LONDON MATHEMATICAL SOCIETY LECTURE NOTE SERIES

Managing Editor: Professor N.J. Hitchin, Mathematics Institute,
University of Oxford, 24–29 St Giles, Oxford OX1 3TG, United Kingdom

The titles below are available from booksellers, or, in case of difficulty, from Cambridge University Press.

46 p-adic Analysis: a short course on recent work, N. KOBLITZ
59 Applicable differential geometry, M. CRAMPIN & F.A.E. PIRANI
66 Several complex variables and complex manifolds II, M.J. FIELD
86 Topological topics, I.M. JAMES (ed)
87 Surveys in set theory, A.R.D. MATHIAS (ed)
88 FPF ring theory, C. FAITH & S. PAGE
89 An F-space sampler, N.J. KALTON, N.T. PECK & J.W. ROBERTS
90 Polytopes and symmetry, S.A. ROBERTSON
92 Representation of rings over skew fields, A.H. SCHOFIELD
93 Aspects of topology, I.M. JAMES & E.H. KRONHEIMER (eds)
96 Diophantine equations over function fields, R.C. MASON
97 Varieties of constructive mathematics, D.S. BRIDGES & F. RICHMAN
98 Localization in Noetherian rings, A.V. JATEGAONKAR
99 Methods of differential geometry in algebraic topology, M. KAROUBI & C. LERUSTE
100 Stopping time techniques for analysts and probabilists, L. EGGHE
104 Elliptic structures on 3-manifolds, C.B. THOMAS
105 A local spectral theory for closed operators, I. ERDELYI & WANG SHENGWANG
107 Compactification of Siegel moduli schemes, C.-L. CHAI
109 Diophantine analysis, J. LOXTON & A. VAN DER POORTEN (eds)
113 Lectures on the asymptotic theory of ideals, D. REES
114 Lectures on Bochner-Riesz means, K.M. DAVIS & Y.-C. CHANG
116 Representations of algebras, P.J. WEBB (ed)
119 Triangulated categories in the representation theory of finite-dimensional algebras, D. HAPPEL
121 Proceedings of *Groups - St Andrews 1985*, E. ROBERTSON & C. CAMPBELL (eds)
128 Descriptive set theory and the structure of sets of uniqueness, A.S. KECHRIS & A. LOUVEAU
130 Model theory and modules, M. PREST
131 Algebraic, extremal & metric combinatorics, M.-M. DEZA, P. FRANKL & I.G. ROSENBERG (eds)
132 Whitehead groups of finite groups, ROBERT OLIVER
133 Linear algebraic monoids, MOHAN S. PUTCHA
134 Number theory and dynamical systems, M. DODSON & J. VICKERS (eds)
137 Analysis at Urbana, I, E. BERKSON, T. PECK, & J. UHL (eds)
138 Analysis at Urbana, II, E. BERKSON, T. PECK, & J. UHL (eds)
139 Advances in homotopy theory, S. SALAMON, B. STEER & W. SUTHERLAND (eds)
140 Geometric aspects of Banach spaces, E.M. PEINADOR & A. RODES (eds)
141 Surveys in combinatorics 1989, J. SIEMONS (ed)
144 Introduction to uniform spaces, I.M. JAMES
146 Cohen-Macaulay modules over Cohen-Macaulay rings, Y. YOSHINO
148 Helices and vector bundles, A.N. RUDAKOV *et al*
149 Solitons, nonlinear evolution equations and inverse scattering, M. ABLOWITZ & P. CLARKSON
150 Geometry of low-dimensional manifolds 1, S. DONALDSON & C.B. THOMAS (eds)
151 Geometry of low-dimensional manifolds 2, S. DONALDSON & C.B. THOMAS (eds)
152 Oligomorphic permutation groups, P. CAMERON
153 L-functions and arithmetic, J. COATES & M.J. TAYLOR (eds)
155 Classification theories of polarized varieties, TAKAO FUJITA
156 Twistors in mathematics and physics, T.N. BAILEY & R.J. BASTON (eds)
158 Geometry of Banach spaces, P.F.X. MÜLLER & W. SCHACHERMAYER (eds)
159 Groups St Andrews 1989 volume 1, C.M. CAMPBELL & E.F. ROBERTSON (eds)
160 Groups St Andrews 1989 volume 2, C.M. CAMPBELL & E.F. ROBERTSON (eds)
161 Lectures on block theory, BURKHARD KÜLSHAMMER
162 Harmonic analysis and representation theory, A. FIGA-TALAMANCA & C. NEBBIA
163 Topics in varieties of group representations, S.M. VOVSI
164 Quasi-symmetric designs, M.S. SHRIKANDE & S.S. SANE
166 Surveys in combinatorics, 1991, A.D. KEEDWELL (ed)
168 Representations of algebras, H. TACHIKAWA & S. BRENNER (eds)
169 Boolean function complexity, M.S. PATERSON (ed)
170 Manifolds with singularities and the Adams-Novikov spectral sequence, B. BOTVINNIK
171 Squares, A.R. RAJWADE
172 Algebraic varieties, GEORGE R. KEMPF
173 Discrete groups and geometry, W.J. HARVEY & C. MACLACHLAN (eds)
174 Lectures on mechanics, J.E. MARSDEN
175 Adams memorial symposium on algebraic topology 1, N. RAY & G. WALKER (eds)
176 Adams memorial symposium on algebraic topology 2, N. RAY & G. WALKER (eds)
177 Applications of categories in computer science, M. FOURMAN, P. JOHNSTONE & A. PITTS (eds)
178 Lower K- and L-theory, A. RANICKI
179 Complex projective geometry, G. ELLINGSRUD *et al*
180 Lectures on ergodic theory and Pesin theory on compact manifolds, M. POLLICOTT
181 Geometric group theory I, G.A. NIBLO & M.A. ROLLER (eds)

182 Geometric group theory II, G.A. NIBLO & M.A. ROLLER (eds)
183 Shintani zeta functions, A. YUKIE
184 Arithmetical functions, W. SCHWARZ & J. SPILKER
185 Representations of solvable groups, O. MANZ & T.R. WOLF
186 Complexity: knots, colourings and counting, D.J.A. WELSH
187 Surveys in combinatorics, 1993, K. WALKER (ed)
188 Local analysis for the odd order theorem, H. BENDER & G. GLAUBERMAN
189 Locally presentable and accessible categories, J. ADAMEK & J. ROSICKY
190 Polynomial invariants of finite groups, D.J. BENSON
191 Finite geometry and combinatorics, F. DE CLERCK *et al*
192 Symplectic geometry, D. SALAMON (ed)
194 Independent random variables and rearrangement invariant spaces, M. BRAVERMAN
195 Arithmetic of blowup algebras, WOLMER VASCONCELOS
196 Microlocal analysis for differential operators, A. GRIGIS & J. SJÖSTRAND
197 Two-dimensional homotopy and combinatorial group theory, C. HOG-ANGELONI, W. METZLER & A.J. SIERADSKI (eds)
198 The algebraic characterization of geometric 4-manifolds, J.A. HILLMAN
199 Invariant potential theory in the unit ball of $\mathbf{C}^n$, MANFRED STOLL
200 The Grothendieck theory of dessins d'enfant, L. SCHNEPS (ed)
201 Singularities, JEAN-PAUL BRASSELET (ed)
202 The technique of pseudodifferential operators, H.O. CORDES
203 Hochschild cohomology of von Neumann algebras, A. SINCLAIR & R. SMITH
204 Combinatorial and geometric group theory, A.J. DUNCAN, N.D. GILBERT & J. HOWIE (eds)
205 Ergodic theory and its connections with harmonic analysis, K. PETERSEN & I. SALAMA (eds)
206 An introduction to noncommutative differential geometry and its physical applications, J. MADORE
207 Groups of Lie type and their geometries, W.M. KANTOR & L. DI MARTINO (eds)
208 Vector bundles in algebraic geometry, N.J. HITCHIN, P. NEWSTEAD & W.M. OXBURY (eds)
209 Arithmetic of diagonal hypersurfaces over finite fields, F.Q. GOUVÊA & N. YUI
210 Hilbert C*-modules, E.C. LANCE
211 Groups 93 Galway / St Andrews I, C.M. CAMPBELL *et al* (eds)
212 Groups 93 Galway / St Andrews II, C.M. CAMPBELL *et al* (eds)
214 Generalised Euler-Jacobi inversion formula and asymptotics beyond all orders, V. KOWALENKO, N.E. FRANKEL, M.L. GLASSER & T. TAUCHER
215 Number theory 1992–93, S. DAVID (ed)
216 Stochastic partial differential equations, A. ETHERIDGE (ed)
217 Quadratic forms with applications to algebraic geometry and topology, A. PFISTER
218 Surveys in combinatorics, 1995, PETER ROWLINSON (ed)
220 Algebraic set theory, A. JOYAL & I. MOERDIJK
221 Harmonic approximation, S.J. GARDINER
222 Advances in linear logic, J.-Y. GIRARD, Y. LAFONT & L. REGNIER (eds)
223 Analytic semigroups and semilinear initial boundary value problems, KAZUAKI TAIRA
224 Computability, enumerability, unsolvability, S.B. COOPER, T.A. SLAMAN & S.S. WAINER (eds)
225 A mathematical introduction to string theory, S. ALBEVERIO, J. JOST, S. PAYCHA, S. SCARLATTI
226 Novikov conjectures, index theorems and rigidity I, S. FERRY, A. RANICKI & J. ROSENBERG (eds)
227 Novikov conjectures, index theorems and rigidity II, S. FERRY, A. RANICKI & J. ROSENBERG (eds)
228 Ergodic theory of $\mathbf{Z}^d$ actions, M. POLLICOTT & K. SCHMIDT (eds)
229 Ergodicity for infinite dimensional systems, G. DA PRATO & J. ZABCZYK
230 Prolegomena to a middlebrow arithmetic of curves of genus 2, J.W.S. CASSELS & E.V. FLYNN
231 Semigroup theory and its applications, K.H. HOFMANN & M.W. MISLOVE (eds)
232 The descriptive set theory of Polish group actions, H. BECKER & A.S. KECHRIS
233 Finite fields and applications, S. COHEN & H. NIEDERREITER (eds)
234 Introduction to subfactors, V. JONES & V.S. SUNDER
235 Number theory 1993–94, S. DAVID (ed)
236 The James forest, H. FETTER & B. GAMBOA DE BUEN
237 Sieve methods, exponential sums, and their applications in number theory, G.R.H. GREAVES, G. HARMAN & M.N. HUXLEY (eds)
238 Representation theory and algebraic geometry, A. MARTSINKOVSKY & G. TODOROV (eds)
239 Clifford algebras and spinors, P. LOUNESTO
240 Stable groups, FRANK O. WAGNER
241 Surveys in combinatorics, 1997, R.A. BAILEY (ed)
242 Geometric Galois actions I, L. SCHNEPS & P. LOCHAK (eds)
243 Geometric Galois actions II, L. SCHNEPS & P. LOCHAK (eds)
244 Model theory of groups and automorphism groups, D. EVANS (ed)
245 Geometry, combinatorial designs and related structures, J.W.P. HIRSCHFELD, S.S. MAGLIVERAS & M.J. DE RESMINI (eds)
246 p-Automorphisms of finite p-groups, E.I. KHUKHRO
247 Analytic number theory, Y. MOTOHASHI (ed)
248 Tame topology and o-minimal structures, LOU VAN DEN DRIES
249 The atlas of finite groups: ten years on, ROBERT CURTIS & ROBERT WILSON (eds)
250 Characters and blocks of finite groups, G. NAVARRO
251 Gröbner bases and applications, B. BUCHBERGER & F. WINKLER (eds)
252 Geometry and cohomology in group theory, P. KROPHOLLER, G. NIBLO, R. STÖHR (eds)
253 The q-Schur algebra, S. DONKIN
254 Galois representations in arithmetic algebraic geometry, A.J. SCHOLL & R.L. TAYLOR (eds)
255 Symmetries and integrability of difference equations, P.A. CLARKSON & F.W. NIJHOFF (eds)
260 Groups St Andrews 1997 in Bath, I, C.M. CAMPBELL *et al* (eds)
261 Groups St Andrews 1997 in Bath, II, C.M. CAMPBELL *et al* (eds)

London Mathematical Society Lecture Note Series. 261

Groups St Andrews 1997 in Bath, II

Edited by

C. M. Campbell
University of St Andrews

E. F. Robertson
University of St Andrews

N. Ruskuc
University of St Andrews

G. C. Smith
University of Bath

CAMBRIDGE UNIVERSITY PRESS
Cambridge, New York, Melbourne, Madrid, Cape Town, Singapore, São Paulo, Delhi

Cambridge University Press
The Edinburgh Building, Cambridge CB2 8RU, UK

Published in the United States of America by Cambridge University Press, New York

www.cambridge.org
Information on this title: www.cambridge.org/9780521655767

First published 1999

A catalogue record for this publication is available from the British Library

ISBN 978-0-521-65576-7 paperback

Transferred to digital printing 2009

CONTENTS

Volume II

CONTENTS OF VOLUME II

Contents of Volume I

Contents of Volume I

INTRODUCTION

An international conference Groups St Andrews 1997 in Bath was held on the campus of the University of Bath during the period 26 July – 9 August 1997. Some 299 mathematicians from 41 countries were involved in the meeting, as well as 82 family members and partners. This was the fifth meeting of the four-yearly Groups St Andrews Conferences, and the series continues to flourish. The shape of the conference was similar to the previous conferences in that the first week was dominated by five series of talks, each surveying an area of rapid contemporary development in group theory. The main speakers were Laszlo Babai (Chicago), Martin Bridson (Oxford), Chris Brookes (Cambridge), Cheryl Praeger (Western Australia) and Aner Shalev (Jerusalem). The second week featured two special days, a Burnside Day and a Lyndon Day. Our thanks are due to Efim Zelmanov (Yale) and Chuck Miller III (Melbourne), respectively, for helping organise the programmes for these days. In addition the week contained a wide variety of research talks. In the evenings throughout the conference, and during the rest periods, there was an extensive social programme, only some of which was disrupted by rain. There was also much extemporised music-making in the Senior Common Room in the evenings.

These Proceedings contain the written evidence of the academic achievements of the conference. The five main speakers have all provided substantial survey articles, giving a wide perspective on their fields. In the case of Laszlo Babai, the article is written jointly with Bob Beals (Arizona), one of the invited speakers of the second week. Sixteen other papers in these Proceedings are written by authors who gave one hour invited lectures. A rigorous journal-style refereeing process was applied to all the research articles and survey articles submitted for publication in the Proceedings.

Less easy to quantify, but much more important, were the exchanges of ideas and joint work that was done, both at the conference and as a result of meetings at the conference. As the largest regular meeting on group theory in the world, this series has provided a continuing stimulus to research in group theory.

There are many who helped to make the conference a memorable occasion; in particular we thank Mrs Nada Harvey for secretarial assistance, and Mr John McDermott, Mr Aaron Wilson and many other students from the Department of Mathematical Sciences of the University of Bath whose help in the day to day running of the conference was invaluable. Many of Geoff Smith's colleagues from the academic staff of the Department of Mathematical Sciences gave valuable help and we particularly thank Professor John Toland for guidance on conference organization, and Dr Fran Burstall for TEX and LATEX assistance to the the conference and its newspaper, The Daily Group Theorist.

We record our thanks also to the then Head of School, Professor Alastair Spence, and to the Pro-Vice Chancellor, Professor I M Jamieson, for smoothing the path of the conference organization. For her constructive attitude and industrious work with the conference accommodation we thank Marian Short from the conference office of the University of Bath.

It is a pleasure to acknowledge the excellent level of support we received from various funding bodies. The Edinburgh Mathematical Society and the London Mathematical Society provided financial help which defrayed the main speakers' travel, registration and accommodation expenses. The London Mathematical Society additionally sponsored three visitors from Moscow State University, Professor A. Yu. Ol'shanskii, and two research students Ivan Arzhantsev and Gulnara Arzhantseva. This funding provided support for travel, accommodation, conference fees and subsistence. The Royal Society of London treated the conference as two meetings, because of its exceptional size, and because it is held only once in four years. They therefore generously supported four visitors from the former Soviet Union.

We also gratefully acknowledge the financial support from the University of Bath Initiative Fund, the Department of Mathematics of the University of Bath and the School of Mathematical and Computational Sciences of the University of St Andrews. We thank the Bath and North-East Somerset Council for hosting a civic reception at the Roman Baths. Finally, it is a pleasure to thank Olga Tabachnikova both for her assistance in countless aspects of conference organization, and for the loan of Geoff Smith.

Colin Campbell
Ed Robertson
Nik Ruškuc
Geoff Smith

GALOIS GROUPS THROUGH INVARIANT RELATIONS[1]

ALEXANDER HULPKE

School of Mathematical and Computational Sciences, University of St Andrews, The North Haugh, St Andrews, Fife KY16 9SS, Scotland

1 Prolegomena

Let $f \in \mathbb{Q}[x]$ be an irreducible polynomial of degree n. Then the splitting field $L \geq \mathbb{Q}$ of f is a normal extension. We want to determine the Galois group $G = \mathrm{Gal}(f) = \mathrm{Gal}(L/\mathbb{Q})$ of f which is the group of all field automorphisms of this extension. This task is basic in computational number theory [Coh93] as the Galois group determines a lot of properties of the field extension defined by f.

Because the index $[L : \mathbb{Q}]$ might be large, however, we do not intend to construct L and thus cannot give explicitly the automorphism action of G on L. Instead we consider the action of G on the roots $\{\alpha_1, \ldots, \alpha_n\}$ of f. As f is defined over the rationals the set of these root must remain invariant under G. This permutation action is faithful, because L can be obtained by adjoining all the α_i to $\mathbb{Q}$. This action has to be transitive because f is irreducible. In other words: For a fixed arrangement of the roots, G can be considered as a transitive subgroup of S_n. We will utilize this embedding without explicitly mentioning it.

While the problem itself is initially number-theoretic, the approaches to solve it are mainly based on commutative algebra and permutation group theory. This paper presents a new approach, approximating the Galois group by its closures (subgroups of S_n that stabilizer orbits of G). This in turn gives rise to questions about permutation groups.

In the course of the paper we shall need a few facts from number theory about p-adic extensions and the relation between extensions of $\mathbb{Q}$ and extensions of $\mathbb{F}_p$. These will be provided in an appendix.

2 Identification tools

We will assume that f is a monic integer polynomial of degree n, that is the roots of f are algebraic integers. Obviously we can always enforce this by a transformation of the type $f(x) \mapsto a^n f(x/a) = \tilde{f}(x)$ which yields a polynomial $\tilde{f}$ defining the same extension as f. The degree n will be typically in the range $n \lessapprox 30$.

2.1 Orbits of elements

By Dedekind's theorem (Appendix, Theorem 5) factorization of f modulo a non-ramifying prime yields the cycle structure of an element of $G = \mathrm{Gal}(f)$. Such factorizations are very cheap and it is feasible to factorize f modulo several hundred

[1]Supported by EPSRC Grant GL/L21013

primes. By this method it is usually very easy to find out whether G is symmetric or alternating [DSar].

Using Tschebotareff's result we may even hope that we have obtained all cycle shapes of G if we look at sufficiently many primes. In practice "sufficiently many" means: Factorize modulo new primes, until $t(n)$ times no new shape emerged (or we can prove already that G contains the alternating group). For small degrees $t(n) = 3n$ seems to be a reasonable choice.

In any case, such factorizations eliminate those groups as candidates for G which do not contain all shapes observed. Unless we are content with a probabilistic result, however, we cannot be certain to have found all shapes and cannot use the frequency of these shapes.

2.2 Orbits of the Galois group

The main tool for the identification of Galois groups is the polynomial ring $\mathcal{R} = \mathbb{Z}[x_1, \ldots, x_n]$ and the specialization homomorphism $\varphi = \varphi_f : \mathcal{R} \to \mathcal{O}(L)$, $h \mapsto h(\alpha_1, \ldots, \alpha_n)$. As a permutation group, G acts on $\mathcal{R}$ by permuting indeterminates; as a Galois group it acts on $\mathcal{O}(L)$. For these two actions φ is a G-module homomorphism, we have that

$$\varphi(r)^g = \varphi(r^g) \qquad \text{for all} \quad r \in \mathcal{R}. \tag{1}$$

The basic idea is to recognize G from G-invariant polynomials in $\mathcal{R}$. G-invariance of an algebraic integer $a \in \mathcal{O}(L)$ implies that $a \in \mathbb{Z}$ and thus G-invariance of $h \in \mathcal{R}$ implies that $\varphi(h) \in \mathbb{Z}$. For example, recall that the discriminant of a monic polynomial with roots $\alpha_1, \ldots, \alpha_n$ is defined as $\operatorname{disc}(f) = \prod_{i<j}(\alpha_i - \alpha_j)^2$. Any transposition of two roots will change a sign of $\sqrt{\operatorname{disc}(f)} = \prod_{i<j}(\alpha_i - \alpha_j)$. Thus this root is invariant under $\operatorname{Gal}(f)$ if and only $\operatorname{Gal}(f)$ consists only of even permutations. Therefore $\operatorname{Gal}(f)$ is a subgroup of A_n if and only if $\operatorname{disc}(f)$ is a square. Otherwise $\mathbb{Q}(\sqrt{\operatorname{disc}(f)})$ is the subfield of L corresponding to the subgroup of even permutations.

In general, the condition $\varphi(h) \in \mathbb{Z}$ is not sufficient to prove G-invariance of h as the following example shows. The polynomial $f = x^4 - 2$ has Galois group $D(4) = \langle (1,2,3,4), (1,3) \rangle$, acting on the roots $\{\sqrt[4]{2}, i\sqrt[4]{2}, -\sqrt[4]{2}, -i\sqrt[4]{2}\}$. Then $h = x_1x_2 + x_3x_4$ is not $D(4)$-invariant, but $\varphi(h) = 0$. It is easy, however, to make the criterion sufficient:

Lemma 2 *Assume that*

$$\varphi(h) \neq \varphi(h') \quad \textit{for all } h' \in h^{S_n}. \tag{3}$$

Then h is G-invariant if and only if $\varphi(h) \in \mathbb{Z}$.

Proof We have to show sufficiency: Assume that $\varphi(h) \in \mathbb{Z}$ and that there is a $g \in G$ such that $h^g \neq h$. Then by (3) we have $\varphi(h) \neq \varphi(h^g)$, in contradiction to $\varphi(h) \in \mathbb{Z}$ and (1). □

2.3 Invariant method

One approach [Sta73, EO95, DF89] uses precomputed invariants h_T for all transitive subgroups of $T \leq S_n$ and filters G from these T by testing which h_T are actually G invariant. This test is done by testing $\varphi(h_T)$ for rationality.

If G and G' are conjugate under S_n, the invariants of G' are S_n-images of the invariants of G, so it is sufficient to store only representatives of the invariants.

A further reduction is obtained by using relative invariants: The set of transitive groups is a semi-lattice with respect to inclusion. Determination of G then proceeds stepwise down this lattice. If G is known to be contained in U and if V is a subgroup of S_n, then every invariant for V can be used to test whether G is contained in $U \cap V$. Such invariants can be substantially simpler than an invariant for $U \cap V$ as a subgroup of G. Furthermore, the number of invariant images to consider is reduced from $[S_n : V]$ to $[U : V]$. This way the identification process steps down through chains of transitive subgroups, stepping from a subgroup U to a maximal subgroup $V < U$.

However, the number of images will still be prohibitive if every subgroup chain from G to S_n includes a step of large index. This typically holds, as (with few exceptions) all transitive maximal subgroups of the symmetric and alternating groups are of large index [LPS87].

A further problem can be approximation accuracy: The roots $\{\alpha_i\}$ are not known exactly but only by approximation. Therefore instead of φ we only know an approximation $\tilde{\varphi}$. Consequentially can only test $\tilde{\varphi}(h)\tilde{\in}\mathbb{Z}$ and $\tilde{\varphi}(a) \approx \tilde{\varphi}(b)$. With numerical approximation of the roots this leads to substantial problems with the needed accuracy to prove approximated numbers to be integers or equal. A better approach is to use p-adic approximation, relying on numerical estimates only to deduce the necessary accuracy of the p-adic approximation as done in [DF89].

Instead of testing integrality of $\varphi(h)$ for a V-invariant h, usually it is checked whether the polynomial $R(x) = \prod_{g \in h^U}(x - \varphi(g)) \in \mathbb{Z}[x]$ has an integral root. This permits to round the coefficients of R to the nearest integer and thus reduce the influence of approximation errors.

In a variant, avoiding approximation, it is shown in [Col95] how to compute R by calculations in $\mathcal{R}$, using only rational operations on the coefficients of f.

2.4 Resolvent method

Another approach [SM85] tries to overcome the problem with approximation of roots and large numbers of invariant images by using only invariants of very restricted type. The approximation of roots in this case is deferred inside a polynomial factorization (which uses p-adic approximation of the factors).

If $A \subset \mathcal{R}$ is an orbit of G, then every elementary symmetric function in the elements of A is obviously G-invariant. Thus the polynomial $\prod_{a \in A}(x - a)$ is G-invariant as well. (We extend the action of G to the polynomial rings $R[x]$ and $L[x]$ by acting on the coefficients.) Thus the polynomial $\prod_{a \in A}(x - \varphi(a))$ is an integer polynomial.

To get A we consider it as a subset of a larger orbit $\widehat{A}$ of the symmetric group.

Suppose that $\widehat{A} = h^{S_n}$. Then $R(h, f) := \prod_{a \in \widehat{A}}(x - \varphi(a))$ is the product of the polynomials of the G-orbits within $\widehat{A}$. We call $R(h, f)$ a *resolvent*. (The name derives from the fact that polynomials of such type were used initially to solve polynomial equations.)

If condition (3) is fulfilled, the $\mathbb{Q}$-irreducible factors of $R(h, f)$ are in bijection to the G-orbits on h^{S_n}. Especially, the factor degrees are the orbit lengths and a factor discriminant is a square if and only if the image of the action on the corresponding orbit is a subgroup of the alternating group.

We define the *parity* of a permutation action to be even if the image of this action is a subgroup of the alternating group and odd otherwise.

By the Galois correspondence we also deduce immediately the same type of correspondence between the orbits of $U \leq G$ and the factorization of $R(g, f)$ over $\mathbb{Q}(\beta)$ when U and $\mathbb{Q}(\beta)$ are Galois correspondents. Such intermediate fields can be obtained by factors of other resolvents or simply by adjoining the root of the discriminant to obtain a field in correspondence to the subgroup of even permutations.

Using this approach one might compute sufficiently many orbit lengths and parities of G and its subgroups to identify up to conjugacy the Galois group uniquely among a list of all transitive subgroups of S_n. A list with the possible orbit lengths and parities therefore has to be prepared a priori. The method will give only the permutational type of the group but not its actual action on the roots. An algorithm of this type has been implemented by the author in GAP 3 [S+97] for $n \leq 15$.

2.5 Computation of resolvents

We note that the coefficients of $R(h, f)$ are symmetric in the root of f. So they can be expressed in the elementary symmetric functions in the $\{\alpha_i\}$ (which are up to a sign the coefficients of f) using only ring operations. Practically this could be done using resultants [Loo82].

As these resultant calculations can be computationally expensive, however, we shall restrict ourselves to the orbits of elements of the type $x_1 + \cdots + x_k$, $x_1 \cdot \cdots \cdot x_k$ or $x_1 + 2x_2 + \cdots + kx_k$. As they correspond to the orbits on k-sets, respectively k-tuples of points, we call the arising resolvents *k-set resolvents* (respectively *k-tuple resolvents*). For computation of the set resolvents $R(x_1 + \cdots + x_k, f)$ and $R(x_1 \cdot \cdots \cdot x_k, f)$, efficient formulae have been published in [CM94].

For a resolvent $R = R(h, f)$, condition (3) simply means that R is square-free. This can be tested easily by checking whether $\gcd(R, R') = 1$ (with $'$ denoting the usual derivative).

If $R(h, f)$ is not square-free, f gets replaced by a Tschirnhaus-transform $\hat{f}$ of f that defines the same field. Then $R(h, \hat{f})$ is computed anew. Unless the preimage of R in $\mathcal{R}$ already contains a square, there always is a transform $\hat{f}$ which renders $R(h, \hat{f})$ square-free [Gir83, theorem 3,(2)]. The transformations given in the proof in [Gir83] however not only might involve resultant calculations to compute $\hat{f}$ but also result in a polynomial $\hat{f}$ with very large coefficients. Thus in practice we restrict ourselves to simple combinations of the transformations $\tau_1: f(x) \mapsto f(x+1)$ and $\tau_2: f(x) \mapsto x^n f(1/x)$. Both together generate (up to a sign) the modular group $PSL(2, \mathbb{Z})$ and so lead to a vast number of possible composite transforms. Most of them, however, have unsuitably large coefficients.

In some cases certain transforms will never yield a square-free resolvent and thus can be discarded immediately. For example if $f(x) = g(x^2)$ roots come in pairs differing by a sign. In such a case resolvents for $h = x_1 + x_2$ are never square-free and τ_1 will not remedy this problem.

Instead of changing f it might be possible as well to change h (without changing the equivalence type of G's action, for example $x_1 + x_2$ and x_1x_2 both correspond to the action on 2-sets) to $\hat{h}$ to obtain a square-free resolvent $R(\hat{h}, f)$.

On a (square-free) $R(h, f)$, similar transforms can be used profitably to get its coefficients smaller again. (The size of the coefficients is a measure for the amount of lifting necessary in the factoring algorithm. Usually in the literature on polynomial factorization only transformations of type τ_2 are suggested.)

2.6 Drawbacks

Both methods mentioned so far rely on explicit lists of all transitive subgroups of the symmetric group of the given degree. Though progress has been made on the determination of such lists [Hul96] it will be hard to compute these lists for further degrees beyond 31. Also the needed determination of properties for all the relevant groups would take substantial time to be spent before running the actual identification algorithms. Finally handling the large lists involved (like for the over 25000 groups needed for degree 24) is a challenge on its own.

Our aim will be to combine both presented methods with a group theoretic approach. From this we obtain an algorithm that is capable of determining the Galois group up to a few possibilities in reasonable time, that will work independently of the degree, provided the Galois group itself is not overly large. It also provides the possibility to obtain the exact Galois group (not just the type but its exact action on approximated roots) if a user is willing to invest further computing resources on this problem. It is understood that there will be cases remaining that are inherently hard to decide. These are basically highly transitive groups which are maximal with very large index in the symmetric or alternating group. The Mathieu groups are a typical examples of this.

3 A new approach

In general the orbits of G convey more information than only their lengths and the permutation actions parities. Therefore the approach from 2.5 does not necessarily make full use of the information obtained by the resolvent factors.

As an example of two potential Galois groups that have the same orbit lengths but different orbits consider $\frac{1}{2}[3{:}2]_c D(4)$ and $\frac{1}{2}[3{:}2]_d D(4)$, the 12th and 15th transitive group of degree twelve. (The names reflect the composition structure corresponding to a block system. Lists of all the groups and a description of the naming scheme used can be found in [CHM]. The same names are also used in GAP [S+97].) On sets of order 2 both have 5 orbits of length 12 and one orbit of length 6. Even the parities for the actions on all those sets are the same for both groups. On the other hand, both groups are stabilizers of their respective partitions of 2-sets within the symmetric group.

To overcome this shortcoming, L. Soicher suggested the following approach [Soi]: Suppose S_n acts on a domain Ω and $A \subset \Omega$ is an orbit of G. Then $G \leq \mathrm{Stab}_{S_n}(A)$ and $G \not\subset \mathrm{Stab}_{S_n}(B)$ for $B \subsetneqq A$. So the intersection of the orbit stabilizers is a subgroup of S_n containing G. By computing these stabilizers we therefore obtain approximations to G without the need to refer to tabulations of subgroups of S_n.

If a is a factor of a (square-free) resolvent $R(g, f)$ the orbit corresponding to a consists of those images $\hat{g}$ of g, for which $\varphi(\hat{g})$ is a root of a. We apply this for approximated roots, using approximation modulo a prime. Let p be a prime which does not divide the discriminant of f or $R(g, f)$ and let π be reduction modulo p as defined in the Appendix. As π is a ring homomorphism, $\pi(\varphi(g))$ is a root of $\pi(a)$.

As in Lemma 2 this is sufficient to distinguish the orbits if $\pi(R(g, f))$ is square-free over $\mathbb{F}_p$. This holds, if $p \not| \mathrm{disc}(R(g, f))$, so there are only finitely many unsuitable primes.

If we identify the roots by their approximations $\{\pi(\alpha_i)\}$ modulo p, we have $\pi(\varphi(h)) = h(\pi(\alpha_1), \ldots, \pi(\alpha_n))$ for any $h \in \mathcal{R}$. Once we have factorized the resolvent $R(g, f)$ obtained from the polynomial $g \in \mathcal{R}$ we compute the reduced values $\pi(\varphi(h))$ for all $h \in g^{S_n}$. Then, under the action of G, $\varphi(h)$ is in the orbit corresponding to the factor a of $R(g, f)$ if $\pi(\varphi(h))$ divides $\pi(a)$. Thus simple evaluation of polynomials if $\mathbb{F}_p$ will yield the orbits of G (with respect to the arrangement of the approximated roots $\pi(\alpha_i)$). Computations of this type can be performed very quickly.

3.1 *k*-closures

Usually, we are considering the action on k-tuples or k-sets of points. For these actions the stabilizer of all the orbits of G are "closures" as defined in [Wie69]: The k closure $G^{(k)}$ of $G \leq S_n$ is the largest subgroup of S_n which has the same orbits on k-tuples of points as G has. Similarly, the $\{k\}$-closure $G^{\{k\}}$ of G is the largest subgroup of S_n which has the same orbits on k-sets as G has. We call k the *level* of the closure. ($G^{(2)}$ also can be interpreted as the stabilizer of all the orbital graphs of G. These are directed graphs with vertices Ω and edges given by one orbit of G on $\Omega \times \Omega$.) Properties of such closures have been studied before in the literature [Wie69, Sie82], we list some of them which we shall need later on:

1) $G \leq G^{(k)} \leq G^{\{k\}}$.
2) $G^{(k+1)} \leq G^{(k)}$.
3) $G^{\{k+1\}} \leq G^{\{k\}}$ if $k < \lfloor \frac{n}{2} \rfloor$.
4) If $p \leq k$ then $p \mid |G|$ if and only if $p \mid |G^{(k)}|$.
5) If G has base length m then $G = G^{(k)}$ $(k > m)$.
6) $(G^{(k)})^{(k)} = G^{(k)}$, $(G^{\{k\}})^{\{k\}} = G^{\{k\}}$.
7) G has the same block systems as $G^{\{2\}}$.

As $G^{(k)} = S_n$ only if G is k-fold transitive, these closures quickly yield a proper subgroup of S_n. For a given permutation group G, closures can be computed as stabilizers by a backtrack search, the 2-closure can also be computed via the orbital graphs. Algorithms for these tasks have been published in [McK81, IAFM94, Leo91, The97] and are available in programs like nauty, COCO, Magma or GAP.

Suppose we have computed a set of closures of G up to the k-closure and $\{k'\}$-closure. Let C be their intersection. (Due to properties 2. and 3. this is in most cases the smallest of the closures computed. Nevertheless it is worthwhile to compute closures of different level, as knowledge of a closure can be very helpful for the factorization of resolvents, needed to compute a closure of higher level. See 5.2.) Then we know that G is a transitive subgroup of C with the following properties:

- G has the same m-closures and $\{m'\}$-closures as C (for all $m \leq k$ and $m' \leq k'$.);
- G contains elements of prescribed cycle types. (Obtained by factoring f, 2.1.)
- One element of G is known. (The Frobenius-Automorphism for the prime chosen to approximate the roots.)

We denote the set of all these groups by $\mathcal{L}$.

3.2 Computation of $\mathcal{L}$

While the computation of $\mathcal{L}$ might be difficult, it is easy to check, whether a given group U is contained in $\mathcal{L}$, as this involves mainly orbit calculations. So the main problem for the computation of $\mathcal{L}$ is to produce a sufficiently small superset of potential candidates.

To do so, we compute the maximal subgroups of C and test which of these are contained in $\mathcal{L}$. For these we compute again the maximal subgroups and so forth. As $\mathcal{L}$ is closed under taking intermediate chains (if $U, W \in \mathcal{L}$ and $U \leq V \leq W$, then also $V \in \mathcal{L}$) this process exhausts all possible groups in $\mathcal{L}$.

3.3 Determination of G

Usually $\mathcal{L}$ will contain more than one group. Nevertheless even this may be sufficient information to, say, show the solvability of the extension or give size bounds for the splitting field. If G is needed exactly, we will make use of the invariant method:

Suppose we know already that $G \leq U$. For each conjugacy class of maximal subgroups of U we compute an relative invariant h and check for all of its images under U whether $\varphi(h^u) \in \mathbb{Z}$. We do this using p-adic approximation of the roots of f as in [DF89].

The use of relative invariants requires knowledge of the action of U on the approximated roots. This is fulfilled as we obtained C not only up to conjugacy but also its action.

Of course in practice one should combine the determination of $\mathcal{L}$ with the downward steps and only compute maximal subgroups of a group U after G has been proven to be contained therein.

4 Group theoretic problems

Algorithms published so far for the determination of Galois groups [AV94, SM85, Sta73] were usually limited to a predetermined set of possible degrees (those degrees for which invariants or orbits information were precomputed). Thus asking for the complexity of these algorithms was not a sensible question. On the other hand the presented approach is, a priori, degree-independent. To estimate its performance, it is crucial, however, to learn more about the set $\mathcal{L}$. Similarly knowledge about $\mathcal{L}$ can determine up to which level closures should be computed.

This leads to purely group-theoretic questions about the set $\mathcal{L}$ of permutation groups. We will state these questions, indicate which type of answer we need for our algorithm to perform well, and give some indication why these hopes are not unreasonable. A full answer seems to be beyond the scope of this article.

Problem 1: How large is $\mathcal{L}$?

As the groups in $\mathcal{L}$ have to be distinguished by invariants, we hope that $\mathcal{L}$ is small.

Experiments with degrees up to 16 show that if C is the intersection of the 2- and the $\{4\}$-closure, that $\mathcal{L}$ is typically of size less then 10 and almost always of size less than 20. The worst case is size 106 in degree 16.

If we have computed k-closures we can (property 5) uniquely identify all groups of base length smaller than k. For example, for primitive, not 2-transitive groups, the base length is limited to $4\sqrt{n}\log n$ [DM96, Theorem 5.3A]. (By property 7., we know whether G is primitive, 2-transitivity is determined by the 2-closure.)

Problem 2: How different are the groups in $\mathcal{L}$?

Or – in other words – which properties of G can we deduce from $\mathcal{L}$? We hope the groups in $\mathcal{L}$ to be of small index in the largest one, as this reduces the number of conjugate maximal subgroups and thus the number of invariants that have to be tested. Again experiments with degrees up to 16 show that for C the intersection of the 2- and the $\{4\}$-closure, the largest and smallest group in $\mathcal{L}$ differ typically by a factor of less than 10 and almost always by less than 100. By far the worst case is index 5040 in degree 12, which happens for M_{12}. Such highly transitive groups however occur rarely and should be considered as exceptions.

For higher degrees however closures of higher level will be needed: If U and V have the same $\{k\}$-closure then for any transitive T the groups $U \wr T$ and $V \wr T$ (acting naturally) have the same $\{k\}$-closure, but $[U \wr T : V \wr T] = [U : V]^{\deg T}$. Therefore the size range for groups with the same $\{k\}$-closure can grow exponentially.

The properties of the closures show that we can obtain small prime divisors of $|G|$ and block systems of G from knowledge of C.

Problem 3: Compute the maximal subgroups

If the group is small, one can simply compute the full subgroup lattice to obtain maximal subgroups iteratively. For solvable groups in general, there is an efficient algorithm to compute maximal subgroups [Eic93]. In the insolvable case such an algorithm (without computing the full lattice first) is still a desideratum. However use of the O'Nan-Scott theorem [DM96, 4.1A] could provide an approach. Finally, for degrees up to 31, the lists of transitive permutation groups computed in [Hul96] can be used.

Problem 4: Determination of invariants

When computing relative invariants for $V < U$ we need *one* invariant of V that is not invariant under U. Furthermore, we want this invariant to be "small", not only to reduce the amount of work needed to evaluate one invariant but also to reduce the magnitude of $|\varphi(h)|$ (which determines the needed p-adic lifting accuracy). As sum of monomials this magnitude is determined primarily by the degree of the monomials. As we use p-adic instead of numerical approximation, error propagation is not a problem. Thus it is not critical to express h in a special way.

The permutation image of a monomial is a monomial of the same degree. So we can assume without loss of generality that we are looking for a homogeneous invariant which is the V-orbit sum $b = \sum_{c \in a^V} c$ of a "defining" monomial a. It is a V-invariant for U if $|a^U| > |a^V|$. If V is normal in U (note that V is maximal in U anyhow) this holds if and only if $\mathrm{Stab}_U(a) \leq V$, otherwise this criterion is sufficient but not necessary.

Classical results on the generation of invariant rings [Noe16] concentrate on generating the full invariant ring and so give an unsuitably large bound (namely $|G|$) for the degree of a.

To restrict the degree we observe the following bounds: If $(b_1, \ldots, b_m)$ is a base of U the monomial $x_{b_1} \cdot x_{b_2}^2 \cdot \cdots \cdot x_{b_m}^m$ has a regular orbit under U and so may serve as an a. So $\deg(a) \leq \frac{m(m+1)}{2}$. Similarly $\deg(a) > k$ if both groups have the same k-closure.

If U and V are small enough to compute their conjugacy classes, one can also compute the Molien series $F_U(\lambda)$ for the permutation action [Sta79]. (It is sufficient to know the conjugacy classes of U to compute this series.) In expanded form $F_U(\lambda) = \sum_{i \geq 0} d(U)_i \lambda^i$ with $d(U)_i$ being the dimension of the space of the i-dimensional homogeneous invariants of U. So the smallest i for which $d(V)_i > d(U)_i$ is the smallest possible degree.

As we are interested only in invariance and not in absolute degrees we can assume that the exponent i occurs in a only if the exponent $i-1$ occurs as well. So we may assume that

$$a = (x_{i_1} \cdots x_{i_{l_1}})(x_{i_{l_1+1}} \cdots x_{i_{l_2}})^2 \cdot \cdots \cdot \left(x_{i_{l_{m-1}+1}} \cdots x_{i_{l_m}}\right)^m \tag{4}$$

with $l_1 \geq l_2 - l_1 \geq \cdots \geq l_m - l_{m-1}$. The stabilizer of such a monomial is an m-times iterated stabilizer of sets of length $l_i - l_{i-1}$. (As the computation of set stabilizers is usually easier for smaller sets, this stabilizer should be computed backwards from m to 1.)

A general strategy to compute invariants is discussed in [Gir87].

It is shown in [Göb95] that a orbit sums of a subset of the monomials of the form (4) of degree at most $\max(n, n(n-1)/2)$ generate the full invariant ring.

5 Factoring resolvents

The hardest subtask of the proposed algorithm usually will be the factorization of the resolvents.

5.1 Difficulties

For the usual factoring approach [Zas69],[DST88, §4.2.2] of Hensel lifting combined with trial quotients of the polynomial with products of the lifted factors, the resolvent polynomials $R(g, f)$ are of the worst possible kind: By Dedekind's theorem (5) the degrees of the factors of $R(g, f)$ modulo any prime divide the order of elements in $\mathrm{Gal}(f)$, but the degrees of the irreducible factors of $R(f, g)$ are orbit lengths of $\mathrm{Gal}(f)$ and so of magnitude $|\mathrm{Gal}(f)|$ instead of $|\sigma|$ ($\sigma \in \mathrm{Gal}(f)$). Thus as soon as the Galois group is not regular, resolvent polynomials will split into many factors modulo any prime while splitting only in few factors over $\mathbb{Q}$. So not only many factors have to be lifted first, but also the exponential combination step might become extremely hard.

An obvious remedy seems to be the use of a polynomial time factoring algorithm as suggested in [LLL82]. Unfortunately this algorithm requires substantially better lifting and thus has a comparatively worse runtime in "small" cases. However, on current computers within reasonable run times only such "small" examples can be computed anyhow. The break-even point of the exponential algorithm can not yet be reached.

5.2 Use of partial information

If we use the fact that we do not factor arbitrary polynomials but resolvents, we can help the factoring algorithm: The partial knowledge about the Galois group and its possible actions will limit the possible orbit lengths and thus the possible factor degrees:

Factoring a polynomial a means to find the orbits of the Galois group on its roots. The first approximation are the orbits of one element (the Frobenius automorphism) of this Galois group. They are given by factorization modulo p: $a \equiv \prod_i a_i \pmod{p}$. This approximation is used in the standard Hensel lifting. Suppose, we have already computed a closure C from the factorization of some resolvents R_i. Reduction of the factors of the R_i modulo this prime p gives (by the save process as described in section 3) the action of C on the roots of f modulo p and in turn (if a is a resolvent) on the roots of a. G is a subgroup of C and so the orbits of G refine the orbits of C. This yields a partition $\mathcal{J}$ of the $\{a_i\}$ (modulo p). When combining the lifted factors, only the $a_i \pmod{p^m}$ in the same block $J \in \mathcal{J}$ of the partition (i.e. $i \in J$) have to be combined. Furthermore this will give better bounds for the degrees of potential factors, probably reducing the necessary lifting bound.

If we even know $\mathcal{L}$ (or a reasonably small superset) we can also test whether modulo p the roots of two factors – say a_i and a_j – always lie in the *same* orbit of U for all $U \in \mathcal{L}$. If this is the case, the factorization will not separate a_i and a_j and we can replace them by their product even before starting the lifting. (Note that Hensel lifting only requires coprime factors. They do not need to be irreducible.)

Both reductions will substantially reduce the work needed for the factorization.

6 Block systems

We have noted already that we obtain the imprimitivity structure of the Galois group G even from the $\{2\}$-closure of G. If the degree n is high, however, factorizing the 2-set resolvent might be initially too hard while computation of block systems [Hul95, KP97] is still feasible. Then we can compute the intersection of the corresponding wreath products and take this as a group C to simplify the polynomial factorization (5.2).

Recall that each block system corresponds to a subfield of $\mathbb{Q}(\alpha)$. We can express a primitive element γ of this subfield as a polynomial $\gamma = h(\alpha)$ and take the minimal polynomial g of γ. So $g(h(\alpha)) = 0$, respectively $f(x)|g(h(x))$.

Then the evaluation of h at any root α_i of f will yield a root of g, which corresponds to a block. So every block consists of those roots α_i of f which evaluate under h to the same number γ_j. Again this test can be done modulo any non-ramifying prime.

If a system of m blocks of size l has been found G is a subgroup of $S_l \wr S_m$. By calling the identification algorithm recursively to determine the action of G on the blocks we can further restrict to the wreath product $S_a \wr \mathrm{Gal}(g)$, its top group acting on the blocks as $\mathrm{Gal}(g)$ does. While the $\{2\}$-closure already determines the block systems, it does not necessarily determine this block action $\mathrm{Gal}(g)$. Therefore this identification may reduce C to a smaller group and improve the discrimination of groups.

Theoretically, by the theorem of Krasner and Kaloujnine [KK51], one could even replace S_a by the action of a block stabilizer on its block. However, the determination of this action requires the identification of Galois groups over algebraic extensions which seems to be too expensive for the gain possible. Also, for obtaining the correct permutation action, this identification would have to take place for any block, as we do not get the action of $\mathrm{Gal}(g)$ on the roots of f.

The method of [KP97] utilizes p-adic approximation of the roots of f. Therefore it might be desirable to use the same prime later on for determining G's action, as the root approximations can be re-used.

We finish this section with the remark that though the best known practical algorithm [KP97] to find block systems is of exponential nature the problem is known to be polynomial in theory [LM85].

Appendix: Reduction modulo a prime

The aim of this appendix is to establish some facts about the relation of field extensions over $\mathbb{Q}$ defined by a polynomial f with extensions of the prime field $\mathbb{F}_p$ defined by a reduction of f modulo p. We shall consider only extensions of the base fields. Generalization to relative extensions is possible, but will not be needed here. For this as well as for proofs we refer to the literature [Cas86, Mar77, Neu92]. We assume knowledge of basic Galois theory. (This appendix proceeds relatively quickly through material which is not needed elsewhere in the paper. The faint-hearted may want to skip it almost completely and only take Theorem 5 and the existence of the homomorphisms established in the last paragraph of this appendix as facts.)

For any prime p there is the p-adic valuation ν_p of the rational numbers given by $\nu_p(\frac{a}{b}) = n$ if $\frac{a}{b} = p^n\frac{c}{d}$ with $p \nmid c, d$. Taking the completion of the rationals with respect to this valuation we obtain the p-adic field $\mathbb{Q}_p$. There is a natural embedding of $\mathbb{Q}$ into $\mathbb{Q}_p$ and we will from now on consider $\mathbb{Q}$ as a subfield of $\mathbb{Q}_p$ without explicitly mentioning this embedding. This implies that we can embed the algebraic closure $\overline{\mathbb{Q}}$ of the rationals into the algebraic closure $\overline{\mathbb{Q}}_p$ of $\mathbb{Q}_p$. If we embed L (as L is a normal extension this image is unambiguous) into $\overline{\mathbb{Q}}_p$ we get a normal extension $L_p = \langle L, \mathbb{Q}_p \rangle$ of $\mathbb{Q}_p$. The Galois group $\mathrm{Gal}(L_p/\mathbb{Q}_p)$ of this extension then naturally embeds into $\mathrm{Gal}(L/\mathbb{Q})$. Note that $[L_p : \mathbb{Q}_p]$ might be strictly smaller than $[L : \mathbb{Q}]$ and the Galois groups be strictly contained in each other, because $\mathbb{Q}_p$ already contains algebraic irrationalities.

As $\mathbb{Q}_p$ is complete there is a unique extension of ν_p to L_p. For $K \in \{\mathbb{Q}_p, L_p\}$ we consider the discrete valuation rings $\mathcal{O}(K) = \{k \in K \mid \nu_p(k) \geq 0\}$ and their maximal ideals $\mathcal{P}(K) = \{k \in K \mid \nu_p(k) > 0\} \lhd \mathcal{O}(K)$. We note that $\mathcal{O}(\mathbb{Q}_p) = \mathbb{Z}_p$. By definition we have $\mathcal{O}(L_p) \cap \mathbb{Q}_p = \mathbb{Z}_p$ and $\mathcal{P}(L_p) \cap \mathbb{Q}_p = \mathcal{P}(\mathbb{Q}_p)$. As $1/k \in \mathcal{O}(K)$ for $k \in K \setminus \mathcal{O}(K)$ (by the definition of a valuation ring), we have $\langle \mathbb{Q}_p, \mathcal{O}(L_p) \rangle = L_p$ while $M := \langle \mathbb{Z}_p, \mathcal{P}(L_p) \rangle$ usually is smaller than $\mathcal{O}(L_p)$. We also note that the ring $\mathcal{O}(L)$ of algebraic integers in L is contained in $\mathcal{O}(L_p)$, because $\mathbb{Z} \subset \mathbb{Z}_p \subset \mathcal{O}(L_p)$ and because valuation rings are integrally closed. Table 1 illustrates this situation.

As $\mathcal{P}(K)$ is a maximal ideal in the integral domain $\mathcal{O}(K)$ the quotient ring $\mathcal{O}(K)/\mathcal{P}(K)$ is a field. This field is finite because the valuation is discrete. We have $\mathbb{Z}_p/\mathcal{P}(\mathbb{Q}_p) \cong \mathbb{F}_p$, the field with p elements. By the isomorphism theorem we get a natural embedding

$$\mathbb{Z}_p/\mathcal{P}(\mathbb{Q}_p) \cong M/\mathcal{P}(L_p) \hookrightarrow \mathcal{O}(L_p)/\mathcal{P}(L_p) =: \mathbb{F}_q,$$

thus $\mathbb{F}_q$ is a finite extension of $\mathbb{F}_p$.

There is a natural homomorphism (action modulo $\mathcal{P}(L_p)$) from $\mathrm{Gal}(L_p/\mathbb{Q}_p)$ onto $\mathrm{Gal}(\mathbb{F}_q/\mathbb{F}_p)$. The kernel of this mapping is usually called the *ramification subgroup*. If it is nontrivial we call p a *ramifying prime*. If L is the splitting field of f, every ramifying prime must divide the discriminant of $\mathbb{Q}(\alpha)/\mathbb{Q}$ and thus divides the discriminant of f. So there are only finitely many ramifying primes and we can avoid these when choosing a prime.

If $L_p/\mathbb{Q}_p$ does not ramify we have $\mathrm{Gal}(L_p/\mathbb{Q}_p) \cong \mathrm{Gal}(\mathbb{F}_q/\mathbb{F}_p)$. Accordingly,

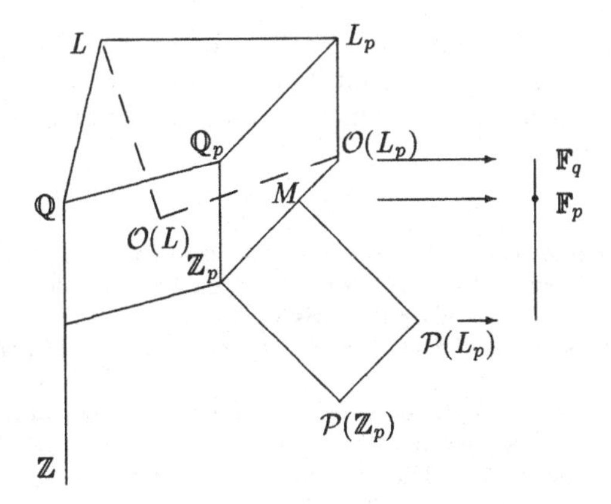

Table 1. Algebraic extensions

there is an element in $\mathrm{Gal}(L/\mathbb{Q})$ that corresponds to the Frobenius automorphism (taking p-th powers) generating $\mathrm{Gal}(\mathbb{F}_q, \mathbb{F}_p)$. We call this element in $\mathrm{Gal}(L/\mathbb{Q})$ the p-Frobenius automorphism. If L is the splitting field of f, the action of the p-Frobenius automorphism on the roots of f is the same as the action of the Frobenius automorphism on the roots of the reduction of f modulo p. The orbits of this automorphism then are in correspondence to the irreducible factors of f modulo p. We have shown:

Theorem 5 (Dedekind's theorem [vdW71, §66]) *For a prime p not dividing the discriminant of an irreducible rational polynomial f the degrees of the irreducible factors of f modulo p give the cycle shape of one element in the Galois group of f.*

Using analytic number theory this result can be strengthened to the Tschebotareff density theorem [Tsc25], by which the density of the non-ramifying primes for which the Frobenius automorphism has a given shape equals the proportion of elements of this shape in the Galois group G.

It can be shown that $\mathrm{Gal}(L/\mathbb{Q})$ is generated by all ramification subgroups. Unfortunately this property cannot be utilized easily, as each ramification subgroup is given by its action on the roots in L_p. However, there is no simple way (without knowing G) to identify the roots in L_p with those in $L_{p'}$ for two different primes p and p'. This identification would be needed to obtain the group they generate together.

Finally, for all powers of $\mathcal{P}(L_p)$ there is a natural ring homomorphism ("reduction modulo p^n"), the approximation map $\pi_i\colon \mathcal{O}(L_p) \to \mathcal{O}(L_p)/\mathcal{P}^i(L_p)$ that restricts to $\pi_i|_{\mathbb{Z}_p}\colon \mathbb{Z}_p \to \mathbb{Z}_p/\mathcal{P}^i(\mathbb{Q}_p) \cong \mathbb{Z}/p^i\mathbb{Z}$. Restricted to $\mathbb{Z}$ this is the well known reduction modulo p^n. We usually abbreviate π_1 to π.

Acknowledgements I'm indebted to Leonard Soicher for suggesting to me the use of stabilizers as described in Section 3 instead of orbit lengths. This collaboration was enabled by an HCM grant of the European Union, whose support I gratefully acknowledge. I would also like to thank Werner Nickel for many helpful comments on a first draft of this paper and Andrew Cutting for rectifying my English.

References

[AV94] Jean-Marie Arnaudies and Annick Valibouze, *Groupes de Galois de polynômes en degré 10 ou 11*, Rapport interne 94.50, Laboratoire informatique théoretique et programmation, Université Paris VI, 1994.

[Cas86] J.W.S. Cassels, *Local fields*, L.M.S. Student Texts, no. 3, Cambridge University Press, 1986.

[CHM] John H. Conway, Alexander Hulpke, and John McKay, *On transitive permutation groups*, to appear in LMS Journal of Computation and Mathematics.

[CM94] David Casperson and John McKay, *Symmetric functions, m-Sets and Galois groups*, Math. Comp. **63** (1994), 749–757.

[Coh93] Henri Cohen, *A course in computational algebraic number theory*, Graduate Texts in Mathematics 138, Springer, Heidelberg, 1993.

[Col95] Antoine Colin, *Formal computation of Galois groups with relative resolvents*, Applied algebra, Algebraic algorithms and Error-correcting Codes (Gérard Cohen, Marc Giustini, and Teo Mora, eds.), Lecture Notes in Computer Science 948, Springer, Heidelberg, 1995, pp. 169–182.

[DF89] Henri Darmon and David Ford, *Computational verification of M_{11} and M_{12} as Galois groups over Q*, Comm. Algebra **17** (1989), 2941–2943.

[DM96] John D. Dixon and Brian Mortimer, *Permutation groups*, Graduate Texts in Mathematics 163, Springer, Heidelberg, 1996.

[DSar] James H. Davenport and Geoff Smith, *Fast recognition of symmetric and alternating galois groups*, to appear.

[DST88] James H. Davenport, Y. Siret, and E. Tournier, *Computer algebra*, Academic Press, 1988.

[Eic93] Bettina Eick, *PAG-Systeme im Computeralgebrasystem GAP*, Diplomarbeit, Lehrstuhl D für Mathematik, Rheinisch-Westfälische Technische Hochschule, Aachen, 1993.

[EO95] Yves Eichenlaub and Michel Olivier, *Computation of Galois groups for polynomials with degree up to eleven*, Preprint, Université Bordeaux I, 1995.

[Gir83] K. Girstmair, *On the computation of resolvents and Galois groups*, Manuscripta Math. **43** (1983), 289–307.

[Gir87] K. Girstmair, *On invariant polynomials and their application in field theory*, Math. Comp. **48** (1987), 781–797.

[Göb95] Manfred Göbel, *Computing bases for rings of permutation-invariant polynomials*, J. Symb. Comput. **19** (1995), 285–291.

[Hul95] Alexander Hulpke, *Block systems of a Galois group*, Experimental Mathematics **4** (1995), no. 1, 1–9.

[Hul96] Alexander Hulpke, *Konstruktion transitiver Permutationsgruppen*, Ph.D. thesis, Rheinisch-Westfälische Technische Hochschule, Aachen, Germany, 1996.

[IAFM94] M. H. Klin I. A. Faradžev and M. E. Muzichuk, *Cellular rings and groups of automorphisms of graphs*, Investigations in algebraic theory of combinatorial objects (I. A. Faradžev, A. A. Ivanov, M. H. Klin, and A. J. Woldar, eds.), Mathematics and its Applications (Soviet Series), vol. 84, Kluwer, 1994, pp. 1–152.

[KK51] Marc Krasner and Leo A. Kaloujnine, *Produit complet des groupes de permutations et problème d'extension de groupes II*, Acta Sci. Math. (Szeged) **14**

(1951), 39–66.

[KP97] Jürgen Klüners and Michael E. Pohst, *On computing subfields*, J. Symb. Comput. **24** (1997), 385–397.

[Leo91] Jeffrey S. Leon, *Permutation group algorithms based on partitions, I: theory and algorithms*, J. Symb. Comput. **12** (1991), 533–583.

[LLL82] A.K. Lenstra, H.W. Lenstra, and L. Lovász, *Factoring polynomials with rational coefficients*, Math. Ann. **261** (1982), 515–534.

[LM85] Susan Landau and Garry Miller, *Solvability by radical is in polynomial time*, J. Comput. System Sci. **30** (1985), 179–208.

[Loo82] Rüdiger Loos, *Computing in algebraic extensions*, Symbolic and Algebraic Computation (Bruno Buchberger, George Edwin Collins, and Rüdiger Loos, eds.), Springer, Wien, 1982, 173–187.

[LPS87] Martin W. Liebeck, Cheryl E. Praeger, and Jan Saxl, *A classification of the maximal subgroups of the finite alternating and symmetric groups*, J. Algebra **111** (1987), 365–383.

[Mar77] Daniel A. Marcus, *Number fields*, Springer, Heidelberg, 1977.

[McK81] Brendan D. McKay, *Practical graph isomorphism*, Congr. Numer. **30** (1981), 45–87.

[Neu92] Jürgen Neukirch, *Algebraische Zahlentheorie*, Springer, Heidelberg, 1992.

[Noe16] Emmy Noether, *Der Endlichkeitssatz der Invarianten endlicher Gruppen*, Math. Ann. **77** (1916), 89–92.

[S+97] Martin Schönert et al., *GAP 3.4, patchlevel 4*, Lehrstuhl D für Mathematik, Rheinisch-Westfälische Technische Hochschule, Aachen, 1997.

[Sie82] Johannes Siemons, *On partitions and permutation groups on unordered sets*, Arch. Math. (Basel) **38** (1982), 391–403.

[SM85] Leonard Soicher and John McKay, *Computing Galois groups over the rationals*, J. Number Theory **20** (1985), 273–281.

[Soi] Leonard H. Soicher, personal communication.

[Sta73] Richard P. Stauduhar, *The determination of Galois groups*, Math. Comp. **27** (1973), 981–996.

[Sta79] Richard P. Stanley, *Invariants of finite groups and their applications to combinatorics*, Bull. Amer. Math. Soc. (N.S.) **1** (1979), 475–511.

[The97] Heiko Theißen, *Eine Methode zur Normalisatorberechnung in Permutationsgruppen mit Anwendungen in der Konstruktion primitiver Gruppen*, Dissertation, Rheinisch-Westfälische Technische Hochschule, Aachen, Germany, 1997.

[Tsc25] Nikolaj Tschebotareff, *Die Bestimmung der Dichtigkeit einer Menge von Primzahlen, welche zu einer gegebenen Substitutionsklasse gehören*, Math. Ann. **95** (1925), 191–228.

[vdW71] Bartel L. van der Waerden, *Algebra, erster Teil*, eighth ed., Heidelberger Taschenbücher 12, Springer, Heidelberg, 1971.

[Wie69] Helmut Wielandt, *Permutation groups through invariant relations and invariant functions*, Lecture notes, Department of Mathematics, The Ohio State University, 1969.

[Zas69] Hans Zassenhaus, *On Hensel factorization I*, J. Number Theory **1** (1969), 291–311.

CONSTRUCTION OF CO_3. AN EXAMPLE OF THE USE OF AN INTEGRATED SYSTEM FOR COMPUTATIONAL GROUP THEORY

ALEXANDER HULPKE[1] and STEVE LINTON

School of Mathematical and Computational Sciences, University of St Andrews, The North Haugh, St Andrews, Fife KY16 9SS, Scotland

1 Introduction

This paper aims to demonstrate, by example, a small sample of the capabilities of the GAP system [S+97] for computational algebra. We specifically focus on the advantages arising from the use of an integrated system such as GAP, which allows the easy combination of techniques from a range of areas, without requiring the user to have a detailed knowledge of the algorithms used.

The sporadic group Co_3 has a faithful permutation representation on 276 points which is unusually small for a group of its size. We want to construct this permutation representation by way of a chain of subgroups of ascending order. In this process we will construct explicitly the sporadic simple groups M_{22} and HS together with associated graphs and codes. Our guide in this is the ATLAS of Finite Simple Groups [CCN+85], which contains a variety of information about the groups of interest, including very terse "constructions" – outlines of settings in which these groups can occur.

We will use GAP (version 3.4, patchlevel 4, including the GRAPE [Soi93] and GUAVA [BCMR] share packages) to realise these constructions. We will see that the integration of many algorithms in one system will permit us to follow the path outlined in theory with concrete constructions.

From a computational stand-point, this is not a very large or difficult computation. It is interesting, however, because it uses a very wide range of techniques, and because it demonstrates one important way in which an integrated system such as GAP can be used.

In giving our example, we give the necessary GAP commands in full, although we sometimes abbreviate the resulting output to save space. The input lines and the full output can be found on the web page

`http://www-gap.dcs.st-and.ac.uk/~ahulpke/paper/bathexample.html`

The complete example session took about 16 minutes on the authors' 200MHz PentiumPro PC running Linux and used about 10MB of GAP workspace.

Some of the calculations rely on random selections. This can make it difficult to reproduce the results. The GAP session presented here results from starting GAP 3.4.4 (which initializes the random number generator to a defined state) and calling exactly the commands listed. If the reader repeats this, she should end up

[1]The first author has been supported by EPSRC Grant GL/L21013

with the same objects. In other circumstances or other versions, input (such as explicit points) that is taken from former output has to be modified accordingly. Though the output of some functions that rely on random methods is abbreviated here (but can be seen in full on the mentioned web page) output relevant for later input is always given.

2 The construction of $M_{22}.2$

We begin with polynomials. By defining an indeterminate, we obtain $\mathbb{F}_2[x]$. We assign a name to this indeterminate so that it will be displayed in a nice way.

```
gap> x:=Indeterminate(GF(2));
X(GF(2))
gap> x.name:="x";;
```

Now we work with polynomials. We define a small degree polynomial (this is where the number 23 comes in) and factor it. Over $\mathbb{Z}$ it would split just into two factors, but over $\mathbb{F}_2$ we get further factors. We take one of them.

```
gap> f:=x^23-1;
Z(2)^0*(x^23 + 1)
gap> Factors(f);
[ Z(2)^0*(x + 1), Z(2)^0*(x^11 + x^10 + x^6 + x^5 + x^
  4 + x^2 + 1), Z(2)^0*(x^11 + x^9 + x^7 + x^6 + x^
  5 + x + 1) ]
gap> f:=First(Factors(f),i->Degree(i)>1);
Z(2)^0*(x^11 + x^10 + x^6 + x^5 + x^4 + x^2 + 1)
```

We next define a code from this polynomial. For this we use the share package GUAVA [BCMR].

```
gap> RequirePackage("guava");
  GUAVA, Version 1.3
  Jasper Cramwinckel, Erik Roijackers, Reinald Baart, Eric Minkes
gap> cod:=GeneratorPolCode(f,23,GF(2));
a cyclic [23,12,1..7]3 code defined by generator polynomial over
GF(2)
```

This code is the binary Golay code, we check a well-known property.

```
gap> IsPerfectCode(cod);
true
```

If we extend the code by a parity bit, we get the extended Golay code. This code has automorphism group M_{24}. The weight distribution for this extended code, that GUAVA can compute for us, gives the well known numbers of subsets in the corresponding Steiner system.

```
gap> ext:=ExtendedCode(cod);
a linear [24,12,8]4 extended code
gap> WeightDistribution(ext);
[ 1, 0, 0, 0, 0, 0, 0, 0, 759, 0, 0, 0, 2576, 0, 0, 0,
  759, 0, 0, 0, 0, 0, 0, 0, 1 ]
```

GUAVA uses an external program desauto [Leo91] to compute the automorphism group of this code. We check that the size of the obtained group is indeed the size of M_{24} and that it acts quintuply transitive.

```
gap> m24:=AutomorphismGroup(ext); m24.name:="m24";
Group( ( 1, 2)( 6,15,24,18)( 7,14,12,19)( 9,17,16,20)
[...]
gap> Size(m24);
244823040
gap> Transitivity(m24,[1..24]);
5
```

$M_{22}.2$ is the stabilizer of a duad (a set of size 2) in M_{24}. As M_{24} acts quintuply transitive it is unimportant which points we select. (The simple group M_{22} itself could be obtained by computing the pointwise stabilizer, using the GAP operation OnTuples instead.) The action of $M_{22}.2$ on the remaining points is faithful.

```
gap> m22a:=Stabilizer(m24,[23,24],OnSets);
Subgroup( m24, [ ( 1, 3)( 5,17)( 6,18)( 7,10)( 8,16)
[...]
gap> Size(m22a);
887040
gap> m22a:=Operation(m22a,[1..22]);m22a.name:="m22a";;
Group( ( 1, 3)( 5,17)( 6,18)( 7,10)( 8,16)( 9,13)(11,20)
[...]
gap> Size(m22a);
887040
```

3 The construction of *HS*

Now we need a little bit of theory. The ATLAS [CCN+85] tells us on page 80:

Graph G.2: the automorphism group of the graph of D.G.Higman 1
and C.C. Sims, a rank 3 graph of valence 22 on 100 vertices. 2
Any given vertex has 22 neighbours (points)*,*
and each of the remaining 77 vertices is joined to 6 of these points 3
and may be labeled by the corresponding hexad. 4
Two of these 77 vertices are joined just
if the corresponding hexads are disjoint. 5

and on the bottom of the page:

```
Order   Index Structure  G.2    Character  Graph
443520    100 M22       : M22:2 1a+22a+77a point     6
```

We deduce that (1) $HS.2$ acts on a graph Γ (the Higman-Sims graph) with 100 vertices (2). There only is one permutation representation of degree 100 for $HS.2$ (6), the point stabilizer is of type $M_{22}.2$. The graph has rank 3 (2), that is the point stabilizer has 3 orbits on the 100 points. A vertex in Γ has 22 neighbours (2) and therefore the point stabilizer must have an orbit of length 22. As it only has 3 orbits there is one remaining orbit of length $100 - 1 - 22 = 77$.

We will construct this graph using the point stabilizer and therefore have to construct $M_{22}.2$ in its intransitive permutation action on 100 points with orbits of lengths 1,22 and 77.

3.1 A representation of $M_{22}.2$ on 77 points

The first step of this construction will be to obtain the transitive action of $M_{22}.2$ on 77 points. We will get it by acting on the cosets of a suitable subgroup. To find this subgroup we use the classification of maximal subgroups of M_{22} given in the ATLAS, in which we find the line:

```
Order   Index  Structure  G.2        Abstract   Mathieu
 5760      77  2^4:A_6    2^4:S_6    N(2A^4)    hexad
```

This tells us:

- M_{22} has a permutation representation of degree 77 which extends to $M_{22}.2$.
- The point stabilizer in this action is a subgroup $H < M_{22}.2$ of type $2^4 : S_6$. This subgroup has a normal subgroup E of type 2^4, of which it must be the normalizer in $M_{22}.2$.
- As it has odd index 77, H must contain a full Sylow 2 subgroup of $M_{22}.2$. Therefore E, which certainly is contained in a Sylow subgroup, must be normal also in the Sylow 2 subgroup.
- As stabilizer of a hexad, H (and therefore E as well) stabilizes a set of 6 points in the permutation representation on 22 points.

To get H we will look for normal subgroups of order 2^4 in a Sylow 2 subgroup. As a Sylow subgroup is polycyclic, we convert it to an AgGroup. This is a special representation for solvable groups in which the computations we want to do run more quickly.

```
gap> s:=SylowSubgroup(m22a,2);;
gap> a:=AgGroup(s);
Group( g1, g2, g3, g4, g5, g6, g7, g8 )
```

We compute all elementary abelian normal subgroups of size 16 in this group, there are 5 of them.

```
gap> n:=Filtered(NormalSubgroups(a),i->Size(i)=16
>         and IsElementaryAbelian(i));
[ Subgroup( Group( g1, g2, g3, g4, g5, g6, g7, g8 ),
    [ g1, g3, g6*g7, g8 ] ), [...]
```

The component bijection of the group a is an homomorphism back into the permutation group. We use it to get these subgroups as permutation groups.

```
gap> n:=List(n,i->Image(a.bijection,i));;
```

To pick the right E, we could now look at the sizes of the normalizers of the subgroups in n and it would turn out that only one of them has the right normalizer size. We happen to know, however, that E acts regularly. (The remaining 6 points form the hexad stabilized by the normalizer.) There only is one group satisfying these conditions. We take this group and compute its normalizer.

```
gap> e:=Filtered(n,i->IsRegular(i,PermGroupOps.MovedPoints(i)));;
gap> Length(e);
1
gap> e:=e[1];;
gap> h:=Normalizer(m22a,e);;
```

Now we compute the action on the cosets.

```
gap> mop:=Operation(m22a,RightCosets(m22a,h),OnRight);;
gap> DegreeOperation(mop,[1..100]);
77
```

It is worth mentioning, that the following command

```
mop:=Operation(m22a,CanonicalRightTransversal(m22a,h),
  OnCanonicalCosetElements(m22a,h));;
```

produces essentially the same result, but usually works much more quickly and uses less space.

The corresponding operation homomorphism is the link between the two permutation representations.

```
gap> ophom:=OperationHomomorphism(m22a,mop);;
```

3.2 Representing $M_{22}.2$ on 100 points

To obtain the action on 100 points we form the direct product of two copies of $M_{22}.2$ acting on 22 and on 77 points respectively and take its diagonal subgroup. We can simply regard 100 as the fixed point.

We get the diagonal subgroup by taking the product of the images of each generator under both embeddings.

```
gap> dp:=DirectProduct(m22a,mop);;
gap> emb1:=Embedding(m22a,dp,1);;
gap> emb2:=Embedding(mop,dp,2);;
gap> diag:=List(m22a.generators,
>            i->Image(emb1,i)*Image(emb2,Image(ophom,i)));;
```

We create the group generated by these diagonal elements and give it a name. (Note that GAP requires us to give the identity here, as **diag** is a list which could be empty.)

```
gap> diag:=Group(diag,());;
gap> diag.name:="M22.2-99";
```

3.3 The Higman-Sims graph

It is now time to construct the graph. For this we use the share package GRAPE.

```
gap> RequirePackage("grape");
Loading  GRAPE 2.31  (GRaph Algorithms using PErmutation groups),
by L.H.Soicher@qmw.ac.uk.
```

GRAPE works with graphs, making maximum use of known automorphisms. We start with an empty graph on 100 vertices on which $M_{22}.2$ acts and adjoin orbits of edges.

```
gap> gamma:=NullGraph(diag,100);
rec( isGraph := true,
     group := M22.2-99,
     representatives := [ 1, 23, 100 ],
[...]
```

Graphs in GRAPE are directed, so we have to add each edge in two directions. The first edge is to connect the point 100 to the orbit of size 22 to make up for valence 22 (2):

```
gap> AddEdgeOrbit(gamma,[1,100]);AddEdgeOrbit(gamma,[100,1]);
```

Now we connect vertices in the orbit of length 77 with vertices in the orbit of length 22, using line (4) of the description from the ATLAS.

The construction of direct products for permutation groups in GAP maps the first factor on the same permutations and the second factor on permutations shifted by the degree of the first factor. Therefore we can use the representative 23 for the orbit of length 77 and know that this point is the point stabilized by the subgroup $H \leq M_{22}.2$. The corresponding hexad (in the first 22 points) must be stabilized by H. We have already observed that H has only one orbit of length 6 in its action on 22 points. We fetch this orbit to obtain the hexad and add the corresponding edges. As the computation of the Sylow subgroup used random methods the reader may get a different hexad as a result. See the comment on random methods at the end of the first section.

```
gap> hexad:=First(Orbits(h,[1..22]),i->Length(i)=6);
[ 2, 8, 17, 15, 22, 6 ]
gap> for i in hexad do AddEdgeOrbit(gamma,[i,23]);
> AddEdgeOrbit(gamma,[23,i]); od;
```

Looking at the neighbourhood of 23 we see that we have not yet reached valence 22.

```
gap> Adjacency(gamma,23);
[ 2, 6, 8, 15, 17, 22 ]
```

Indeed, we still have to consider the next rule ($_5$), to determine edges within the orbit of size 77. Because we have a group acting, it is sufficient to test for each representative of the orbits of the stabilizer of 23, whether it is joined to 23. The other edges in the orbits will be added automatically by GRAPE.

```
gap> stab:=Stabilizer(diag,23);;
gap> orbs:=Orbits(stab,[24..99]);;
gap> orbreps:=List(orbs,i->i[1]);
[ 24, 39 ]
```

The hexad corresponding to a point a in the 77-point orbit consists (we constructed the graph that way) of the points in the first orbit adjacent to a. The hexad corresponding to 23 is already known as **hexad**.

```
gap> Intersection(hexad,Adjacency(gamma,24));
[ 15, 17 ]
gap> Intersection(hexad,Adjacency(gamma,39));
[ ]
```

This tells us that 23 must be joined only to 39. We add this edge orbit.

```
gap> AddEdgeOrbit(gamma,[23,39]); AddEdgeOrbit(gamma,[39,23]);
```

Now we have completed the graph we want. We check that it is indeed a simple graph, look at the adjacency of 23 and find out that it is distance regular.

```
gap> IsSimpleGraph(gamma);
true
gap> Adjacency(gamma,23);
[ 2, 6, 8, 15, 17, 22, 39, 42, 46, 49, 52, 56, 63, 68, 70, 76, 80,
  81, 86, 90, 93, 94 ]
gap> IsDistanceRegular(gamma);
true
```

Finally, we compute the automorphism group of **gamma**. Once again an external program (nauty, [McK90]) is used, but we do not see it:

```
gap> aug:=AutGroupGraph(gamma);;
gap> Size(aug);
88704000
```

Indeed the size is correct for $HS.2$. We can confirm this by looking at the composition series. GAP identifies simple composition factors by their size and the degree of a primitive permutation action, using the classification of finite simple groups.

```
gap> DisplayCompositionSeries(aug);
(G) (3 gens, size 88704000)
 | Z(2)
(S) (2 gens, size 44352000)
 | HS
(1) (0 gens)
```

We get HS as the derived subgroup of $HS.2$.

```
gap> hs:=DerivedSubgroup(aug);;
```

4 Getting Co_3

Now we want to construct Co_3. The ATLAS tells us (page 134) that Co_3 has a permutation representation on 276 points in which the point stabilizer is $McL{:}2$. Furthermore HS is a maximal subgroup of Co_3. If we can construct HS in this permutation representation on 276 points, then, together with one further suitably constructed element it will generate Co_3. The construction is not mentioned explicitly in the ATLAS but this "amalgamation" of subgroups is a standard technique for the construction of sporadic groups [PW90].

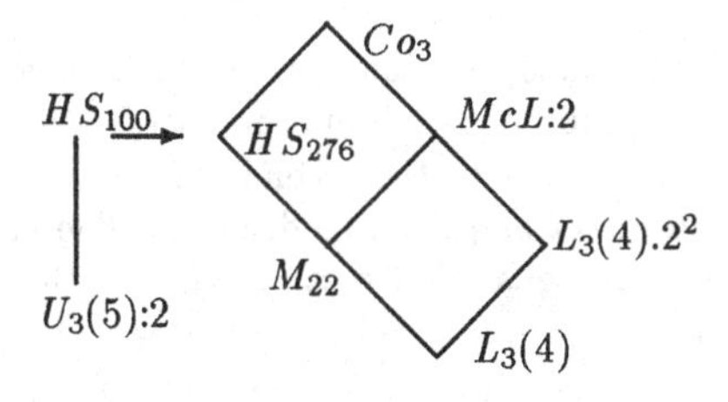

The picture will illustrate the construction process. The first step will be to construct HS as a permutation group acting on 276 points. The smallest degrees of faithful permutation representations of HS are 100, 176 and 1100. Therefore, as a subgroup of Co_3, HS has two orbits of lengths 100 and 176 respectively. (HS cannot have any fixed points because it is not contained in $McL{:}2$.)

The stabilizer in HS of a point in an orbit of length 176 is of type $U_3(5){:}2$, but there is no description of how to find such a subgroup $U \leq HS$ of index 176 when only the representation of HS on 100 points is known. Instead we will try to find elements by random search in the permutation representation of HS on 100 points that generate such a subgroup U. As the index $[HS : U]$ is relatively small we have good chances. We just have to decide which of the element orders occurring in $U_3(5){:}2$ to try to find a suitable U.

To determine our chances of success we use the character tables of $U_3(5)$, $U_3(5){:}2$ and HS. We can look them up in the ATLAS but they are more conveniently obtained from a GAP data base.

```
gap> ct:=CharTable("U3(5)");;ct2:=CharTable("U3(5).2");;
gap> cths:=CharTable("hs");;
gap> ct.orders;ct2.orders;cths.orders;
[ 1, 2, 3, 4, 5, 5, 5, 5, 6, 7, 7, 8, 8, 10 ]
[ 1, 2, 3, 4, 5, 5, 5, 6, 7, 8, 10, 2, 4, 6, 8, 10, 12, 20, 20 ]
[ 1, 2, 2, 3, 4, 4, 4, 5, 5, 5, 6, 6, 7, 8, 8, 8, 10, 10,
11, 11, 12, 15, 20, 20 ]
```

The character tables show that $U_3(5)$:2 contains elements of order 12 which $U_3(5)$ does not. Thus an element of order 12 will ensure that we find U and not its derived subgroup U'. Furthermore HS contains only one class of elements of order 12 and this class is self-centralizing. Therefore there is a chance of 1/12 that a random element of HS has order 12 and we know that there is a conjugate of $U_3(5)$:2 in HS that contains this element. We search for such an element of order 12 which we denote by `e1` and compute its 6th power `e2`.

```
gap> repeat e1:=Random(hs);until OrderPerm(e1)=12;
gap> e2:=e1^6;;
```

As 6 is divisible by 2, `e2` must lie in $U_3(5)$ and is therefore in the first class of elements of order 2. This class in $U_3(5)$:2 has 525 elements

```
gap> ct2.classes[2];
525
```

Along the lines of [Lin95] we can now estimate how much random search is required to find a subgroup of HS isomorphic to $U_3(5):2$. In HS, the 6th power of an element of order 12 lies in the second class. (GAP stores only prime power maps from which all power maps can be deduced. Therefore we compute the sixth power as the cube of the square.) We can read from the table that the class of `e2` in HS is of order 5775.

```
gap> cths.orders[21];
12
gap> cths.powermap[3][cths.powermap[2][21]];
2
gap> cths.classes[2];
5775
```

If we take random conjugates of `e2` about every 10th conjugate will lie in a maximal subgroup of type $U_3(5)$:2 which contains also `e1`.

A similar count of pairs for the maximal subgroups of $U_3(5)$:2 (which is left to the reader) establishes that at least every second conjugate of `e2` in $U_3(5)$:2 together with `e1` generates the full group $U_3(5)$:2. Therefore we can assume that in the mean at least one of 20 conjugates of `e2` in HS together with `e1` will generate a maximal subgroup U isomorphic to $U_3(5)$:2 and we can find it by a random search. (As U operates transitively on 100 points the `until` statement has to use the index and cannot check for orbit lengths which otherwise would be quicker.)

```
gap> cnt:=0;;repeat u:=Subgroup(aug,[e1,e2^Random(hs)]);
> cnt:=cnt+1;until Index(hs,u)=176;cnt;
26
```

We were a little bit unlucky in that we had to try 26 random subgroups. (Anyhow this random search took only about 7 seconds on the authors' system.)

We now construct the operation of HS on the cosets of U. Again we use the special operation on canonical coset representatives (which saves over 40MB of memory and several minutes of runtime). We test that its image is indeed primitive. (So U is indeed a maximal subgroup and we don't have to rely on the list of maximal subgroups being complete.) Afterwards we construct a diagonal action of HS on 100+176 points by mirroring the process used above for $M_{22}.2$.

```
gap> hsop:=Operation(hs,CanonicalRightTransversal(hs,u),
> OnCanonicalCosetElements(hs,u));;
gap> IsPrimitive(hsop,[1..176]);
true
gap> ophom:=OperationHomomorphism(hs,hsop);;
gap> dp:=DirectProduct(hs,hsop);;
gap> emb1:=Embedding(hs,dp,1);;
gap> emb2:=Embedding(hsop,dp,2);;
gap> diag:=List(hs.generators,
>              i->Image(emb1,i)*Image(emb2,Image(ophom,i)));;
gap> diag:=Group(diag,());;diag.name:="hs-276";;
```

4.1 Search for a further element

The strategy for finding a further element is as follows: Such elements exist in $McL{:}2$. At this point, however, we don't have $McL{:}2$ in an appropriate action. We may choose conjugates such that $HS \cap McL{:}2 = M_{22}$ ($McL{:}2$ is a stabilizer of a point in the action of Co_3 on 276 points and M_{22} is a point stabilizer in the orbit of length 100 of HS).

Assume that T is a subgroup of HS such that $N_{McL:2}(T) \not\leq HS$. In this case there are elements in $N_{McL:2}(T)$ which are not in HS and we can take such an element as a further generator for Co_3. We will see that (with appropriate choice of T) such elements can be found directly from the automorphism action they induce on T. We shall obtain this action from the automorphism group of T.

A look at the ATLAS list of maximal subgroups of $McL{:}2$ shows that the outer automorphism of M_{22} cannot be realized in $McL{:}2$ and therefore we cannot simply choose $T = M_{22}$. If we look at the point stabilizer in the action of M_{22} on 22 points, we have more luck: This group is of type $L_3(4)$ and while HS contains a maximal subgroup of type $L_3(4).2_1$, $McL{:}2$ contains $L_3(4).2^2$. So we can choose T to be this subgroup $L_3(4).2_1$.

Furthermore, T is just the stabilizer in HS of an edge of the Higman-Sims graph and is therefore easily obtainable. Again we use that the direct product construction mapped the copy of HS acting on 100 points to the same permutations, so we can simply use the graph we have constructed already and do not need to do any point translation.

```
gap> adj:=Adjacency(gamma,1);
[ 79, 80, 81, 82, 83, 84, 85, 86, 87, 88, 89, 90, 91,
92, 93, 94, 95, 96, 97, 98, 99, 100 ]
gap> t:=Stabilizer(diag,[1,adj[1]],OnSets);;
```

Because we computed T as stabilizer it might have many generators. For the computation of the automorphism group of T, however, it will be helpful to have a subgroup with fewer generators. Therefore we look at subgroups generated by two random elements of T and find quickly two elements which generate a subgroup that equals T and use this generated subgroup as T further on.

```
gap> cnt:=0;;repeat s:=Subgroup(diag,[Random(t),Random(t)]);
> cnt:=cnt+1;until Size(s)=Size(t);cnt;t:=s;
1
```

We know (well, have strong indications. For a proof we would have to classify subgroups of McL isomorphic to $L_3(4)$) that there is an element $g \in McL{:}2$ which normalizes T and is not contained in HS. We want to obtain this element from its automorphism action on T.

4.2 Finding suitable automorphisms

Now we start by computing the automorphism group of T. (In principle it should be possible to use the command **AutomorphismGroup** here. There is, however, a problem with GUAVA that we used before: it overwrites the dispatcher function for **AutomorphismGroup**).

```
gap> aus:=t.operations.AutomorphismGroup(t);;
gap> Size(aus);
241920
```

As we want an outer automorphism, we take the subgroup of inner automorphisms of T (GAP has already computed a generating set and stored it) and consider a transversal in the outer automorphism group. As we are looking for an automorphism of order two, we select those representatives from this transversal whose order modulo the inner automorphisms is 2. (It might be worth mentioning that these few calculations in the automorphism group, together with the calculation of the automorphism group itself, actually take up more than 65 percent of the total runtime!)

```
gap> inner:=Subgroup(aus,aus.innerAutomorphisms);;
gap> Index(aus,inner);
6
gap> rt:=RightTransversal(aus,inner);;
```

```
gap> automs:=Filtered(rt,i->i^2 in inner and not i in inner);;
gap> Length(automs);
3
gap> List(automs,i->Order(aus,i));
[ 2, 2, 2 ]
```

We were lucky in that all the coset representatives we got have order 2 (and that is a proof that every coset contains a representative of order 2. The reader starting with different initial data may end up with representatives of composite order. If this is the case, she can multiply these with inner automorphisms to obtain representatives of order 2 for the same cosets. This may require some random searching).

There are three possible representatives. We will continue with all three of them for now. It will turn out later that only one of them is a suitable candidate.

If we look at the orbits of T

```
gap> ot:=Orbits(t,[1..276]);;
gap> List(ot,Length);
[ 2, 56, 42, 56, 120 ]
```

we see that the only orbits of T that its normalizer in Co_3 may swap are those of length 56, the other orbits must remain fixed. (On the other hand, as this normalizer together with HS generates Co_3 it then must swap these to obtain a transitive action.) For each orbit of length $\neq 56$ and for the fused orbit of length 112 we then construct a permutation in the symmetric group on this orbit which by conjugation induces the same action on the image of T projected on this orbit as the automorphism does. Except for the orbit of length 2 this permutation on each (fused) orbit is unique, as the image of the action of T on the orbit has trivial centralizer in the full symmetric group.

4.3 Constructing a permutation

We shall need a small GAP function to compute such a permutation. For this we use a theorem (see for example [DM96, Theorem 4.2B]) which states that an automorphism of a permutation group is induced by conjugation with a permutation in the symmetric group if and only if it maps a point stabilizer to another point stabilizer. In this case, the automorphism maps cosets of the one stabilizer to cosets of the other. This can be used to define a permutation, which then induces this automorphism.

The GAP function `MappingPermListList` creates a permutation which maps one list onto another. From this we will build another function that will try to realize automorphisms on orbits. This function gets a permutation group `grp`, an automorphism `aut` of `grp` and a domain `dom` on which a permutation action will be computed. If the group has two orbits on this domain, the function will create a permutation that will swap both orbits. If the action cannot be induced by a permutation, `false` is returned.

```
PermutationByAutomorphism := function(grp,aut,dom)
local op,oh,p,s,sim,simp,rt,rtim,extelm,l1,l2;
 # We compute the action on the given domain and transfer
 # the automorphism to this permutation action
 op:=Operation(grp,dom);
 oh:=OperationHomomorphism(grp,op);
 aut:=GroupHomomorphismByImages(op,op,op.generators,
      List(op.generators,i->Image(oh,Image(aut,
                                 PreImagesRepresentative(oh,i)))));
 aut.isMapping:=true;
 # just to save time (otherwise GAP will test that
 # it is indeed a homomorphism, but we know this already)
 # compute stabilizer and images
 s:=Stabilizer(op,1);
 sim:=Image(aut,s);

 # is the image a stabilizer? It is if it has an orbit of length 1
 simp:=Filtered(Orbits(sim,[1..Length(dom)]),i->Length(i)=1);
 if Length(simp)=0 then
  return false; # it cannot be induced by a permutation action
 fi;

 # the permutation can be obtained by the induced action on
 # the right cosets.
 simp:=simp[1][1]; #image base point
 rt:=RightTransversal(op,s);
 rtim:=List(rt,i->Image(aut,i));
 l1:=List(rt,i->1^i);l2:=List(rtim,i->simp^i);

 # we got the images, make a permutation from them.
 if Length(Orbits(grp,dom))=1 then
  extelm:=MappingPermListList(l1,l2);
 else
  # if we have two orbits, we have to ensure they get swapped
  extelm:=MappingPermListList(Concatenation(l1,l2),
                              Concatenation(l2,l1));
 fi;
 # test whether the computed element indeed fulfills the
 # specifications (This is a safety test only)
 if ForAny(op.generators,i->i^extelm<>i^aut) then
  Error("something went wrong");
 fi;
 # finally move the points acted on to the original domain.
 return extelm^MappingPermListList([1..Length(dom)],dom);
end;
```

It turns out, that of the three candidates we had, only one is induced by a permutation on the orbit of length 120.

```
gap> lo:=First(ot,i->Length(i)=120);;
gap> automs:=Filtered(automs,
>             i->PermutationByAutomorphism(t,i,lo)<>false);;
gap> Length(automs);
1
gap> autom:=automs[1];;
```

Now we simply build a permutation using our function. We get unique permutations for all orbits of length $\notin \{2, 56\}$, and one permutation to swap the two orbits of length 56.

The product of all these is the permutation we need.

```
gap> pos:=Filtered([1..Length(ot)],i->Length(ot[i])=56);
[ 2, 4 ]
gap> perms:=List(ot{Difference([1..5],pos)},
>              i->PermutationByAutomorphism(t,autom,i));;
gap> element:=Product(perms)*PermutationByAutomorphism(t,
>                            autom,Concatenation(ot{pos}));
```

The permutation `element` is unique but for the orbit of length 2. The element we are looking for might swap the two points in this orbit or it might not and so we will have to try both possibilities. In this situation, however, it turns out that swapping both points gives the right permutation and we only show this part of the calculation, for which we change the permutation.

```
gap> ot[1];
[ 1, 79 ]
gap> 1^element;
1
gap> element:=element*(1,79);;
```

Now we can generate Co_3. Again we check that everything went well

```
gap> co3:=Group(Concatenation(diag.generators,[element]),());;
gap> Size(co3);
495766656000

gap> DisplayCompositionSeries(co3);
<G> (3 gens, size 495766656000)
 | Co(3)
<1> (0 gens)
```

Using the same techniques we could now continue and form Co_1 on 98280 points by adding an outer automorphism to $HS < Co_3$. The large number of points however makes this rather elaborate and unsuitable for a presentation in this form.

5 Concluding remarks

5.1 Summary of example and points to note

We started our calculation with a polynomial over $\mathbb{F}_2$ and obtained a code from its factors. From the code we constructed M_{24} and hence $M_{22}.2$. In $M_{22}.2$ acting on 22 points, we found a subgroup of index 77 and so obtained an action on 100 points. We constructed a graph on 100 vertices, on which $M_{22}.2$ acts. The full automorphism group of this graph is $HS.2$. We formed its derived subgroup HS as a subgroup of Co_3 on 276 points and, within HS, a subgroup $L_3(4).2$. From an automorphism of this subgroup we obtained a permutation contained in Co_3, but not in HS. Together with generators of HS, this permutation generates Co_3.

While these constructions were known before in theory, and had been performed relatively explicitly by hand, most of the objects involved are not given explicitly in a book or article. Using the very terse descriptions in the ATLAS, informed by a general knowledge of what computations are likely to be feasible, we have turned a piece of mathematics relatively painlessly into a program. In this presentation, we have, of course, omitted some false starts. Nevertheless, the interactive nature of GAP makes it relatively easy to develop programs of this kind. At every stage, the full structure of all the objects is exposed, so that we can ask GAP mathematical questions about each code, graph or group that we construct.

As well as its considerable library of built-in algorithms, especially for group theory, a key role of GAP in this work was to provide common interfaces and data structures for techniques from a number of mathematical areas, so that we could, for example, construct M_{24} using tools from coding theory, manipulate it as a group and apply the results in graph theory.

Another point to note is the pattern of working. The combination of interactive capabilities and programmability of GAP allows a middle road between purely interactive exploration and programming. Purely interactive work would not (for example) have allowed us to find the "extra" element of Co_3, because the conversion from automorphisms to permutations was not built into GAP. On the other hand, a non-interactive "design – program – test" model would have been too cumbersome for the main thread of the construction, which is more easily found by interactive experiment. A normal way of working with GAP is to explore interactively, keeping a log file as a record, and stopping, from time to time to program and debug functions that seem necessary or useful, which thereafter act as library extensions.

5.2 Further information

Further information about Computational Group Theory in general can be found in excellent survey articles [Neu95, Ser97]. The GAP system itself, and a great deal of associated information, can be found on the GAP World Wide Web site (`http://www-gap.dcs.st-and.ac.uk/~gap`). In particular the site includes the GAP distribution itself, details about the authors of GAP and of the contributed Share Packages and information about the forthcoming version 4 of the system.

Acknowledgements The development of GAP has been supported by grants (listed under `http://www-gap.dcs.st-and.ac.uk/~gap/Info/funding.html`) for which we are indebted to the granting bodies. We would also like to thank the referee for many helpful comments.

References

[BCMR] Reinalt Baart, Jasper Cramwinckel, Eric Minkes, and Erik Roijackers, *GUAVA: Users manual.*

[CCN+85] John H. Conway, Robert T. Curtis, Simon P. Norton, Richard A. Parker, and Robert A. Wilson, ATLAS *of finite groups*, Oxford University Press, 1985.

[DM96] John D. Dixon and Brian Mortimer, *Permutation groups*, Graduate Texts in Mathematics, vol. 163, Springer, Heidelberg, 1996.

[Leo91] Jeffrey S. Leon, *Partition backtrack programs, users manual*, University of Illinois at Chicago, Mathematics Dept., m/c 249, 1991.

[Lin95] Steve A. Linton, *The art and science of computing in large groups*, Proceedings of CANT '92 (Wieb Bosma and Alf J. van der Poorten, eds.), Kluwer, 1995, 91–109.

[McK90] B. D. McKay, *nauty user's guide (version 1.5), technical report tr-cs-90-02*, Australian National University, Computer Science Department, ANU, 1990.

[Neu95] Joachim Neubüser, *An invitation to computational group theory*, Groups '93 Galway/St. Andrews (C. M. Campbell, T. C. Hurley, E. F. Robertson, S. J. Tobin, and J. J. Ward, eds.), London Mathematical Society Lecture Note Series, vol. 212, Cambridge University Press, 1995, 457–475.

[PW90] Richard A. Parker and Robert A. Wilson, *The computer construction of matrix representations of finite groups over finite fields*, J. Symb. Comput. **9** (1990), 583–590.

[S+97] Martin Schönert et al., *GAP 3.4, patchlevel 4*, Lehrstuhl D für Mathematik, Rheinisch-Westfälische Technische Hochschule, Aachen, 1997.

[Ser97] Ákos Seress, *An introduction to computational group theory*, Notices Amer. Math. Soc. **44** (1997), 671–679.

[Soi93] Leonard H. Soicher, *GRAPE: a system for computing with graphs and groups*, Groups and Computation (Larry Finkelstein and William M. Kantor, eds.), DIMACS: Series in Discrete Mathematics and Theoretical Computer Science, vol. 21, American Mathematical Society, Providence, RI, 1993, pp. 287–291.

EMBEDDING SOME RECURSIVELY PRESENTED GROUPS

D. L. JOHNSON

Department of Mathematics and Computer Science, UWI Mona, Kingston 7, Jamaica

Abstract

We seek to illustrate the Higman Embedding Theorem by finding actual embeddings of various popular recursively presented groups in finitely presented ones, and are successful in at least one case.

1 Finite symmetric groups

The symmetric group $S_n \cong \mathrm{Sym}\,\{1,2,\ldots,n\}$ of degree $n \in \mathbf{N}$ is generated by symbols x_i corresponding to the transposition $(i\; i+1)$, $1 \le i \le n-1$, and defined by the relations

$$\left.\begin{array}{lll} {x_i}^2 & , & 1 \le i \le n-1, \\ (x_i x_{i+1})^3 & , & 1 \le i \le n-2, \\ (x_i x_j)^2 & , & 1 \le i \le n-3,\;\; i+2 \le j \le n-1. \end{array}\right\} \quad (1)$$

This is just the Coxeter system A_{n-1} [1], and we call it the *standard presentation of S_n on the generators x_i, $1 \le i \le n-1$* [5]. For example, the relators in the standard presentation for S_4 on x, y, t are

$$x^2, y^2, t^2, (xy)^3, (yt)^3, (xt)^2. \quad (2)$$

We aim to produce certain non-standard presentations of S_n, beginning with the case $n = 4$.

Consider the effect on the relators (2) of the Tietze transformation that replaces t by the new generator $z = yty$ corresponding to the transposition (24). Substituting $t = yzy$ into the relators (2) (and into the equation $z = yty$, which gives a consequence of $y^2 = e$ and so can be ignored), we get

$$x^2, y^2, (yzy)^2, (xy)^3, (y^2zy)^3, (xyzy)^2$$

or, equivalently,

$$x^2, y^2, z^2, (xy)^3, (yz)^3, (xyzy)^2 \quad (3)$$

as relators defining S_4 on the generators x, y, z. Since the map $\alpha : x \mapsto y \mapsto z \mapsto x$ induces an automorphism of this group, some consequences of (3) can be got by applying α to its members. In particular, we have the following lemma, a direct proof of which makes an interesting exercise.

Lemma 1 *The relator $(xz)^3$ is a consequence of (3).*

Returning to the general case, first take standard presentations of S_n and S_m on generators $x_1, \ldots, x_{n-1}$ and $y_1, \ldots, y_{m-1}$, respectively. Next, form the standard presentation $P(n, m)$ of their direct product [5] by taking the union of the generating sets, the union of the sets of relators, and the commutators $[x_i, y_i]$, $1 \leq i \leq n-1, 1 \leq j \leq m-1$. Finally, adjoin a new generator t to $P(n, m)$ and new relators

$$\left.\begin{array}{lll} t^2, (tx_k)^3, & (x_{k-1}x_k t x_k)^2, & (ty_1)^3, \\ (tx_i)^2, 1 \leq i \leq n-1, & i \neq k-1 \text{ or } k, & (ty_j)^2, 2 \leq j \leq m-1, \end{array}\right\} \quad (4)$$

where k is a new parameter, $2 \leq k \leq n-1$. Call this presentation $P(n, m, k)$.

The key step is now to replace t in $P(n, m, k)$ by the new generator $z = x_k t x_k$. The generators and relators in $P(n, m)$ are unchanged, while modulo these substitution of $t = x_k z x_k$ in (4) yields

$$\left.\begin{array}{llll} z^2, & (x_k z)^3, & (x_{k-1}z)^2, & (zy_1)^3, \\ (zx_i)^2, 1 \leq i \leq n-1, & & i \neq k-1, k, k+1, & (zy_j)^2, 2 \leq j \leq m-1, \\ (x_k z x_k x_{k+1})^2, & & & \end{array}\right\} \quad (5)$$

where the last relator is only present when $k \leq n-2$. When $k = n-1$, (5) is just the standard presentation of S_{m+n} on the generators $x_1, \ldots, x_{n-1}, z, y_1, \ldots, y_{m-1}$. But if $k \leq n-2$, we can apply Lemma 1 twice, first to adjoin the relator $(zx_{k+1})^3$, then to remove $(x_k z)^3$, which gives the presentation $P(n, m, k+1)$. A simple induction now yields the following result.

Lemma 2 *The presentation $P(n, m, k)$ described above defines S_{m+n}.*

An obvious change of notation yields the following recipe as a special case.

Lemma 3 *Let P_n be the presentation obtained by taking the direct product of three standard presentations of S_n, on generators x_i, y_i, z_i, respectively, $1 \leq i \leq n-1$, and adjoining new generators x, y, z and new relations (3) along with*

$$(xx_1)^3, (yy_1)^3, (zz_1)^3$$

and the commutators of each of x, y, z with all other x_i, y_i, z_i. Then P_n is a presentation of S_{3n+1}.

Remark 1 Just as the standard presentations of the symmetric groups correspond in an obvious way to trees with no vertices of degree three or more, those given in Lemma 2 correspond to trees with just one vertex of degree three and no others of degree more than two. It might be interesting to find presentations for the S_n corresponding to more complicated graphs. This could lead to a more elegant proof of Lemma 3 than the above, which looks rather clumsy.

2 A direct limit

If we are given a sequence of groups G_n and inclusions $i_n : G_n \to G_{n+1}$, then the union $G = \bigcup_{n \in \mathbb{N}} G_n$ can be made into a group in an obvious way. If in addition we have presentations $G_n = < X_n \mid R_n >$ with inclusions $X_n \subseteq X_{n+1}$, $R_n \subseteq R_{n+1}$ inducing the i_n, then putting $X = \bigcup_{n \in \mathbb{N}} X_n$, $R = \bigcup_{n \in \mathbb{N}} R_n$ we get the following result.

Lemma 4 *With the above notation, $P = < X \mid R >$ is a presentation for G.*

Proof For each n there is a commutative diagram

$$\begin{array}{ccc} P_n & \rightarrow & P \\ \downarrow & & \downarrow \\ G_n & \rightarrow & G \end{array}$$

in which the map on the left is an isomorphism and the lower map a monomorphism. It follows easily that the map on the right is also an isomorphism, as required. □

It follows from this that the removal of all bounds involving n from (1) yields a presentation of the finitary symmetric groups $S = \mathrm{Sym}^f\ \mathbb{N}$ consisting of all permutations of $\mathbb{N}$ that move only a finite number of points. The same applies to the presentation P_n of Lemma 3, and we now describe the resulting presentation P of S.

With a small change in notation (write $x = x_0$, $y = y_0$, $z = z_0$, then add 1 to all subscripts), we take as generators the set

$$X = \{x_n, y_n, z_n \mid n \in \mathbb{N}\}.$$

The relators in Lemma 3 can then be partitioned as follows.

$R_1 \;:\; {x_n}^2,\ {y_n}^2,\ {z_n}^2,\ n \in \mathbb{N}.$

$R_2 \;:\; (x_n x_{n+1})^3,\ (y_n y_{n+1})^3,\ (z_n z_{n+1})^3,\ n \in \mathbb{N}.$

$R_3 \;:\; (x_m x_{m+n+1})^2,\ (y_m y_{m+n+1})^2,\ (z_m z_{m+n+1})^2,\ m,\ n \in \mathbb{N}.$

$R_4 \;:\; (x_1 y_1)^3,\ (z_1 y_1)^3,\ (x_1 y_1 z_1 y_1)^2.$

$R_5 \;:\; (x_m y_n)^2,\ (y_m z_n)^2,\ (z_n z_m)^2,\ \forall m, n \in \mathbb{N}, m + n \neq 2.$

We can now summarise all that we have so far achieved into the following lemma.

Lemma 5 *With the above notation, $S = < X \mid R_1, R_2, R_3, R_4, R_5 >$.*

3 A group "with three ends"

To obtain an embedding of S in a finitely presented group, we need briefly to name the points of the set Ω on which it acts. We take

$$\Omega = \left\{n, n', n'' \mid n \in \mathbb{N}\right\} \cup \{0\}$$

and put

$$x_n = (n-1\ n),\ y_n = ((n-1)'\ n'),\ z_n = ((n-1)\ n''),\ n \in \mathbf{N},$$

where $0'' = 0' = 0$. Then the presentation in Lemma 4 defines $\mathrm{Sym}^f\ \Omega$. The key to our embedding is supplied by the following two non-finitary permutations a, b of Ω:

$$\begin{array}{lll} na = n+1 = nb, & & n \geq 0, \\ n'a = (n-1)', & n'b = n', & n \geq 1, \\ n''a = n'', & n''b = (n-1)'', & n \geq 1. \end{array}$$

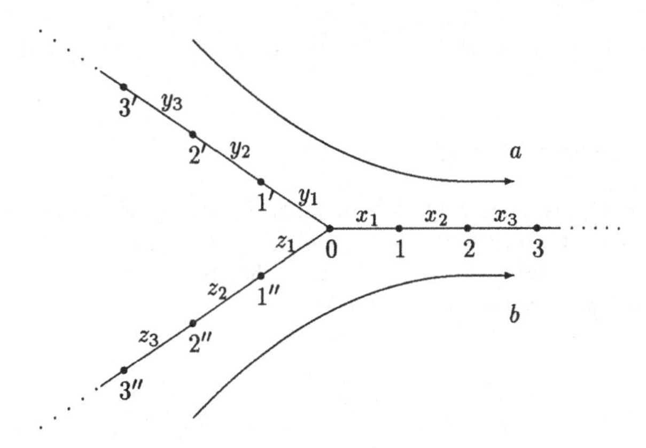

Putting $H = \langle a, b \rangle \leq \mathrm{Sym}\ \Omega$, we get the following remarkable result.

Lemma 6 $H' = S$.

Proof H' is generated by the conjugates of $a^{-1}b^{-1}ab = [a, b] = (01) = x_1$, and

$$x_1^{a^n} = x_{n+1},\ x_1^{a^{-n}} = y_n,\ x_1^{b^{-n}} = z_n,\ n \in \mathbf{N}. \tag{6}$$

Hence, $S = \langle X \rangle \leq H'$. On the other hand,

$$y_1^{\ b} = y_1^{\ x_1},\ z_1^{\ a} = z_1^{\ x_1},\ \text{and} \tag{7}$$

$$x_1^{\ b^n} = x_{n+1},\ y_{n+1}^{\ \ b} = y_{n+1},\ z_{n+1}^{\ \ a} = z_{n+1},\ n \in \mathbf{N}, \tag{8}$$

and it follows that $\langle X \rangle$ is normal in H, whence $H' \leq X$. □

Since the permutation $a^m b^n$ sends k' to $(k-m)'$ and k'' to $(k-n)''$ for all sufficiently large k, it is clear that H/H' is isomorphic to the free abelian group $Z^2 = \langle a, b \mid [a, b] \rangle$ of rank 2. Hence, H is a non-split extension of S by Z^2, and the obvious guess at a presentation for H turns out to be correct.

Lemma 7 *The group $H = \langle a, b\rangle \leq Sym\,\Omega$ has the presentation*

$$Q = \langle X, a, b \mid R_1, R_2, R_3, R_4, R_5, [a,b]\,{x_1}^{-1}, R\rangle \tag{9}$$

where X and the R_i are as in Lemma 5 and R consists of the relators corresponding to the action (6), (7), (8).

Proof We have a commutative diagram with exact rows and obvious maps.

$$\begin{array}{ccccccccc} 1 & \longrightarrow & \langle X\rangle & \longrightarrow & Q & \longrightarrow & Q/\langle X\rangle & \longrightarrow & 1 \\ & & \downarrow & & \downarrow & & \downarrow & & \\ 1 & \longrightarrow & H' & \longrightarrow & H & \longrightarrow & Z^2 & \longrightarrow & 1 \end{array}$$

As the outer vertical maps are both isomorphisms, so is the inner one by the five-lemma (see [5]). □

4 A finite presentation for H

We seek to reduce the presentation Q of $H = \langle a, b\rangle$. First reduce the generating set to $a, b, x = x_1$ using (6). Since this is a subset of $X \cup \{a, b\}$, defining relations are obtained by simply substituting (6) in all the other relators in (9). Ignoring obvious superfluities, this reduces them as follows.

R_1' : x^2.

R_2' : $(xx^a)^3, \left(x^{a^{-1}}x^{a^{-2}}\right)^3, \left(x^{b^{-1}}x^{b^{-2}}\right)^3$.

R_3' : $\left(xx^{a^{n+1}}\right)^2, \left(x^{a^{-1}}x^{a^{-n-2}}\right)^2, \left(x^{b^{-1}}x^{b^{-n-2}}\right)^2, n \in \mathrm{N}$.

R_4' : $\left(xx^{a^{-1}}\right)^3, \left(x^{b^{-1}}x^{a^{-1}}\right)^3, \left(xx^{a^{-1}}x^{b^{-1}}x^{a^{-1}}\right)^2$.

R_5' : $\left(x^{a^{m-1}}x^{a^{-n}}\right)^2, \left(x^{a^{-m}}x^{b^{-n}}\right)^2, \left(x^{b^{-n}}x^{a^{m-1}}\right)^2 m, n \in \mathrm{N},\ m+n \neq 2$.

R_6' : $[a,b]\,x^{-1}$.

R_7' : $\left[x^{a^{-1}},\ bx^{-1}\right], \left[x^{b^{-1}},\ ax^{-1}\right]$.

R_8' : $[x, b^n a^{-n}], \left[x^{a^{-n-1}},\ b\right], \left[x^{b^{-n-1}}, a\right], n \in \mathrm{N}$.

Our main result asserts that just five of these are sufficient to define the group.

Theorem 1 *On the generators a, b, x, the group H is defined by the relations*

$$x^2 = e, \quad (xx^a)^3 = e, \quad x \sim x^{a^2}, \quad x = [a,b], \quad x^a = x^b, \tag{10}$$

where $\sim$ denotes commutation and e the identity.

Proof These correspond to the relators $R_1{}'$, $R_6{}'$ and the first relators in $R_2{}'$, $R_3{}'$, $R_8{}'$ in the order given above, and so they all hold in H. To deduce the others, we first get them from (10) together with

$$x \sim b^{a^2}, \quad x \sim a^{b^2} \tag{11}$$

$$x^{a^n} = x^{b^n}, \quad x \sim x^{a^{n+1}}, \quad n \geq 1, \tag{12}$$

and then deduce (12) from (10) and (11), and finally obtain (11) from (10). Call the relations in (10) $r_1, \ldots, r_5$, and those in (11) r_6, r_7, in the order given.

We assume (10), (11), (12) and deduce $R_i{}'$ for $i = 2, 3, 4, 5, 8$ in turn. For $R_2{}'$ conjugate the second and third relations by a^2 and b^2 respectively to see that they follow from r_2 and r_7. Because of r_1, the first relation in $R_3{}'$ follows from (12), the second is a conjugate, and so is the third by (12) again. The first relation in $R_4{}'$ is a conjugate of r_2, and its conjugate by $a^{-1}b^{-1}a$ is

$$\left(x^{a^{-1}b^{-1}a}\, x^{a^{-2}b^{-1}a}\right)^3 = \left(x^{b^{-1}} x^{a^{-1}}\right)^3,$$

using r_4 and r_5. By (12), $x^{b^{-1}} \sim x^b = x^a$, and so

$$x \sim x^{b^{-1}a^{-1}} = x^{b^{-2}aba^{-1}} = x^{b^{-1}axa^{-1}} = x^{a^{-1}} x^{b^{-1}} x^{a^{-1}}$$

using r_6, r_4, r_1, which finishes $R_4{}'$. The first and third relations in $R_5{}'$ obviously follow from (12), while the second also uses the second and third relations in $R_8{}'$ (m and n cannot both be equal to 1). The first relation in $R_8{}'$ is part of (12), and the second and third follow by inductions based on r_6 and r_7 respectively. For example, for $n \geq 1$, if $x^{a^{-n-1}} \sim b$, then $x^{a^{-n-1}} \sim b^{-1}$ and so

$$x^{a^{-n-2}} \sim ab^{-1}a^{-1} = a\,[b, a]\,a^{-1}b^{-1} = x^{a^{-1}}b^{-1},$$

using r_1 and r_4. But $x^{a^{-n-2}} \sim x^{a^{-1}}$ already by (12) and this completes the induction.

We must now deduce (12) from (10) and (11), and this is by simultaneous induction. Note that r_7, r_3 provide the base $n = 1$ and that the first relation holds trivially when $n = 0$. So let $n \geq 2$, assume

$$x^{a^{n-2}} = x^{b^{n-2}}, \quad x^{a^{n-1}} = x^{b^{n-1}}, \quad x \sim x^{a^n}, \tag{13}$$

and seek to deduce (12) in the presence of (10) and (11). Using (13) and r_4,

$$x^{a^n} = x^{a^n x} = x^{b^{n-1}ax} = x^{b^{n-2}ab} = x^{a^{n-1}b} = x^{b^n},$$

which is the first equation. For the second, r_3 and r_6 say that $x^{a^{-2}}$ commutes with both x and b, whence

$$x^{a^{-2}} \sim x^{b^{n-1}} = x^{a^{n-1}},$$

as required.

Finally, we obtain r_6 and r_7 from (10). From r_2 using r_1

$$xx^a x = x^a x x^a \Rightarrow x^{ax} = x^{a^{-1}xa} \Rightarrow x^{a^{-1}x} = x^{axa^{-1}} = x^{bxa^{-1}} = x^{a^{-1}b},$$

by r_5 and r_1, Hence,

$$x^{a^{-1}} \sim bx^{-1} = a^{-1}ba \Rightarrow x \sim b^{a^2},$$

which is r_6. Similarly, using r_1,

$$r_2 + r_5 \Rightarrow x^{bx} = x^{b^{-1}xb} \Rightarrow x^{b^{-1}x} = x^{bxb^{-1}} = x^{axb^{-1}} = x^{b^{-1}a},$$

whence, using r_1 again,

$$x^{b^{-1}} \sim ax = b^{-1}ab \Rightarrow x \sim a^{b^2},$$

which is r_7, as required. □

5 Conclusion

From what we have proved it follows a fortiori that any countable direct power of finite groups embeds in a finitely presented group. We had hoped to construct such an embedding for the countably infinite direct power of any finitely presented group via the obvious wreath product, but our "proof" broke down. Our main aim, of embedding in a finitely presented group the additive group of rational numbers, continues to elude us.

I am grateful to Marston Conder and his colleagues in the Mathematics Department at the University of Auckland for their hospitality during the preparation of this article. Thanks are also due to Bob Burns (private communication), who drew attention to another proof [2] of the main result above and independently confirmed it: (he got six relations whereas I originally had seven), and (10) is the intersection of these sets, and to C F Miller III for sending details of another embedding (in unpublished work of G Higman and B H Neumann) of S in a finitely presented group using *HWN extraneous*. I am also grateful to Graham Higman for raising the question studied here, and to Jim Wiegold for telling me the answer.

References

[1] C. T. Benson and L. C. Grove, *Finite reflection groups*, Bogden and Quigley, Tarrytown 1971.

[2] K. S. Brown, *Finiteness properties of groups*, J. Pure Appl. Algebra **44**:1-3 (1987), 45-75.

[3] G. Higman, *Subgroups of finitely presented groups*, Proc. Royal Soc. London Ser. A **262** (1961), 455-475.

[4] C. Houghton, *The first cohomology of a group with permutation module coefficients*, Arch. Math. (Basel) **31:3** (1978/79), 254-258.

[5] D. L. Johnson, *Presentations of groups*, 2nd ed., Cambridge University Press, Cambridge 1997.

THE DEDEKIND-FROBENIUS GROUP DETERMINANT: NEW LIFE IN AN OLD PROBLEM

KENNETH W. JOHNSON

The Pennsylvania State University, Abington Campus, 1600 Woodland Road, Abington, PA 19001, U.S.A.

1 Introduction

The following remarks appear in the introduction to the first edition of Burnside's group theory book (1897).

"Cayley's dictum that 'a group is defined by means of the laws of combination of its symbols' would imply that, in dealing purely with the theory of groups, no more concrete mode of representation should be used than is absolutely necessary. It may then be asked why, in a book which professes to leave all applications aside, a considerable space is devoted to substitution groups; while other particular modes of representation, such as groups of linear transformations, are not even referred to. My answer to this question is that while, in the present state of our knowledge, many results in the pure theory are arrived at most readily by dealing with the properties of substitution groups, *it would be difficult to find a result that could be directly obtained by consideration of groups of linear transformations*" (italics inserted by the present author).

The mathematics which initiated this work provides a diametrically opposite point of view to Burnside's interpretation of Cayley's dictum. A problem motivated by number theory and posed in terms of determinants, which Dedekind thought might lead to new systems of "hypercomplex numbers", was solved and incidentally gave rise to both group character theory and group representation theory thanks to the virtuosity of Frobenius. The original methods are difficult to follow for those not intimately familiar with 19th century determinant theory but were quickly supplanted by those of Burnside, Dickson and Schur, so that it is possible to be an expert in representation theory without any acquaintance with the original work in the area.

Recently it has happened that modern work has drawn upon 19th century mathematics which has been called "bumbling" by at least one reviewer to reveal results which have either anticipated "new" concepts or suggested intricate solutions to current problems. For examples we may refer to the book [13] which reproduces an 1848 Cayley paper containing a "Koszul"complex, the paper [11] on determinantal varieties and the inaugural address by Deligne for the new mathematics building at the Institute of Advanced Study (1993) which connected the theory of n-logarithms (associated with Chern classes in Grassmannians) with ideas in motivic cohomology. The work presented here provides another example of how some of the intricate ideas of our forefathers can suggest different problems and new ideas in the modern context.

Since it is difficult nowadays to divorce character theory from representation theory it is surprising that matrix representations are completely absent from Frobenius' papers [8] and [10] in which character theory was first introduced. Most of the statements and proofs are in terms of determinants, and the problem that Frobenius addressed was that of the factorisation of the group determinant, which is the determinant Θ_G of a matrix X_G encoding the multiplication of the group G. Since his proofs are difficult to follow it is not surprising that after Burnside, Molien and Dickson produced a good part of his results by alternative methods and Schur gave an account of representation theory which is more or less equivalent to modern treatments the group determinant languished.

There is however a paper by Poincaré [29] in 1903 on linear differential equations which gives a criterion for such an equation to have a rational solution in terms of the group determinant (and which incidentally points out that Frobenius' and Cartan's work could "mutuellement éclairer", as well as indicating a proof of the Wedderburn theorems on algebras). Also in [6] Dickson examines the group determinant factorisation over a field of finite characteristic and there are several other mainly expository papers by him, but until recently few other papers have appeared which mention the group determinant and these have remained well outside the mainstream.

In 1986 the paper [23] was published examining the analogue of the group determinant for latin squares and in it the question of whether non-isomorphic groups could have the same determinant was raised. In [24] it was pointed out that Frobenius had introduced "k-characters", which are invariants of a finite group with a cohomological flavour, in an algorithm to produce an irreducible factor of the group determinant, and a start was made on examining their properties. It was shown in [8] , using ideas put forward by Hoehnke that the group determinant is unique to the group, a result which caused surprise in some quarters, but more remarkably in [19] it was shown that a much smaller amount of information, namely the 3-characters, determine the group. This result is an application of the work on norm-type forms of Hoehnke and the school of A. Bergmann.

Since then there have been developments in different directions, some of which have already been surveyed at Groups 93 Galway/St Andrews and in [25]. In this discussion a survey of results is given which are either subsequent to [25] or were discovered in the literature after [25] appeared . One such is the realisation that the "Weak Cayley Table" or WCT of a group, which gives for each pair (g, h) of elements the class containing the product gh, and which arises from the discussion of 2-characters, can be used to simplify proofs involving pairs of groups with the same character tables. This is developed much more fully in [26] (and the talk of Mattarei at the conference). There is also a group of WCT-preserving maps first discussed by Humphries which arises naturally, and it seems that interesting problems for infinite groups can arise at this stage.

A feature of Frobenius' work on group determinant factors is the formal similarity between it and the construction of the Chern classes of a Lie group representation using curvature matrices. The Chern classes of a representation of finite groups have proved difficult to construct and understand. Here, as mentioned above, the idea of a "curvature" of a finite group is introduced and is used to set out a way to obtain the total Chern class of a representation from the corresponding determinant factor. One tempting consequence of this connection between groups and geometry may be the further application of the extensive results on norm-type forms on algebras and related topics, which has perhaps not received adequate attention, to cohomology theory and geometry.

A summary is given of work of Sjogren ([31]), in which the factorisation of the symmetrised group determinant is used to obtain results in graph theory, and of Humphries who used Frobenius' work on the group matrix to calculate the cogrowth of a group presentation, and prove results on amenability.

This article is set out as follows. In Section 2 a brief summary of some of Frobenius' results is given for the convenience of the reader and to establish notation. A brief account of developments subsequent to 1986 which have been presented in [25] is given in Section 3, with additional comments arising from subsequent work. Chern classes of a group are discussed in Section 4, and the definition of curvature of a finite group is given together with the construction of the Chern class via a group determinant factor. Other developments not discussed in [25] are given in Section 5.

In this year of the centennial of Frobenius' seminal work on representation theory (although not character theory which came one year earlier) it would appear to be appropriate to comment on the great technical skill which he displayed. Not only did he produce a surprisingly complete character theory in the space of three months, but he did this with what would nowadays be called the theory of association schemes. In particular his calculation of the character table of $PSL(2,p)$ in [9] is a computational tour de force. Mention should also be made of the insight of Dedekind whose suggestion initiated Frobenius' work and who also indicated the connection with matrix representations. For those of us who like to explore the "highways and byways" of mathematics, it is interesting to read the diffident quotes of Dedekind in [16]. I give one example (in a letter to Frobenius):

"Do hypercomplex numbers with non-commutative multiplication also intrude in your research? But I do not at all wish to bother you with the request for an answer, which I would best obtain through your work?"

In [17] some discussion is made of the apparent rivalry between Frobenius and Burnside. To quote from a letter from Frobenius to Dedekind:

" This is the same Herr Burnside who annoyed me several years ago by quickly rediscovering all the theorems I had published on the theory of groups, in the same order and without exception: first my proof of Sylow's Theorem, then the theorem on groups with square-free orders, on groups of order $p^\alpha q$, on groups whose order is a product of four or five prime numbers, etc., etc. In any case, a very remarkable and amazing example of intellectual harmony, probably as is possible only in England and perhaps America."

Burnside was shortly to reprove many of the Frobenius results discussed here, but Hawkins points out that this work relies on that of Frobenius and cannot be considered independent discovery.

No claim is made that this paper is in any way a historical article per se, since sources used are for the most part secondary, the quotes from letters of Frobenius to Dedekind in [17] being especially useful. Although an attempt has been made to make this contribution as self-contained as possible, the reader is referred to [16] and [17] for much of the historical background. For a fuller account of some aspects the reader may also wish to refer to [24], [25], [19] and [20].

We assume that the field over which factors are taken is $\mathbf{C}$ and that all groups are finite unless otherwise stated.

A significant part of the work described here was carried out in separate collaborations with H.-J. Hoehnke and Surinder K. Sehgal, for whose input many thanks are due.

2 Frobenius' group determinant work

Let G be a finite group. Let $\{x_g\}_{g\in G}$ be a set of variables. The *group matrix* X_G is the matrix whose rows and columns are indexed by the elements of G and whose $(g,h)^{\text{th}}$ entry is $x_{gh^{-1}}$. The *group determinant* $\Theta(G)$ is defined to be $\det(X_G)$. It will usually be denoted by Θ if no ambiguity occurs.

Example Let G be S_3, the symmetric group on 3 objects. Let the elements be $\{e, (123), (132), (12), (23), (13)\}$ be abbreviated by 1,...,6 in the order given. Then X_G is

$$\begin{pmatrix} x_1 & x_3 & x_2 & x_4 & x_5 & x_6 \\ x_2 & x_1 & x_3 & x_5 & x_6 & x_4 \\ x_3 & x_2 & x_1 & x_6 & x_4 & x_5 \\ x_4 & x_5 & x_6 & x_1 & x_3 & x_2 \\ x_5 & x_6 & x_4 & x_2 & x_1 & x_3 \\ x_6 & x_4 & x_5 & x_3 & x_2 & x_1 \end{pmatrix}$$

and

$$\Theta = (x_1+x_2+x_3+x_4+x_5+x_6)(x_1+x_2+x_3-x_4-x_5-x_6)$$
$$(x_1^2+x_2^2+x_3^2-x_1x_2-x_2x_3-x_3x_1-x_4^2-x_5^2-x_6^2+x_4x_5+x_5x_6+x_6x_4)^2.$$

Any two "group matrices" X_G and Y_G $(=(y_{gh^{-1}}))$ have the property that

$$X_G Y_G = Z_G$$

where Z_G is also a group matrix with entries $z_g = \sum_{hk=g} x_h y_k$. This means that any power of X_G (including its inverse) has the same form, with the entries x_g replaced by suitable polynomials or rational functions.

Frobenius established the following (see for example [24]).

(1) There is a 1:1 correspondence between irreducible factors of Θ and irreducible characters of G.

(2) There is an algorithm to produce a factor ψ_χ of Θ from a character χ of G as follows. The *k-character* $\chi^{(k)} : G^k \to \mathbf{C}$ is defined for $k = 1, \ldots, \deg\chi$ by

$$\chi^{(1)}(g) = \chi(g)$$

$$\begin{aligned}\chi^{(k)}(g_1,g_2,\ldots,g_k) &= \chi(g_1)\chi^{(k-1)}(g_2,g_3,\ldots,g_k) \\ &\quad -\chi^{(k-1)}(g_1g_2,g_3,\ldots,g_k) - \chi^{(k-1)}(g_2,g_1g_3,\ldots,g_k) \\ &\quad -\cdots - \chi^{(k-1)}(g_2,g_3,\ldots,g_1g_k). \qquad (1)\end{aligned}$$

(3) Suppose f is the degree of the character χ. The factor ψ_χ of θ is given by

$$\psi_\chi = \frac{1}{f!}\sum \chi^{(k)}(g_1,g_2,\ldots,g_f)x_{g_1}x_{g_2}\cdots x_{g_f}$$

where the summation runs over all f-tuples of elements of G.

(4) If the $\rho(g)$ is a set of representing matrices for G corresponding to χ then ψ_χ is the determinant of

$$\sum_{g\in G} \rho(g)x_g.$$

Using these results it has been possible to factorise the determinants of all groups of orders at most 27 by hand, and to obtain a description of the factorisations of certain families such as the dihedral groups.

3 First developments

The following remarkable theorem of Frobenius is behind the Formanek-Sibley proof that Θ_G determines G.

Theorem 3.1 *Let X be an $n\times n$ matrix with entries which are algebraically independent variables, and let the entries in the $n\times n$ matrix Y be linear combinations of those in X with constant coefficients. If det(X) and det(Y) differ by a non-zero constant scalar multiple then either $Y = AXB$ or $Y = AX'B$ where A and B are constant matrices and if $n > 1$ only one of the above cases can occur, and A and B are determined up to a scalar factor.*

This theorem appears to have had many reincarnations in the literature.

Question 3.2 Is there an analogous theorem for the hyperdeterminants (i.e. determinants of multidimensional matrices) in [13], or for the non-commutative determinants in [14]?

A *norm-type form* N on an algebra A with basis $\{u_1,\ldots,u_n\}$ over a field K is a homogeneous polynomial in the coordinates x_i of an element $x = x_1u_1+\ldots+x_nu_n$ with values in K, or some polynomial extension of K. It is *multiplicative* if $N(xy) = N(x)N(y)$. The group determinant is a multiplicative norm-type form (this arises from the symmetry property of X_G given in Section 2). A fundamental result on norm-type forms is the following ([18]).

Theorem 3.3 *Let N be a multiplicative norm-type form on the algebra A with basis $\{u_1, \ldots, u_n\}$ and structure constants a^k_{ij} defined by*

$$u_i u_j = \sum a^k_{ij} u_k.$$

Let

$$N(\lambda - x) = \lambda^m - s_1(x)\lambda^{m-1} + \ldots + (-1)^m s_m(x)$$

where λ is an indeterminate and x is a generic element,

$$x = x_1 u_1 + \ldots + x_n u_n.$$

If the discriminant of N is non-zero (which is true when N is the group determinant) it follows that from the knowledge of $s_1(x)$, $s_2(x)$ and $s_3(x)$
(a) *the $s_k(x)$ are determined for all k, and*
(b) *the "symmetrised" structure constants $a^k_{(ij)} = a^k_{ij} + a^k_{ji}$ are determined.*

See [34] for an account of generalisations of this result. The question of how much information is lost in the transition from group to determinant was raised in [24]. Define polynomials $f(x_1, x_2, \ldots, x_n)$ and $g(x_1, x_2, \ldots, x_n)$ to be *equivalent* if there exists σ in S_n such that

$$f(x_1, x_2, \ldots, x_n) = \pm g(x_{\sigma(1)}, x_{\sigma(2)}, \ldots, x_{\sigma(n)}).$$

The following results have already been surveyed in [25].

(1) If G and H have equivalent determinants they must be isomorphic ([8]).

(2) If G and H have equivalent determinant factors of degree 2 which arise from a faithful representation then G isomorphic to H ([24]).

(3) The 3-character of the regular representation determines the group, or equivalently the 3-characters of the irreducible representations determine the group. This result was originally proved using the theory of norm-type forms ([19]) but after it was announced more direct proofs have appeared.

(4) The 2-characters of the irreducible representations do not determine the group ([27], see also [28] and [26]).

It may be remarked that the results mentioned above imply that if the characteristic equation of X_G is

$$det(\lambda I - X_G) = \lambda^n - s_1 \lambda^{n-1} + \ldots + (-1)^n s_n \tag{2}$$

then X_G is a matrix which can be constructed from the form s_3. This has caused surprise for those working in the theory of matrices with polynomial entries.

A 2-character table and 3-character table of a group are defined in [24] and [20]. The columns are indexed by the symmetrised orbits of G acting on $G \times G$ by simultaneous conjugation.

The problem of whether latin squares are determined up to the obvious equivalence by their determinants is addressed in [7]. It turns out that the first case of non-equivalent classes with equivalent determinants is for 8×8 squares, but of the

842,227 classes all but 37 have inequivalent determinants. Each of the exceptional classes consist of squares which are of the form

$$\begin{pmatrix} A & B \\ C & D \end{pmatrix}$$

where A, B, C and D are 4×4 group matrices. It is surprising that the exceptional squares should have this regular form and this calculation gives an indication that we do not have a very good understanding of these symbolic determinants.

4 Chern classes of representations

Given a representation $\rho : G \to \mathrm{GL}(m, \mathbf{C})$ of the finite group G there are defined Chern classes $c_i, i = 1, \ldots, m$, with $c_i \in H^{2i}(G, \mathbf{Z})$. Often it is useful to consider the total Chern class $1 + \sum_{i=1}^{m} c_i$. The problem of characterising Chern classes of representations was posed by Atiyah in [1], (see [32], [33]), and Fulton and MacPherson essentially solved this in [12], placing the problem in a much wider context. It appears that these Chern classes are not well understood. They may be defined via a vector bundle $X \to BG$ arising from ρ, where BG is the classifying space of G.

However, for Lie groups there is an alternative definition using concepts from differential geometry. For any m-dimensional vector bundle $X \to M$ and a connection θ on X one can obtain an $m \times m$ curvature matrix M_θ by standard means, the entries in M_θ being 2-forms (in the sense of differential geometry), see [15], p.400. Define the elementary invariant polynomials P^i by

$$det(A + tI) = \sum_{k=0}^{m} P^{m-k} t^k.$$

For each P^i, $P^i(M_\theta)$ is a well-defined global $2i$-form on M, and the cohomology class represented by this is independent of the connection θ.

Define

$$c_k(M_\theta) = P^k((i/2\pi)M_\theta)$$

and the Chern class $c_k(X)$ to be the class of $P^k((i/2\pi)M_\theta)$ in $H^{2i}_{DR}(M)$ (the De Rham cohomology of M). It can be shown that $c_k(X)$ lies in $H^{2i}(X, \mathbf{Z})$, and the total Chern class $c(\rho)$ may be obtained as the class of $\det(I + (i/2\pi)M_\theta) \in H^*(G, \mathbf{Z})$.

Now consider G finite. Frobenius proved that X_G is similar to a block diagonal matrix, the distinct diagonal blocks being $B_1, B_2, \ldots, B_r$, with B_i a square matrix with n_i rows, and B_i occurring n_i times. Moreover, $\sum_{i=1}^{r} n_i^2 = |G|$ and the $|G|$ entries in the B_i which are linear combinations of the x_g are algebraically independent (this appears to be equivalent to a formulation of the Peter-Weyl theorem in the finite case).

Consider a typical B_k. Let M_θ be the curvature matrix of $\mathrm{GL}(n, \mathbf{C})$. The set of matrix equations $B_k = M_\theta$, $k = 1, \ldots, n$ reduces to a set of linear equations of the form

$$p_i = \psi_i$$

where ψ_i may be thought of as a 2-form for $GL(m, \mathbf{C})$ with m large enough. This system may be regarded as a set of linear equations for the x_g, and using the algebraic independence of the p_i it follows that the matrix of coefficients is invertible and hence there is a unique solution $x_g = \theta(g)$ where $\theta(g)$ is a 2-form.

Define $\theta : G \to GL(m, \mathbf{C})$ to be the curvature of G. Now let ϕ be an irreducible factor of Θ_G of degree t corresponding to the representation ρ. If x_e is replaced by $1 + (i/2\pi)\theta(e)$ and each x_g, $g \neq e$ replaced by $(i/2\pi)\theta(g)$ in ϕ there is obtained a form with values in the Lie algebra of $GL(m, \mathbf{C})$ and hence an element of $\overset{*}{H}_{DR}(GL(m, \mathbf{C}), \mathbf{Z})$. It is claimed that the pullback of this along $\rho : G \to GL(m, \mathbf{C})$ is the total Chern class of ρ.

Question 4.1 If the Chern classes c_1, c_2 and c_3 are known for all irreducible representations of a group, can the higher classes be constructed?

Question 4.2 Are the vector bundle maps f_*^k defined in [12] related to k-characters?

5 More developments

(a) The work of Poincaré.

In [29] a discussion of a linear differential equation

$$\sum_{i=1}^{n} P_i(x) \frac{d^i y}{dx^i} = 0$$

with algebraic solutions is given. To such an equation there is associated a monodromy group G which of finite order m. Poincaré obtains the following.

Theorem 5.1 *The equation*

$$\sum_{i=1}^{n} P_i(x) \frac{d^i y}{dx^i} = 0$$

has a rational solution if and only if there exist $b_1, b_2, \ldots, b_m$ in $\mathbf{C}$ such that the group matrix X_G after the substitution $x_{g_i} = b_i$ is made has rank $m - n$.

A consequence of the above result is that if all the forms $s_k(x_{g_1}, \ldots, x_{g_n})$ defined in equation (2) vanish for $k > m - n$ after the substitution $x_{g_i} = b_i$ is made, the equation has a rational solution, or in terms of k-characters, if π is the character of the regular representation and $\pi^k(b_1, \ldots, b_n)$ vanishes for $k > m - n$ the equation has a rational solution.

It is not clear whether the above conditions can be translated into more conventional representation theory.

(b) Bergmann's interpretation of k-characters.

In the communication [2] Bergmann has indicated that the k-characters can be given in terms of "reduced functions" as follows. If F is any form of degree m, and

$$F(\sum_{i=1}^{m} \lambda_i z_i) = \sum \lambda_1^{\mu_1} \dots \lambda_n^{\mu_n} F_{\mu_1,\dots,\mu_n}(z_1, \dots, z_n)$$

this defines the multihomogeneous polynomials $F_{\mu_1,\dots,\mu_n}$ of multidegree $(\mu_1, \dots, \mu_n) \in \mathbb{N}_0^m$. In the literature ([3], [4], [5]) these are called *reduced functions* and are denoted by $\mathrm{red}F^{(\mu_1,\dots,\mu_r)}$ (excluding μ_i if it is 0). Then

$$\chi^r(g_1, \dots, g_r) = \mathrm{red}\phi^{(1,\dots,1,m-r)}.$$

where ϕ is the group determinant factor corresponding to χ. We refer also to [30], [35] and [36].

(c) The work of Sjogren ([31]).

He considers the factorisation of the determinant of a square symmetric matrix consisting of blocks of symmetrised group matrices (obtained by setting $x_g = x_{g^{-1}}$) for a group G of odd order n. Examples of such matrices occur as adjacency and Kirchoff matrices of a graph Γ on which G acts as a group of regular automorphisms. Information on the spectrum and Laplacian of such graphs are obtained using results derived from those of Frobenius. Some of the results involve the form s_{m-1} defined in equation 2. The reader is referred to [31] for more details.

(d) Weak Cayley Tables.

Since the 3-characters determine G and the 2-characters do not the question arises as to what properties of G can be determined by the irreducible 2-characters. It turns out that the knowledge of the 2-characters is equivalent to that in the *Weak Cayley Table* (WCT) of G, which is the array with $(i, j)^{\text{th}}$ entry consisting of the conjugacy class containing the product $g_i g_j$. Define G, H to *have the same WCT* if there exists $f : G \to H$ preserving conjugacy classes such that $f(g)f(h)$ is conjugate to $f(gh)$ for all $g, h \in G$. Pairs of groups having the same WCT necessarily have the same ordinary character table, but for example the dihedral and quaternion groups of order 8 do not have the same WCT, whereas the pair of nonabelian groups of order p^3, p odd have the same WCT. (It is well-known that all these pairs have the same character tables). On examination, many of the pairs of groups constructed as "Brauer pairs" or to provide answers to questions in character theory can be shown to have the same WCT without the need to construct character tables, which often provides significantly simpler proofs. The question of what information is contained in the WCT remains open, and we refer to the talk of Mattarei at the conference and [26] for further details.

(e) Cogrowth of groups ([21]).

Let

$$1 \to N \to F \to G \to 1$$

be a presentation for the group G where $F = \langle a_1, \dots, a_n \rangle$ is a free group. Put $A = \{a_1, \dots, a_n, a_1^{-1}, \dots, a_n^{-1}\}$ and call n the *degree* of the presentation. For $g \in G, c \in A$ let $W(k, g, c)$ be the set of words $W \in F$ having freely reduced length k, which end on the right in c and which represent the element $g \in G$. Let $w(k,g,c) = |W(k,g,c)|$. The function

$$\Gamma(k) = \sum_{c \in A} w(k, id_G, c)$$

is called the *cogrowth function* for this presentation. Let $\gamma(k) = \sum_{j=0}^{k} \Gamma(j)$ and $\gamma = lim_{k \to \infty} \gamma(k)^{1/k}$. Then $\frac{log(\gamma)}{log(2n-1)}$ is called the *cogrowth* of the presentation. By a result of Cohen a group is amenable if and only if $\gamma = 2n - 1$.

The above presentation gives rise to a recurrence relation for the variables $w(k,g,c)$ with recurrence matrix R, where R is a block matrix of the form $(H_{gh^{-1}})$ where the H_g are (0,1) matrices of a certain type. Thus R exhibits some of the symmetry of the group matrix X_G. The diagonalisation of X_G by Frobenius is then used in calculations of cogrowth.

(f) WCT functions and WCT groups ([22]).

Let G, H be groups (not necessarily finite) Then a weak Cayley table function (WCTF) is a bijection $f : G \to H$ such that
(i) for all $x, y \in G$ $f(xy)$ is conjugate to $f(x)f(y)$
(ii) if C is a conjugacy class in G then $f(C)$ is a conjugacy class of H. A group G is *indecomposable* if it cannot be written as a free product $G = G_1 * G_2$ of non-trivial groups. G is *co-Hopfian* if every monomorphism from G to G is an automorphism. Humphries has obtained the following.

Theorem 5.2 *Let G_i, H_i, $i = 1,2$ be non-trivial groups with H_1 indecomposable and such that H_1 is not isomorphic to a subgroup of G_1 or G_2. Then there are no WCTFs*

$$f : G = G_1 * G_2 \to H = H_1 * H_2.$$

Theorem 5.3 *Let $G_i, H_i, i = 1,2$ be non-trivial non-isomorphic co-Hopfian groups with Hi indecomposable. Then there are no WCTFs*
$f : G = G_1 * G_2 \to H = H_1 * H_2$.

Theorem 5.4 *Let $f : G \to H$ be a WCTF. Then G/G' is isomorphic to H/H'.*

Corollary 5.5 *If $f : F_n \to F_m$ is a WCTF between free groups F_n and F_m then $n = m$.*

The set of WCTF's from a group G to itself form a group $W(G)$ which contains the automorphism group of G, and the map $\iota : g \to g^{-1}$. If $W_0(G) =< Aut(G), \iota >$ we say that $W(G)$ is trivial if $W(G) = W_0(G)$. Let S_n be the symmetric group on n symbols.

Theorem 5.6 *For $n > 2$ $W(S_n)$ is trivial.*

Theorem 5.7 *Let G and H be non-abelian infinite groups. Then $W(G * H)$ is trivial.*

Corollary 5.8 *$W(F_n)$ is trivial if $n > 3$.*

It may also be proved that the WCTF group for a finite dihedral group is trivial. There are examples where $W(G)$ is much larger than $W_0(G)$, such as the non-abelian group of order 21. In fact the Heisenberg group over the integers has a $\mathbf{Z}^\infty$ as a subgroup.

References

[1] M.F. Atiyah, "Characters and cohomology of finite groups", *Inst. Hautes Études Sci, Publ. Math.* 9 (1961) 2-64.

[2] A. Bergmann "k- characters and forms", *Private communication* (1994).

[3] A. Bergmann, "Formen auf Moduln über kommutativen Ringen beliebiger Charakteristic", *J.Reine Angew. Math.***219** (1965) 113-156.

[4] A. Bergmann, "Hauptnorm und Struktur von Algebren", *J.Reine Angew. Math.* **222** (1966) 160-194.

[5] A. Bergmann, "Reduzierte Normen und Theorie von Algebren", *Algebra Tagung Halle 1986, Tagungsband* 29-57.

[6] L.E. Dickson, "Modular theory of group-matrices", *Trans. Amer. Math. Soc.* (1907) 389-398.

[7] D. Ford and K.W. Johnson, "Determinants of latin squares of order 8", *Experimental Mathematics.* **5** (1996) 317-325.

[8] E. Formanek and D. Sibley, "The group determinant determines the group", *Proc. Amer. Math. Soc* **112** (1991) 649-656.

[9] G. Frobenius, "Über Gruppencharaktere", *Sitzungsber. Preuss. Akad. Wiss. Berlin* (1896) 985-1021. (Gesammelte Abhandlungen, (Springer-Verlag 1968), 1-37).

[10] G. Frobenius, "Über die Primfactoren der Gruppendeterminante", *Sber. Preuss Akad. Wiss. Berlin* (1896) 1343-1382. (Gesammelte Abhandlungen, (Springer-Verlag 1968), 38-77).

[11] W. Fulton, "Flags, Schubert polynomials and determinantal bundles", *Duke Math. J.* **65** (1992) 381-420.

[12] W. Fulton, R. MacPherson, "Characteristic classes of direct image bundles for covering maps", *Annals of Mathematics* **125** (1987) 1-92.

[13] I. M. Gelfand, M.M. Kapranov, A.V. Zelevinsky, "Determinants, resultants and multidimensional determinants", Birkhäuser, Boston, (1994).

[14] I. M. Gelfand, V. S. Retakh, "Determinants of matrices over non-commutative rings", *Functional Anal. Appl.* **26** (1992), 231–246.

[15] P. Griffiths and J. Harris, "Principles of algebraic geometry", *John Wiley and Sons* , New York, 1978.

[16] T. Hawkins, "The origins of the theory of group characters", *Arch. Hist. Exact Sci.* **7** (1971) 142-170.

[17] T. Hawkins, "New light on Frobenius' creation of the theory of group characters", *Arch. Hist. Exact Sci.* **12** (1974) 217-243.

[18] H.-J. Hoehnke, "Über komponierbare Formen und konkordante hyperkomplexe Grossen", *Math. Zeitschr.* **70** (1958) 1-12.

[19] H.-J. Hoehnke and K.W. Johnson, "The 3-characters are sufficient for the group determinant", Proceedings of the Second International Conference on Algebra, *Contemporary Mathematics* **184** (1995) 193-206.

[20] H.-J. Hoehnke and K.W. Johnson, "K-characters and group invariants", *Comm. Algebra* 26 (1998) 1-27.

[21] S. P. Humphries, "Cogrowth of groups and the Dedekind-Frobenius group determi-

nant", *Math. Proc. Cambridge Phil. Soc.* **121** (1997) 193-217.

[22] S. P. Humphries, "Weak Cayley table groups", Preprint (1997).

[23] K.W. Johnson, "Latin square determinants", in Algebraic, extremal and metric combinatorics, London Math. Soc. Lecture Notes Series 131 Cambridge Univ. Press, Cambridge (1988) 146-154.

[24] K.W. Johnson, "On the group determinant", *Math. Proc. Cambridge Phil. Soc.* **109** (1991) 299-311.

[25] K.W. Johnson, "The group determinant and k-characters, a survey and a conjecture", *Proceedings of the biennial Ohio-State-Denison Conference, Eds. Surinder K. Sehgal, R. Solomon, World Sci. Press, Singapore* (1993) 181-197.

[26] K.W. Johnson, S. Mattarei, and Surinder K. Sehgal, "Weak Cayley tables", Preprint.

[27] K.W. Johnson and Surinder K. Sehgal, "The 2-character table does not determine a group", *Proc. Amer. Math. Soc.* **119** (1993) 1021-1027.

[28] K.W. Johnson and Surinder K. Sehgal, "The 2-characters and the group determinant", *Europ. J. Combinatorics* **16** (1995) 623-631.

[29] H. Poincaré, "Sur l'intégration algébrique des équations linéaires et les périodes des integrales abéliennes", *Journal de Mathématiques* 5 Ser. **2** (1903) 107-168.

[30] R. Schmähling, "Separable Algebren über kommutativen Ringen", *J. Reine Angew. Math.* **262/63**, (1973) 307-322.

[31] J.A. Sjogren, "Connectivity and the spectrum in a graph with a rectangular automorphism group of odd order", *Internat. J. Algebra and Computation.* **4** (1994) 529-560.

[32] C.B.Thomas, "Chern classes of representations", *Bull. London Math. Soc* . **18** (1986) 225-240.

[33] C.B.Thomas, "Characteristic classes and the cohomology of finite groups", Cambridge University Press, Cambridge, (1986).

[34] W.C. Waterhouse, "Automorphism group schemes of basic matrix invariants", *Trans. Amer. Math. Soc.* **347** (1995) 3859-3872.

[35] D. Ziplies, "Divided powers and multiplicative polynomial laws", *Comm. Algebra* **14** (1986) 49-108.

[36] D. Ziplies, "A characterization of the norm of an Azumaya algebra of constant rank through the divided powers algebra of an algebra", *Beiträge zur Algebra und Geometrie* **22** (1986) 53-70.

GROUP CHARACTERS AND π-SHARPNESS[1]

KENNETH W. JOHNSON* and EIRINI POIMENIDOU†

*Pennsylvania State University, Abington Campus, 1600 Woodland Road, Abington, PA 19001, U.S.A.

†New College of USF, 5700 N.Tamiami Trail, Sarasota, FL 34243, U.S.A.

Abstract

Let G be a finite group and χ a generalized character of G of degree n. Suppose that $f_\chi(x)$ is the monic polynomial of least degree whose roots are the distinct values of χ on non-identity elements. The quotient $b_\chi = f_\chi(n)/|G|$ is a rational integer ([4], [7]). If $b_\chi = 1$, then Cameron and Kiyota defined χ to be *sharp.* This paper discusses the situation where π is a set of primes and $f_\chi^\pi(x)$ is the monic polynomial of least degree whose roots are the values of χ on non-identity π-elements. We show that $b_\chi^\pi = f_\chi^\pi(n)/|G|_\pi$ is also a rational integer and if $b_\chi^\pi = 1$ we define χ to be π-*sharp.* We discuss examples and relate the existence of π-sharp generalized characters to properties of the prime graph of G.

1991 Mathematics Subject Classification: Primary 20C15; Secondary 20C20.

1 Introduction

Let χ be a generalized character of a finite group G, with $\chi(1) = n$ and let $L_\chi = \{\chi(g) | g \in G, g \neq 1\}$. Let also $f_\chi(x)$ be the monic polynomial of least degree whose set of roots is L_χ, i.e. $f_\chi(x) = \prod_{l \in L_\chi}(x - l)$.

The result that $f_\chi(n)$ is an integer divisible by $|G|$ was proved in a modern context by Cameron and Kiyota in [7]. When $b_\chi = f_\chi(n)/|G|$ is 1, χ is said to be *sharp.* It was remarked in [7] that the result that b_χ is an integer is already contained in an early paper of Blichfeldt [4] and Matsuhisa and Yamaki [12] have indicated that Blichfeldt proved a more general result which leads to the definition of a sharp triple. Work on sharp characters has focussed on the characterization of groups admitting a sharp ordinary character χ with L_χ of a certain type (see for example [7], [1], [6], [2], [3], [11]), but in [9] a relationship is exhibited between sharp generalized characters and the prime graph of a group, and in [10] sharpness is discussed for quasigroup characters. It remains unclear whether the integer b_χ has a more intrinsic interpretation. The result above may be rephrased as: for any generalized character χ $f_\chi(\chi)$ is an integral multiple of ρ where ρ is the character of the regular representation. Alvis has remarked that this is a strengthening of the well-known result of Burnside that every irreducible character appears as a constituent of at least one of χ, χ^2, ... χ^k, where k is the number of distinct values taken on by χ.

[1]The authors wish to thank the office of the Dean and Warden and the New College Foundation for their support while working on this project.

This paper discusses the situation where π is a set of primes which divide $|G|$, χ is a generalized character and $L_\chi^\pi = \{\chi(g) | g \text{ a } \pi\text{-element in } G, g \neq 1\}$. It follows directly that if π consists of a single prime or if G is π-separable that $b_\chi^\pi = f_\chi^\pi(n)/|G|_\pi$ is an integer, where $f_\chi^\pi(x)$ is the monic polynomial of least degree whose set of roots is L_χ^π, essentially by using the restriction of χ to a Hall π-subgroup and using the Blichfeldt result. It is interesting to note that the result remains true for any group. This is the content of Theorem 2.2 below. Thus for each character χ we have a set of integer invariants b_χ^π, for all sets π of primes dividing $|G|$.

A definition of π-sharpness follows naturally: the generalized character χ is π-*sharp* if $\chi(g) \neq \chi(1)$ for g a π-element and $b_\chi^\pi = 1$. A sharp character need not be π-sharp, and examples are readily available to show that if $\pi \subset \pi' \subset \pi''$ a character may be π -sharp and π'' -sharp but not π' -sharp. However the dihedral groups of order $2n$, n odd, give examples of groups for which all faithful characters are π-sharp for all π.

In the context of modular representations it is mentioned in [5] that Kataoka and Kiyota considered sharpness for the Brauer character corresponding to a faithful representation in characteristic $p > 0$. This notion of sharpness is equivalent to the Brauer character being p'-sharp in our sense. We show that the Brauer character of the natural representation of $Sl(2, p^r)$ in characteristic p is p'-sharp.

In [9] Iiyori used the prime graph of a group to construct a non-trivial generalized sharp character when this graph is disconnected. In Section 4 we show that a result of Frobenius may be used to drastically shorten his discussion. We introduce the π-graph of a group and use Frobenius' result to prove the existence of non-trivial π-sharp characters when the π-graph is disconnected.

2 Character values and π-sharpness

Theorem 2.1 *Let χ be a generalized character of a finite group G, with $\chi(1) = n$ and let $L_\chi^p = \{\chi(g) | g$ a p-element $g \neq 1\}$. Let also $f_\chi^p(x)$ be the monic polynomial of least degree whose set of roots is L_χ^p,i.e.*

$$f_\chi^p(x) = \prod_{l \in L_\chi^p} (x - l).$$

Then $|G|_p$ divides $f_\chi^p(n)$, where $|G|_p$ denotes the p-part of the group order.

Proof Let P be a Sylow p-subgroup of G and let χ_P denote restriction of χ to P. Let

$$L_{\chi_P} = \{\chi_P(g) | g \in P, g \neq 1\}.$$

Since every p-conjugacy class of G intersects P non-trivially, it follows that $L_{\chi_P} = L_\chi^p$. We therefore have that $f_\chi^p(x) = f_{\chi_P}(x)$. Applying Blichfeldt's result to χ_P and $f_{\chi_P}(x)$, we have that $|P| = |G|_p$ divides $f_{\chi_P}(n) = f_\chi^p(n)$ and result follows. □

If π is a set of primes and G is π-separable, the above argument may be modified to produce an analogous result. However, this result is true without any restriction on G, and this is the content of the theorem below.

Theorem 2.2 *Let π be a set of primes. Let χ be a generalized character of a finite group G, with $\chi(1) = n$ and let $L_\chi^\pi = \{\chi(g) | g$ a π-element, $g \neq 1\}$. Let also $f_\chi^\pi(x)$ be the monic polynomial of least degree whose set of roots is L_χ^π, i.e*

$$f_\chi^\pi(x) = \prod_{l \in L_\chi^\pi} (x - l).$$

Then $|G|_\pi$ divides $f_\chi^\pi(n)$.

Proof Let ω be a $|G|_\pi$-th root of unity. For each π-element $g \in G$, we have that $\chi(g) = T(\omega)$, where $T(x)$ is a polynomial with integer coefficients. Since the algebraic conjugates of $T(\omega)$ are precisely $T(\omega^r)$, where $(r, |G|_\pi) = 1$, it follows that L_χ^π contains a full set of algebraic conjugates of $\chi(g)$, for all π-elements $g \in G$. Let θ_r be any Galois automorphism of $\mathbf{Q}(\omega)$ over $\mathbf{Q}$. Then θ_r has the form $\theta_r(\omega) = \omega^r$, where $(r, |G|_\pi) = 1$. It therefore follows that the set L_χ^π is fixed by θ_r and that $f_\chi^\pi(n)$ is fixed by $Gal(\mathbf{Q}(\omega)/\mathbf{Q})$ and is thus a rational number. Since $f_\chi^\pi(n)$ is also an algebraic integer it has to be an integer.
If $p \in \pi$ is a prime, then $L_\chi^p \subseteq L_\chi^\pi$ and hence $f_\chi^\pi(x)/f_\chi^p(x)$ is a polynomial with algebraic integer coefficients. Thus $f_\chi^\pi(n)/f_\chi^p(n)$ is an algebraic integer and a rational number and therefore a rational integer. By Theorem 2.1 we have that $|G|_p$ divides $f_\chi^\pi(n)$, for all $p \in \pi$. Therefore $\prod_{p \in \pi} |G|_p = |G|_\pi$ divides $f_\chi^\pi(n)$. □

It is not difficult to find examples to show that sharpness does not imply π-sharpness for π a proper subset of $\pi(G)$, the set of primes dividing $|G|$.

Example 1 Consider the permutation character χ of S_5. Then $L_\chi = \{0, 1, 2, 3, 4\}$ and $f_\chi(5) = 5! = |S_5|$ and χ is sharp. If $\pi = \{2, 3\}$ then $L_\chi^\pi = \{0, 1, 2, 3\}$ and $f_\chi^\pi(5) = 5 \cdot 4 \cdot 3 \cdot 2 = 5! \neq 24 = |S_5|_\pi$, hence χ is not $\{2, 3\}$-sharp. If however $\pi = \{2, 5\}$, then $L_\chi^\pi = \{0, 1, 3\}$ and $f_\chi^\pi(5) = 5 \cdot 4 \cdot 2 = 40 = |S_5|_\pi$ and χ is $\{2, 5\}$-sharp.

Example 2 illustrates that π-sharpness is not in general transitive.

Example 2 Consider the permutation character χ of S_7. Then χ is $\{2, 3, 5, 7\}$-sharp. It is not $\{3, 5, 7\}$-sharp but it is $\{5, 7\}$-sharp. This can be shown directly or by applying Lemma 2.4.

Example 3 Let $G = D_{2n}$ where n is odd. Then each faithful irreducible character χ is sharp. Suppose that π is a set of primes dividing $2n$, and $2 \in \pi$. The union of the π-classes together with 1 form a dihedral group of order $2m$, and the restriction of χ to this subgroup is sharp, from which we deduce that χ is π-sharp. Now suppose that $2 \notin \pi$. The π-classes along with the class of involutions and 1 again form a dihedral subgroup H of order $2m$, and it is sufficient to show that χ is π-sharp on this. Now, χ takes the value 0 only on the class of involutions and thus $f_{\chi_H}^\pi(2) = f_{\chi_H}(2)/2$ which along with the sharpness of χ on H demonstrates that χ is π-sharp on H.

Lemma 2.3 *Let χ be the permutation character of S_n and let π be a set of primes such that $p > \frac{n}{2}$ for all $p \in \pi$. Then χ is π-sharp.*

Proof By hypothesis, S_n has a unique conjugacy class of p-elements for each $p \in \pi$ and hence it follows that $\chi(c) = n - p$ for each p-cycle c. Hence $L_\chi^\pi = \{n-p | p \in \pi\}$. Therefore $f_\chi^\pi(n) = \prod_{p\in\pi} p = |S_n|_\pi$ and the result follows. □

Lemma 2.4 *Let χ be the permutation character of S_n and let π be a set of primes such there exists $p \in \pi$ with $p \leq \frac{n}{k}$ and k an integer. If there exists a prime $q \leq k$ with $q \notin \pi$, then χ is not π-sharp.*

Proof Using the hypothesis, there are elements in S_n which are products of i disjoint p-cycles for $i = 1, \cdots, k$. It therefore follows that $(n - (n - p)(n - (n - 2p) \cdots (n - (n - kp) = k!p^k$ divides $vert f_\chi^\pi(n)$, which in turn implies that $f_\chi^\pi(n) \neq |S_n|_\pi$ since q divides $f_\chi^\pi(n)$ but not $|S_n|_\pi$. □

3 Sharp Brauer characters

If $\zeta : G \to GL(n, F)$ is a representation of a group G over a field F whose characteristic divides the order of the group, then the Brauer character corresponding to ζ may be regarded as a generalized character of G.

Theorem 3.1 *In $Sl(2, q)$, the Brauer character corresponding to the natural representation in characteristic p is p'-sharp, where q is a power of p.*

We need the following lemma for the proof of Theorem 3.1

Lemma 3.2 *Let n be an even positive integer. Then*

$$\prod_{k=1}^{\frac{n}{2}} (2 - 2\cos\frac{2\pi}{n}k) = 2n.$$

Proof We have

$$\begin{aligned}\prod_{k=1}^{\frac{n}{2}} (2 - 2\cos\frac{2\pi}{n}k) &= \prod_{k=1}^{\frac{n}{2}} 4\sin^2\frac{\pi}{n}k = (\prod_{k=1}^{\frac{n}{2}} 2\sin\frac{\pi}{n}k) \cdot (\prod_{k=1}^{\frac{n}{2}} 2\sin\frac{\pi}{n}(n-k)) \\ &= (\prod_{k=1}^{n-1} 2\sin\frac{\pi}{n}k) \cdot 2 = 2n\end{aligned}$$

as required. □

Proof of Theorem 3.1 Let χ be the Brauer character corresponding to the natural representation of $Sl(2,q)$ over $GF(q)$. The group $Sl(2,q)$ is of order $q(q^2-1)$.Each of the elements of $Sl(2,q)$ is similar to a matrix of the following type:

$$A_1:\begin{pmatrix}\xi^a & 0\\ 0 & \xi^a\end{pmatrix}, A_2:\begin{pmatrix}\xi^a & 0\\ 1 & \xi^a\end{pmatrix}, A_3:\begin{pmatrix}\xi^a & 0\\ 0 & \xi^b\end{pmatrix}_{a\neq b}, B_1:\begin{pmatrix}\sigma^a & 0\\ 0 & \sigma^{aq}\end{pmatrix}_{a\not\equiv 0\,(\mathrm{mod}\, q+1)}$$

where ξ,σ are primitive elements of $GF(q), GF(q^2)$ respectively. Clearly $\chi(1)=2$. To find $L_\chi^{p'}$ we examine the p-regular elements of the above types.

- Type A_1: $\begin{pmatrix}1 & 0\\ 0 & 1\end{pmatrix}, \begin{pmatrix}-1 & 0\\ 0 & -1\end{pmatrix}$.
- Type A_3: $\begin{pmatrix}\xi^a & 0\\ 0 & \xi^{-a}\end{pmatrix}$.
- Type B_1: $\begin{pmatrix}\tau^k & 0\\ 0 & \tau^{qk}\end{pmatrix}$ where $\tau=\sigma^{q-1}$.

The character values obtained in the p-regular classes are $\{-2\}\cup L_{A_3}\cup L_{B_1}$ where

$$L_{A_3}=\{2\cos\frac{2\pi}{q-1}a \mid a=1,2,\ldots\frac{q-1}{2}\}$$

and

$$L_{B_1}=\{2\cos\frac{2\pi}{q+1}k \mid k=1,2,\ldots\frac{p+1}{2}\}.$$

Hence

$$\begin{aligned}
f_\chi^{p'}(2) &= (2-(-2))\cdot\prod_{a=1}^{\frac{q-1}{2}}(2-2\cos\frac{2\pi}{q-1}a)\cdot\prod_{k=1}^{\frac{q-1}{2}}(2-2\cos\frac{2\pi}{q+1}k)\\
&= \prod_{a=1}^{\frac{q-1}{2}}(2-2\cos\frac{2\pi}{q-1}a)\cdot\prod_{k=1}^{\frac{q+1}{2}}(2-2\cos\frac{2\pi}{q+1}k)\\
&= (q-1)(q+1)=q^2-1.
\end{aligned}$$

Thus χ is p'-sharp. □

4 π-prime graphs

Let G be a finite group. The *prime graph* $\Gamma(G)$ of G, see [13], is the graph with vertices $\mathcal{V}(\Gamma(G))=\pi(G)$ and edges $\mathcal{E}(\Gamma(G))$ such that $(x,y)\in\mathcal{E}(\Gamma(G))$ if and only if there exists a $g\in G$ such that xy divides $o(g)$. With a slight abuse of notation we will denote the i-th component of $\Gamma(G)$ by π_i, where π_i is the set of primes representing the vertices of the i-th component. Let $\Gamma_i=\{g\in G|g \text{ a } \pi_i\text{-element}, g\neq 1\}$. We define the *natural sharp class function* Ψ of G to be the function

$$\Psi(g)=\begin{cases}0, & \text{if } g=1\\ -|G|_{\pi_i}, & \text{if } g\in\Gamma_i.\end{cases}$$

Theorem 4.1 *The natural sharp class function of a finite group G is a generalized character.*

A proof of Theorem 4.1 is implicit in [9]. We give a much shorter proof using the following result of Frobenius ([8], page 428).

Theorem 4.2 *If G is a group and m is a divisor of $|G|$, then the class function θ defined by $\theta(g) = |G|/m$ if $g^m = e$ and $\theta(g) = 0$ otherwise is a generalized character of G.*

Proof of Theorem 4.1 For a fixed component π_i of $\Gamma(G)$, we define

$$\Psi_i(g) = \begin{cases} -|G|_{\pi_i} & \text{if } g \in \Gamma_i \\ 0 & \text{otherwise.} \end{cases}$$

Now $\Psi = \sum_i \Psi_i$, where the sum is over the components of $\Gamma(G)$, so it suffices to prove that Ψ_i is a generalized character of G. From Theorem 4.2 with $m = |G|/|G|_{\pi_i}$, the function

$$\mu_i(g) = \begin{cases} |G|_{\pi_i} & \text{if } g^m = 1 \\ 0 & \text{otherwise.} \end{cases}$$

is a generalized character. Thus $\Psi_i = |G|_{\pi_i} 1_G - \mu_i$ is a generalized character. □

It is easily seen that a generalized character χ of G is sharp if and only if $f_\chi(\chi) = \rho$, where $f_\chi(\phi) = \prod_{l_i \in L_\chi}(\phi - l_i 1_G)$ and ϕ is a generalized character of G. If χ is π-sharp, then $f_\chi^\pi(\chi) = \prod_{l_i \in L_\chi^\pi}(\chi - l_i 1_G)$ takes the value $|G|_\pi$ on 1 and 0 on the π-elements. Let ρ^π be the π-class function defined by $\rho^\pi(1) = |G|_\pi$ and $\rho^\pi(g) = 0$ when g is a π-element. It follows from Theorem 4.2 with $m = |G|_{\pi'}$ that ρ^π is always the restriction of a generalized character of G to the π-classes and in a sense ρ^π is analogous to the regular character. Thus there always exists (trivially) a π-sharp generalized character. Note that if G is π-separable then ρ^π is the restriction of an ordinary character to the π-classes, namely $(1_H)^G$ where H is a Hall π'-subgroup. We may restate Theorem 2.2 as: $f_\chi^\pi(\chi)$ is an integral multiple of ρ^π. The set of characters $f_\chi^\pi(\chi)$ for χ irreducible seem to be interesting.

For a set of primes π, we define naturally the *π-graph* $\Gamma_\pi(G)$ of G to be the full subgraph of $\Gamma(G)$ with vertices in π. Let $\pi_1, \pi_2, \ldots \pi_r$ represent the connected components of $\Gamma_\pi(G)$ and $G_i = \{g \in G | g \text{ a } \pi_i\text{-element } g \neq 1\}$. We define the *natural π-sharp function* Ψ^π of G to be the π-class function

$$\Psi^\pi(g) = \begin{cases} 0, & \text{if } g = 1 \\ -|G|_{\pi_i}, & \text{if } g \in G_i \end{cases}.$$

Theorem 4.3 *Ψ^π is the restriction of a generalized character of G to the π-classes.*

Proof For a fixed component π_i of $\Gamma_\pi(G)$, we define the π-class function

$$\Psi_i^\pi(g) = \begin{cases} -|G|_{\pi_i} & \text{if } g \in G_i \\ 0 & \text{otherwise.} \end{cases}$$

It is again a consequence of 4.2 that Ψ_i^π is the restriction of a generalized character to the π-classes and the theorem follows immediately. □

Corollary 4.4 *If $\Gamma_\pi(G)$ is disconnected, there exists a non-trivial π-sharp generalized character of G.*

References

[1] D. Alvis, "On finite groups admitting certain sharp characters with irrational values", *Comm. Algebra* **21** (1993) 535-554.

[2] D. Alvis, M. Kiyota, H. W. Lenstra Jr., S. Nozawa, "Sharp characters with only one rational value ", *Comm. Algebra* **22** (1994) 95-115.

[3] D. Alvis, S. Nozawa, "Sharp characters with irrational values", (preprint).

[4] H.W. Blichfeldt, "A theorem concerning the invariants of linear homogeneous groups with some applications to substitution groups", *Trans. Amer. Math. Soc.* **5** (1904) 461-466.

[5] P.J. Cameron, "Extremal problems from Theorems of Maillet and Blichfeldt", Unpublished Notes (private correspondence) (1994).

[6] P.J. Cameron,T. Kataoka, M. Kiyota, "Sharp characters of finite groups of type $\{-1,1\}$", *J. Algebra* **152** (1992) 248-258.

[7] P.J. Cameron, M. Kiyota, "Sharp characters of finite groups", *J. Algebra* **115** (1988) 125-143.

[8] G. Frobenius, "Über einen Fundamentalsatz der Gruppentheorie II", *S'Ber Akad. Wiss. Berlin* (1907) 428-437.

[9] N. Iiyori, "Sharp characters and prime graphs of finite groups", *J. Algebra* **163** (1994) 1-8.

[10] K. W. Johnson, "Sharp Characters of quasigroups ", *Europ. J. Combin.* **14** (1993) 103-112.

[11] M. Kiyota, S. Nozawa, "Sharp characters whose values on non-identity elements are 0 and a family of algebraic conjugates", to appear in *Comm. Algebra.*

[12] T. Matsuhisa, H. Yamaki, "A class of certain groups admitting certain sharp characters II", *Proc. Amer. Math. Soc.* **110** (1990) 1-5.

[13] J.S. Williams, "Prime graph components of finite groups", *J. Algebra* **69** (1981) 487-513.

PERMUTATION GROUP ALGORITHMS VIA BLACK BOX RECOGNITION ALGORITHMS

WILLIAM M. KANTOR* and ÁKOS SERESS†

*University of Oregon, Eugene, OR 97403, U.S.A.
†The Ohio State University, Columbus, OH 43210, U.S.A.

Abstract

If a black box simple group is known to be isomorphic to a classical group over a field of known characteristic, a Las Vegas algorithm is used to produce an explicit isomorphism. This is used to upgrade all nearly linear time Monte Carlo permutation group algorithms to Las Vegas algorithms when the input group has no composition factor isomorphic to an exceptional group of Lie type or a 3–dimensional unitary group.

Key words and phrases: computational group theory, black box groups, classical groups, matrix group recognition
1991 Mathematics Subject Classification: Primary 20B40, 20G40; Secondary: 20P05, 68Q25, 68Q40

1 Introduction

There is a large library of nearly linear time permutation group algorithms [BCFS, BS, CF, LS, Mo, Ra, SchS, Ser]. Most of these are Monte Carlo (which means that the algorithm can return an incorrect answer, although the probability of that can be made as small as desired). The main result of this note is that Monte Carlo can be upgraded to Las Vegas (which means that the output is always correct, but the algorithm may also report failure, although the probability of that can be made as small as desired), whenever there are suitable recognition algorithms for the simple groups occurring as composition factors.

There is a growing literature of recognition algorithms for quasisimple groups of Lie type. The first of these, due to Neumann and Praeger [NP], solved the following problem: given a group $G \leq \mathrm{GL}(d,q)$ by a set of generating matrices, decide whether G contains $\mathrm{SL}(d,q)$. Subsequently, this result was generalized to the other classical groups [NiP1, NiP2, CLG1, CLG2, Ce, Pr], also assuming that matrices were given of the desired size over the desired field. A much more general setting is where a quasisimple matrix group is given only as a black box group. Recently, Cooperman, Finkelstein and Linton [CFL] studied the case $G \cong \mathrm{PSL}(d,2)$, providing a methodology for handling many such questions simultaneously. We extended their result in [KS] to all classical groups over all finite fields. In [BCFL] black box groups isomorphic to $\mathrm{PSL}(d,q)$ will also be dealt with for any q in a manner similar to [CFL].

A *black box group* G is a group whose elements are encoded as 0-1 strings of uniform length N, and the group operations are performed by an oracle (the "black box"). Given strings representing $g, h \in G$, the black box can compute strings representing gh and g^{-1}, and decide whether or not $g = h$. Note that $|G| \leq 2^N$: we have an upper bound on $|G|$. Algorithms usually try to exploit the specific features of the representation of the group they work with. By contrast, a black box group algorithm does not rely on specific features of the group representation or on particulars of how the group operations are performed. It turns out that this is a critical aspect of our uses for these algorithms (cf. Section 2.3).

We state our results about classical groups in a more general setting so their applications for permutation groups can be extended easily when recognition algorithms for additional groups become available.

Definition 1.1 Let $\mathcal{F}$ be a family of simple groups and $f: \mathcal{F} \to \mathbb{R}$ a function taking positive values. We say that $\mathcal{F}$ is *black box f-recognizable* if, whenever a group $G = \langle \mathcal{S} \rangle$ isomorphic to a member of $\mathcal{F}$ is given as a black box group encoded by strings of length N and, in the case of Lie-type G, the characteristic of G is given, there are Las Vegas algorithms for the following:

(i) Find the isomorphism type of G.

(ii) Find a new set $\mathcal{S}^*$ of size $O(N)$ generating G, and a presentation of length $O(N^2)$ in terms of $\mathcal{S}^*$. (This presentation proves that G has the isomorphism type determined in (i).)

(iii) Given $g \in G$, find a straight-line program of length $O(N)$ from $\mathcal{S}^*$ to g.

Moreover,

(iv) The algorithms for (i)–(iii) run in time $O((\xi + \mu)f(G)N^c)$, where ξ is an upper bound on the time requirement per element for the construction of independent, (nearly) uniformly distributed random elements of G, μ is an upper bound on the time required for each group operation in G, and c is an absolute constant.

A *straight-line program of length m* reaching some $g \in G$ can be thought of as a sequence of group elements $(g_1, \ldots, g_m)$ such that $g_m = g$ and for each i one of the following holds: $g_i \in \mathcal{S}^*$, or $g_i = g_j^{-1}$ for some $j < i$, or $g_i = g_j g_k$ for some $j, k < i$. More precisely, since we do not want to store the group elements themselves, the straight-line program reaching g is a sequence of expressions $(w_1, \ldots, w_m)$ such that, for each i, either w_i is a symbol for some element of $\mathcal{S}^*$, or $w_i = (w_j, -1)$ for some $j < i$, or $w_i = (w_j, w_k)$ for some $j, k < i$, such that if the expressions are evaluated the obvious way then the value of w_m is g. This more abstract definition not only requires less memory, but also *enables us to construct a straight-line program in one representation of G and evaluate it in another*, which is an important feature of the algorithms.

We shall prove the following two theorems. Let $\mathcal{G}$ denote the family of all finite simple groups, and let $m: \mathcal{G} \to \mathbb{R}$ be the function such that $m(G)$ is the degree of the smallest faithful permutation representation of G.

Theorem 1.2 *Given a permutation group $G = \langle \mathcal{S} \rangle \leq S_n$ such that all nonabelian composition factors of G are from a black box m-recognizable family $\mathcal{F}$, a base and* *strong generating set for G can be computed in nearly linear Las Vegas time.*

Bases and strong generating sets (SGS) are the basic data structures in algorithms for permutation groups; we shall define them in Section 2.1. We call an algorithm *nearly linear* if its running time is of the form $O(n|S|\log^k|G|)$ for some constant k. We shall justify the name and elaborate more on this notion in Section 2.2.

The novelty in Theorem 1.2 is that the base and SGS construction is Las Vegas. Earlier nearly linear time algorithms used the Monte Carlo construction of [BCFS]. All currently known nearly linear Monte Carlo algorithms can be modified so that after an initial base and SGS computation, all further steps of the algorithm are deterministic or Las Vegas. Thus, *for the groups described in* Theorem 1.2, *we can upgrade the entire nearly linear time library to Las Vegas.*

The algorithm in Theorem 1.2 differs significantly from the traditional SGS constructions [Si1, Si2]; by the time we have found $|G|$ we have also constructed a composition series for G. In this respect, the algorithm resembles the parallel handling of permutation groups [BLS1] and the current fastest deterministic algorithms for computing strong generating sets [BLS2, BLS3].

The second theorem is a constructive version of a result from [BGKLP] about short presentations of groups.

Theorem 1.3 *There is a nearly linear Las Vegas algorithm which, when given a permutation group G satisfying the composition factor restriction of* Theorem 1.2, *computes a presentation of length $O(\log^3|G|)$ for G.*

Using the terminology of this paper, the main result of [KS] can be stated as:

Theorem 1.4 [KS] *The classical simple groups, with the possible exception of the 3-dimensional unitary groups, comprise a black box m-recognizable family.*

We shall also use a similar result for the alternating groups.

Theorem 1.5 [BLNPS] *The alternating groups comprise a* 1*-recognizable family (i.e., one can take $f(G) = 1$ for all alternating groups G).*

It is easy to check that cyclic groups of prime order are m-recognizable and, obviously, sporadic simple groups are 1-recognizable. Hence, combining the previous two theorems with Theorem 1.2 we obtain the

Corollary 1.6 *Given a permutation group $G \leq S_n$ with no exceptional Lie type or 3-dimensional unitary composition factors, all known nearly linear time algorithms dealing with G can be upgraded to Las Vegas algorithms.*

We note that in [KS] a new generating set S^* satisfying Definition 1.1(iii) was found within the required time bound in 3-dimensional unitary groups as well, but it is an open problem whether these groups have a presentation of length $O(\log^2|G|)$ as needed in 1.1(ii).

Actually, in [KS] we prove more than Theorem 1.4: an isomorphism λ with a group of matrices in the correct dimension is constructed, defined by the images of generators, together with procedures to compute the image of any element of G under λ or of any element of $G\lambda$ under λ^{-1}. The analogous isomorphism for alternating groups is constructed in [BLNPS]. These procedures are very useful for further computations with G, such as the construction of Sylow subgroups [Ka, Mo]. For possible applications of [KS] in matrix recognition algorithms see [BB] and [Pr] in these Proceedings.

In [KS] there are also more precise timings of algorithms than required in Theorem 1.4. For example, we show that the family of classical groups, with the possible exception of 3-dimensional unitary groups, is black box f-recognizable for the function

$$f(G) = \begin{cases} q & \text{if } G \cong \mathrm{PSL}(d,q) \text{ for some } d \\ q^2 & \text{for all other } G \text{ defined on a vector space over } \mathrm{GF}(q). \end{cases}$$

It seems very likely that the set of *all* groups of Lie type is a black box f–recognizable family with $f(G) \leq m(G)$. Research is presently under way on the groups of Lie rank ≥ 2 other than ${}^2F_4(q)$. Possibly the biggest obstacle is condition (ii) of Definition 1.1 in the case of rank 1 groups: finding $O(\log^c |G|)$-length presentations for $\mathrm{PSU}(3,q)$, ${}^2B_2(q)$ and ${}^2G_2(q)$ has been a very annoying open problem for several years (cf. [BGKLP]).

2 The proofs

2.1 Bases, strong generating sets, and Schreier trees

Fundamental data structures for computing with permutation groups were introduced by Sims in [Si1, Si2]. A *base* for a permutation group $G \leq \mathrm{Sym}(\Omega)$ of degree n is a sequence $B = (\beta_1, \ldots, \beta_M)$ of points from Ω such that the pointwise stabilizer $G_B = 1$. The *point-stabilizer chain* of G relative to B is the chain of subgroups

$$G = G^{(1)} \geq G^{(2)} \geq \cdots \geq G^{(M+1)} = 1,$$

where $G^{(i)} = G_{(\beta_1,\ldots,\beta_{i-1})}$. The base B is called *nonredundant* if there is strict inclusion $G^{(i)} > G^{(i+1)}$ for all $1 \leq i \leq M$; then $(\log|G|)/(\log n) \leq |B| \leq \log|G|$. A *strong generating set* (SGS) for G relative to B is a set $\mathcal{S}$ of generators of G with the property that

$$\langle \mathcal{S} \cap G^{(i)} \rangle = G^{(i)} \text{ for } 1 \leq i \leq M+1.$$

Let $B = (\beta_1, \ldots, \beta_M)$ be a base of the group G, let $G = G^{(1)} \geq G^{(2)} \geq \cdots \geq G^{(M+1)} = 1$ be the corresponding point-stabilizer chain, and let R_i denote a transversal for $G^{(i+1)}$ in $G^{(i)}$, $1 \leq i \leq M$. Such a transversal can be computed from the SGS by a standard orbit computation of $\beta_i^{G^{(i)}}$, keeping track of the group elements sending β_i to the points of the orbit. Each $g \in G$ can be written uniquely in the form

$$g = r_M r_{M-1} \cdots r_2 r_1, \; r_i \in R_i. \tag{2.1}$$

The process of factoring g in this form is called *sifting* or *stripping.* Note that the order of G can be obtained easily as $|G| = \prod_{i=1}^{M} |R_i|$.

In practical computation, the transversals R_i usually are not computed and stored explicitly; rather, they are encoded in a Schreier-tree data structure. Suppose that a base B and an SGS $\mathcal{S}$ for G relative to B are given. A *Schreier-tree data structure* for G is a sequence of pairs (S_i, T_i) called *Schreier trees*, one for each base point β_i, $1 \leq i \leq M$, where T_i is a directed labeled tree with all edges directed toward the root β_i, and with edge-labels selected from the set $S_i := \mathcal{S} \cap G^{(i)} \subseteq G^{(i)}$. The nodes of T_i are the points of the orbit $\beta_i^{G^{(i)}}$. The labels satisfy the condition that, for each directed edge from γ to δ with label h, $\gamma^h = \delta$. If γ is a node of T_i, then the sequence of the edge-labels along the path from γ to β_i in T_i is a word in the elements of S_i such that the product of these permutations moves γ to β_i. Thus each Schreier tree (S_i, T_i) defines *inverses* of the elements of the transversal R_i for $G^{(i+1)}$ in $G^{(i)}$.

Given an arbitrary SGS $\mathcal{S}$ relative to B, an algorithm in [BCFS] constructs a new SGS $\mathcal{T}$ in $O(nM|\mathcal{S}|\log^2|G|)$ deterministic time such that the depth of each Schreier tree defined by $\mathcal{T}$ is at most $2\log|G|$. We call a Schreier-tree data structure *shallow* if the depth of each tree is at most $2\log|G|$. A shallow Schreier-tree data structure supports membership testing in $O(nM\log|G|)$ time. We will assume that all bases computed in our algorithms are nonredundant and that all Schreier-tree data structures we consider are shallow.

2.2 Nearly linear time algorithms

In groups of current interest for implementations, it frequently happens that the degree of $G = \langle S \rangle$ is in the tens of thousands or even higher, so even a $\Theta(n^2)$ algorithm may not be practical. On the other hand, $\log|G|$ is often small. Therefore, a recent trend is to search for algorithms with running time of the form $O(n|S|\log^k|G|)$. More precisely, given a constant c, a family $\mathcal{G}$ of permutation groups is called a family of *small-base groups* if all $G \in \mathcal{G}$ of degree n admit bases of size $O(\log^c n)$; or, equivalently, if there is a constant c' such that $\log|G| = O(\log^{c'} n)$ for each $G \in \mathcal{G}$ of degree n. For example, all classical simple groups, in all of their permutation representations, comprise a small-base family (with $c = 2$).

We call a permutation group algorithm a *nearly linear time algorithm* if its running time for any $G = \langle S \rangle \leq S_n$ is $O(n|S|\log^k|G|)$. The name is justified by the fact that, if G is a member of a small-base family then the running time is a nearly linear, $O(n|S|\log^{c''}(n|S|))$, function of the input length. We will require the following algorithms of this sort:

Theorem 2.2 *There are Monte Carlo nearly linear time algorithms which, when given $G \leq S_n$, find the following:*

(i) [BCFS] *A base, strong generating set, and a shallow Schreier-tree data structure for G;*

(ii) *As a consequence of* (i): *given a homomorphism $\varphi: G \to S_n$ specified by the images of generators, data structures which enable the nearly linear time* *computation of $\varphi(g)$ for any $g \in G$ and a preimage of any $g \in \varphi(G)$;*

(iii) [BS] *A composition series $G = N_1 \rhd N_2 \rhd \cdots \rhd N_m = 1$ and, for each $1 \leq i \leq m-1$, a homomorphism $\varphi_i: N_i \to S_n$ with* $\ker \varphi_i = N_{i+1}$.

A large part of the permutation group library in GAP [Sch+] consists of implementations of nearly linear algorithms.

2.3 Permutation groups as black box groups

Suppose that a base $B = (\beta_1, \ldots, \beta_M)$, a strong generating set $\mathcal{S}$ with respect to B, and a shallow Schreier-tree data structure $\mathcal{ST} = \{(S_i, T_i) \mid 1 \leq i \leq M\}$ are given for some $G \leq \mathrm{Sym}(\Omega)$, where the sum of the depths of these M Schreier trees is $t = O(\log^2 |G|)$. We may assume that the SGS $\mathcal{S}$ is closed under taking inverses.

Any $g \in G$ can be written uniquely in the form (2.1) for elements r_i of the transversals whose inverses were coded by $\mathcal{ST}$. Each such inverse can be written as a word in the strong generators $\mathcal{S}$, following the path in the appropriate Schreier tree. Taking the inverse of this word and using the fact that $\mathcal{S} = \mathcal{S}^{-1}$, we obtain the r_i, and so g, as a word in $\mathcal{S}$ in a well-defined way. The length of the word representing g is at most t; this word is called the *standard word representing g*. We note the following:

Lemma 2.3 *In deterministic $O(t|B|)$ time, given an injection $f: B \to \Omega$, it is possible to find a standard word representing some $g \in G$ with $B^g = f(B)$ or to determine that no such element of G exists.*

This algorithm *relies on a base, SGS and Schreier-tree data structure.* However, those inputs are computed by Monte Carlo algorithms, and hence may not be correct. Therefore, it is possible that the preceding algorithm returns an incorrect answer — though with small probability.

Now *we show how to consider G as a black box group H*; this will be crucial for Theorems 1.2 and 1.3. The elements of H are defined to be the standard words representing the elements of G; these are strings over the alphabet $\mathcal{S}$, of length at most t. Of course, we can write the elements of H as 0-1 strings of uniform length N: $\mathcal{S}$ can be coded by $\lceil \log(|\mathcal{S}| + 1) \rceil$–long 0-1 sequences for the numbers $1, 2, \ldots, |\mathcal{S}|$; every standard word can be padded by 0's to length t. Since $|\mathcal{S}|$ and t are $O(\log^2 |G|)$, N is $O(\log^2 |G| \log\log |G|)$. As customary in the analysis of permutation group algorithms, we assume that small numbers can be read in $O(1)$ time, and therefore we shall ignore the factor $\log\log |G|$ above; this is at worst $\log n$, and hence is appropriate within the nearly linear time context.

Formally, we have an isomorphism $\psi: G \to H$ with the following properties. Each $g \in G$ defines an injection $f: B \to \Omega$ by $f(\beta_i) := \beta_i^g$, and then Lemma 2.3 can be used to compute $g\psi$ in $O(t|B|) = O(\log^3 |G|)$ time. Conversely, given $h \in H$, $h\psi^{-1}$ can be computed in $O(n \log^2 |G|)$ time, by multiplying out the product of the elements of h as a permutation.

Each $h \in H$ is represented by a unique string, so comparison of group elements can be performed in $O(\log^2 |G|)$ time. In order to take the product of $h_1, h_2 \in H$, we concatenate these two words, and define a function $f\colon B \to \Omega$ by $f(\beta_i) := \beta_i^{h_1 h_2}$. Then the standard word representing $h_1 h_2$ can be obtained by Lemma 2.3. This procedure runs in $O(\log^3 |G|)$ time. Similarly, to take the inverse of some $h \in H$, we take the inverse of the word h. This defines an injection $f\colon B \to \Omega$, and again we use Lemma 2.3 in $O(\log^3 |G|)$ time. Hence, *we have a black box oracle which performs the black box group operations in* $O(\log^3 |G|)$ *time*, which is potentially faster than the ordinary permutation multiplication. In particular, if $G \leq S_n$ is a member of a small-base family, then, in the notation of Definition 1.1, $\mu = O(\log^c n)$ for some constant c. Recall, however, that this oracle can give incorrect answers if our base, SGS or Schreier-tree data structure was incorrect.

(Nearly) random elements of $G\psi$ and of subgroups of $G\psi$ can be constructed, by a remarkable algorithm of Babai [Ba], in $O(\mu \log^5 |G|)$ time. (An apparently practical heuristic algorithm for this purpose is given in [CLMNO].)

Summarizing, we can construct an isomorphism between a permutation group $G \leq S_n$ and a black box group H such that the word length N of the encoding of H, as well as the time requirement for the group operations in H and constructing random elements in H are bounded from above by a polylogarithmic function of $\log |G|$. Therefore, we can perform $O(n)$ group operations in H within nearly linear time. Note that if we considered G as a black box group, with the original permutation multiplication as black box group operation, then $O(n)$ group operations would result in an $O(n^2)$ algorithm.

2.4 Proofs of Theorems 1.2 and 1.3

Proof of Theorem 1.2 Let $G = \langle T \rangle \leq S_n$ be a permutation group. Compute a base and strong generating set, and a composition series $G = N_1 \rhd N_2 \rhd \cdots \rhd N_m = 1$, by the nearly linear Monte Carlo algorithms in Theorem 2.2. The composition series algorithm also provides homomorphisms $\varphi_i\colon N_i \to S_n$ with $\ker \varphi_i = N_{i+1}$, for $1 \leq i \leq m-1$. We also compute strong generating sets for all N_i with respect to the base of G. We will verify the correctness of the base and strong generating sets for the subgroups N_i by induction on $i = m, m-1, \ldots, 1$.

Suppose that we already have verified an SGS for N_{i+1}. Using Theorem 2.2, we compute a base, SGS, shallow Schreier-tree data structure, and an isomorphism ψ_i with a black box group for the image $N_i\varphi_i$ (cf. Section 2.3), which is a subgroup of S_n and is allegedly isomorphic to a simple group. Our first goal is to *obtain in nearly linear Las Vegas time a presentation of length* $O(\log^2 |N_i\varphi_i|)$ *for* $N_i\varphi_i$, using a generating set S_i^* such that a straight-line program of length $O(\log |N_i\varphi_i|)$ from S_i^* to any given element of $N_i\varphi_i$ can be obtained in nearly linear Las Vegas time.

As a consequence of the classification of finite simple groups, we know that there are no three pairwise nonisomorphic simple groups of the same order. So we have at most two candidate simple groups C for the isomorphism type of $N_i\varphi_i$, and in the ambiguous cases we try both possibilities. Also, if $|N_i\varphi_i| > 8!/2$ then $|N_i\varphi_i|$ determines whether $N_i\varphi_i$ is of Lie type, and if it is, its characteristic. Hence, using that $N_i\varphi_i$ is from an m-recognizable family, we can obtain $S_i^*\psi_i$, a presentation, and straight-line programs in $N_i\varphi_i\psi_i$, in nearly linear Las Vegas time. By Theorem 2.2(ii), the preimage S_i^* of $S_i^*\psi_i$ can also be obtained in nearly linear time.

Now the correctness of the SGS for N_i can be proved the following way. Let $\mathcal{T}_i$ be the set of generators of N_i computed by the composition series algorithm (Theorem 2.2(iii)). We check that (i) $N_{i+1} \lhd N_i$ and $N_i \neq N_{i+1}$; (ii) $\langle \mathcal{S}_i^* \varphi_i^{-1} \rangle N_{i+1}/N_{i+1}$ satisfies the presentation computed for $N_i\varphi_i$; and (iii) $\mathcal{T}_i \subset \langle \mathcal{S}_i^* \varphi_i^{-1} \rangle N_{i+1}$, where $h\varphi_i^{-1}$ denotes a lift (i.e., an arbitrary preimage) of $h \in \mathcal{S}_i^* \subset N_i\varphi_i$. Checking (i)–(iii) shows that $|N_i| = |N_{i+1}||N_i\varphi_i|$. If $|N_i|$ is equal with the value for $|N_i|$ computed from the alleged SGS of N_i then the SGS construction is correct: it is known that the alleged order of a group obtained from the Monte Carlo SGS construction is not greater than the true order, with equality if and only if the SGS construction is correct.

For (i), conjugate the generators of N_{i+1} by the elements of $\mathcal{T}_i$, and check that the resulting permutations are in N_{i+1} (since the correctness of N_{i+1} is already known, membership testing giving guaranteed correct results is available for that group). Also, check that not all elements of $\mathcal{T}_i$ are in N_{i+1}. For (ii), multiply out the relators that were written in terms of $\mathcal{S}_i^*$, using the permutations in $\mathcal{S}_i^* \varphi_i^{-1}$; then check that the resulting permutations are in N_{i+1}. Finally, for (iii) write straight-line programs from $\mathcal{S}_i^*$ to $\mathcal{T}_i\varphi_i$, and for each $t \in \mathcal{T}_i$ evaluate it starting from $\mathcal{S}_i^* \varphi_i^{-1}$ (this is where we use our unusually precise definition of straight-line programs). This produces some $t^* \in \langle \mathcal{S}_i^* \varphi_i^{-1} \rangle$; check that $t^* t^{-1} \in N_{i+1}$. By (ii) and (iii), we have checked that the factor group $N_i/N_{i+1} \cong C$.

At the end of the induction, we have obtained a correct SGS for the group $N_1 = \langle \mathcal{T}_1 \rangle$ which was output by the composition series algorithm. After that, we verify that $G = N_1$ by sifting the elements of the original generating set $\mathcal{T}$ in N_1.

To justify the nearly linear running time of the entire algorithm, note that m is $O(\log |G|)$ so it is enough to show that the ith step of the induction runs in nearly linear time. We have already seen that the constructions of both $\mathcal{S}_i^*$ and the presentation of $N_i\varphi_i$ are within this time bound. Since both $|\mathcal{T}_i|, |\mathcal{S}_i^*|$ are $O(\log |G|)$, while the length of the presentation is $O(\log^2 |G|)$ and the Schreier-tree data structure of N_{i+1} is shallow, the number of permutation multiplications in (i)–(iii) is bounded from above by a polylogarithmic function of $|G|$.

We note that we have to require that calls to the algorithms in Theorems 1.4, 1.5, and 2.2 fail with probability $< 1/(c \log |G|)$, since during the induction, $O(\log |G|)$ such calls may be made; however, this multiplies the running time only by a $\log \log |G|$ factor. □

Proof of Theorem 1.3 The following result is contained in [BGKLP, Sec. 8]. *If each composition factor H_i of the finite group G has a presentation of length $O(\log^C |H_i|)$ for some $C \geq 2$, then G has a presentation of length $O(\log^{C+1} |G|)$.* The proof in [BGKLP] proceeds by the following steps; we need to show that these can be handled in nearly linear time.

(i) Let L be a lifting of the generators of the composition factors to G. Let M be a subset of L of size $O(\log |G|)$ which also generates G.

(ii) Let S be a subset of G such that any element of G can be reached from S by a straight-line program of length $O(\log |G|)$. Write straight-line programs from M to S.

(iii) Write a presentation for G, and simultaneously write straight-line programs from S to $O(\log^C |G|)$ elements of G. Roughly, these elements are those whose membership in N_{i+1} was tested in (i)–(iii) in the proof of Theorem 1.2. Now the presentation in [BGKLP] is obtained in $O(\log^{C+1} |G|)$ deterministic time.

We saw in the proof of Theorem 1.2 that presentations of the composition factors H_i of a black box group satisfying the restriction of Theorem 1.2 can be obtained having length $O(\log^2 |H_i|)$, using a nearly linear time algorithm. Hence, we shall apply the result of [BGKLP] with the value $C = 2$. Moreover, the generating sets S_i^* of the composition factors constructed in the proof of Theorem 1.2 are such that $\bigcup_i S_i^* \varphi_i^{-1}$ has $O(\log |G|)$ elements. In Proposition 2.4 we shall show that any given $g \in G$ can be reached from $\bigcup_i S_i^* \varphi_i^{-1}$ by a straight-line program of length $O(\log |G|)$, and such a straight-line program can be computed in nearly linear time. This means that we can choose $S := L = M$ in (i) and (ii), so that a presentation of G of length $O(\log^3 |G|)$ can indeed be written in nearly linear time, as indicated in (iii) and as required in the theorem.

Proposition 2.4 *Let $G \leq S_n$ be a permutation group, and suppose that the following have already been computed by Las Vegas algorithms, as in the proof of* Theorem 1.2*: a composition series $G = N_1 \rhd N_2 \rhd \cdots \rhd N_m = 1$, homomorphisms $\varphi_i : N_i \to S_n$ with $\ker \varphi_i = N_{i+1}$, and presentations using generating sets $S_i^* \subset N_i \varphi_i$. Then any $g \in G$ can be reached from $\bigcup_i S_i^* \varphi_i^{-1}$ by a straight-line program of length $O(\log |G|)$, and such a straight-line program can be computed in nearly linear time.*

Proof By induction on $i = 1, 2, \ldots, m$, we will construct a straight-line program of length $O(\log(|G|/|N_i|))$ to some $g_i \in G$ such that $gg_i^{-1} \in N_i$. Let $g_1 := 1$. If g_i has already been obtained for some i, then write a straight-line program of length $O(\log |N_i/N_{i+1}|)$ from S_i^* to $(gg_i^{-1})\varphi_i$. In the case when N_i/N_{i+1} is cyclic or sporadic, this can be done by brute force. In the other cases, we use the isomorphism ψ_i between $N_i\varphi_i$ and a black box group, as in the proof of Theorem 1.2, and the fact that $N_i\varphi_i\psi_i$ is black box m-recognizable. Evaluate this straight-line program starting from $S_i^* \varphi_i^{-1}$, producing an element $h_i \in N_i$. Here, $gg_i^{-1}h_i^{-1} \in N_{i+1}$, and we can define $g_{i+1} := h_i g_i$. Finally, we notice that the procedure runs in nearly linear time, since $m(N_i/N_{i+1}) \leq n$. □

References

[Ba] L. Babai, Local expansion of vertex-transitive graphs and random generation in finite groups, pp. 164–174 in: Proc. ACM Symp. on Theory of Computing 1991.

[BB] L. Babai and R. Beals, A polynomial-time theory of matrix groups and black box groups, in these Proceedings.

[BCFL] S. Bratus, G. Cooperman, L. Finkelstein and S. Linton (in preparation).

[BCFS] L. Babai, G. Cooperman, L. Finkelstein and Á Seress, Nearly linear time algorithms for permutation groups with a small base, pp. 200–209 in: Proc. Int. Symp. Symbolic and Algebraic Computation, ACM 1991.

[BGKLP] L. Babai, A. J. Goodman, W. M. Kantor, E. M. Luks and P. P. Pálfy, Short presentations for finite groups, J. Algebra 194 (1997), 79–112.

[BLNPS] R. Beals, C. R. Leedham-Green, A. C. Niemeyer, C. E. Praeger and Á. Seress, A mélange of black box algorithms for recognising finite symmetric and alternating groups (in preparation).

[BLS1] L. Babai, E. M. Luks and Á. Seress, Permutation groups in NC, pp. 409–420 in: Proc. ACM Symp. on Theory of Computing 1987.

[BLS2] L. Babai, E. M. Luks and Á. Seress, Fast management of permutation groups I. SIAM J. Computing 26 (1997).

[BLS3] L. Babai, E. M. Luks and Á. Seress, Fast management of permutation groups II (in preparation).

[BS] R. Beals and Á. Seress, Structure forest and composition factors for small base groups in nearly linear time, pp. 116–125 in: Proc. ACM Symp. on Theory of Computing 1992.

[Ce] F. Celler, Matrixgruppenalgorithmen in GAP. Ph. D. thesis, RWTH Aachen 1997.

[CF] G. Cooperman and L. Finkelstein, A random base change algorithm for permutation groups, J. Symb. Comp. 17 (1994), 513–528.

[CFL] G. Cooperman, L. Finkelstein and S. Linton, Recognizing $GL_n(2)$ in non-standard representation, pp. 85–100 in [FK].

[CLG1] F. Celler and C. R. Leedham-Green, A non-constructive recognition algorithm for the special linear and other classical groups, pp. 61–67 in [FK].

[CLG2] F. Celler and C. R. Leedham-Green, A constructive recognition algorithm for the special linear group (to appear in Proc. ATLAS Conference).

[CLMNO] F. Celler, C. R. Leedham-Green, S. H. Murray, A. C. Niemeyer and E. A. O'Brien, Generating random elements of a finite group. Comm. Alg. 23 (1995) 4931–4948.

[FK] L. Finkelstein and W. M. Kantor, editors, Groups and Computation II, DIMACS Series in Discrete Math. and Theoretical Computer Science, vol. 28, AMS 1997.

[Ka] W. M. Kantor, Sylow's theorem in polynomial time. J. Comp. Syst. Sci. 30 (1985) 359–394.

[KS] W. M. Kantor and Á. Seress, Black box classical groups (submitted).

[LS] E. M. Luks and Á. Seress, Computing the Fitting subgroup and solvable radical of small-base permutation groups in nearly linear time, pp. 169–181 in [FK].

[Mo] P. Morje, A nearly linear algorithm for Sylow subgroups of permutation groups. Ph.D. thesis, The Ohio State University 1995.

[NiP1] A. C. Niemeyer and C. E. Praeger, Implementing a recognition algorithm for classical groups, pp. 273–296 in [FK].

[NiP2] A. C. Niemeyer and C. E. Praeger, A recognition algorithm for classical groups over finite fields (to appear in Proc. London Math. Soc.).

[NP] P. M. Neumann and C. E. Praeger, A recognition algorithm for special linear groups. Proc. London Math. Soc. (3) 65 (1992), 555–603.

[Pr] C. E. Praeger, Primitive prime divisor elements in finite classical groups, in these Proceedings.

[Ra] F. Rákóczi, Fast recognition of nilpotency of permutation groups, pp. 265–269 in: Proc. Int. Symp. Symbolic and Algebraic Computation, ACM 1995.

[Sch+] M. Schönert et. al., GAP: Groups, Algorithms, and Programming, Lehrstuhl D für Mathematik, RWTH Aachen, 1994.

[SchS] M. Schönert and Á. Seress, Finding blocks of imprimitivity in small-base groups in nearly linear time, pp. 144–147 in: Proc. Int. Symp. Symbolic and Algebraic Computation, ACM 1994.

[Ser] Á. Seress, Nearly linear time algorithms for permutation groups: an interplay

between theory and practice (to appear in Acta Appl. Math).
[Si1] C. C. Sims, Computational methods in the study of permutation groups, 169–183 in: Computational problems in abstract algebra (ed. J. Leech), Pergamon Press 1970.
[Si2] C. C. Sims, Computation with permutation groups, pp. 23–28 in: Proc. 2nd Symp. on Symb. and Alg. Manipulation (ed. S. R. Petrick), ACM 1971.

NONABELIAN TENSOR PRODUCTS OF GROUPS: THE COMMUTATOR CONNECTION

LUISE-CHARLOTTE KAPPE

SUNY at Binghamton, New York, NY 13902-6000, U.S.A.

Abstract

This is a progress report on some of the developments in nonabelian tensor products of groups since the appearance of the paper "Some Computations of Non-Abelian Tensor Products of Groups" by Brown, Johnson and Robertson, ten years ago.

In the spring of 1988 Ronnie Brown came to Binghamton and gave a talk about nonabelian tensor products, in particular about his paper with Johnson and Robertson [7] which had just appeared. I fell in love with tensor products on first sight and started my student Michael Bacon on this topic for his dissertation, and since then others have joined in these investigations.

This talk is an invitation for others to join in this research. There are many interesting and accessible problems and it appears likely that there are interesting applications to group theory, in the same way as regular tensors have been applied.

All this is provided you do not immediately get thrown off by the notation. In context with nonabelian tensor products left actions are used. Early on I contemplated switching to right action but decided against it. That would be like insisting on driving on the right in a country where everyone else drives on the left.

We use the following notation. For elements g, g', h, h' in a group G we set ${}^h g = hgh^{-1}$ for the conjugate of g by h, and $[h, g] = hgh^{-1}g^{-1}$ for the commutator of h and g. The familiar expansion formulas using left action appear as follows:

$$\begin{aligned} [gg', h] &= [{}^g g', {}^g h][g, h], \\ [g, hh'] &= [g, h][{}^h g, {}^h h']. \end{aligned}$$

Now we define the nonabelian tensor product of two groups, preceded by the definition of a compatible action which is intimately connected with nonabelian tensor products.

Definition 1 *Let G and H be a pair of groups acting upon each other in a compatible way, that is*

$$^{({}^g h)}g' = {}^g({}^h({}^{g^{-1}}g')) \text{ and } ^{({}^h g)}h' = {}^h({}^g({}^{h^{-1}}h'))$$

for $g, g' \in G$ and $h, h' \in H$, and acting upon themselves by conjugation.

Then the nonabelian tensor product $G \otimes H$ is the group generated by the symbols $g \otimes h$ with defining relations

$$\begin{aligned} gg' \otimes h &= ({}^g g' \otimes {}^g h)(g \otimes h), \\ g \otimes hh' &= (g \otimes h)({}^h g \otimes {}^h h') \end{aligned}$$

for $g, g' \in G, h, h' \in H$.

If $G = H$ *and all actions are conjugation, then* $G \otimes G$ *is called the nonabelian tensor square of* G.

As the reader might have already realized, a slight change in notation gets you from the commutator expansion formulas to the defining relations of the nonabelian tensor product. It was this visual similarity that lead me to look into this topic. Of course, there is a conceptual connection between commutators and nonabelian tensors, but this will remain in the background here. The idea is to explore tensor products (and squares) as group theoretical objects, and to explicitly determine nonabelian tensor squares and products for whole classes of groups, by using techniques similar to commutator calculus.

To illustrate my point, I will discuss a classical group theoretical problem originating with Levi [23] and a nonabelian tensor analogue. Levi considered groups in which the commutator operation is associative. Here is a characterization of this class of groups based on Levi's results.

Theorem 2 *For a group* G *the following conditions are equivalent:*

(i) $[[x, y], z] = [x, [y, z]]\ \forall\, x, y, z \in G$;

(ii) $[[x, y], z] = 1\ \forall\, x, y, z \in G$;

(iii) $[xy, z] = [y, z][x, z]$ *and* $[x, yz] = [x, y][x, z]\ \forall\, x, y, z \in G$;

(iv) $G' \subseteq Z(G)$.

Bacon [3] looked at an analogue of Levi's result for nonabelian tensors. To formulate the analogue we need to consider the tensor center of a group which is a characteristic and central subgroup of every group and is defined as $Z^{\otimes}(G) = \{g \in G; g \otimes x = 1_{\otimes}\ \forall\, x \in G\}$. Replacing the center by the tensor center we may ask now for a characterization of groups in which the commutator subgroup is contained in the tensor center. We arrive at the following five equivalent conditions.

Theorem 3 *The following are equivalent for a group* G *and its tensor square:*

(i) $[x, y] \otimes z = x \otimes [y, z]\ \forall\, x, y, z \in G$;

(ii) $[x, y] \otimes z = 1_{\otimes}\ \forall\, x, y, z \in G$;

(iii) $xy \otimes z = (y \otimes z)(x \otimes z)$ *and* $x \otimes yz = (x \otimes y)(x \otimes z)\ \forall\, x, y, z \in G$;

(iv) $G' \subseteq Z^{\otimes}(G)$;

(v) $G \otimes G \cong G/G' \otimes G/G'$.

The first four conditions bear a striking resemblance to Levi's characterization of groups in which the commutator operation is associative. In fact, the methods for their proofs are very similar. The fifth condition comes somewhat as a surprise and arises out of the question whether there are nonabelian groups whose nonabelian tensor square is isomorphic to the tensor square of its abelianization. The answer is yes, and such groups are exactly those groups whose commutator subgroup is contained in the tensor center. This follows immediately from a result due to Graham Ellis [13].

Proposition 4 *Let G be a group and $N \lhd G$. Then $G \otimes G \cong G/N \otimes G/N$ if and only if $N \leq Z^{\otimes}(G)$.*

The following example shows that the conditions of Theorem 3 can be satisfied in a nontrivial way.

Example 5 Let p be a prime and $G = \langle a \rangle \rtimes \langle b \rangle$, where $|a| = p^{\alpha}, |b| = p^{\beta}, |[a,b]| = p^{\gamma}, \alpha \geq 2\gamma, \beta \geq \gamma \geq 1, \alpha - \gamma \geq \beta$. Then G satisfies the conditions of Theorem 3.

More could be said in this context, but I want to move on and look briefly at the origins and the history. Nonabelian tensor products have their roots in algebraic K-theory as well as topology. Everyone agrees that some ideas can already be found in Whitehead's work [31]. The nonabelian tensor square appears in essence but not in name in the work of Keith Dennis [10] and is based on ideas of Miller [25]. Independently, Lue in [24] defines nonabelian tensor products in the setting of nilpotent groups. He extends earlier work by Ganea [16]. Brown and Loday in [8] and [9] can lay claim to be the inventors of the nonabelian tensor product of groups. It is a direct outgrowth of their involvement with generalized Van Kampen theorems. The nonabelian tensor squares of K-theory origin are a special case of nonabelian tensor products.

This brings us up to the 1987-paper of Brown, Johnson and Robertson [7], the starting point for looking at nonabelian tensor products as group theoretical objects. Almost simultaneously several other papers on this topic appear, e.g. by Aboughazi [1], Ellis [11] and Johnson [21]. I will come back to some of them later.

Here was the starting point for us in 1988. Initially, our research was guided by the 8 open problems at the end of [7]. I will report next on the progress that has been made during the past ten years in solving these problems.

1. *Let G and H be finite groups acting compatibly on each other. Then is it true that $G \otimes H$ is finite?*

Already Brown and Loday in [9] establish that the tensor square, $G \otimes G$, is finite for finite G. Ellis settled this question for tensor products affirmatively in [11], but no purely algebraic proof is known. In addition he shows that the tensor product $G \otimes H$ is of p-power order if G and H are of p-power order. Rocco in [27] gives a bound for the order of $G \otimes G$ if G has order p^n. In [15], Ellis and McDermott improve Rocco's bound and extend it to the case of nonabelian tensor products $G \otimes H$ of prime-power groups G and H. In particular it is shown that if G has order p^n and d is the minimal number of generators of G, then the order of $G \otimes G$ does not exceed p^{dn}.

These finiteness results open the door for determining tensor squares and products with the help of computers. Already in [7] the nonabelian tensor squares of all groups up to order 30 are computed in this way, using just the definition of a tensor square. In [14], Ellis and Leonard give a computer algorithm capable of determining the tensor squares of much larger groups, e.g. the Burnside group $B(2,4)$ which has order 4096.

2. *Let $d(G)$ be the minimal number of generators for a group G. Can any general estimate of $d(G \otimes G)$ be found when G is finite?*

This question was apparently prompted by the following result:

Proposition 6 ([7]) *Let F be free of rank 2, then $d(F \otimes F)$ is countably infinite.*

More generally, one can now ask under what conditions on G does $d(G)$ finite imply $d(G \otimes G)$ finite. As a sampler of some of our results related to this question I provide one due to Bacon [2]:

Theorem 7 *Let G be a group of nilpotency class two with $d(G) = n$, then*

$$d(G \otimes G) \leq \frac{n(n^2 + 3n - 1)}{3}.$$

The bound given in the above theorem is sharp. It is attained for the free group of nilpotency class 2 and rank n.

The third and fourth open problems in [7] address the solvability length and nilpotency class of $G \otimes G$, given such information about G.

3. *If G is solvable of derived length $l(G)$, then $G \otimes G$ is solvable and $l(G \otimes G) = l(G)$ or $l(G) - 1$. Is there any intrinsic characterization of solvable groups of either type?*

4. *If G is nilpotent of class $cl(G)$, then $G \otimes G$ is nilpotent and $cl(G \otimes G) = cl(G')$ or $cl(G') + 1$. Can either of these types be characterized internally?*

Here are some of our contributions to Problem 4.

Theorem 8 ([4]) *Let G be nilpotent of class 2. Then $G \otimes G$ is abelian.*

The nilpotency of $G \otimes G$ only depends on the nilpotency of the commutator subgroup G' of G as the following result shows.

Theorem 9 ([5]) *Let G be a group. If the derived subgroup G' is nilpotent of class $cl(G')$, then $G \otimes G$ is nilpotent with $cl(G \otimes G) = cl(G')$ or $cl(G') + 1$.*

As a byproduct of our investigations in [5] we have the following result.

Proposition 10 *Let T_n be the free nilpotent group of class 3 and rank n. Then $T_2 \otimes T_2$ is abelian and $T_n \otimes T_n$ is nilpotent of class 2 precisely for $n \geq 3$.*

The answer to Problem 4 seems not only related to the nilpotency class of G but also to the number of generators. Having now more examples gives hope that we will soon get an answer at least in the case of metabelian groups.

Before going into the specifics of the next three open problems in [7], all dealing with determinations of tensor squares, let me report on the second big topic addressed in our research, namely the explicit computation of tensor squares and products for whole classes of groups. The key to the method are crossed pairings (see [7]).

Definition 11 Let G and L be groups. A function $\Phi : G \times G \to L$ is called a crossed pairing if

$$\begin{aligned}\Phi(gg', h) &= \Phi({}^g g', {}^g h)\Phi(g, h),\\ \Phi(g, hh') &= \Phi(g, h)\Phi({}^h g, {}^h h')\end{aligned}$$

for $g, g', h, h' \in G$.

Proposition 12 *A crossed pairing Φ determines a unique homomorphism of groups $\Phi^* : G \otimes G \to L$ such that $\Phi^*(g \otimes h) = \Phi(g, h)$ for all $g, h \in G$.*

This method involves conjecturing L and Φ. It seems to look like an improbable task, but can be successfully managed if L can be shown to be abelian, as in the case of groups of nilpotency class 2. The key element in the analysis is the expansion of $g \otimes h$ which is obtained in a similar fashion as a commutator expansion. In [4] a complete classification of nonabelian tensor squares of 2-generator p-groups of class 2 was obtained in the case $p \neq 2$. We lacked at the time a classification of these groups for $p = 2$.

Sarmin recently obtained a classification of 2-generator 2-groups of class 2. We have now completed the computation of the tensor squares of these groups and are in the process of preparing a paper on these results [22].

If L is not commutative, conjecturing L is much more difficult and the calculations for the expansion of $g \otimes h$ and checking the crossed pairing are much more involved. Recently we were able to compute $\mathcal{E} \otimes \mathcal{E}$, a nonabelian group, where $\mathcal{E}$ is the free 3-generator 2-Engel group [5]. The group L was conjectured from our knowledge of the tensor square of $B(3,3)$, the 3-generator Burnside group of exponent 3, which is a finite homomorphic image of $\mathcal{E}$. The computations for $B(3,3) \otimes B(3,3)$ were done by Ellis using the algorithm in [14]. The extensive calculations for the expansion of $g \otimes h$ and the verification of the crossed pairing were done with the help of symbolic calculations in GAP [28].

In this context I want to mention a paper by Hartl [20]. Extending ideas from [1], he develops a method to determine nonabelian tensor squares for groups of nilpotency class two and computes some of them using this method.

I want to return to the remaining open problems from [7], the next three addressing the determination of nonabelian tensor squares.

5. *Examine the behavior of $G \otimes G$ under the formation of free products.*

As already mentioned in [7] this was done by Gilbert in [17].

6. *Complete the evaluation of $G \otimes G$ for all metacyclic groups G.*

Some results for split extensions can already be found in [7]. In [21], Johnson settled the question for finite metacyclic split extensions completely. This leaves us with the cases of infinite metacyclic groups and those which are not split extensions. James Beuerle, a Ph.D. student who just started working with me in the fall of 1996 has made good progress and we hope to have our results ready for publication soon [6]. All tensor squares of metacyclic groups are abelian. In addition to crossed pairings we also use the fact that the tensor square of a homomorphic image of a group G is a homomorphic image of $G \otimes G$ (see [7]).

7. *Compute the tensor square of $GL(2,p)$ and other linear groups.*

This problem was solved by Hannebauer [19] before we ever got a chance to look at it.

The first seven problems in [7] are of a group theoretical nature. The last problem concerns itself with the diagram (1) in [7] which we reproduce here for the convenience of the reader.

$$
\begin{array}{ccccccccc}
 & & & & 0 & & 0 & & \\
 & & & & \downarrow & & \downarrow & & \\
H_3(G) & \longrightarrow & \Gamma(G^{ab}) & \xrightarrow{\psi} & J_2(G) & \longrightarrow & H_2(G) & \longrightarrow & 0 \\
\| & & \| & & \downarrow & & \downarrow & & \\
H_3(G) & \longrightarrow & \Gamma(G^{ab}) & \xrightarrow{\psi} & G \otimes G & \longrightarrow & G \wedge G & \longrightarrow & 1 \\
 & & & & \downarrow{\scriptstyle \kappa} & & \downarrow{\scriptstyle \kappa'} & & \\
 & & & & G' & \xrightarrow{=} & G' & & \\
 & & & & \downarrow & & \downarrow & & \\
 & & & & 1 & & 1 & &
\end{array}
\qquad (1)
$$

From results in [8] and [9] it follows that the above diagram is commutative with exact rows and central extensions as columns. Here Γ is Whitehead's quadratic functor [31], $G \wedge G$ is a generalized exterior product, and $H_n(G)$ is the n-th Eilenberg-MacLane homology of G with trivial integer coefficients. For further details see [7]. The last in the list of open problems in [7] is the following:

8. *Give an algebraic proof of the exactness of the top row of diagram (1):*

$$H_3(G) \to \Gamma(G^{ab}) \xrightarrow{\psi} J_2(G) \to H_2(G) \to 0$$

We suspect that this is an edge exact sequence of a spectral sequence of algebraic origin.

A purely algebraic proof of the exactness of the sequence of Problem 8 has been given by Ellis in [12]. The material in this paper has since been developed further by Parashvili in [26].

Most of the open problems in [7] deal with nonabelian tensor squares, and as a consequence, so does this progress report. In the concluding paragraphs I want to take a look at general tensor products. Compared to the good progress made with tensor squares, the information on products is still very sparse. In [18], Gilbert and Higgins compute the first nonabelian tensor product. They establish that

$\mathbb{Z} \otimes \mathbb{Z} \cong \mathbb{Z} \oplus \mathbb{Z}$ where the action is mutual inversion. In [14], Ellis and Leonard developed a method to compute $N \otimes M$, where N and M are embedded as normal subgroups in a group G and implemented it by computer for $|G| \leq 14$.

The largest class so far addressed are cyclic groups. M. Visscher is preparing these results for publication [29]. The methods employed are again crossed pairings $\Phi : G \times H \to L$ (see [7]). Let $G = \mathbb{Z}_n$ and $H = \mathbb{Z}_m$, where $\mathbb{Z}_n$ and $\mathbb{Z}_m$ are cyclic, and not necessarily finite, groups of order n and m. Then $d(\mathbb{Z}_n \otimes \mathbb{Z}_m) \leq 2$, and this bound is sharp even in the case that $\mathbb{Z}_n$ and $\mathbb{Z}_m$ are both finite.

So far we do not have a single example for a nonabelian tensor product $G \otimes H$ where G and H are each nonabelian with nontrivial compatible actions. Compatability is still an enigma to us. We hope that we can say more about it in the near future.

A natural extension of Problems 3 and 4 in [7] is to consider questions about solvability length and nilpotency class in the setting of nonabelian tensor products. What can one say about the solvability length and nilpotency class of $G \otimes H$ if such information is given on G and H ? Visscher addresses these questions in [30]. We conclude with a summary of these results.

Definition 13 Let H and G be groups and let H act on G. We define $D_H(G) = \langle g^h g^{-1} \mid g \in G, h \in H \rangle$ as the derivative of H in G.

Proposition 14 *If H and G act compatibly on each other, then $D_H(G) \triangleleft G$ and $D_G(H) \triangleleft H$.*

Theorem 15 *Let G and H be groups acting compatibly on each other. If $D_H(G)$ is solvable then so are $G \otimes H$ and $D_G(H)$. Furthermore if $l(D_H(G))$ is the derived length of $D_H(G)$ then $l(D_H(G)) \leq l(G \otimes H) \leq l(D_H(G)) + 1$, and $l(D_H(G)) - 1 \leq l(D_G(H)) \leq l(D_H(G)) + 1$.*

Theorem 16 *Let G and H be groups acting compatibly on each other. If $D_H(G)$ is nilpotent then so are $G \otimes H$ and $D_G(H)$. Furthermore, if $cl(D_H(G))$ is the nilpotency class of $D_H(G)$, then $cl(D_H(G)) \leq cl(G \otimes H) \leq cl(D_H(G)) + 1$, and $cl(D_H(G)) - 1 \leq cl(D_G(H)) \leq cl(D_H(G)) + 1$.*

References

[1] R. Aboughazi, *Produit tensoriel du group d'Heisenberg*, Bull. Soc. Math. France **115** (1987), 95-106.

[2] M. Bacon, *The nonabelian tensor square of a nilpotent group of class 2*, Glasgow Math. J. **36** (1994), 291-296.

[3] M. Bacon, *The associative law in tensor squares*, in preparation.

[4] M. Bacon, L.C. Kappe, *The nonabelian tensor square of a 2-generator p-group of class 2*, Arch. Math. **61** (1993), 508-516.

[5] M. Bacon, L.C.Kappe, R.F.Morse, On the nonabelian tensor square of a 2-Engel group, Arch. Math. 69 (1997), 353-364.

[6] J. Beuerle, L.C. Kappe, *Metacyclic groups and their nonabelian tensor squares*, in preparation.

[7] R. Brown, D.L. Johnson, E.F. Robertson, *Some Computations of Non-Abelian Tensor Products of Groups*, J. Algebra **111** (1987), 177-202.
[8] R. Brown, J.-L. Loday, *Excision homotopique en basse dimension*, C.R. Acad. Sci. Ser.I Math. Paris **298** (1984), 353-356.
[9] R. Brown, J.-L. Loday, *Van Kampen theorems for diagrams of spaces*, Topology **26** (1987), 311-335.
[10] R.K. Dennis, *In search of new "homology" functors having a close relationship to K-theory*, Preprint, Cornell University, Ithaca, NY 1976.
[11] G. Ellis, *The Nonabelian Tensor Product of Finite Groups is Finite*, J. Algebra **111** (1987), 203-205.
[12] G. Ellis, *An algebraic derivation of a certain exact sequence*, J. Algebra **127** (1989), 178-181.
[13] G. Ellis, *Tensor Products and q-Crossed Modules*, J. London Math. Soc. (2) **51** (1995), 243-258.
[14] G. Ellis, F. Leonard, *Computing Schur Multipliers and Tensor Products of Finite Groups*, Proc. Royal Irish Acad. **95A** (1995), 137-147.
[15] G. Ellis, A. McDermott, *Tensor products of prime power groups*, J. Pure Applied Algebra, to appear.
[16] T. Ganea, *Homologie et extensions centrales des groupes*, C.R. Acad. Sci. Paris **266** (1968), 556-568.
[17] N.D. Gilbert, *The non-abelian tensor square of a free product of groups*, Arch. Math. **48** (1987), 369-375.
[18] N.D. Gilbert, P.J. Higgins, *The non-abelian tensor product of groups and related constructions*, Glasgow Math. J. **31** (1989), 17-29.
[19] T. Hannebauer, *On nonabelian tensor squares of linear groups*, Arch. Math. **55** (1990), 30-34.
[20] M. Hartl, *The Nonabelian Tensor Square and Schur Multiplicator of Nilpotent Groups of Class 2*, J. Algebra **179** (1996), 416-440.
[21] D.L. Johnson, *The nonabelian tensor square of a finite split metacyclic group*, Proc. Edinburgh Math. Soc. **30** (1987), 91-96.
[22] L.C. Kappe, N. Sarmin, M. Visscher, *Two-generator 2-groups of class 2 and their nonabelian tensor squares*, Glasgow Math.J., to appear.
[23] F.W. Levi, *Groups in which the commutator operation satisfies certain algebraic conditions*, J. Indian Math. Soc. **6** (1942), 87-97.
[24] A.S.-T. Lue, *The Ganea map for nilpotent groups*, J. London Math. Soc. **14** (1976), 309-312.
[25] C. Miller, *The second homology group of a group; relations among commutators*, Proc. Amer. Math. Soc. **3** (1952), 588-595.
[26] T. Pirashvili, *A universal coefficient theorem for quadratic functors*, preprint, A.M.R. Math. Inst. Tbilisi, 1993.
[27] N.R. Rocco, *On a construction related to the nonabelian tensor square of a group*, Bol. Soc. Brasil, Mat. **22** (1991), 63-79.
[28] M. Schönert, et al., *GAP-Groups, Algorithms, and Programming* (3rd ed.), Lehrstuhl D für Mathematik, RWTH, Aachen, 1993.
[29] M. Visscher, *The classification of nonabelian tensor products of a pair of cyclic groups*, in preparation.
[30] M. Visscher, *On the Nilpotency Class and Solvability Length of Nonabelian Tensor Products of Groups*, in preparation.
[31] J.H.C. Whitehead, *A certain exact sequence*, Ann. of Math. **51** (1950), 51-110.

SIMPLE SUBALGEBRAS OF GENERALIZED WITT ALGEBRAS OF CHARACTERISTIC ZERO

NAOKI KAWAMOTO

Maritime Safety Academy, 5-1, Wakaba, Kure 737, Japan

1 Introduction

In this paper we give a short survey on infinite-dimensional Witt type Lie algebras and their simple subalgebras mainly over a field of characteristic zero. We give a list of papers related to this subject, but we did not intend to make a complete list.

We denote by k the ground field of any characteristic unless otherwise specified.

2 Some history

Let k be a field of characteristic $p > 0$, and W be a vector space over k with basis $\{D_i \mid 0 \leq i < p\}$. Define the multiplication by $[D_i, D_j] = (i-j)D_{i+j}$, and this makes W a Lie algebra. W is called a Witt algebra. We denote this algebra by $W_{\mathbf{Z}/p\mathbf{Z}}$, and the following result is well known [Se67].

Theorem *If* $p \neq 2$ *then* $W_{\mathbf{Z}/p\mathbf{Z}}$ *is a simple Lie algebra.*

This result was the starting point for later research, and the finite cyclic group $\mathbf{Z}/p\mathbf{Z}$ was replaced by several groups.

Kaplansky [K54] has generalized in the following form: Let V be a vector space over k, and G be an additive subgroup of the dual space V^* of V. Let I be an index set of a basis of V, and we denote an element a of V^* by $a = (a_i)_{i\in I}$, where $a_i \in k$. Assume that G is a total additive group, that is, the only element $\alpha = (\alpha_i)_{i\in I}$, where $\alpha_i = 0$ except for finite i, such that $\sum_i a_i\alpha_i = 0$ for any $a = (a_i)_{i\in I} \in G$ is the zero-element. Let L be a vector space over k with basis $\{(x,\alpha) \mid x \in V, \alpha \in G\}$. Define the multiplication by $[(x,\alpha),(y,\beta)] = (\alpha(y)x - \beta(x)y, \alpha+\beta)$ $(x, y \in V, \alpha, \beta \in G)$. Then L is a Lie algebra.

Theorem ([K54]) *Except when* $\dim V = 1$ *and* $\operatorname{char} k = 2$ *the Lie algebra* L *is simple.*

3 Generalizations

Block [B58] has considered another generalization. Let G be a non-zero abelian group, $g : G \longrightarrow k$ be an additive map, and $f : G \times G \longrightarrow k$ be a biadditive and alternative map. Then a vector space $L(G, g, f)$ with basis $\{u_\alpha \mid \alpha \in G\}$, and the multiplication defined by $[u_\alpha, u_\beta] = (f(\alpha,\beta) + g(\alpha-\beta))u_{\alpha+\beta}$ $(\alpha, \beta \in G)$ is a Lie algebra.

Theorem ([B58]) *$L(G,g,f)$ is simple if and only if* $\operatorname{char} k \neq 2$, $g \not\equiv 0$, $f(\alpha,\beta) = g(\alpha)h(\beta) - g(\beta)h(\alpha)$ *for some additive map* $h : G \longrightarrow k$, *and* $g(\delta) = 0$, $h(\delta) = 0$ *or* 2 *implies* $\delta = 0$.

Block has also given a condition for two simple Lie algebras $L(G,g,f)$ and $L(G',g',f')$ to be isomorphic, and shown if $\operatorname{rank} G = 1$ then $\operatorname{Aut} L$ is a semi-direct product of $\operatorname{Hom}(G,k^*)$ and $\operatorname{Aut} G$.

Amayo and Stewart [AS74] have defined a generalized Witt algebra W_G: Let G be an additive subgroup of k, and W_G be a vector space with basis $\{ w_g \mid g \in G\}$. Define the multiplication by $[w_g, w_h] = (g-h)w_{g+h}$ $(g,h \in G)$. Then W_G is a Lie algebra. This is a special case of Block's algebras. W_G is simple if $\operatorname{char} k \neq 2$ and $G \neq 0$. For example $W_{\mathbf{Z}}$ and $W_{\mathbf{R}}$ are simple Lie algebras. The algebra $W_{\mathbf{Z}}$ have a very interesting property, that is, it satisfies the maximal condition for subalgebras shown by Amayo [A75] and Kubo [Ku76].

4 Generalizations over a field of characteristic zero

From now on we assume that the characteristic of k is zero. The author has considered a Lie algebra $W(G,I)$ [Ka86] inspired by the result of Kaplansky. Let $I \neq \phi$ be an index set, and G be a nonzero subgroup of $\prod_{i\in I} k_i^+$, where k_i^+ is a copy of k. $W(G,I)$ is a vector space with basis $\{w(a,i) \mid a \in G, i \in I\}$ and multiplication $[w(a,i), w(b,j)] = a_j w(a+b,i) - b_i w(a+b,j)$ $(a,b \in G,\ i,j \in I)$, where $a = (a_i)_{i\in I}, b = (b_i)_{i\in I} \in G$.

Theorem ([Ka86]) *$W(G,I)$ is a semi-direct product of R and S, where R is a unique maximal abelian ideal of $W(G,I)$, and S is a nonabelian simple subalgebra S of $W(G,I)$. S is isomorphic to some $W(H,J)$.*

$W(G,I)$ is simple if and only if the radical $R = 0$ or G is "total". If $|I| = n$, $G = \mathbf{Z}^n$ then $W(G,I)$ is a simple infinite dimensional Lie algebra isomorphic to the derivation algebra of the Laurent polynomial ring in indeterminates $x_1, \cdots, x_n$ [Ka86]. We write $W_{\mathbf{Z}^n}$ instead of $W(G,I)$. The algebra $W(\mathbf{R}^n, I)$ is also simple.

Đoković and Zhao [DZ95] have considered in the more general setting : Let A be an abelian group, T be a vector space over k, $kA = \langle t^x \mid x \in A\rangle$ is a group algebra of A. Let $W = kA \otimes T$, and a map $\varphi : T \times A \longrightarrow k$ is linear in T and additive in A. Define the multiplication of W by $[t^x\partial_1, t^y\partial_2] = t^{x+y}(\partial_1(y)\partial_2 - \partial_2(x)\partial_1)$ $(x,y \in A,\ \partial_1,\partial_2 \in T)$, where, $t^x\partial = t^x \otimes \partial$, $\partial(x) = \varphi(\partial,x) = \langle \partial, x\rangle$. Then $W = W(A,T,\varphi)$ is a Lie algebra. For φ let $A_0 = \{x \in A \mid \langle\partial,x\rangle = 0 \text{ for any } \partial \in T\}$ be a right kernel, and $T_0 = \{\partial \in T \mid \langle \partial, x\rangle = 0 \text{ for any } x \in A\}$ be a left kernel.

Theorem ([DZ95]) *$W(A,T,\varphi)$ is simple if and only if $A \neq 0$ and φ is nondegenerate, that is, $A_0 = 0, T_0 = 0$.*

Yu has considered classification of G-graded Lie algebras with 1-dimensional homogeneous components [Y95, Y96].

5 Cartan type Lie algebras

There are well known infinite-dimensional simple Lie algebras called Cartan type Lie algebras (see for example [Fu86, R69]). Let $R = \mathbf{C}[[x_1, \cdots, x_n]]$ be a formal power series ring. Let $W_n = \{\sum_i f_i \frac{\partial}{\partial x_i} \mid f_i \in R\}$, and define the multiplication by $[\sum_i f_i \frac{\partial}{\partial x_i}, \sum_i g_i \frac{\partial}{\partial x_i}] = \sum_i (\sum_j f_j \frac{\partial g_i}{\partial x_j} - \sum_j g_j \frac{\partial f_i}{\partial x_j}) \frac{\partial}{\partial x_i}$. Then W_n is a simple Lie algebra, and the subalgebra $S_n = \{\sum_i f_i \frac{\partial}{\partial x_i} \mid \sum_i \frac{\partial f_i}{\partial x_i} = 0\}$ is simple. If n is even then Hamiltonian algebra H_n is defined, and if n is odd then contact algebra K_n is defined. These Lie algebras are subalgebras of $W_{\mathbf{Z}^n}$.

Dzhumadil'daev has considered derivation algebras of formal Laurent power series rings corresponding to the above Cartan type Lie algebras and their central extensions [Dz92].

Osborn has extended W_n, S_n, H_n, K_n to $W(G, I)$, $|I| = n$, denoted by W^*, S^*, H^*, K^*, and investigated simplicity, derivations and automorphisms of these algebras [O96a, O96b, O96c]. Further, Osborn and Zhao have investigated generalized Cartan type H Lie algebras and generalized Cartan type K Lie algebras [OZ96, OZ97a, OZ97b, OZ97c]. Đoković and Zhao have considered generalized Cartan type S Lie algebras [DZ97].

6 Automorphism groups

Block has determined automorphism groups of his Lie algebras [B58]. Amayo and Stewart have shown $W_G \simeq W_H$ $(G, H \leq k^+)$ if and only if there exists $\alpha \in k \setminus \{0\}$ such that $G = \alpha H$ [AS74].

For $G, H \leq k^n$ Ikeda has shown if $G = HX$ for $X \in GL(n, k)$ then $W(G, I_n) \simeq W(H, I_n)$ [I91]. The converse was a question, and has been solved by Đoković and Zhao for $W(A, T, \varphi)$ [DZ95].

The author has shown that the G-graded automorphism group of $W_{\mathbf{Z}^n}$ is $GL(n, \mathrm{Z}) \times (k \setminus \{0\})^n$ [Ka95]. Đoković and Zhao have determined the automorphism groups of $W(A, T, \varphi)$ and shown the above result for $W_{\mathbf{Z}^n}$ without the restriction of "G-graded" [DZ95].

7 Derivation algebras

Many authors have considered the derivation algebras of generalizations of Witt algebra. It is well known that finite dimensional simple Lie algebras over $\mathbf{C}$ have no outer derivation. But the situation is different for generalized Witt algebras, and they usually have outer derivations [B58, Fa88, IKa90, I91, Dz92, OP95, O96b].

8 Simple subalgebras

Nam [N96a] has considered a different kind of Lie algebras. Let $W(0, 1, 0)$ be a vector space over k of characteristic 0 with basis $\{e^{ax} x^i \partial \mid a, i \in \mathbf{Z}\}$, where $\partial = \frac{d}{dx}$. The multiplication is defined by

$$[e^{ax} x^i \partial, e^{bx} x^j \partial] = (b - a) e^{ax+bx} x^{i+j} \partial + (j - i) e^{ax+bx} x^{i+j-1} \partial.$$

Then $W(0,1,0)$ is a simple Lie algebra, and he has considered generalizations of $W(0,1,0)$ [N96b].

Ikeda and the author have observed that $W(0,1,0)$ is isomorphic to a subalgebra of $W_{\mathbf{Z}^2}$. Related to this subalgebra Ikeda and Mitsukawa [IM97] have shown that $W_{\mathbf{Z}^2}$ has a family of infinite-dimensional simple subalgebras: $\langle w(a+n\delta_1,1)+w(a+m\delta_2,2) \mid a \in \mathbf{Z}^2\rangle$, where $\delta_1 = (1,0), \delta_2 = (0,1)$, is simple if $n = 0, m \neq 0$ or $n \neq 0, m = 0$.

Mathieu [M92] has classified simple **Z**-graded Lie algebras of finite growth over an algebraically closed field of characteristic 0, but simple subalgebras of $W_{\mathbf{Z}}$ or $W_{\mathbf{Z}^2}$ are not classified.

References

[A75] R. K. Amayo, A construction for algebras satisfying the maximal condition for subalgebras, Compositio Math. **31**(1975), 31–46.

[AS74] R. K. Amayo and I. Stewart, Infinite-dimensional Lie Algebras, Noordhoff, Leyden, 1974.

[B58] R. Block, On torsion-free abelian groups and Lie algebras, Proc. Amer. Math. Soc. **9**(1958), 613-620.

[DZ95] D. Z. Đoković and K. Zhao, Derivations, isomorphisms, and second cohomology of generalized Witt algebras, Proc. Amer. Math. Soc., to appear.

[DZ96] D. Z. Đoković and K. Zhao, Some infinite-dimensional simple Lie algebras in characteristic 0 related to those of Block, Preprint.

[DZ97] D. Z. Đoković and K. Zhao, Generalized Cartan type S Lie algebras in characteristic zero, J. Algebra **193**(1997), 144-179.

[Dz92] A. S. Dzhumadil'daev, Central extensions of infinite-dimensional Lie algebras, Functional Anal. Appl. **26**(1992), no. 4, 21–29, English transl. 246–253.

[Fa88] R. Farnsteiner, Derivations and central extensions of finitely generated graded Lie algebras, J. Algebra **118**(1988), 33-45.

[Fu86] D. B. Fuks, Cohomology of Infinite-dimensional Lie Algebras, Consultants Bureau, New York, 1986.

[I91] T. Ikeda, Derivations and central extensions of a generalized Witt algebra, Nonassociative Algebras and Related Topics, 47–57, World Scientific, Singapore, 1991.

[IKa90] T. Ikeda and N. Kawamoto, On the derivations of generalized Witt algebras over a field of characteristic zero, Hiroshima Math. J. **20**(1990), 47–55.

[IKa94] T. Ikeda and N. Kawamoto, On derivation algebras of group algebras, Non-Associative Algebra and its Applications, 188-192, Kluwer Acad. Publ., Dordrecht, 1994.

[IM97] T. Ikeda and A. Mitsukawa, Lie algebras with a basis indexed by $\mathbf{Z}^2$, Preprint.

[K54] I. Kaplansky, Seminar on simple Lie algebras, Bull. Amer. Math. Soc. **60**(1954). 470–471.

[Ka86] N. Kawamoto, Generalizations of Witt algebras over a field of characteristic zero, Hiroshima Math. J. **16**(1986), 417–426.

[Ka95a] N. Kawamoto, On G-graded automorphisms of generalized Witt algebras, Second International Conference on Algebra, 225-230, Contemp. Math., 184, Amer. Math. Soc., Providence, RI, 1995.

[Ku76] F. Kubo, On an infinite-dimensional Lie algebra satisfying the maximal condition for subalgebras, Hiroshima Math. J. **6**(1976), 485–487.

[Ku77b] F. Kubo, A note on Witt algebras, Hiroshima Math. J. **7**(1977), 473–477.

[M92] O. Mathieu, Classification of simple graded Lie algebras of finite growth, Invent.

Math. **108**(1992), 455-519.

[N96a] K-B. Nam, New proof of derivations of Witt algebras and applications to the derivations of generalized Witt algebras, Preprint.

[N96b] K-B. Nam, Simple Lie algebras which generalizes Witt algebras, Preprint.

[O96a] J. M. Osborn, New simple infinite dimensional Lie algebras of characteristic 0, J. Algebra **185**(1996), 820-835.

[O96b] J. M. Osborn, Derivations and isomorphisms of Lie algebras of characteristic 0, Preprint.

[O96c] J. M. Osborn, Automorphisms of the Lie algebras W^* in characteristic 0, Preprint.

[OP95] J. M. Osborn and D. S. Passman, Derivations of skew polynomial rings, J. Algebra **176**(1995), 417-448.

[OZ96] J. M. Osborn and K. Zhao, Generalized Poisson brackets and Lie algebras of type H in characteristic 0, Preprint.

[OZ97a] J. M. Osborn and K. Zhao, Generalized Cartan type K Lie algebras in characteristic 0, Preprint.

[OZ97b] J. M. Osborn and K. Zhao, Infinite dimensional Lie algebras of type L, Preprint.

[OZ97c] J. M. Osborn and K. Zhao, Infinite dimensional Lie algebras of generalized Block type, Preprint.

[R69] A. N. Rudakov, Groups of automorphisms of infinite-dimensional simple Lie algebras, Math USSR-Izv. **3**(1969), 707–722.

[Se67] G. B. Seligman, Modular Lie algebras, Ergebnisse der Mathematik und ihrer Grenzgebiete, 40, Springer, Berlin, 1967.

[Y95] R. W. T. Yu, Algèbres de type Witt sur les réseaux, C. R. Acad. Sci. Paris Sér. I Math. **320**(1995), 907-910.

[Y96] R. W. T. Yu, Algèbres de Lie de type Witt, Comm. Algebra, to appear.

APPLICATIONS OF THE BAKER-HAUSDORFF FORMULA IN FINITE p-GROUPS

E. I. KHUKHRO

Institute of Mathematics, Novosibirsk-90, 630090, Russia

School of Mathematics, University of Wales, College of Cardiff, CF2 4YH, Wales

In memory of A. I. Mal'cev (1909–1967)

1 Introduction

The Baker–Hausdorff Formula $H(x, y)$ is defined by the equality $e^x e^y = e^{H(x,y)}$ for formal power series in non-commuting variables. This formula is an important instrument in the theory of Lie groups giving a local correspondence between a Lie group and its Lie algebra, but we shall not discuss Lie groups in this paper.

The Mal'cev Correspondence makes use of the Baker–Hausdorff Formula to provide a global correspondence (a so-called equivalence of categories) between nilpotent Lie $\mathbb{Q}$-algebras and discreet nilpotent $\mathbb{Q}$-powered (that is, torsion-free and divisible) groups. Although finite p-groups are neither torsion-free nor divisible, we shall show how the Mal'cev Correspondence can be applied in the theory of finite p-groups.

Under some rather restrictive conditions (like the nilpotency class to be less than p) an analogous correspondence can be established between finite p-groups and Lie rings (the Lazard Correspondence). As G. Higman remarked in his talk at the Congress of Mathematicians in Edinburgh, 1958, these conditions are "*... too severe to be used..., ... the sort of thing one wants in the conclusion of one's theorem, rather than in the hypothesis.*" Nevertheless, we shall also give examples of applications of the Lazard Correspondence. In particular, it can be used for faster reductions to Lie rings and for constructing certain examples, which may be easier for Lie rings.

As a proving ground for applications of the Baker–Hausdorff Formula, we shall discuss results on automorphisms with few fixed points, regular and almost regular ones. The results are stated mainly for finite groups, although they usually have natural extensions to locally finite or locally nilpotent groups.

2 The Mal'cev correspondence

We feel it necessary to remind the reader of some basic definitions (see, for example, [6, 32, 33]). Let A be a free nilpotent associative $\mathbb{Q}$-algebra on free generators $x_1, x_2 \ldots$. The bracket multiplication $[a, b] = ab - ba$ defines the Lie $\mathbb{Q}$-algebra $A^{(-)}$ on the additive group of A. The x_i freely generate a free nilpotent Lie subalgebra L of $A^{(-)}$. Adjoining the outer unity 1 to A, we define the formal exponent e^a for $a \in A$. The formal logarithm is defined satisfying $\log e^a = a$ and $e^{\log(1+b)} = 1 + b$ for $a, b \in A$. The elements e^{x_i} freely generate a free nilpotent group F. The Baker–Hausdorff Formula $H(x_1, x_2)$ in two variables is defined as $H(x_1, x_2) = \log(e^{x_1} \cdot e^{x_2})$. Often an important theorem stating that $H(x_1, x_2) \in L$ is referred to as the Baker–Hausdorff Formula. It follows that the set $e^L = \{e^l \mid l \in L\}$ is a group, which is a nilpotent $\mathbb{Q}$-powered group containing F.

It is obvious that the operations $a * b = H(a, b)$ and $a^{*r} = ra$, $r \in \mathbb{Q}$, define the structure of a $\mathbb{Q}$-powered group on the set L isomorphic to e^L under the isomorphism $a \to e^a$, $a \in L$.

It can be actually shown that e^L coincides with $\sqrt{F}$, the set of all roots of element of F, and that $\sqrt{F}$ is a free nilpotent $\mathbb{Q}$-powered group freely generated by the e^{x_i}. It follows that $e^{x_1+x_2} = h_1(x_1, x_2)$ and $e^{[x_1,x_2]} = h_2(x_1, x_2)$ for some $\mathbb{Q}$-powered group words h_1 and h_2 in e^{x_1} and e^{x_2} (the Inverse Baker–Hausdorff Formulae). These formulae define the structure of a nilpotent $\mathbb{Q}$-algebra on the same set $\sqrt{F}$ isomorphic to L.

The above constructions can be applied to any nilpotent $\mathbb{Q}$-powered group H (or any nilpotent $\mathbb{Q}$-algebra M), because the appropriate laws hold for any elements in H (or M), since they hold for the free generators e^{x_i} of $\sqrt{F}$ (or x_i of L). These transformations are inverses of one another: if L_H is the Lie algebra defined via the Inverse Baker–Hausdorff Formulae on the same set H, and G_M is the $\mathbb{Q}$-powered group defined via the Baker–Hausdorff Formula on the same set M, then $L_{(G_M)} = M$ and $G_{(L_H)} = H$. This is the Mal'cev Correspondence between nilpotent $\mathbb{Q}$-powered groups and nilpotent Lie $\mathbb{Q}$-algebras. Every statement in terms of the $\mathbb{Q}$-powered group H can be translated into the language of the Lie $\mathbb{Q}$-algebra L_H and vice versa. Note, however, that in the $\mathbb{Q}$-powered group H only the $\mathbb{Q}$-powered (divisible) subgroups correspond to the $\mathbb{Q}$-subalgebras of L_H, only $\mathbb{Q}$-powered normal subgroups correspond to ideals, etc. With such caution in mind, one can prove that the derived length of H is equal to that of L_H. Every automorphism of H as an abstract group is also an automorphism of H as a $\mathbb{Q}$-powered group and is also an automorphism of L_H (acting on the same set in the same way).

3 The Lazard correspondence

One can easily see that if the nilpotency class of A (which coincides with the nilpotency class of L, or F, or $\sqrt{F}$) is less than a given prime p, then the denominators of the coefficients of the Baker–Hausdorff Formula $H(x, y)$ and its inverses $h_1(x, y)$ and $h_2(x, y)$ are not divisible by p. If p' is the set of all primes other than p, then p-groups are p'-divisible and p'-torsion-free. Hence the same formulae give (a special case of) the Lazard Correspondence [25] between nilpotent p-groups of class $\leq p-1$ and nilpotent Lie rings of class $\leq p-1$ with additive p-group. Here the situation is even better since every subgroup of such a group H is automatically an "allowed" one, corresponding to a subring of L_H.

4 Lie ring methods

Applications of the Baker–Hausdorff Formula in Group Theory belong to the Lie ring methods. Any Lie ring method consists of three major steps. First, a hypothesis on a group is translated into a hypothesis on a Lie ring that is constructed from the group in some way. Then a theorem on Lie rings is proved (or used). Finally, a result on the group must be recovered from the Lie ring information obtained.

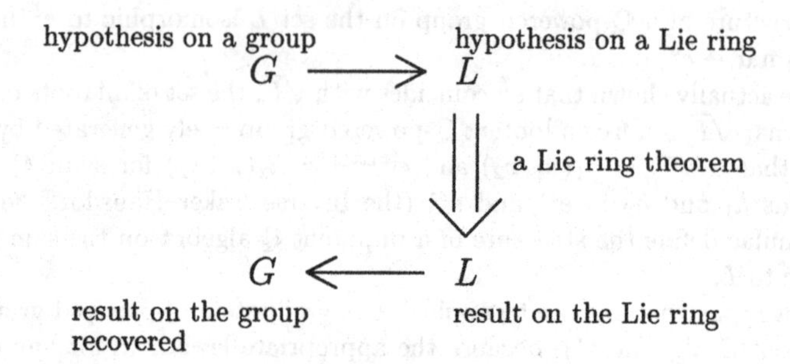

The advantage of this scheme lies in the fact that Lie rings are "more linear" objects, which may be easier to study. For example, one can extend the ground ring in order to decompose the Lie algebra into the sum of eigenspaces with respect to an automorphism regarded as its linear transformation. Sometimes both overpasses, from groups to Lie rings and back, may be quite non-trivial; sometimes, of course, the Lie ring theorem is the main part of the proof. One of the best-known triumphs of the Lie ring methods is the positive solution of the Restricted Burnside Problem for groups of prime exponent (Kostrikin) and of prime-power exponent (Zel'manov). Note that the reduction to Engel Lie algebras for groups of prime exponent by Magnus [29] and Sanov [39] was based on the Baker–Hausdorff Formula. The reduction to Engel Lie algebras for groups of prime-power exponent was completed by Zel'manov.

5 Regular automorphisms

In the above scheme of a Lie ring method, results on Lie rings may be regarded as an inspiration for proving analogous theorems on groups. We shall discuss the group-theoretic analogues of the following classical theorems on Lie rings. Recall that an automorphism φ of a group (or a Lie ring) G is *regular*, if it has no non-trivial fixed points: $C_G(\varphi) = 1$ (or $C_G(\varphi) = 0$).

5.1 Higman's Theorem *If a Lie ring L admits a regular automorphism φ of prime order p, then L is nilpotent of class at most $h(p)$, where $h(p)$ depends on p only.*

Higman [10] only proved the existence of his function $h(p)$; later Kreknin and Kostrikin [23, 24] found a new effective proof with an explicit upper bound for $h(p)$. Simple examples show that there exist non-nilpotent Lie rings with regular automorphisms of composite order. But they must be soluble anyway, as proved by Kreknin [23].

5.2 Kreknin's Theorem *If a Lie ring L admits a regular automorphism φ of finite order n, then L is soluble of derived length at most $k(n)$, where $k(n)$ depends on n only.*

Kreknin's work [23] gives $2^n - 2$ as an upper estimate for Kreknin's function $k(n)$. As a typical example of a Lie ring method, we derive the following consequence of Higman's Theorem, which is also due to Higman [10].

5.3 Corollary *If a finite nilpotent group G admits a regular automorphism φ of prime order p, then G is nilpotent of class at most $h(p)$, where $h(p)$ is Higman's function.*

Proof $G \to L$: We consider the so-called associated Lie ring

$$L(G) = \bigoplus_{i=1}^{\infty} \gamma_i(G)/\gamma_{i+1}(G)$$

whose additive group is the direct sum of the factors of the lower central series and whose multiplication is defined via taking group commutators. The automorphism φ acting on the $\gamma_i(G)/\gamma_{i+1}(G)$ induces an automorphism of $L(G)$ (denoted by the same letter). It can be shown that $C_G(\varphi) = 1$ implies $C_{L(G)}(\varphi) = 0$.

$L \Rightarrow L$: Then, by Higman's Theorem 5.1, $L(G)$ is nilpotent of class $\leq h(p)$.

$G \leftarrow L$: It is well-known that the nilpotency class of the associated Lie ring $L(G)$ is exactly the same as that of G. Hence G is nilpotent of class $\leq h(p)$. □

5.4 Remark Of course, by the celebrated theorem of Thompson [44], the word "nilpotent" can be dropped from the hypothesis.

5.5 Remark Unlike theorems on regular automorphisms of Lie rings, there must be some additional conditions in analogous theorems on groups. For example, a free 2-generator group $F = \langle x, y \rangle$ admits the automorphism $\psi : x \leftrightarrow y$, which has order 2 and is regular: $C_F(\psi) = 1$.

As for Kreknin's Theorem on regular automorphisms of arbitrary finite order, there is still no analogous result for groups.

5.6 Conjecture *An analogue of Kreknin's Theorem for a locally finite group G: if G admits a regular automorphism φ of finite order n, then G is soluble of derived length bounded in terms of n only.*

By the classification of finite simple groups and by a work of Thompson, the proof of Conjecture 5.6 is reduced to the case where G is a finite nilpotent group. (In the works of Berger, Gross, Hartley, Isaacs, Kurzweil, Shult, Turull and others, better bounds for the nilpotency length of G were obtained, even under weaker hypotheses.) One can also state a similar conjecture for G being a locally nilpotent group. But until now, apart from Higman's case of an automorphism of prime order, Conjecture 5.6 is proved only for $|\varphi| = 4$ (L. Kovács [22]). Using the associated Lie ring $L(G)$ fails to produce the required result because the derived length of $L(G)$ may well be smaller than that of G. Although in the direction "$G \to L$" we may assert that φ induces a regular automorphism of order n on $L(G)$, and $L(G)$ is then soluble of derived length $\leq k(n)$ by Kreknin's Theorem 5.2, this does not say much about the group G, that is, the overpass "$G \leftarrow L$" does not work here.

6 Almost regular automorphisms

However, a great progress was achieved in the "modular" case, where a finite p-group P admits a p-automorphism φ of order p^n, say, with p^m fixed points. Of course, a p-automorphism of a finite p-group can never be regular. But one can regard φ as an "almost regular" automorphism of P, the problem being to obtain restrictions on the structure of P in terms of the number of fixed points and the order of the automorphism. In theorems analogous to Higman's and Kreknin's, it was proved that P is "almost" nilpotent or soluble, having a subgroup of bounded index, which has bounded nilpotency class or derived length. For $|\varphi| = p$ such results were obtained by Alperin [1], Khukhro [13] and Medvedev [35, 36, 37], and for $|\varphi| = p^n$ by Shalev [40] and Khukhro [15], plus a sharper result of Makarenko [31] for $|\varphi| = 4$. These results can be regarded as generalizations of some theorems on p-groups of maximal class and on p-groups of given coclass, where one of the main cases amounts to the situation $|C_P(\varphi)| = p$, which was dealt with in the works of Shepherd [43], and Leedham-Green and McKay [27] (for $|\varphi| = p$) and McKay [34] and Kiming [21] (for $|\varphi| = p^n$). In the works of Shalev and Zel'manov [42] and Shalev [41] theorems on regular automorphisms of Lie rings and on almost regular p-automorphisms of finite p-groups were used to prove some of the Coclass Conjectures.

We mention in passing that in the "ordinary", coprime situation there is the following theorem, which is a natural generalization of Higman's Theorem 5.1 to almost regular automorphisms of prime order. It is convenient to say that a value is (a, b)-bounded, say, if this value is bounded by a function depending on a and b only.

6.1 Theorem (Khukhro [14]) *If a finite nilpotent group G admits an automorphism φ of prime order p with exactly m fixed points, then G has a subgroup of (p, m)-bounded index which is nilpotent of p-bounded class.*

The "coprime part" of this theorem is much more difficult and involves an analogous theorem on Lie rings with an almost regular automorphism of prime order. The associated Lie rings are used in the overpasses from groups to Lie rings and back. Note the difficulty in recovering the group-theoretic result in the direction "$G \leftarrow L$": a subring of $L(G)$ does not usually correspond to a subgroup of G (with the same nilpotency class). By a reduction of Fong [8] and Hartley and Meixner [9], one can drop the nilpotency assumption on the group G in the hypothesis; alternatively, by Medvedev [37], one can drop the word "finite" (preserving "nilpotent").

There is also some progress for an almost regular automorphism of order four: a generalization of Kovács' result [22] was proved for Lie rings by Makarenko and

Khukhro in [18, 20], while in [19] they proved a somewhat weaker result for groups.

We single out the "modular" component of Theorem 6.1 that will be needed later.

6.2 Theorem *If a finite p-group P admits an automorphism φ of order p with exactly p^m fixed points, then P has a subgroup of (p,m)-bounded index which is nilpotent of class at most $h(p)$, the value of Higman's function.*

Alperin [1] proved that P has (p,m)-bounded derived length; Khukhro [13] obtained a subgroup of (p,m)-bounded index which is nilpotent of class $\leq h(p)+1$, and Makarenko [30] improved the bound for the nilpotency class to the best possible value $h(p)$ (if required to depend on the order of the automorphism only).

7 Centralizers and ranks

The main advantage in the "ordinary" case, when the order of the automorphism $\varphi \in \operatorname{Aut} G$ is coprime to the order of a (finite) group G, is in the fact that $C_{G/N}(\varphi) = C_G(\varphi)N/N$ for any normal φ-invariant subgroup N. This implies that $|C_{L(G)}(\varphi)| = |C_G(\varphi)|$ for the induced automorphism of the associated Lie ring; if φ is regular on G, then φ induces a regular automorphism of $L(G)$.

Such an equality for the centralizers may not hold in general. This gives rise to certain difficulties in translating the hypothesis from groups to Lie rings: the number of fixed points on the associated Lie ring, say, may become much greater than on the group. There is, however, a weaker assertion, the following well-known folklore lemma.

7.1 Lemma *If φ is an automorphism of a finite group G, then $|C_{G/N}(\varphi)| \leq |C_G(\varphi)|$ for any normal φ-invariant subgroup N.*

The following lemma reveals an advantage, and a really powerful one, of the modular case.

7.2 Lemma *Suppose that φ is an automorphism of order p^n of a finite p-group P with exactly p^m fixed points. Then P can be generated by at most mp^n elements.*

This lemma can be proved using Lemma 7.1 and the Jordan normal form of the linear transformation induced by φ on $P/\Phi(P)$ regarded as a vector space over $\mathbb{F}_p$.

All characteristic subgroups of the group P satisfying Lemma 7.2 can be generated by at most mp^n elements. According to the theory of powerful p-groups of A. Lubotzky and A. Mann [28] (anticipated also by M. Lazard [26]), then P contains a characteristic subgroup H of (p,m,n)-bounded index which is a powerful p-group (that is, $H^p \geq [H,H]$ for $p \neq 2$, or $H^4 \geq [H,H]$ for $p=2$).

7.3 Corollary *Suppose that φ is an automorphism of order p^n of a finite p-group P with exactly p^m fixed points. Then P has a characteristic powerful subgroup of (p,m,n)-bounded index.*

Thus, proving the existence of a subgroup of (p, m, n)-bounded index which is nilpotent of bounded class or soluble of bounded derived length, we may assume that P is a powerful p-group. Powerful p-groups have many nice linear properties which make it easier to apply various Lie ring methods.

8 Summary of the applications of the Baker–Hausdorff formula discussed.

8.1 When the Mal'cev Correspondence can be applied, it works perfectly, as in the proof of the following result, which is an analogue of Kreknin's Theorem for torsion-free locally nilpotent groups.

Theorem (Folklore) *Suppose that a torsion-free locally nilpotent group G admits a regular automorphism φ of finite order n. Then G is soluble of derived length at most $k(n)$, the value of Kreknin's function.*

8.2 The Lazard Correspondence for nilpotent p-groups of class $\leq p-1$ can be used for

8.2.1 constructing certain examples, which may be easier for Lie rings;

8.2.2 fast efficient reductions to Lie rings, as, for example, in the proof of the following theorem of Yu. Medvedev.

Theorem (Medvedev [35, 36]) *If a finite p-group P admits an automorphism φ of prime order p with exactly p^m fixed points, then P has a subgroup of index bounded in terms of p and m which is nilpotent of class bounded in terms of m only.*

8.2.3 proving results involving p-groups of class 2 for odd p in certain minimal situations.

8.3 The Mal'cev Correspondence is an essential ingredient in the proof of the following theorem, which naturally matches Kreknin's Theorem for p-automorphisms of finite p-groups.

Theorem (Khukhro [15]) *If a finite p-group P admits an automorphism φ of order p^n with exactly p^m fixed points, then P has a subgroup of (p, m, n)-bounded index which is soluble of derived length bounded in terms of p^n only.*

Earlier Shalev [40] proved that the derived length of P is bounded in terms of p, m and n.

9 Proof of folklore's Theorem 8.1

Let G be a torsion-free locally nilpotent group and φ its automorphism of finite order n, which has only trivial fixed points: $C_G(\varphi) = 1$. Proving that G is soluble of derived length $\leq k(n)$, we may assume that G is generated by $n2^{k(n)}$ elements (the corresponding identity depends on $2^{k(n)}$ elements and we take their orbits under $\langle\varphi\rangle$, each of length $\leq n$, to make the subgroup φ-invariant); hence G is nilpotent. It is well-known that G can be embedded in the so-called Mal'cev completion $\hat{G}$, which is a nilpotent $\mathbb{Q}$-powered group consisting of the roots of the elements of G. Then φ extends to an automorphism of $\hat{G}$ denoted by the same letter φ. Clearly, φ is regular on $\hat{G}$ too: if $g \in C_{\hat{G}}(\varphi)$, then, for some $k \in \mathbb{N}$, we have $g^k \in G \cap C_{\hat{G}}(\varphi) = C_G(\varphi) = 1$, whence $g = 1$ since $\hat{G}$ is torsion-free along with G.

Let L be the Lie $\mathbb{Q}$-algebra on the same set $L = \hat{G}$ that is in the Mal'cev Correspondence with $\hat{G}$. Since the operations in L are expressed by fixed formulae in the operations of $\hat{G}$, the mapping φ acting in the same way on the same set becomes an automorphism of L with $C_L(\varphi) = 0$. By Kreknin's Theorem, L is then soluble of derived length $\leq k(n)$. Since the Mal'cev Correspondence is known to preserve the derived length, $\hat{G}$ is also soluble of derived length $\leq k(n)$, and hence so is G.

10 Constructing examples

The advantage of using the Lazard Correspondence for constructing examples is in the fact that it is usually easier to check the Jacobi identity (on a basis, say), rather than the laws of a group operation.

10.1 Example In 1976 Leedham-Green and McKay asked in [27] whether there exist finite p-groups of maximal class of arbitrarily large derived length (of course, for different p, in view of the existence of a subgroup of p-bounded index that is nilpotent of class 2 proved in [27, 43]). A positive answer was given in 1980 in [38] by the following example. Let $e_1, \dots, e_p$ be a basis of the Lie algebra L over $\mathbb{F}_p$ with structural constants

$$[e_i, e_j] = \begin{cases} (i-j)e_{i+j}, & \text{if } i+j \leq p \\ 0, & \text{if } i+j > p \end{cases}.$$

It is easy to see that L has order p^p and is nilpotent of class $p-1$ and that the derived length of L is $\approx \log_2 p$. The Lazard Correspondence defines a finite p-group P on the same set with the same nilpotency class and the same derived length.

10.2 Example Let p be an odd prime and let e_1, e_2, e_3 be linearly independent generators of the additive group of the Lie ring ($\mathbb{Z}$-algebra) M with structural constants

$$[e_1, e_2] = pe_3; \quad [e_2, e_3] = pe_1; \quad [e_3, e_1] = pe_2.$$

It is easy to see that $\gamma_s(M) = p^{s-1}M$ and $M^{(d)} = p^{2^d-1}M$. The factor ring $M/p^{p-1}M$ is nilpotent of class $p-1$ and its additive group is a p-group. Hence we can apply the Lazard Correspondence to obtain a nilpotent p-group P on the same set. It is easy to see that the linear transformations α_1, α_2 defined by

$$\alpha_1: \; e_1 \to -e_1, \quad e_2 \to -e_2, \quad e_3 \to e_3;$$

$$\alpha_2: \; e_1 \to e_1, \quad e_2 \to -e_2, \quad e_3 \to -e_3$$

are automorphisms of M and hence induce automorphisms of $M/p^{p-1}M$. The group $A = \langle \alpha_1, \alpha_2 \rangle$ is elementary abelian of order 4, and it is easy to see that $C_{M/p^{p-1}M}(A) = 0$. The same A is a group of automorphisms of P, with $C_P(A) = 1$. Note that the derived length of P, which is equal to that of $M/p^{p-1}M$, grows to infinity with the growth of p, that is, it is not bounded by any constant ("depending" only on A). This shows that there is no direct analogue of Kreknin's Theorem for non-cyclic regular groups of automorphisms of nilpotent p-groups.

Similarly, a simple three-dimensional Lie $\mathbb{Q}$-algebra L admits A as a group of automorphisms with $C_L(A) = 0$. But Example 10.2 shows that even nilpotency and finiteness of the group do not help (as they do for cyclic regular groups of automorphisms, compare with Remark 5.5).

There are some other examples constructed via the Lazard Correspondence. Higman [10] produced finite nilpotent groups of class $(p^2 - 1)/4$ with a regular automorphism of prime order p by applying the Lazard Correspondence to the similar examples of Lie rings (thus proving that $h(p) \geq (p^2-1)/4$). Similarly, Khukhro [12] produced examples of certain subdirect products of unbounded nilpotency class.

11 p-groups of nilpotency class 2 for odd p

Baer [3] discovered the special case of the Lazard Correspondence in a classification of odd p-groups that are nilpotent of class 2.

Using this special case of the Lazard Correspondence, Bender [4] extended Thompson's [45] theorem on so-called signalizers; in the minimal situation in the proof the normal p-subgroup, p odd, has nilpotency class 2.

Thompson's work [45] involved a theorem on fixed points of a p-automorphism of a p-group of class 2, for odd p. Using the Lazard Correspondence E. I. Khukhro [11] extended this theorem of Thompson to p-groups of class $p - 1$.

12 Fast reduction to Lie rings

Recently, Medvedev proved the following theorem, which is another generalization of the results on p-groups of maximal class.

12.1 Theorem (Medvedev [35, 36]) *If a finite p-group P admits an automorphism φ of prime order p with exactly p^m fixed points, then P has a subgroup of (p, m)-bounded index which is nilpotent of m-bounded class.*

Our purpose is to show how the Lazard Correspondence can be used for a very fast and efficient reduction of the proof of Theorem 12.1 to Lie rings (so that the bulk of the proof, which is rather difficult, is about Lie rings).

By the Alperin–Khukhro Theorem 6.2, we may assume P to be nilpotent of class $\leq h(p)$. Then we may use induction on the nilpotency class. By Lemma 7.2 and by the induction hypothesis, $P/Z(P)$ has a subgroup of (p, m)-bounded index which is nilpotent of m-bounded class $g(m)$. Hence we may assume that P is nilpotent of class $\leq g(m) + 1$. We may also assume that p is large enough with respect to m: more precisely, we may assume that $g(m) + 1 \leq p - 1$. Otherwise we could simply take $h(p)$ for the value of $g(m)$. Now the nilpotency class of P is less than p, and hence the Lazard Correspondence translates the problem on P into the corresponding problem on the Lie ring defined on the same set via the Baker–Hausdorff Formula, with the same automorphism φ.

13 An essential use of the Mal'cev correspondence

Now we show how the Mal'cev Correspondence is used in the proof of the following result.

13.1 Theorem (Khukhro [15]) *If a finite p-group P admits an automorphism φ of order p^n with exactly p^m fixed points, then P has a subgroup of (p, m, n)-bounded index which is soluble of p^n-bounded derived length.*

Scheme of proof (a) By Corollary 7.3, P has a characteristic powerful subgroup of (p, m, n)-bounded index; hence we may assume P to be powerful from the outset. Let us fix notation $k = k(p^n)$ for the value of Kreknin's function. Some calculations in uniformly powerful sections of P, based on Kreknin's Theorem, yield that $P^{(k)}$, the kth derived subgroup of P, is nilpotent of (p, m, n)-bounded class.

Here an idea from Shalev's paper [40] is used. Note that instead of the "p-adic" Lie ring construction in [40], we used the usual associated Lie ring in these calculations. Another important tool in our calculations is the interchanging property

$$[M^p, N] = [M, N]^p \tag{13.2}$$

for powerfully embedded subgroups M, N, whose final form is due to Shalev [40].

The result obtained at this stage is a kind of "weak" result, since the bound on the nilpotency class of $P^{(k)}$ depends on all parameters, including m.

(b) Using the Mal'cev Correspondence and a combinatorial form of Kreknin's Theorem, we prove another kind of "weak" result, where the conclusion depends on the nilpotency class.

13.3 Proposition (Khukhro [15]) *Suppose that G is a nilpotent group of class c, and φ is its automorphism of finite order t. Then, for some (c, t)-bounded number $N = N(c, t)$, the $k(t)$th derived subgroup of the subgroup generated by the Nth powers is contained in the normal closure of the centralizer of φ in G, that is, $(G^N)^{(k(t))} \leq \langle C_G(\varphi)^G \rangle$.*

Scheme of proof The result for an arbitrary group satisfying the hypothesis follows as a homomorphic image of the required inclusion for a free nilpotent $\langle \varphi \rangle$-group (which is simply a free nilpotent group whose free generators are freely permuted by φ). Hence we may assume G to be a free nilpotent $\langle \varphi \rangle$-group of class c.

Then, as before, we embed G into its Mal'cev completion $\hat{G}$, a $\mathbb{Q}$-powered nilpotent group of class c which consists of all roots of elements of G. The extension of φ to an automorphism of $\hat{G}$ is denoted by the same letter.

Let L be the Lie $\mathbb{Q}$-algebra on the same set $L = \hat{G}$ that is in the Mal'cev Correspondence with $\hat{G}$. As before, the mapping φ acting in the same way on the same set becomes an automorphism of L. By Kreknin's Theorem in a combinatorial form, the $k(t)$th derived subring of L is contained in the ideal generated by $C_L(\varphi)$, that is, $L^{(k(t))} \leq {}_{\rm id}\langle C_L(\varphi)\rangle$.

Translating back into the language of the $\mathbb{Q}$-powered group $\hat{G}$ via the Mal'cev Correspondence we obtain the equivalent inclusion $\hat{G}^{(k(t))} \leq \left\langle C_{\hat{G}}(\varphi)^{\hat{G}}\right\rangle$. However, although the left-hand side certainly contains $G^{(k(t))}$, the right-hand side is larger than required. An application of certain technical arguments, typical for nilpotent groups, allows us to replace the right-hand side by $\langle C_G(\varphi)^G\rangle$, at the expense of taking a suitable power G^N on the left: $(G^N)^{(k(t))} \leq \langle C_G(\varphi)^G\rangle$, as required. □

(c) Finally, we have a kind of miracle: two "weak" results produce the required strong one, with a subgroup of bounded index which is soluble of derived length bounded in terms of $|\varphi|$ only. By part (a), P is a powerful p-group whose kth derived subgroup $P^{(k)}$ is nilpotent of (p, m, n)-bounded class $c = c(p, m, n)$. Then, by Proposition 13.3, we have

$$\left((P^{(k)})^{p^u}\right)^{(k)} \leq \left\langle C_{P^{(k)}}(\varphi)^{P^{(k)}}\right\rangle$$

for some (p, m, n)-bounded number $u = u(c, p^n)$. The subgroup on the right is generated by the conjugates of the elements from $C_{P^{(k)}}(\varphi)$ all having order dividing p^m. This subgroup is also nilpotent of (p, m, n)-bounded class $\leq c$, being a subgroup of $P^{(k)}$. Hence its exponent is bounded by p^{mc}, which is a (p, m, n)-bounded number. As a result,

$$\left(\left((P^{(k)})^{p^u}\right)^{(k)}\right)^{p^v} = 1$$

for some (p, m, n)-bounded number $v = v(p, m, n)$. Using the interchanging property (13.2) for powerfully embedded subgroups, we deduce that

$$\left(P^{p^w}\right)^{(2k)} = 1$$

for some (p, m, n)-bounded number $w = w(p, m, n)$. The index of P^{p^w} is (p, m, n)-bounded too, because of the bounds for the ranks of all abelian sections of P given by Lemma 7.2. Thus, P^{p^w} is a required subgroup of (p, m, n)-bounded index which is soluble of derived length $2k(p^n)$. □

14 Some other applications of the Baker–Hausdorff formula to finite p-groups

We note the work of T. Weigel [46], where the Baker–Hausdorff Formula is used to establish a correspondence similar to Mal'cev's (or Lazard's) for the so-called uniformly powerful p-groups and Lie rings.

The Baker–Hausdorff Formula is used in the theory of p-adic analytic groups as described in [7]; although pro-p-groups are not finite, they are, after all, inverse limits of finite p-groups.

Alperin and Glauberman [2] use the Lazard Correspondence to prove certain generalizations of the Thompson–Glauberman replacement theorem for finite p-groups of class $\le p-1$.

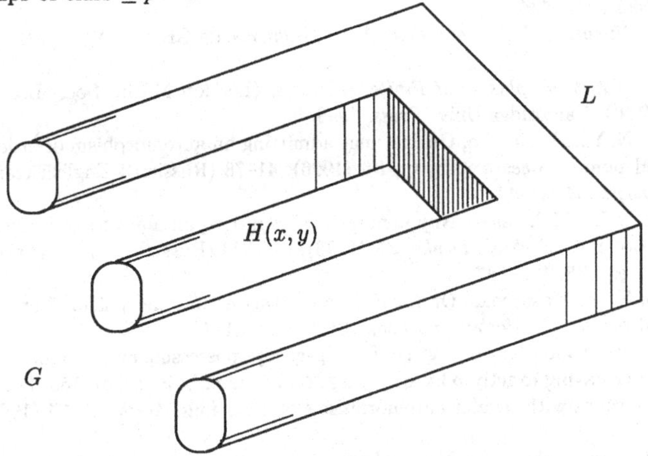

References

[1] J. Alperin, Automorphisms of solvable groups, *Proc. Amer. Math. Soc.*, **13** (1962), 175–180.

[2] J. Alperin and G. Glauberman, Limits of abelian subgroups of finite p-groups, *Preprint*, Univ. of Chicago, 1997.

[3] R. Baer, Groups with abelian central quotient groups, *Trans. Amer. Math. Soc.*, **44** (1938), 357–386.

[4] H. Bender, Über den grössten p'-Normalteiler in p-auflösbaren Gruppen, *Arch. Math. (Basel)*, **18** (1967), 15–16.

[5] N. Blackburn, On a special class of p-groups, *Acta Math.*, **100** (1958), 45–92.

[6] N. Bourbaki, *Lie groups and algebras, Chapters 1–3*, Springer, Berlin, 1980.

[7] J. D. Dixon, M. P. F. du Sautoy, A. Mann, D. Segal, *Analytic pro-p-groups* (London Math. Soc. Lecture Note Series, **157**), Cambridge Univ. Press., 1991.

[8] P. Fong, On orders of finite groups and centralizes of p-elements, *Osaka J. Math.*, **13** (1976), 483–489.

[9] B. Hartley, T. Meixner, Finite soluble groups containing an element of prime order whose centralizer is small, *Arch. Math. (Basel)*, **36** (1981), 211–213.

[10] G. Higman, Groups and rings which have automorphisms without non-trivial fixed elements, *J. London Math. Soc. (2)*, **32** (1957), 321–334.

[11] E. I. Khukhro, Verbal commutativity and fixed points of p-automorphisms of finite p-groups, *Mat. Zametki*, **25** (1979), 505–512 (Russian); English transl. *Math. Notes*, **25**, 262–265.

[12] E. I. Khukhro, On nilpotent subdirect products, *Siberian Mat. J.*, **23** (1982), No 6, 178–180 (Russian).

[13] E. I. Khukhro, Finite p-groups admitting an automorphism of order p with a small number of fixed points, *Mat. Zametki*, **38** (1985), 652–657 (Russian); English transl. *Math. Notes*, **38** (1986), 867–870.

[14] E. I. Khukhro, Groups and Lie rings admitting an almost regular automorphism of prime order, *Mat. Sbornik*, **181** (1990), 1207–1219 (Russian); English transl. *Math.*

USSR Sbornik, **71** (1992), 51–63.

[15] E. I. Khukhro, Finite p-groups admitting p-automorphisms with few fixed points, *Mat. Sbornik*, **184** (1993), 53–64 (Russian); English transl. *Russian Acad. Sci. Sbornik Math.*, **80** (1995), 435–444.

[16] E. I. Khukhro, *Nilpotent Groups and their Automorphisms*, de Gruyter–Verlag, Berlin, 1993.

[17] E. I. Khukhro, *p-Automorphisms of Finite p-Groups*, (London Math. Soc. Lecture Note Series, **246**), Cambridge Univ. Press, 1998.

[18] E. I. Khukhro, N. Yu. Makarenko, On Lie rings admitting an automorphism of order 4 with few fixed points, *Algebra i Logika*, **35** (1996), 41–78 (Russian); English transl. *Algebra and Logic*, **35**, 21–43.

[19] E. I. Khukhro, N. Yu. Makarenko, Nilpotent groups admitting an almost regular automorphism of order 4, *Algebra i Logika*, **35** (1996), 314-333 (Russian); English transl. *Algebra and Logic*, **35**, 176–187.

[20] E. I. Khukhro, N. Yu. Makarenko, On Lie rings admitting an automorphism of order 4 with few fixed points. II, *Algebra and Logic*, **35** (1997), 21–43.

[21] I. Kiming, Structure and derived length of finite p-groups possessing an automorphism of p-power order having exactly p fixed points, *Math. Scand.*, **62** (1988), 153–172.

[22] L. G. Kovács, Groups with regular automorphisms of order four, *Math. Z.*, **75** (1961), 277–294.

[23] V. A. Kreknin, The solubility of Lie algebras with regular automorphisms of finite period, *Dokl. Akad. Nauk SSSR*, **150** (1963), 467–469 (Russian); English transl. *Math. USSR Doklady*, **4** (1963), 683–685.

[24] V. A. Kreknin, A. I. Kostrikin, Lie algebras with regular automorphisms, *Dokl. Akad. Nauk SSSR*, **149** (1963), 249–251 (Russian); English transl. *Math. USSR Doklady*, **4**, 355–358.

[25] M. Lazard, Sur les groupes nilpotents et les anneaux de Lie, *Ann. Sci. École Norm. Supr.*, **71** (1954), 101–190.

[26] M. Lazard, Groupes analytiques p-adiques, *Publ. Math. Inst. Hautes Études Sci.*, **26** (1965), 389–603.

[27] C. R. Leedham-Green, S. McKay, On p-groups of maximal class. I, *Quart. J. Math. Oxford Ser. (2)*, **27** (1976), 297–311.

[28] A. Lubotzky, A. Mann, Powerful p-groups. I: finite groups, *J. Algebra*, **105** (1987), 484–505; II: p-adic analytic groups, *ibid.*, 506–515.

[29] W. Magnus, A connection between the Baker–Hausdorff formula and a problem of Burnside, *Ann. of Math. (2)*, **52** (1950), 111–126; Errata, *Ann. of Math. (2)*, **57** (1953), 606.

[30] N. Yu. Makarenko, Almost regular automorphisms of prime order, *Sibirsk. Mat. Zh.*, **33** (1992), No 5, 206–208 (Russian); English transl. *Siberian Math. J.*, **33**, 932–934.

[31] N. Yu. Makarenko, Finite 2-groups that admit automorphisms of order 4 with a small number of fixed points, *Algebra i Logika*, **32** (1993), 402–427 (Russian); English transl. *Algebra and Logic*, **32**, 215–230.

[32] A. I. Mal'cev, Nilpotent groups without torsion, *Izv. Akad. Nauk SSSR, Ser. Mat.*, **13** (1949), 201–212 (Russian).

[33] A. I. Mal'cev, *Selected works. Vol. 1, Classical algebra; Vol. 2, Mathematical logic and the general theory of algebraic systems*, Nauka, Moscow, 1976 (Russian).

[34] S. McKay, On the structure of a special class of p-groups, *Quart. J. Math. Oxford Ser. (2)*, **38** (1987), 489–502.

[35] Yu. Medvedev, p-Groups, Lie p-rings and p-automorphisms, *Preprint*, Univ. Wisconsin, Madison, 1994.

[36] Yu. Medvedev, p-Divided Lie rings and p-groups, *Preprint*, Univ. Wisconsin, Madison,

1994.

[37] Yu. Medvedev, Groups and Lie rings with almost regular automorphisms, *J. Algebra*, **164** (1994), 877–885.

[38] B. A. Panfërov, On nilpotent groups with lower central factors of minimal ranks, *Algebra i Logika*, **19** (1980), 701–706 (Russian); English transl. *Algebra and Logic*, **19** (1981), 455–458.

[39] I. N. Sanov, Establishment of a connection between periodic groups with period a prime number and Lie rings, *Izv. Akad. Nauk SSSR Ser. Mat.*, **16** (1952), 23–58 (Russian).

[40] A. Shalev, On almost fixed point free automorphisms, *J. Algebra*, **157** (1993), 271–282.

[41] A. Shalev, The structure of finite p-groups: effective proof of the coclass conjectures, *Invent. Math.*, **115** (1994), 315–345.

[42] A. Shalev, E. I. Zelmanov, Pro-p-groups of finite coclass, *Math. Proc. Cambridge Philos. Soc.*, **111** (1992), 417–421.

[43] R. Shepherd, *Ph. D. Thesis*, Univ. of Chicago, 1971.

[44] J. G. Thompson, Finite groups with fixed-point-free automorphisms of prime order, *Proc. Nat. Acad. Sci. U.S.A.*, **45** (1959), 578–581.

[45] J. G. Thompson, Fixed points of p-groups acting on p-groups, *Math. Z.*, **86** (1964), 12–13.

[46] T. Weigel, Exp *and* Log *functors in the categories of powerful p-central groups and Lie algebras*, Habilitationsschrift, Freiburg i. Br., 1994.

GENERALIZATIONS OF THE RESTRICTED BURNSIDE PROBLEM FOR GROUPS WITH AUTOMORPHISMS

E.I. KHUKHRO

Institute of Mathematics, Novosibirsk-90, 630090, Russia
School of Mathematics, University of Wales, College of Cardiff, Senghennydd Road, Cardiff, CF2 4YH, Wales

To the 100th anniversary of W. Burnside's book

1 Introduction

The Restricted Burnside Problem (RBP) asks whether the order of a d-generator finite group G of exponent n is bounded in terms of d and n only. By the classification of the finite simple groups, G may be assumed soluble, and, by the reduction theorem of P. Hall and G. Higman [9], the exponent of G may be assumed to be a power of a prime p, say: $n = p^k$. A positive solution of the RBP was obtained for prime exponent by A. I. Kostrikin [46] and for any prime-power exponent by E. I. Zelmanov [79, 80] as a consequence of their theorems on Engel Lie algebras. The reduction to Lie algebras was performed for prime exponent by W. Magnus [55] and A. I. Sanov [58] and for prime-power exponent by E. I. Zelmanov [78].

Bounding the order of a finite d-generator group G of prime-power exponent $n = p^k$ in terms of d and n is equivalent to bounding the nilpotency class of G in terms of d and n. Hence the positive solution of the RBP for groups of prime-power exponent p^k can be stated as following: the nilpotency class of a nilpotent d-generator group of exponent p^k is bounded in terms of d and p^k. This fact can also be stated in terms of varieties. Let $\mathfrak{B}_n$ denote the variety of all groups of exponent n, and let $LN\mathfrak{C}$ denote the subclass of all locally nilpotent groups from the class $\mathfrak{C}$. Then the positive solution of the RBP for exponent p^k means that $LN\mathfrak{B}_{p^k}$ is a variety.

From the above "nilpotent" viewpoint, it is natural to say that the RBP has a positive solution for a class of groups $\mathfrak{C}$ if $LN\mathfrak{C}$ is a variety. We shall discuss certain generalizations of the RBP for groups with automorphisms, and the word "variety" in the preceding sentence may mean a variety of groups with operators. Informally speaking, we shall consider the RBP-type problems for groups satisfying certain conditions that are obtained by "diluting" the law $x^n = 1$ by automorphisms.

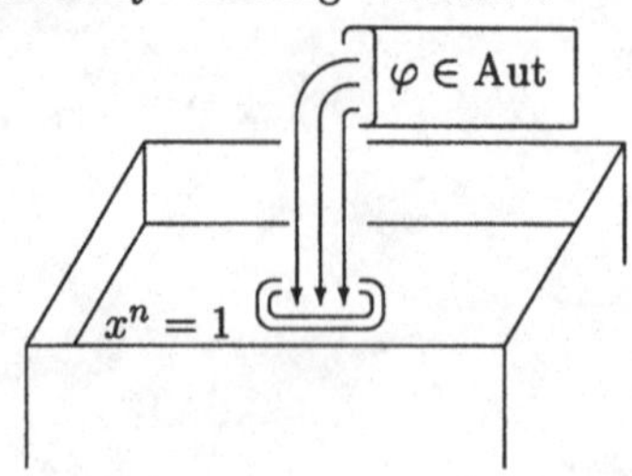

First in § 2–4 we recall some older results on the RBP for groups with the so-called splitting automorphisms of prime order (which are related to the Hughes Problem and some others), if only to prepare the further generalizations. These generalizations concern the RBP for arbitrary operator groups (§ 5), groups with coset identities (§ 6) and torsion compact group (§ 7). Finally, in § 8–9 our joint result with P. Shumyatsky is discussed, on a non-cyclic abelian group of automorphisms all of whose non-trivial elements have fixed points of bounded exponent.

2 Splitting automorphisms of prime order

Let p be a prime and let $\mathfrak{S}_p$ denote the variety of groups with operator φ satisfying the laws $x^{\varphi^p} = 1$ and

$$x \cdot x^{\varphi} \cdot x^{\varphi^2} \cdots x^{\varphi^{p-1}} = 1. \tag{2.1}$$

In other words, for a fixed cyclic group $\langle\varphi\rangle$ of order p the class $\mathfrak{S}_p$ consists of the $\langle\varphi\rangle$-groups G (acted upon by $\langle\varphi\rangle$ as automorphisms, not necessarily faithfully) satisfying the identity with operators (2.1) for all $x \in G$. Then φ is a *splitting automorphism of order* p of G (with some abuse of terminology, here we do not exclude the case where φ acts trivially on G).

Note that if φ acts trivially on $G \in \mathfrak{S}_p$, then G has prime exponent p (just put $\varphi = 1$ in (2.1)). On the other hand, a periodic p'-group G belongs to $\mathfrak{S}_p$ if and only if $C_G(\varphi) = 1$. By J. G. Thompson [65], every finite p'-group in $\mathfrak{S}_p$ is nilpotent. By D. R. Hughes and J. G. Thompson [21], every finite group in $\mathfrak{S}_p$ is soluble and moreover, by O. H. Kegel [27], nilpotent. The following theorem gives a positive solution to the RBP for $\mathfrak{S}_p$.

Theorem 2.2 (E. I. Khukhro [36]) *The nilpotency class of a d-generator nilpotent group from $\mathfrak{S}_p$ is bounded in terms of d and p (in other words, $LN\mathfrak{S}_p$ is a variety of groups with operators).*

By G. Higman [15], the nilpotency class of a nilpotent p'-group in $\mathfrak{S}_p$ is bounded in terms of p only; thus, the proof of Theorem 2.2 is reduced to the case of a p-group $G \in \mathfrak{S}_p$. For such G a generalization of the Magnus–Sanov Theorem is proved: the associated Lie ring of the semidirect product $G \rtimes \langle\varphi\rangle$ is a Lie algebra over $\mathbb{F}_p$ satisfying all but some particular consequences of the $(p-1)$-Engel identity. Then an application of Kostrikin's Theorem yields that the number of fixed points of φ on G is bounded in terms of d and p. This, in turn, implies that G is soluble of bounded derived length and even has a subgroup of bounded index which is nilpotent of p-bounded class (J. Alperin [1] and E. I. Khukhro [33]). Hence the proof of Theorem 2.2 concludes by the following theorem.

Theorem 2.3 (E. I. Khukhro [30]) *A soluble group from $\mathfrak{S}_p$ of derived length s is nilpotent of class at most $(p^s-1)/(p-1)$ (in other words, $\mathfrak{A}^s \cap \mathfrak{S}_p \subseteq \mathfrak{N}_{(p^s-1)/(p-1)}$).*

Due to G. Higman [15] and V. A. Kreknin and A. I. Kostrikin [49], this result was known for p'-groups, so the proof of Theorem 2.3 is essentially about p-groups. Theorem 2.3 may be regarded also as a generalization of P. J. Higgins' result [13] on soluble Engel Lie algebras (applied to soluble groups of exponent p).

3 The RBP for p-groups with a partition

We recall two definitions.

Definition 3.1 A group G has a non-trivial *partition* if G is the set-theoretic union $G = \bigcup_i G_i$ of some of its proper subgroups $G_i < G$ with pairwise trivial intersections: $G_i \cap G_j = 1$ for $i \neq j$.

Definition 3.2 For a prime p, the *Hughes H_p-subgroup* of a group G is $H_p(G) = \langle x \in G \mid x^p \neq 1 \rangle$, the smallest subgroup outside which all elements have order p.

The following proposition establishes connections between splitting automorphisms of prime order, partitions and the Hughes subgroup in finite p-groups, it can be found in R. Baer [2].

Proposition 3.3 *For a finite p-group P the following are equivalent:*

(1) *P has a non-trivial partition;*

(2) *$P \neq H_p(P)$;*

(3) *P is a semidirect product $P = P_1 \rtimes \langle \varphi \rangle$, where $|\varphi| = p$ and (2.1) holds for all $x \in P_1$, that is, φ is a splitting automorphism of P_1 of order p (or $P_1 \in \mathfrak{S}_p$).*

By Proposition 3.3, Theorems 2.2 and 2.3 have consequences for finite p-groups with partitions and for finite p-groups distinct from the Hughes subgroup, consequences that may be regarded as a positive solution of the RBP for such groups with respect to the nilpotency class.

Corollary 3.4 *If a finite d-generator p-group P has a non-trivial partition or, equivalently, is distinct from its Hughes subgroup, then P has a subgroup of index p which is nilpotent of class bounded in terms of d and p.*

Corollary 3.5 *If a finite p-group P has a non-trivial partition or, equivalently, is distinct from its Hughes subgroup, then P has a subgroup P_1 of index p which is nilpotent of class at most $(p^s - 1)/(p - 1)$, where s is the derived length of P_1.*

4 The RBP for anti-Hughes groups and a positive solution of the Hughes problem for almost all finite p-groups.

D. R. Hughes [20] conjectured that if $G \neq H_p(G) \neq 1$ for a finite group G, then $|G : H_p(G)| = p$. This conjecture was proved for $p = 2$ in [19], for $p = 3$ in [64], for metabelian groups in [18], and for p-groups of class $2p - 2$ in [54]. Using J. G. Thompson's [65] fundamental theorem on Thompson's subgroup, D. R. Hughes and J. G. Thompson [21] proved that the Hughes conjecture is true for a finite group G if G is not a p-group. But among finite p-groups counterexamples were found by G. E. Wall [69], for $p = 5$ with $|P : H_p(P)| = p^2$ and $H_p(P) \neq 1$. In fact, as shown by G. E. Wall [70], such a counterexample exists for a prime p if the associated Lie ring $L(B(\infty, p))$ of the free countably generated group of prime exponent p satisfies an identity of degree $2p-1$ which is not a consequence of the $(p-1)$-Engel identity mod p. But so far such identities are known to be really new in this sense only for $p = 5$, 7 and 11, due to computer-aided calculations [6] (the case $p = 5$ was dealt with "by hand" in [70]). Moreover, M. R. Vaughan-Lee [67] discovered certain multilinear identities of degrees $1 + k(p-1)$, $k \in \mathbb{N}$, that hold in $L(B(\infty, p))$, and by G. E. Wall [72], if they are "really new" for $k = 1, 2, \ldots, k_0$, say, then there exists a counterexample with $|P : H_p(P)| = p^{k_0}$ and $H_p(P) \neq 1$. But such "really new" identities are known only in the above-mentioned case of $k_0 = 2$ for $p = 5, 7, 11$. There exists even a 2-generator counterexample with $H_p(P) = \Phi(P) \neq 1$ constructed in E. I. Khukhro [32] for $p = 7$ with the aid of computer. Such a counterexample exists for a prime p if there is a "really new" relation of degree $2p$ in $L(B(2, p))$, which is the case for $p = 7$ (and may well be for all $p \geq 7$). Such a p-group is a kind of monster among finite p-groups: all elements outside the Frattini subgroup have order p, while the group is not of exponent p. This group serves as a counterexample to some other problems on finite p-groups from [17, 24, 53].

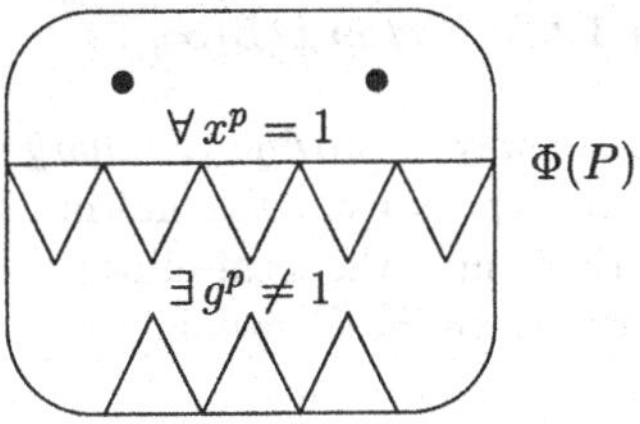

$\forall x^p = 1$

$\Phi(P)$

$\exists g^p \neq 1$

Nevertheless, in spite of the existence of counterexamples, the Hughes conjecture is valid for almost all finite p-groups. The following corollary may be regarded as a positive solution of the RBP for anti-Hughes groups in the usual sense, with respect to the order.

Corollary 4.1 (E. I. Khukhro [37]) *Suppose that $|P : H_p(P)| \geq p^2$ for a d-generated finite p-group P. Then*

(a) *the exponent of P is at most $p^{f(p)}$, where $f(p)$ depends on p only;*

(b) *$|P| \leq g(d, p)$, where $g(d, p)$ depends on d and p only.*

(Part (b) was deduced directly from Corollary 3.4, before Zelmanov's Theorem.) One can say that, although monsters do exist, they are caged.

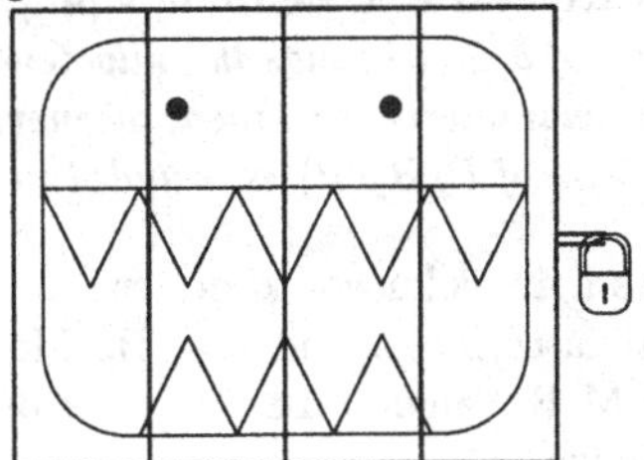

As a consequence of Corollary 4.1(b), there are universal anti-Hughes groups. Let us abbreviate to *p-counterexample* the title of "a finite p-group that is a counterexample to the Hughes conjecture".

Corollary 4.2 *If, for a prime p, there exist p-counterexamples, then, for every $d \geq 3$, there is the largest d-generator p-counterexample $C(d,p)$ such that any other d-generator p-counterexample is a homomorphic image of $C(d,p)$. If there is a 2-generator p-counterexample, then there is the largest 2-generator p-counterexample $C(2,p)$ such that any other 2-generator p-counterexample is a homomorphic image of $C(2,p)$.*

In [39] we proved a partial converse of G. E. Wall's theorem, bounding the index of a non-trivial Hughes p-subgroup under the condition that all identities of $L(B(\infty,p))$ are consequences of finitely many multilinear identities. This result can be reformulated in another "unconditional" form.

Theorem 4.3 (E. I. Khukhro [43, 7.4.17]) *If $|P : H_p(P)| = p^k$ in a finite p-group P, then the associated Lie ring $L(P)$ satisfies all multilinear identities of degrees $\leq (k-1)(p-1)+1$ that hold in $L(B(\infty,p))$.*

This theorem gives another way of proving Corollary 4.1(b), since for $k = 2$ the hypothesis implies the identities $px = 0$ and the multilinear equivalent of the $(p-1)$-Engel identity, which together imply the local nilpotency identities by Kostrikin's Theorem. In addition we can state the following.

Theorem 4.4 *If P is a d-generator p-counterexample, then P is nilpotent of class $k(d,p)$, where $k(d,p)$ is the bound for the nilpotency class of a d-generator $(p-1)$-Engel Lie algebra of characteristic p.*

As a consequence of the proof of Theorem 4.3 we have the following.

Theorem 4.5 *Suppose that $H_p(P) \neq 1$ in a finite p-group P, and let m and c be, respectively, the minimal number of generators and the nilpotency class of the factor-group $P/H_p(P)$. Then the nilpotency class of P is at least $(p-1)(m-1+c(c+1)/2)+1$.*

This technical result can be used to prove the following theorem.

Theorem 4.6 (E. I. Khukhro and E. I. Zelmanov [43, 7.5.1]) *If, for a prime p, there exists a linear function of d that bounds the nilpotency class of a d-generated $(p-1)$-Engel Lie algebra of characteristic p, then, whenever $H_p(P) \neq 1$ in a finite p-group P, the nilpotency class of $P/H_p(P)$ is bounded in terms of p only.*

The hypothesis of Theorem 4.6 is known to be true for $p = 5$ due to G. Higman [14] (apart from the trivial cases $p = 2$ and $p = 3$). Moreover, as calculated by G. Havas, M. Newman and M. R. Vaughan-Lee [11], the precise function for $p = 5$ is such that the following holds.

Corollary 4.7 *For any 5-group P, the factor-group $P/H_5(P)$ is abelian whenever $H_5(P) \neq 1$.*

Remarks 4.8 The new identities in $L(B(\infty,p))$ were also used in E. I. Khukhro [34] to construct so-called secretive p-groups of large rank (larger than that of G. E. Wall [71]), which refuted a conjecture of N. Blackburn and A. Espuelas [5]. As a consequence, in [34] finite soluble non-nilpotent groups were constructed with large index of the generalized Hughes subgroup $H_{p^2}(G) = \left\langle x \mid x^{p^2} \neq 1 \right\rangle$, which contrasts with the result of D. R. Hughes and J. G. Thompson [21] for $H_p(G)$. We note also that C. Scoppola [59] showed that any finite p-group H can occur as a section of $G/H_{p^k}(G)$ for some finite p-group G and for some k depending on H.

5 The RBP-type problems for varieties of groups with operators

Theorems 2.2 and 2.3 on splitting automorphisms were the prototypes of certain general results on nilpotency and local nilpotency in varieties of groups with operators. For a fixed group Ω we consider Ω-groups, which are groups acted upon by Ω as automorphisms, but not necessarily faithfully. A free Ω-group $F = \langle X \rangle$ is just an abstract free group with free generators x^ω, $x \in X$, $\omega \in \Omega$, admitting Ω as a group of automorphisms acting freely (regularly) on the generators: $(x^\omega)^\alpha = x^{\omega\alpha}$ for $\alpha \in \Omega$. A subset $W \subseteq F$ of Ω-group words defines the variety $\mathfrak{V}_W$ of Ω-groups on which all words from W take only trivial values.

Definition 5.1 The *projection* $\overline{W}$ of a set of Ω-words $W \subseteq F$ is the set of group words obtained from the Ω-words in W by replacing all operators by the trivial one.

Example 5.2 If $\Omega = \langle \varphi \rangle$ and $W = \{x \cdot x^\varphi \cdot x^{\varphi^2} \cdots x^{\varphi^{p-1}}\}$ for a prime p, then $\mathfrak{V}_W = \mathfrak{S}_p$ and $\overline{W} = \{x^p\}$.

We denote by $\mathfrak{V}_{\overline{W}}$ the variety of abstract groups defined by the projection set of group words $\overline{W}$. (In Example 5.2 we have $\mathfrak{V}_{\overline{W}} = \mathfrak{B}_p$, which is the variety of all groups of exponent p.) The main idea of the following results is that if there is a positive solution of the RBP for $\mathfrak{V}_{\overline{W}}$ with respect to the nilpotency class (and in certain multilinear sense), then there is a kind of a positive solution of the RBP for some groups in $\mathfrak{V}_W$.

Theorem 5.3 (E. I. Khukhro [42]) *Suppose that Ω is a finite group of order n and W is a set of Ω-group words. Suppose that the RBP has a positive multilinear solution for the variety $\mathfrak{V}_{\overline{W}}$ in the following sense: the associated Lie ring of the free countably generated group of the variety $\mathfrak{V}_{\overline{W}}$ satisfies some system of multilinear identities such that the variety of Lie rings $\mathfrak{L}$ defined by this system is locally nilpotent, with function $f(d)$ bounding the nilpotency class of a d-generator Lie ring in $\mathfrak{L}$.*

If, for an Ω-group $G \in \mathfrak{V}_W$, the semidirect product $G \rtimes \Omega$ is locally nilpotent, then G belongs to a locally nilpotent variety, in which a d-generator group is nilpotent of class at most $f(d(n^n - 1)/(n - 1))$.

Note that the condition on the semidirect product $G \rtimes \Omega$ to be locally nilpotent is automatically satisfied if both Ω and G are locally finite p-groups. The hypothesis of Theorem 5.3 is satisfied for Example 5.2: there the associated Lie ring of the free group of the variety $\mathfrak{V}_{\overline{W}} = \mathfrak{B}_p$ satisfies multilinear identities $px = 0$ and $\sum_{\pi \in S_{p-1}} [x_0, x_{\pi(1)}, \dots, x_{\pi(p-1)}] = 0$, which imply local nilpotency identities by Kostrikin's Theorem [46]. Hence the result of Theorem 2.2 for p-groups now follows by Theorem 5.3 directly from Kostrikin's Theorem, and probably with better bounds for the nilpotency class. Moreover, we can state a more general corollary.

Corollary 5.4 *Suppose that a locally finite p-group P admits a finite p-group of automorphisms Ω of order n, and suppose that, for some fixed elements $\omega_1, \dots, \omega_p \in \Omega$, the identity $x^{\omega_1} \cdot x^{\omega_2} \cdots x^{\omega_p} = 1$ holds for all $x \in P$. Then P belongs to a locally nilpotent variety in which the nilpotency class of a d-generator group is at most $k(d(n^n - 1)/(n-1))$, where $k(s)$ is the upper bound for the nilpotency class of an s-generator $(p-1)$-Engel Lie algebra of characteristic p.*

Problem 5.5 Can one get rid of the dependence on the order of Ω in the bound for the nilpotency class in the conclusion of Theorem 5.3? Can one drop the assumption of the finiteness of Ω in Theorem 5.3?

So far, we know a positive answer to Problem 5.5 in the special case, where all nilpotent groups in the variety $\mathfrak{V}_{\overline{W}}$ have bounded nilpotency class.

Theorem 5.6 (E. I. Khukhro [41]) *Suppose that Ω is an arbitrary group and W is a set of Ω-group words. Suppose that all nilpotent groups from the variety $\mathfrak{V}_{\overline{W}}$ are nilpotent of class $\le c$. If, for an Ω-group $G \in \mathfrak{V}_W$, the semidirect product $G \rtimes \Omega$ is locally nilpotent, then G is nilpotent of class at most c.*

Corollary 5.7 *Suppose that a soluble p-group P of derived length s admits a finite p-group of automorphisms Ω such that, for some fixed elements $\omega_1, \dots, \omega_p \in \Omega$, the identity $x^{\omega_1} \cdot x^{\omega_2} \cdots x^{\omega_p} = 1$ holds for all $x \in P$. Then P is nilpotent of class at most $((p-1)^s - 1)/(p-2)$.*

Problem 5.8 Find a connection of Theorems 5.3 and 5.6 with a certain construction, similar to the connection of Theorems 2.2 and 2.3 with p-groups with a partition or distinct from the Hughes subgroup.

6 Coset identities

May be, a connection required in Problem 5.8 could be found in terms of coset identities. At least, the prototype examples of the splitting automorphisms can be interpreted using this notion.

Definition 6.1 Let $w = w(x_1, \dots, x_n)$ be a group word. A group G satisfies a *coset identity* w on the fixed cosets $g_1H, \dots, g_nH$ of a subgroup $H \le G$ if $w(y_1, \dots, y_n) = 1$ for any $y_i \in g_iH$, $i = 1, \dots, n$.

For example, by Proposition 3.3, for $P = P_1 \rtimes \langle\varphi\rangle$ we have $P_1 \in \mathfrak{S}_p$ (that is, φ is a splitting automorphism of P_1 of order p) if and only if the coset $P_1\varphi$ satisfies the identity $x^p = 1$ (this is, obviously, equivalent to $H_p(P) \leq P_1$).

J. S. Wilson and E. I. Zelmanov [74] proved that if a finitely generated pro-p-group G satisfies a coset identity as in Definition 6.1 for an open subgroup H, then the Lazard's Lie $\mathbb{F}_p$-algebra $L_p(G)$ (see definition in § 9) is PI, that is, satisfies a polynomial identity. See E. I. Zelmanov's survey [82] for a thorough discussion of the importance of $L_p(G)$ to be PI. (We shall also use in § 9 a powerful theorem of E. I. Zelmanov on PI Lie algebras with ad-nilpotent commutators in generators.)

Problem 6.2 (A. Shalev [66, 12.95]) Suppose that a finitely generated pro-p-group G satisfies some coset identity as in Definition 6.1 for an open subgroup H. Does it follow that G satisfies some non-trivial identity? A positive answer would imply that an analogue of the Tits Alternative holds for finitely generated pro-p-groups.

Recently L. Levai gave a positive answer to Problem 6.2 in the special case of the so-called λ-analytic groups.

L. Levai and L. Pyber proved that if a finitely generated pro-p-group P satisfies a coset identity (for an open subgroup) then P has no dense free (abstract) subgroups. This result belongs to the "probabilistic" approach in the study of pro-p-groups.

7 Torsion compact groups

These are precisely torsion (periodic) profinite groups. Note that a torsion profinite group always has an open subset of elements of given order n, that is, satisfies the coset identity $x^n = 1$ for some open subgroup. There is a positive solution of a kind of RBP for these groups.

Theorem 7.1 (E. I. Zelmanov [81]) *Every torsion pro-p-group is locally finite.*

By J. Wilson's reduction theorem [73], it follows that every torsion compact group is locally finite. But the following problem remains open.

Problem 7.2 Is every torsion compact group a group of bounded exponent?

Problem 7.2 is reduced to pro-p-groups. The following consequence of Theorem 2.2 gives a positive answer in a special case.

Corollary 7.3 (E. I. Khukhro [38]) *If a profinite group G has an open subset consisting of elements of prime order p, then G has a subgroup of finite index which belongs to a locally nilpotent variety of bounded exponent.*

Proving the following conjecture may turn to be useful in dealing with Problem 7.2 for arbitrary torsion pro-p-groups.

Conjecture 7.4 Suppose that a finite d-generated p-group P admits a splitting automorphism φ of order p^k, that is, such that $x \cdot x^{\varphi} \cdot x^{\varphi^2} \cdots x^{\varphi^{p^k-1}} = 1$ for all $x \in P$. Then the derived length of P is bounded in terms of p, k and d.

While Theorem 2.2 gives even a stronger result in the case a splitting automorphism of prime order, even the simplest case of composite order, $|\varphi| = 4$, remains open. We mention that E. Jabara [22] proved that any finite group G with a splitting automorphism of order 4 is soluble, and that the nilpotency class of $[G, G]$ is bounded in terms of the derived length [23].

8 Non-cyclic abelian group of automorphisms all of whose non-trivial elements have fixed points of bounded exponent

This section reflects a recent joint work of the author with P. Shumyatsky. Let G be a finite group and let A be a non-cyclic abelian group of automorphisms $A \leq \mathrm{Aut}\, G$ of coprime order, $(|A|, |G|) = 1$. It is a well-known fact that then G is generated by the centralizers of the non-trivial elements of A:

$$G = \left\langle C_G(a) \mid a \in A^{\#} \right\rangle, \tag{8.1}$$

where $A^{\#} = A \setminus \{1\}$. Suppose that in this situation n is the least common multiple of the exponents of the centralizers $C_G(a)$, $a \in A^{\#}$. If G is abelian, then, of course, the exponent of G is n too. P. Shumyatsky raised a question: can one bound the exponent of G in terms of n and A if G is an arbitrary finite group? This may be regarded as another way of diluting the Burnside identity by automorphisms. A positive answer is given by the following result.

Theorem 8.2 (E. I. Khukhro and P. Shumyatsky) *Suppose that A is a non-cyclic abelian group of automorphisms of a finite group G of coprime order, $(|A|, |G|) = 1$, and let n be the least common multiple of the exponents of the centralizers $C_G(a)$ of non-trivial elements $a \in A^{\#}$. Then the exponent of G is bounded in terms of n and q, where q is the least prime divisor of $|A|$ such that the Sylow q-subgroup of A is non-cyclic.*

The proof of Theorem 8.2 is sketched in § 9 below; it involves using powerful p-groups, a "finite" version of Lazard's criterion for a pro-p-group to be p-adic analytic, a deep theorem of E. I. Zelmanov on PI Lie algebras with ad-nilpotent commutators in the generators, and a theorem of Yu. A. Bahturin and M. V. Zaicev on PI centralizers of automorphisms of Lie algebras.

Theorem 8.2 has an obvious extension to the case where G is a locally finite group and A is a finite abelian non-cyclic group of automorphisms of G such that $(|A|, |g|) = 1$ for every $g \in G$: if all centralizers $C_G(a)$, $a \in A^{\#}$, are groups of bounded exponent, then G is a group of bounded exponent too.

Theorem 8.2 can also be applied to Problem 7.2 in the following special case.

Corollary 8.3 (E. I. Khukhro and P. Shumyatsky) *Suppose that a torsion compact group Q has an elementary abelian subgroup with finite centralizer. Then Q is a group of bounded exponent.*

Of course, the proof of Corollary 8.3 begins with an application of Zelmanov's Theorem [81] by which Q is a locally finite group. Clearly, $C_H(A) = 1$ for some normal subgroup H of finite index. If $|A|$ is a prime, then H is locally nilpotent by J. G. Thompson [65] and, furthermore, nilpotent by G. Higman [15]. For nilpotent groups the result follows easily. If $|A|$ is not a prime then, by induction on $|A|$, all centralizers $C_H(a)$, $a \in A^{\#}$, are of bounded exponent, and the result follows by Theorem 8.2.

In view of the examples of H. Heineken and I. J. Mohamed [12], a locally finite group all of whose nilpotent subgroups have finite exponents need not be of finite exponent. The following corollary provides a sufficient condition for such a group to be of finite exponent.

Corollary 8.4 (E. I. Khukhro and P. Shumyatsky) (a) *Suppose that G is a locally finite group such that every nilpotent subgroup of G has finite exponent. Suppose that G admits a finite elementary abelian group of automorphisms A such that $(|A|, |g|) = 1$ for every $g \in G$. If the fixed-point subgroup of A in G is finite, then G is a group of bounded exponent.*

(b) *Suppose that G is a periodic almost locally soluble group such that every nilpotent subgroup of G has finite exponent. Suppose that G has an elementary abelian subgroup B with finite centralizer. Then G is a group of bounded exponent.*

The result is deduced from Theorem 8.2 similarly to the proof of Corollary 8.3, with the results on fixed-point-free automorphisms replaced by the generalizations on almost fixed-point-free ones. (Namely, if a locally finite group G admits an almost fixed-point-free automorphism of prime order, then G has a nilpotent subgroup of finite index. This follows from the classification of the finite simple groups and from the works of P. Fong [8], B. Hartley and T. Meixner [10], and E. I. Khukhro [40].)

9 Scheme of proof of Theorem 8.2.

Without loss of generality we may, obviously, assume that A is a non-cyclic group of order q^2 for a prime q (which does not divide $|G|$). Since for every prime p dividing $|G|$ there is an A-invariant Sylow p-subgroup P of G, we may assume G to be a finite p-group (for a prime $p \neq q$). By (8.1), we have, of course, $p \leq n$. We may also assume G to be generated by q^2 elements, since every element $g \in G$ is contained in the A-invariant subgroup $\langle g^a \mid a \in A \rangle$.

The following well-known lemma is used several times.

Lemma 9.1 *Suppose that B is a group of automorphisms of a finite group H of coprime order, $(|B|, |H|) = 1$. Then $C_{H/N}(B) = C_H(B)N/N$ for any normal B-invariant subgroup N.*

9.2 The Lazard's Lie algebra and powerful p-groups. Given a prime p, the *Lazard–Zassenhaus p-central series* D_i of a group G can be defined as follows:

$$D_i = \left\langle [a_1, \ldots, a_s]^{p^t} \mid sp^t \geq i,\ a_k \in G \right\rangle.$$

(In particular, $D_1 = G$, and $D_2 = G^p[G,G]$ is the Frattini subgroup if G is a finite p-group.) The D_i form a filtration in the sense that $[D_i, D_j] \leq D_{i+j}$ for all i, j. These inclusions imply that a Lie ring $LD(G)$ can be defined on the additive group $\bigoplus_i D_i/D_{i+1}$, with Lie products defined for elements $aD_{i+1} \in D_i/D_{i+1}$, $bD_{j+1} \in D_j/D_{j+1}$ as $[aD_{i+1}, bD_{j+1}] = [a,b]D_{i+j+1}$ (on the right is the image of the group commutator $[a,b]$ in D_{i+j}/D_{i+j+1}). In fact, $LD(G)$ is a Lie algebra over $\mathbb{F}_p$, since all of the D_i/D_{i+1} have exponent p. The *Lazard Lie algebra* $L_p(G)$ is defined as the subalgebra of $LD(G)$ generated by the additive subgroup D_1/D_2. The homogeneous component $L_p(G)_i$ of degree i (with respect to the generating set D_1/D_2) is an additive subgroup (a subspace) of D_i/D_{i+1}.

A theorem of M. Lazard [50] states that a finitely generated pro-p-group P is p-adic analytic if and only if $L_p(P)$ is nilpotent. To be p-adic analytic for P is equivalent to have a subgroup H of finite index which is *powerful*, that is, $H^p \geq [H,H]$ for $p \neq 2$ (or $H^4 \geq [H,H]$ for $p = 2$).

Powerful finite p-group are defined in the same way, they have many nice linear properties, of which we need the following: if a powerful p-group T is generated by elements of exponent p^e, then the exponent of T is p^e too (see A. Lubotzky and A. Mann [52]). By (8.1), our group G is generated by elements of exponent n. Thus, it is sufficient to reduce the proof of Theorem 8.2 to the case of powerful p-groups.

We shall prove that $L_p(G)$ is nilpotent of (q,n)-bounded class. This will imply the required reduction to powerful p-groups in the proof of Theorem 8.2 via the following quantitative version of Lazard's theorem for finite p-groups.

Proposition 9.2.1 *Suppose that P is a d-generated finite p-group such that the Lazard Lie algebra $L_p(P)$ is nilpotent of class c. Then P has a powerful characteristic subgroup of (p,c,d)-bounded index.*

Proof Let $\rho_1, \ldots, \rho_r$ be all simple commutators of weight $\leq c$ in the generators of P; here r is a (d,c)-bounded number. The nilpotency of $L_p(P)$ of class c implies that for any $s \in \mathbb{N}$ every element $g \in P$ can be written in the form $g = \rho_1^{k_1} \cdots \rho_r^{k_r} \lambda$, where $\lambda \in D_s$, $k_i \in \mathbb{Z}$ (see E. I. Zelmanov [82]). Since P is a finite p-group, $D_s = 1$ for some $s \in \mathbb{N}$. Hence every element $g \in P$ can be written in the form $g = \rho_1^{k_1} \cdots \rho_r^{k_r}$. Since $\rho_i^{p^m} \in P^{p^m}$, it follows that for every $m \in \mathbb{N}$ every element $g \in P$ can be written in the form

$$g = \rho_1^{k_1} \cdots \rho_r^{k_r} \varkappa, \qquad 0 \leq k_i \leq p^m - 1 \text{ for all } i, \quad \varkappa \in P^{p^m}.$$

Hence $|P/P^{p^m}| \leq p^{rm}$ for any $m \in \mathbb{N}$.

Let V be the intersection of the kernels of all homomorphisms of P into $GL_r(\mathbb{F}_p)$. Put $W = V$ if $p \neq 2$ (or $W = V^2$ if $p = 2$). The exponent of the Sylow p-subgroup of $GL_r(\mathbb{F}_p)$ is a (p,r)-bounded number. Then $P^{p^a} \leq W$ for some (p,r)-bounded number a, which is also (p,c,d)-bounded, since r is (c,d)-bounded. There is a (p,c,d)-bounded number $u \geq a$ such that $|P^{p^u}/P^{p^{u+1}}| \leq p^r$, for otherwise the inequality $|P/P^{p^m}| \leq p^{rm}$ would be violated for some m. Then $P^{p^u} \leq P^{p^a} \leq W$, and P^{p^u} is r-generated since $|P^{p^u}/\Phi(P^{p^u})| \leq |P^{p^u}/P^{p^{u+1}}| \leq p^r$. Now, by [7, Proposition 2.12], P^{p^u} is a powerful subgroup. The index of P^{p^u} is at most p^{ur} and hence is (p,c,d)-bounded. □

Proving that $L_p(G)$ is nilpotent of bounded class relies on one of the most general forms of the positive solutions of the Restricted–Burnside–type problems for Lie algebras by E. I. Zelmanov [82]. We shall need the following lemma to check one of the hypothesis of that theorem. An element y of a Lie algebra M is said to be *ad-nilpotent of index k* if $[x, \underbrace{y, \ldots, y}_{k}] = 0$ for all $x \in M$.

Lemma 9.2.2 (M. Lazard [50]) *If $g^{p^m} = 1$ for some $g \in D_i$, then the element $\bar{g} = gD_{i+1}$ of the Lie algebra $LD(G)$ is ad-nilpotent of index p^m.*

9.3 PI centralizers of automorphisms of Lie algebras Another hypothesis in E. I. Zelmanov's theorem is that the Lie algebra satisfies a polynomial identity (in short, *is PI*). To check that this is the case for our Lie algebra, we use the following recent result.

Theorem 9.3.1 (Yu. A. Bahturin and M. V. Zaicev [4]) *Suppose that B is a finite soluble group of automorphisms of a Lie algebra M over a field k of characteristic coprime to $|B|$. If the fixed-point subalgebra $C_M(B)$ is PI, then M is also PI.*

(There is a similar result for associative algebras of V. K. Kharchenko [28] and S. Montgomery [56] without assuming B soluble; in [51] V. Linchenko also removed the solubility hypothesis from Theorem 9.3.1, but we shall need it only with B cyclic.) The identity that holds on the Lie algebra M may be chosen to depend only on B, on the identity of $C_M(B)$, and on the ground field. This follows from Theorem 3 by a standard "universal" argument.

We apply this to the Lazard algebra of our group G.

Proposition 9.3.2 *The Lie algebra $L_p(G)$ satisfies a polynomial identity that depends only on q and n.*

Proof We pick an arbitrary element $a \in A^{\#}$ and regard a as an automorphism of $L_p(G)$. By the Bahturin–Zaicev Theorem 9.3.1, it is sufficient to show that $C_{L_p(G)}(a)$ satisfies a polynomial identity that depends only on q and n. This is done in the following proposition, which will be used also for another purpose later. □

Proposition 9.3.3 *The subring $C_{L_p(G)}(a)$ satisfies the identity*

$$\sum_{\pi \in \mathbb{S}_{n-1}} [x_0, x_{\pi(1)}, x_{\pi(2)}, \ldots, x_{\pi(n-1)}] = 0. \tag{9.3.4}$$

This follows from Lemma 9.1 and from the fact that the group $C_G(a)$ has exponent n, which is here a power of p. "Multilinear" argument must be applied similar to those of G. Higman [16] proving the same identity mod p in the associated Lie ring of a group of exponent n.

9.4 Ad-nilpotent commutators in generators We shall complete the proof of Theorem 8.2 by applying the following deep result.

Theorem 9.4.1 (E. I. Zelmanov [82]) *Let L be a Lie algebra over a field k that is generated by finitely many elements $a_1, \dots, a_d$. Suppose that L is PI and every commutator in the a_i is ad-nilpotent. Then the Lie algebra L is nilpotent.*

A standard "universal" argument gives us at once the following.

Corollary 9.4.2 *Suppose that under the hypothesis of Theorem 9.4.1 all commutators in the a_i are ad-nilpotent of bounded index, m, say. Then L is nilpotent of class depending only on m, d, on the polynomial identity of L, and on the ground field.*

By Proposition 9.3.2 the Lazard algebra $L_p(G)$ of our group G is PI; hence we shall be able to use Zelmanov's Theorem, if we show that $L_p(G)$ is generated by q^2 elements such that all commutators in these generators are ad-nilpotent (and moreover, nilpotent of (q, n)-bounded index). The ith homogeneous component $L_p(G)_i$ of $L_p(G)$ is an A-invariant additive subgroup of D_i/D_{i+1}. Since the decomposition of type (8.1) holds for every $L_p(G)_i$, each of the $L_p(G)_i$ is spanned by elements $x \in C_{L_p(G)}(a)$, $a \in A^{\#}$. By Lemma 9.1, every such x is the image in D_i/D_{i+1} of some element $y \in C_G(a)$, $a \in A^{\#}$. Since $y^n = 1$ by the hypothesis, by Lazard's Lemma 9.2.2 we immediately have the following.

Lemma 9.4.3 *For any $i \in \mathbb{N}$ and for any $a \in A^{\#}$, every element $x \in C_{L_p(G)_i}(a)$ is an ad-nilpotent element of $L_p(G)$ of index n.*

In particular, $L_p(G)$ is generated by ad-nilpotent elements from $C_{L_p(G)_1}(a)$, $a \in A^{\#}$. But we cannot claim that every Lie commutator in these generators is again in some of the centralizers $C_{L_p(G)_i}(b)$, $b \in A^{\#}$, and hence is ad-nilpotent too. To overcome this difficulty, we extend the ground field of $L_p(G)$ by a primitive qth root of unity ω, forming $L = L_p(G) \otimes \mathbb{F}_p[\omega]$. The idea is to replace $L_p(G)$ by L and to prove that L is nilpotent of (q, n)-bounded class, which will, of course, imply the same nilpotency result for $L_p(G)$. Before that we translate the properties of $L_p(G)$ into the language of L.

It is a well-known fact that every polynomial identity has multilinear consequences. Hence $L_p(G)$ satisfies some multilinear identity that depends only on q and n. Then the Lie algebra L satisfies the same multilinear identity. It is natural to identify $L_p(G)$ with the $\mathbb{F}_p$-subalgebra $L_p(G) \otimes 1$ of L. We note that if an element $x \in L_p(G)$ is ad-nilpotent of index m, say, then the "same" element $x = x \otimes 1$ is ad-nilpotent in L of the same index m. Put $L_1 = L_p(G)_1 \otimes \mathbb{F}_p[\omega]$;

then $L = \langle L_1 \rangle$, since $L_p(G) = \langle L_p(G)_1 \rangle$, and L is the direct sum of the homogeneous components $L_i = L_p(G)_i \otimes \mathbb{F}_p[\omega]$. Since G is generated by q^2 elements, the $\mathbb{F}_p$-space $L_p(G)_1 = D_1/D_2 = G/\Phi(G)$ is spanned by q^2 elements, and hence so is the $\mathbb{F}_p[\omega]$-space L_1.

The group A acts naturally on L, and $C_L(a) = C_{L_p(G)}(a) \otimes \mathbb{F}_p[\omega]$ for any $a \in A$. Since A is abelian, and the ground field is now a splitting field for A, every L_i decomposes in the direct sum of common eigenspaces for A. In particular, L_1 is spanned by common eigenvectors for A, and it requires at most q^2 of them to span L_1, since L_1 has dimension at most q^2. Hence L is generated by q^2 common eigenvectors for A from L_1. Every common eigenspace is contained in the centralizer $C_{L_i}(a)$ for some $a \in A^\#$, since A is non-cyclic. The main advantage of extending the ground field now becomes clear: any commutator in common eigenvectors is again a common eigenvector. Hence our preparation for the application of Zelmanov's Theorem will be completed by the following lemma.

Lemma 9.4.4 *For any $i \in \mathbb{N}$ and any $a \in A^\#$ every element $y \in C_{L_i}(a)$ is ad-nilpotent of (q, n)-bounded index.*

Proof We have

$$y = x_0 + \omega x_1 + \omega^2 x_2 + \cdots + \omega^{q-2} x_{q-2}$$

for some $x_s \in C_{L_p(G)_i}(a)$, so that each of the summands $\omega^i x_i$ is ad-nilpotent of index n by Lemma 9.4.3. A sum of ad-nilpotent elements need not be ad-nilpotent in general. But in our case all summands belong to a nilpotent subalgebra, so that the following elementary lemma will finish the proof.

Lemma 9.4.5 *Suppose that M is a Lie algebra, H is a subalgebra of M generated by s elements $h_1, \ldots, h_s$ such that all commutators in the h_i are ad-nilpotent of index t. If H is nilpotent of class u, then for some (s, t, u)-bounded number v we have $[M, \underbrace{H, \ldots, H}_{v}] = 0$.*

Proof We apply to a sufficiently long (but of (s, t, u)-bounded length) commutator

$$[m, h_{i_1}, h_{i_2}, \ldots]$$

a collecting process whose aim is to rearrange the h_i (and emerging by the Jacobi identity commutators in the h_i) in an ordered string after m, where all occurrences of a given element (h_i or a commutator in the h_i) would form an unbroken segment. Since H is nilpotent, this process terminates at a linear combination of commutators with sufficiently long segments of equal elements. All these commutators are equal to 0 because all commutators in the h_i are ad-nilpotent by the hypothesis. □

To finish the proof of Lemma 9.4.4, we set $H = \langle x_0, \omega x_1, \ldots, \omega^{q-2} x_{q-2} \rangle$. Note that $H \subseteq C_L(a)$, since $\omega^j x_j \in C_L(a)$ for all j. A commutator of weight k in the $\omega^j x_j$ has the form $\omega^t x$ where $x \in C_{L_p(G)_{ki}}(a)$. By Lemma 9.4.3, such an x

is ad-nilpotent of index n and hence so is $\omega^t x$. Since the identity (9.3.4), which holds on $C_{L_p(G)}(a)$ by Proposition 9.3.3, is multilinear, the same identity holds on $C_L(a) = C_{L_p(G)}(a) \otimes \mathbb{F}_p[\omega]$, and hence on H. By Corollary 9.4.2 of Zelmanov's Theorem, H is nilpotent of (q, n)-bounded class. Now Lemma 9.4.5 implies that $y \in H$ is ad-nilpotent of (q, n)-bounded index. □

9.4.6 Completion of the proof of Theorem 8.2 The Lie algebra $L_p(G)$ satisfies a polynomial identity that depends only on q and n. Hence the Lie algebra $L = L_p(G) \otimes \mathbb{F}_p[\omega]$ satisfies some multilinear polynomial identity that depends only on q and n. As shown above, L is generated by q^2 elements all commutators in which are ad-nilpotent of (q, n)-bounded index. Hence L is nilpotent of (q, n)-bounded class by Corollary 9.4.2 of Zelmanov's Theorem. Then, of course, $L_p(G)$ is nilpotent of the same class, which implies that G has a powerful subgroup of (q, n)-bounded index by Proposition 9.2.1. This completes the proof of Theorem 8.2, as shown in § 9.2. □

Remarks 9.5 A similar combination of the Bahturin–Zaicev Theorem 9.3.1 and Zelmanov's Theorem 9.4.1 as in the proof of Theorem 8.2 is applied by A. Shalev in [60] (where earlier results of N. Rocco and P. Shumyatsky [57] and P. Shumyatsky [61, 62] are generalized). In [63] P. Shumyatsky uses Theorem 9.4.1 and Linchenko's [51] extension of Theorem 9.3.1 to arbitrary, not necessarily soluble, group of automorphisms to generalize A. Shalev's results from [60].

References

[1] J. Alperin, Automorphisms of solvable groups, *Proc. Amer. Math. Soc.*, **13** (1962), 175–180.

[2] R. Baer, Partitionen endlicher Gruppen, *Math. Z.*, **75** (1960), 333–372.

[3] R. Baer, Einfachepartitionen endlicher Gruppen mit nicht-trivialer Fittingscher Untergruppe, *Arch. Math. (Basel)*, **12** (1961), 81–89.

[4] Yu. A. Bakhturin, M. V. Zaitsev, Identities of graded algebras, *J. Algebra*, to appear.

[5] N. Blackburn, A. Espuelas, The power structure of metabelian p-groups, *Proc. Amer. Math. Soc.*, **92** (1984), 478–484.

[6] J. Cannon, Some combinatorial and symbol manipulation programs in group theory, *in: Computational Problems in Abstract Algebra (Proc. Conf., Oxford, 1967)*, Pergamon Press, 1970, 199–203.

[7] J. D. Dixon, M. P. F. du Sautoy, A. Mann, D. Segal, *Analytic pro-p-groups* (London Math. Soc. Lecture Note Series, **157**), Cambridge Univ. Press, 1991.

[8] P. Fong, On orders of finite groups and centralizes of p-elements, *Osaka J. Math.*, **13** (1976), 483–489.

[9] P. Hall, G. Higman, The p-length of a p-soluble group and reduction theorems for Burnside's problem, *Proc. London Math. Soc. (3)*, **6** (1956), 1–42.

[10] B. Hartley, T. Meixner, Finite soluble groups containing an element of prime order whose centralizer is small, *Arch. Math. (Basel)*, **36** (1981), 211–213.

[11] G. Havas, M. Newman, M. R. Vaughan-Lee, A nilpotent algorithm for graded Lie rings, *J. Symb. Comput.*, **9** (1990), 653–664.

[12] H. Heineken, I. J. Mohamed, A group with trivial centre satisfying the normalizer condition, *J. Algebra*, **10** (1968), 368–376.

[13] P. J. Higgins, Lie rings satisfying the Engel condition, *Math. Proc. Cambridge Philos. Soc.*, **50** (1954), 8–15.

[14] G. Higman, On finite groups of exponent five, *Math. Proc. Cambridge Philos. Soc.*, **52** (1956), 381–390.
[15] G. Higman, Groups and rings which have automorphisms without non-trivial fixed elements, *J. London Math. Soc. (2)*, **32** (1957), 321–334.
[16] G. Higman, Lie ring methods in the theory of finite nilpotent groups, *in: Proc. Intern. Congr. Math. Edinburgh, 1958*, Cambridge Univ. Press, 1960, 307–312.
[17] C. R. Hobby, Nearly regular p-groups, *Canad. J. Math.*, **19** (1967), 520–522.
[18] G. T. Hogan, W. P. Kappe, On the H_p-problem for finite p-groups, *Proc. Amer. Math. Soc.*, **20** (1969), 450–454.
[19] D. R. Hughes, Partial difference sets, *Amer. J. Math.*, **78** (1956), 650–677.
[20] D. R. Hughes, A research problem in group theory, *Bull. Amer. Math. Soc.*, **63** (1957), 209.
[21] D. R. Hughes, J. G. Thompson, The H_p-problem and the structure of H_p-groups, *Pacific J. Math.*, **9** (1959), 1097–1101.
[22] E. Jabara, Solvability of finite groups equipped with a splitting automorphism of order 4, *Boll. Unione Mat. Ital. B (7)*, **8** (1994), 915–928 (Italian).
[23] E. Jabara, On soluble finite groups with a splitting automorphism of order four, *Preprint*, 1994 (Italian).
[24] W. Kappe, Properties of groups related to the second center, *Math. Z.*, **101** (1967), 356–368.
[25] O. H. Kegel, Nicht einfache Partitionen endlicher Gruppen, *Arch. Math. (Basel)*, **12** (1961), 170–175.
[26] O. H. Kegel, Aufzälung der Partitionen endlicher Gruppen mit trivialer Fittingscher Untergruppen, *Arch. Math. (Basel)*, **12** (1961), 409–412.
[27] O. H. Kegel, Die Nilpotenz der H_p-Gruppen, *Math. Z.*, **75** (1960), 373–376.
[28] V. K. Kharchenko, Galois extensions and rings of quotients, *Algebra i Logika*, **13** (1974), 460–484; English transl. *Algebra and Logic*, **13** (1975), 265–281.
[29] E. I. Khukhro, A soluble group admitting a regular splitting automorphism of prime order is nilpotent, *Algebra i Logika*, **17** (1978), 611–618 (Russian); English transl. *Algebra and Logic*, **17** (1979), 402–405.
[30] E. I. Khukhro, Nilpotency of soluble groups admitting a splitting automorphism of prime order, *Algebra i Logika*, **19** (1980), 118–129 (Russian); English transl. *Algebra and Logic*, **19** (1981), 77–84.
[31] E. I. Khukhro, On a connection between the Hughes conjecture and relations in finite groups of prime exponent, *Mat. Sbornik*, **116** (1981), 253–264 (Russian); English transl. *Math. USSR Sbornik*, **44** (1983), 227–237.
[32] E. I. Khukhro, On the associated Lie ring of a free 2-generator group of prime exponent and on the Hughes conjecture for 2-generator p-groups, *Mat. Sbornik*, **118** (1982), 567–575 (Russian); English transl. *Math. USSR Sbornik*, **46** (1983), 571–579.
[33] E. I. Khukhro, Finite p-groups admitting an automorphism of order p with a small number of fixed points, *Mat. Zametki*, **38** (1985), 652–657 (Russian); English transl. *Math. Notes*, **38** (1986), 867–870.
[34] E. I. Khukhro, Finite p-groups close to groups of prime exponent, *Algebra i Logika*, **25** (1986), 227–240 (Russian); English transl. *Algebra and Logic*, **25** (1987), 143–153.
[35] E. I. Khukhro, A new identity in the Lie ring of a free group of prime exponent, *Izv. Akad. Nauk SSSR Ser. Mat.*, **50** (1986), 1308–1325 (Russian); English transl. *Math. USSR Izvestiya*, **29** (1987), 659–676.
[36] E. I. Khukhro, Locally nilpotent groups admitting a splitting automorphism of prime order, *Mat. Sbornik*, **130** (1986), 120–127 (Russian); English transl. *Math. USSR Sbornik*, **58** (1987), 119–126.
[37] E. I. Khukhro, On the Hughes problem for finite p-groups, *Algebra i Logika*, **26** (1987), 642–646 (Russian); English transl. *Algebra and Logic*, **26** (1988), 398–401.

[38] E. I. Khukhro, A remark on periodic compact groups, *Sibirsk. Mat. Zh.*, **30** (1989), 187–190 (Russian); English transl. *Siberian Math. J.*, **30** (1990), 493–496.
[39] E. I. Khukhro, On the structure of finite p-groups admitting a partition, *Sibirsk. Mat. Zh.*, **30** (1989), 208–218 (Russian); English transl. *Siberian Math. J.*, **30** (1990), 1010–1019.
[40] E. I. Khukhro, Groups and Lie rings admitting an almost regular automorphism of prime order, *Mat. Sbornik*, **181** (1990), 1207–1219 (Russian); English transl., *Math. USSR Sbornik*, **71** (1992), 51–63.
[41] E. I. Khukhro, Nilpotency in varieties of groups with operators, *Mat. Zametki*, **50** (1991), 142–145 (Russian); English transl. *Math. Notes*, **50** (1992), 869–871.
[42] E. I. Khukhro, Local nilpotency in varieties of groups with operators, *Mat. Sbornik*, **184** (1993), 137–160; English transl. *Russian Acad. Sci. Sb. Math.*, **78** (1994), 379–396.
[43] E. I. Khukhro, *Nilpotent Groups and their Automorphisms*, de Gruyter–Verlag, Berlin, 1993.
[44] E. I. Khukhro, *p-Automorphisms of Finite p-Groups* (London Math. Soc. Lecture Note Series, **246**), Cambridge Univ. Press., 1998.
[45] E. I. Khukhro and P. Shumyatsky, Bounding the exponent of a finite group with automorphisms, submitted to *J. Algebra*, 1997.
[46] A. I. Kostrikin, On a problem of Burnside, *Izv. Akad. Nauk SSSR Ser. Mat.*, **23** (1959), 3–34 (Russian); English transl. *Transl. II Ser. Amer. Math. Soc.*, **36** (1964), 63–99.
[47] A. I. Kostrikin, *Around Burnside*, Nauka, Moscow, 1986 (Russian); English transl. Springer–Verlag, 1990.
[48] V. A. Kreknin, The solubility of Lie algebras with regular automorphisms of finite period, *Dokl. Akad. Nauk SSSR*, **150** (1963), 467–469 (Russian); English transl. *Math. USSR Doklady*, **4** (1963), 683–685.
[49] V. A. Kreknin, A. I. Kostrikin, Lie algebras with regular automorphisms, *Dokl. Akad. Nauk SSSR*, **149** (1963), 249–251 (Russian); English transl. *Math. USSR Doklady*, **4** (1963), 355–358.
[50] M. Lazard, Groupes analytiques p-adiques, *Publ. Math. Inst. Hautes Études Sci.*, **26** (1965), 389–603.
[51] V. Linchenko, Identities of Lie algebras with actions of Hopf algebras, *Comm. Algebra*, to appear.
[52] A. Lubotzky, A. Mann, Powerful p-groups. I: finite groups, *J. Algebra*, **105** (1987), 484–505; II: p-adic analytic groups, *ibid.*, 506–515.
[53] I. D. Macdonald, The Hughes problem and others, *J. Austral. Math. Soc. Ser. A*, **10** (1969), 457–459.
[54] I. D. Macdonald, Solution of the Hughes problem for finite p-groups of class $2p-2$, *Proc. Amer. Math. Soc.*, **27** (1971), 39–42.
[55] W. Magnus, A connection between the Baker–Hausdorff formula and a problem of Burnside, *Ann. of Math. (2)*, **52** (1950), 111–126; Errata, *Ann. of Math. (2)*, **57** (1953), 606.
[56] S. Montgomery, *Fixed Rings of Finite Automorphism Groups of Associative Rings* (Lecture Notes in Math., **818**), Springer, Berlin, 1980.
[57] N. Rocco, P. V. Shumyatsky, On periodic groups having almost regular 2-elements, *in Proc. Edinburgh Math. Soc.*, to appear.
[58] I. N. Sanov, Establishment of a connection between periodic groups with period a prime number and Lie rings, *Izv. Akad. Nauk SSSR Ser. Mat.*, **16** (1952), 23–58 (Russian).
[59] C. M. Scoppola, Groups of prime-power order as Frobenius–Wielandt complements, *Trans. Amer. Math. Soc.*, **325** (1991), 855–874.

[60] A. Shalev, Centralizers in residually finite torsion groups, *Proc. Amer. Math. Soc.*, to appear.
[61] P. Shumyatsky, On groups having a four-subgroup with finite centralizer, *Quart. J. Math.*, to appear.
[62] P. Shumyatsky, Nilpotency of some Lie algebras associated with p-groups, *Preprint of the Univ. of Brasilia*, 1996.
[63] P. V. Shumyatsky, Centralizers in groups with finiteness conditions, *J. Group Theory*, to appear.
[64] E. G. Straus, G. Szekeres, On a problem of D. R. Hughes, *Proc. Amer. Math. Soc.*, **9** (1958), 157–158.
[65] J. G. Thompson, Finite groups with fixed-point-free automorphisms of prime order, *Proc. Nat. Acad. Sci. U.S.A.*, **45** (1959), 578–581.
[66] *Unsolved Problems in Group Theory: The Kourovka Notebook, 13th ed.* (V. D. Mazurov, E. I. Khukhro, editors), Novosibirsk, 1995.
[67] M. R. Vaughan-Lee, The restricted Burnside problem, *Bull. London Math. Soc.*, **17** (1985), 113–133.
[68] M. R. Vaughan-Lee, *The restricted Burnside problem*, 2nd ed., Oxford, Clarendon Press, 1993.
[69] G. E. Wall, On Hughes' H_p-problem, *in: Proc. Int. Conf. Theory of Groups, Canberra, 1965*, New York, Gordon and Breach, 1967, 357–362.
[70] G. E. Wall, On the Lie ring of a group of prime exponent, *in: Proc. Second Int. Conf. Theory of Groups, Canberra, 1973* (Lecture Notes in Math., **372**), Springer, 1974, 667–690.
[71] G. E. Wall, Secretive prime-power groups of large rank, *Bull. Austral. Math. Soc.*, **12** (1975), 363–369.
[72] G. E. Wall, On the multilinear identities which hold in the Lie ring of a group of prime-power exponent, *J. Algebra*, **104** (1986), 1–22.
[73] J. Wilson, On the structure of compact torsion groups, *Monatsh. Math.*, **96** (1983), 57–66.
[74] J. S. Wilson, E. I. Zelmanov, Identities for Lie algebras of pro-p-groups, *J. Pure Appl. Algebra*, **81** (1992), 103–109.
[75] H. Zassenhaus, Ein Verfahren, jeder endlichen p-Gruppe einen Lie-Ring mit der Charakteristik p zuzuordnen, *Abh. Math. Sem. Univ. Hamburg*, **13** (1939), 200–207.
[76] H. Zassenhaus, Liesche Ringe mit Primzahlcharacteristik, *Abh. Math. Sem. Univ. Hamburg*, **13** (1939), 1–100.
[77] E. I. Zelmanov, On Engel Lie algebras, *Sibirsk. Mat. Zh.*, **29** (1988), No 5, 112–117 (Russian); English transl. *Siberian Math. J.*, **29** (1989), 777–781.
[78] E. I. Zelmanov, On some problems of the theory of groups and Lie algebras, *Mat. Sbornik*, **180** (1989), 159–167 (Russian); English transl. *Math. USSR Sbornik*, **66** (1990), 159–168.
[79] E. I. Zelmanov, A solution of the Restricted Burnside Problem for groups of odd exponent, *Izv. Akad. Nauk SSSR Ser. Mat.*, **54** (1990), 42–59 (Russian); English transl. *Math. USSR Izvestiya*, **36** (1991), 41–60.
[80] E. I. Zelmanov, A solution of the Restricted Burnside Problem for 2-groups, *Mat. Sbornik*, **182** (1991), 568–592 (Russian); English transl. *Math. USSR Sbornik*, **72** (1992), 543–565.
[81] E. I. Zelmanov, On periodic compact groups, *Israel J. Math.*, **77** (1992), 83–95.
[82] E. I. Zelmanov, Lie ring methods in the theory of nilpotent groups, *in: Proc. Groups '93 Galway / St Andrews, vol. 2* (London Math. Soc. Lecture Note Ser., **212**), Cambridge Univ. Press, 1995, 567–585.

THE Σ^m–CONJECTURE FOR A CLASS OF METABELIAN GROUPS

DESSISLAVA H. KOCHLOUKOVA

Trinity College, Cambridge, England

1 Introduction to the Σ^m–Conjecture

Suppose G is a finitely generated group. In [B-R] R. Bieri and B. Renz attached to every (left) $\mathbb{Z}G$–module A a series of homological invariants $\{\Sigma^m(G,A)\}_{m\in\mathbb{N}}$ defined by

$$\Sigma^m(G,A) = \{[\chi] \in S(G) \mid A \text{ is of type } FP_m \text{ over } \mathbb{Z}G_\chi\}$$

where $S(G) = \{[\chi] = \mathbb{R}_{>0}.\chi \mid \chi \in Hom(G,\mathbb{R}) \setminus \{0\}\}$ and $G_\chi = \{g \in G \mid \chi(g) \geq 0\}$. We note that $\Sigma^1(G,A) \supseteq \ldots \supseteq \Sigma^m(G,A) \supseteq \Sigma^{m+1}(G,A) \supseteq \ldots$ and $\Sigma^1(G,\mathbb{Z}) = \Sigma_{G'}(G)$, where if N is a left G–operator group $\Sigma_N(G)$ consists of all points $[\chi] \in S(G)$ such that N is finitely generated as an operator group over a finitely generated submonoid of G_χ. The invariant $\Sigma_N(G)$ was first introduced in [B-N-S] for right operator groups. The version of $\Sigma_N(G)$ for left operator groups generalizes the invariant $\Sigma_A(Q) = \{[\chi] \in S(Q) \mid A \text{ is finitely generated as a left } \mathbb{Z}Q_\chi - \text{ module}\}$ defined in [B-S 1] for finitely generated $\mathbb{Z}Q$–modules A. As shown in [B-R] the invariants $\{\Sigma^m(G,A)\}_{m\in\mathbb{N}}$ generalize some properties of the invariants defined in [B-S 1] and [B-N-S].

Theorem ([B-R], Thm A) *$\Sigma^m(G,A)$ is an open subset of $S(G)$ for every finitely generated group G and every $\mathbb{Z}G$–module A.*

The next theorem shows the importance of the invariants $\{\Sigma^m(G,A)\}_{m\in\mathbb{N}}$.

Theorem ([B-R], Thm B) *Let G be a finitely generated group, N a subgroup of G containing G' and A a $\mathbb{Z}G$–module. Then A is of type FP_m over $\mathbb{Z}N$ if and only if $\Sigma^m(G,A)$ contains the great subsphere $S(G,N) = \{[\chi] \in S(G) \mid \chi(N) = 0\}$.*

We note that the higher geometric invariant $\Sigma^m(G,A)$ behaves nicely with respect to subgroups of finite index in G.

Proposition ([B-S 2], Corollary B4.13) *Let H be a subgroup of finite index in G, A a $\mathbb{Z}G$–module of type FP_m and $\chi : G \to \mathbb{R}$ a non-zero character. Then $[\chi\mid_H] \in \Sigma^m(H,A)$ if and only if $[\chi] \in \Sigma^m(G,A)$.*

The condition that A is of type FP_m as a $\mathbb{Z}G$–module in the above proposition is a natural one. If $\Sigma^m(G,A)$ is non-empty then A should be of type FP_m over $\mathbb{Z}G$.

In the case when the group G is of type F_m i.e. G has an Eilenberg-MacLane complex $K(G,1)$ with finite m–skeleton, it is possible to define a homotopical version $\Sigma^m(G)$ of $\Sigma^m(G,\mathbb{Z})$. In order to define $\Sigma^m(G)$ we first discuss height functions on CW–complexes acted on by G. If χ is a real character of G, a χ–equivariant height function on X is a continuous map $h = h_\chi : X \to \mathbb{R}$ with the property $h(gx) = \chi(g) + h(x)$ for all $g \in G, x \in X$. Let $X_h^{[\lambda,\infty)}$ denote the maximal subcomplex of X contained in $h^{-1}([\lambda,\infty))$. We say that $X_h^{[\lambda,\infty)}$ is essentially k–connected, for some $k \geq -1$, if there is a real positive number d such that the map

$$\pi_i(X_h^{[\lambda,\infty)}, *) \longrightarrow \pi_i(X_h^{[\lambda-d,\infty)}, *),$$

induced by inclusion, is trivial for $i \leq k$ and every choice of a base point $* \in X_h^{[\lambda,\infty)}$. Similarly $X_h^{[\lambda,\infty)}$ is essentially k–acyclic, for some $k \geq -1$, if there is a real positive number d such that the map

$$H_i(X_h^{[\lambda,\infty)}) \longrightarrow H_i(X_h^{[\lambda-d,\infty)}),$$

induced by inclusion, is trivial for $i \leq k$.

Finally we come to the definition of the invariant $\Sigma^m(G)$. As before $\Sigma^m(G)$ is a subset of the unit sphere $S(G)$, though some authors as Meinert [M 2] and Gehrke [Ge] prefer using subsets of $V(G) = Hom(G,\mathbb{R})$ instead of the projections to $S(G)$ of their non-trivial elements. By definition the class $[\chi]$ of a non-trivial character $\chi \in V(G)$ belongs to $\Sigma^m(G)$ if and only if there exists a $K(G,1)$ complex K with finite m–skeleton and a χ–equivariant height function $h = h_\chi : \widetilde{K} \to \mathbb{R}$ on the universal covering complex of K such that $\widetilde{K}_h^{[0,\infty)}$ is $(m-1)$–connected. Furthermore $[\chi]$ belongs to $\Sigma^m(G,\mathbb{Z})$ if and only if there exists a $K(G,1)$–complex K with finite m–skeleton and a χ–equivariant height function $h = h_\chi : \widetilde{K} \to \mathbb{R}$ such that $\widetilde{K}_h^{[0,\infty)}$ is $(m-1)$–acyclic but this approach works only if G is of type F_m. Therefore for groups G of type F_m $\Sigma^m(G) \subseteq \Sigma^m(G,\mathbb{Z})$. It turns out that $\Sigma^m(G)$ can be characterized by a weaker condition than the one included in the above definition.

Proposition ([M 2], Corollary 2.6) *Let K be any $K(G,1)$–complex with finite m–skeleton and χ be a non-trivial real character of G. If $h : \widetilde{K} \to \mathbb{R}$ is a χ–equivariant height function then $[\chi] \in \Sigma^m(G)$ if and only $\widetilde{K}_h^{[0,\infty)}$ is essentially $(m-1)$–connected.*

For general groups G it is known that $\Sigma^1(G) = \Sigma^1(G,\mathbb{Z})$ and if G is of type F_m then $\Sigma^m(G) = \Sigma^2(G) \cap \Sigma^m(G,\mathbb{Z})$ for $m \geq 2$. In this paper we deal with a special case of the following conjecture due to R. Bieri.

Σ^m–Conjecture *Let G be a metabelian group of type FP_m. Then*

$$conv_{\leq m}\Sigma^1(G,\mathbb{Z})^c = \Sigma^m(G,\mathbb{Z})^c = \Sigma^m(G)^c.$$

As usual upper index c denotes taking complement in $S(G)$. If T is a subset of the unit sphere $S(G)$ we denote by $conv_{\leq m}T$ the projection to $S(G)$ of all non-trivial elements of the union of the convex hulls of all m–subsets of $\mathbb{R}_{>0}T$.

There is another conjecture due to R. Bieri. The FP_m–Conjecture states that G is of type FP_m if and only if the union of the convex hulls of all m–subsets of $\mathbb{R}_{>0}.\Sigma^1(G,\mathbb{Z})^c$ does not contain the trivial character. The FP_m–Conjecture is still an open problem though some cases of the conjecture were answered positively in [Å], [B-G 1],[N], [K 1].

We note that the Σ^m–Conjecture has not been not so much explored as the FP_m–Conjecture. In [M 1] and [M 2] H. Meinert shows that the Σ^m–Conjecture holds in the case when G is a metabelian group of finite Prufer rank. In [Ge] R. Gehrke proves that $conv_{\leq 2}\Sigma^1(G,\mathbb{Z})^c \subseteq \Sigma^2(G,\mathbb{Z})^c \subseteq \Sigma^2(G)^c$ provided G is a metabelian group of type F_2 (which is equivalent to type FP_2 in this case) and if further $rk(G/G') = 2$ then all the equalities hold. A proof of the Σ^2–Conjecture for split extension metabelian groups can be found in [K 2].

Our goal is to prove the Σ^m–Conjecture for a class of metabelian groups. We show that the methods used in the proof of Theorem A and Theorem B from [K 1] can be used to prove some cases of the Σ^m–Conjecture. We use the main results of [M 1] and [M 2] where the case of groups of finite Prufer rank is considered. Our main results are the following theorems.

Theorem A *If $A \to G \to Q$ is a short exact sequence of groups with A, Q abelian, A $\mathbb{Z}$–torsion, Krull dimension one and G is of type FP_m, then*

$$\Sigma^m(G)^c \subseteq conv_{\leq m}\Sigma^1(G,\mathbb{Z})^c.$$

Theorem B *Let $A_1 \to G_1 \xrightarrow{\mu} Q_1$ be a short exact sequence of groups, A_1, Q_1 abelian and G_1 of type FP_m. Furthermore we assume that the extension is split or A_1 is $\mathbb{Z}$–torsion. Then $conv_{\leq m}\Sigma^1(G_1,\mathbb{Z})^c \subseteq \Sigma^m(G_1,\mathbb{Z})^c$.*

As a corollary of the above results, the fact that $\Sigma^m(G) \subseteq \Sigma^m(G,\mathbb{Z})$ for groups of type F_m i.e. groups of type FP_m which are finitely presented, and the classification of the finitely presented metabelian groups given in [B-S 1] we deduce that the Σ^m–Conjecture holds for a particular type of metabelian groups.

Corollary C *If $A \to G \to Q$ is a short exact sequence of groups with A, Q abelian, A $\mathbb{Z}$–torsion, Krull dimension one as a $\mathbb{Z}Q$–module and G is of type FP_m then $conv_{\leq m}\Sigma^1(G,\mathbb{Z})^c = \Sigma^m(G,\mathbb{Z})^c = \Sigma^m(G)^c$.*

2 Proof of Theorem A

2.1 Review of Meinert's and Åberg's approaches

In this section we review one of the main results in [M 1] i.e. if $A \to G \to Q$ is a short exact sequence of groups, A, Q abelian, G of type FP_m and of finite Prufer rank then $\Sigma^m(G,\mathbb{Z})^c \subseteq conv_{\leq m}\Sigma^1(G,\mathbb{Z})^c$. The following theorems for arbitrary groups are some of the basic tools used in [M 1].

Theorem ([M 1], Thm 3.3) *Suppose G is an arbitrary group acting on a $(m-1)$–acyclic CW–complex X by permuting the cells and X has finite m–skeleton modulo*

G. Let G_e denote the stabilizer in G of the cell e. If $\chi\mid_{G_e}\neq 0$ and $[\chi\mid_{G_e}]\in\Sigma^{m-p}(G_e,\mathbb{Z})$ for every p–cell e of X with $p\le m$ then G is of type FP_m if and only if $[\chi]\in\Sigma^m(G,\mathbb{Z})$.

Theorem ([M 1], Thm 3.8) *Let G be an arbitrary group acting on a $(m-1)$–acyclic CW–complex X by permuting the cells and X has finite m-skeleton modulo G. Assume that X admits a χ–equivariant height function $h=h_\chi : X\to\mathbb{R}$ for some real non-trivial character χ of G. Assume further that for every p–cell e, $0\le p\le m$, the stabilizer G_e is contained in the kernel of χ, G_e is of type FP_{m-p} and e is fixed pointwise by G_e. Then $[\chi]\in\Sigma^m(G,\mathbb{Z})$ if and only if $X_h^{[0,\infty)}$ is essentially $(m-1)$–acyclic.*

The main idea in [M 1] is to combine the above theorems with a construction of H. Åberg [Å] of a $(m-1)$–connected CW–complex X on which a subgroup of finite index in $\tilde{G}$ acts cocompactly with polycyclic cell stabilizers that fix the cells pointwise. Here $\tilde{G}$ is a metabelian group of finite Prufer rank that contains the original metabelian group G of finite Prufer rank as a subgroup of finite index. The point is that $\tilde{G}$ has all the properties of the original group G plus the additional one that it is a split extension of abelian groups. In [M 1], Thm 6.15, Meinert proves that if $\chi\in\mathbb{R}$–span $\Sigma^1(G,\mathbb{Z})^c\setminus\mathbb{R}_{>0}.conv_{\le m}\Sigma^1(G,\mathbb{Z})^c$ there exists a χ–equivariant height function $h:X\to\mathbb{R}$ such that for all $r\in\mathbb{R}$, $X_h^{[-r,\infty)}$ is essentially $(m-1)$–connected. Then the second Meinert's theorem implies that $[\chi]\in\Sigma^m(G,\mathbb{Z})$.

2.2 Some cases of the homological part of the Σ^m–Conjecture: an application of the generalized complex of Åberg's type

In this section we prove that in the conditions of Theorem A $\Sigma^m(G,\mathbb{Z})^c$ is a subset of $conv_{\le m}\Sigma^1(G,\mathbb{Z})^c$. Note that this is a weaker statement than the conclusion of Theorem A. We assume that $[\chi]\notin conv_{\le m}\Sigma^1(G,\mathbb{Z})^c$ and aim to prove that $[\chi]\in\Sigma^m(G,\mathbb{Z})$.

The proof is based on the construction of the generalized complex of Åberg's type defined in [K 1]. In [K 1] we constructed a $(m-1)$–connected CW–complex Y on which G acts cocompactly with polycyclic stabilizers that fix the cells pointwise. The fact that the space Y is $(m-1)$–connected is a consequence of [K 1], Theorem B, which implies that A is m–tame as a $\mathbb{Z}Q$–module. The construction of the space Y resembles the one in [Å] i.e. Y is a 'special' subspace of a finite product of trees $\prod_{v\in V}X_v$. Every tree X_v corresponds to a character v whose equivalence class $[v]$ is in $\Sigma^1(G,\mathbb{Z})^c$ and X_v is built by gluing copies of the real line $\mathbb{R}$ with common end $-\infty$. X_v is equipped with a v–equivariant height function $f_v : X_v\to\mathbb{R}$. We note that the set of characters V defined in [K 1] has the property that $\{[\varphi]\mid\varphi\in V\}\simeq\Sigma^c_A(Q)=\Sigma^1(G,\mathbb{Z})^c$. The latter equality holds because in general if $A\to G\to Q$ is a short exact sequence of groups with A,Q abelian and G finitely generated then the group homomorphism $G/G'\to Q$ induces a bijection $\Sigma^c_A(Q)\simeq\Sigma^c_{G'}(G/G')=\Sigma^1(G,\mathbb{Z})^c$.

As in [M 1] we distinguish two cases depending on whether χ belongs or not to the $\mathbb{R}$–span of $\Sigma^1(G,\mathbb{Z})^c$.

Step 1. If $\chi \notin \mathbb{R}$-span $\Sigma^1(G,\mathbb{Z})^c$ then for any p–cell $e, 0 \leq p \leq m$, the restriction of χ on G_e is non-trivial. Furthermore since G_e is polycyclic all invariants $\Sigma^k(G_e,\mathbb{Z}), k \geq 0$, coincide with the unit sphere $S(G_e)$. Therefore we can use the first of Meinert's theorems quoted in Section 2.1 to deduce $[\chi] \in \Sigma^m(G,\mathbb{Z})$.

Step 2. Suppose $\chi = \sum_{v\in V} \alpha_v.v$ for some $\alpha_v \in \mathbb{Z}$. Then as in [M 1], Thm 6.6.1, there exists a χ–equivariant height function $h = h_\chi : Y \to \mathbb{R}$ given by $h = H \circ f$, where $f = \prod f_v : \prod_{v\in V} X_v \to \prod_{v\in V} \mathbb{R}_v$ with $\mathbb{R}_v = \mathbb{R}$ and H is the $\mathbb{R}$-linear map $\prod_{v\in V} \mathbb{R}_v \to \mathbb{R}$ given by $H(\prod_{v\in V} r_v) = \sum_{v\in V} \alpha_v r_v$. We note that the restriction of h on every cell of Y attains its extreme values on the boundary of the cell.

The similarities between the original Åberg's complex and the generalized complex of Åberg's type show that the topological result [M 1], Thm 6.15, applies for the space Y i.e. if $[\chi] \notin conv_{\leq m}\Sigma^1(G,\mathbb{Z})^c$ then $Y \cap h_\chi^{-1}[-r,\infty)$ is $(m-1)$-connected for all $r \in \mathbb{R}$. Then by [M 1], Lemma 3.5, $Y_{h_\chi}^{[-r,\infty)}$ is essentially $(m-1)$–acyclic (furthermore using the proof of [M 1], Lemma 3.5, we see that $Y_{h_\chi}^{[-r,\infty)}$ is essentially $(m-1)$–connected). This combined with the second Meinert's result quoted in the previous section shows that $[\chi] \in \Sigma^m(G,\mathbb{Z})$.

2.3 The homotopical part of the Σ^m–Conjecture

In this section we consider the homotopical version of the results included in the previous section. In the introduction we discussed a characterization of $\Sigma^m(G)$ in terms of equivariant height functions and essentially $(m-1)$–connected subspaces of universal covers of $K(G,1)$–complexes. It turns out that it is possible to use non-free actions to determine $\Sigma^2(G)$. The following theorems are the homotopical analogues for $m = 2$ of the theorems reviewed in Section 2.1.

Theorem ([M 2], Thm 4.3) *Suppose that a group G acts cocompactly on a 1–connected, 2–complex X and χ is a real character of G. If $0 \neq \chi\,|_{G_\sigma}$ and $[\chi\,|_{G_\sigma}] \in \Sigma^{2-dim\sigma}(G_\sigma)$ for every cell σ and cell stabilizer G_σ of σ then $[\chi] \in \Sigma^2(G)$.*

Theorem ([M 2], Thm 4.1) *Suppose a group G acts cocompactly on a 1–connected 2–complex X such that all vertex stabilizers are finitely presented and all edge stabilizers are finitely generated. Let χ be a non-trivial real character of G and $h = h_\chi : X \to \mathbb{R}$ be a χ–equivariant height function. Then $[\chi] \in \Sigma^2(G)$ if and only if $X_h^{[0,\infty)}$ is essentially 1–connected.*

We assume that G is a group satisfying the assumptions of Theorem A. Using the above theorems we can repeat the argument of Section 2.2 for $m = 2$ replacing $\Sigma^2(G,\mathbb{Z})$ with $\Sigma^2(G)$. This implies that if $[\chi] \notin conv_{\leq 2}\Sigma^1(G,\mathbb{Z})^c$ then $[\chi] \in \Sigma^2(G)$ i.e. $\Sigma^2(G)^c \subseteq conv_{\leq 2}\Sigma^1(G,\mathbb{Z})^c$. By [Re] if G is of type F_m and $m \geq 2$ we have $\Sigma^m(G) = \Sigma^2(G) \cap \Sigma^m(G,\mathbb{Z})$ and hence $\Sigma^m(G)^c = \Sigma^2(G)^c \cup \Sigma^m(G,\mathbb{Z})^c$. Therefore

$$\Sigma^m(G)^c \subseteq conv_{\leq m}\Sigma^1(G,\mathbb{Z})^c.$$

3 Proof of Theorem B

3.1 The main construction in the proof of Theorem B

In this section we show how the main idea of the proof of Theorem B in [K 1] can be adapted to the Σ^m–Conjecture. Note that we cannot use Meinert's approach of [M 1], because it is based on embedding metabelian groups of finite Prufer rank into constructible metabelian groups.

Using the fact that Theorem B is invariant with respect to subgroups of finite index in G_1 we can assume that $Q_1 \simeq \mathbb{Z}^{n+1}$ and so Q_1 is endowed with inner product $(\, , \,)$ inherited from $\mathbb{R}^{n+1}$. Now we assume Theorem B does not hold and hence there exist non-trivial real characters $\chi, \chi_1^*, \ldots, \chi_m^*$ of G_1 such that $\chi = \chi_1^* + \ldots + \chi_m^*$, $[\chi_1^*], \ldots, [\chi_m^*] \in \Sigma^1(G_1, \mathbb{Z})^c$ and $[\chi] \in \Sigma^m(G_1, \mathbb{Z})$. Since $\Sigma^1(G_1, \mathbb{Z})^c = \Sigma^c_{[G_1,G_1]}(G_1/[G_1, G_1])$, we deduce by [B-G 2], Thm E, that $\Sigma^1(G_1, \mathbb{Z})^c$ is a rational spherical polyhedron. Using that $\Sigma^m(G_1, \mathbb{Z})$ is open in $S(G_1)$ without loss of generality we can assume that $\chi_1^*, \chi_2^*, \ldots, \chi_m^*$ are discrete characters.

We note that because $[\chi_i^*] \in \Sigma^c_{[G_1,G_1]}(G_1/[G_1, G_1]) \simeq \Sigma^c_{A_1}(Q_1)$ we have $A_1 \subseteq Ker\chi_i^*$ for all $1 \leq i \leq m$. Then $A_1 \subseteq Ker\chi$ and from now on we view $\chi_1^*, \ldots, \chi_m^*$, χ as characters of Q_1.

We define χ_i to be the restriction of χ_i^* to $Ker\chi$ and then $\chi_1 + \ldots + \chi_m = 0$. By [K 1], Thm B, $\chi_1^*, \ldots, \chi_m^*$ lie in an open halfspace of $Hom(Q_1, \mathbb{R}) \simeq \mathbb{R}^{n+1}$ and since $\chi = \chi_1^* + \ldots + \chi_m^*$ we deduce $\chi \notin \mathbb{R}_{<0}.\chi_i^*$. Since $[\chi] \in \Sigma^m(G_1, \mathbb{Z}) \subseteq \Sigma^1(G_1, \mathbb{Z})$ and $[\chi_i^*] \in \Sigma^1(G_1, \mathbb{Z})^c$ we have $\chi \notin \mathbb{R}_{>0}.\chi_i^*$. Then $\chi \notin \mathbb{R}\chi_i^*$, so all the characters $\chi_1, \ldots, \chi_m$ are non-trivial.

By induction we can assume without loss of generality that Theorem B holds for $m-1$ i.e. $\Sigma^{m-1}(G_1, \mathbb{Z}) \cap (conv_{\leq m-1}\Sigma^1(G_1, \mathbb{Z})^c) = \emptyset$. Then the characters $\chi_1^*, \ldots, \chi_m^*$ span a subspace of $Hom(Q_1, \mathbb{R}) \simeq \mathbb{R}^{n+1}$ of dimension m, otherwise we deduce that $\chi \in conv_{\leq m-1}\{\chi_1^*, \ldots, \chi_m^*\} \cap \Sigma^m(G_1, \mathbb{Z})$, a contradiction. Hence the $\mathbb{R}$-span of $\chi_1, \ldots, \chi_m$ is a space of dimension $m-1$.

We define $Q = Ker\chi \leq Q_1$ and by multiplying $\chi_1^*, \ldots, \chi_m^*$ by a positive integer if necessary we can assume that there exist elements $q_{\chi_1}, \ldots, q_{\chi_m} \in Q$ such that $\chi_i(q) = (q, q_{\chi_i})$ for all $q \in Q$. Let Q' be the subgroup of Q generated by q_{χ_1}, $\ldots, q_{\chi_m}$ and H be the orthogonal complement of Q' in Q. We choose an element $q^* \in Q_{1,\chi} \setminus Ker\chi$ where $Q_{1,\chi} = \{q \in Q_1 \mid \chi(q) \geq 0\}$ with the additional property that $\chi_i^*(q^*) > 0$ for all $1 \leq i \leq m$. Note that this is possible because $\chi_1^*, \ldots, \chi_m^*$ lie in an open halfspace of $Hom(Q_1, \mathbb{R}) \simeq \mathbb{R}^{n+1}$. We define Q'' to be the submonoid of Q_1 generated by H and q^* and hence the subgroup of Q_1 generated by Q'' and Q' has finite index in Q_1. Then, by substituting G_1 if necessary with a subgroup of finite index, we can assume that $Q_{1,\chi} = Q' \times Q''$.

We note that for every $\mathbb{Z}G_{1,\chi}$–module M there is a first quadrant spectral sequence $E^2_{p,q} \underset{p}{\Longrightarrow} Tor^{\mathbb{Z}G_{1,\chi}}_{p+q}(\mathbb{Z}, M)$ where $E^2_{p,q} = Tor_p^{\mathbb{Z}Q_{1,\chi}}(\mathbb{Z}, Tor_q^{\mathbb{Z}A_1}(\mathbb{Z}, M))$. This spectral sequence is an obvious modification of Lyndon-Serre spectral sequence in homology of groups and can be deduced as an easy consequence of Grothendieck spectral sequence [Rot], Thm 11.39. If $M = \prod \mathbb{Z}G_{1,\chi}$ the spectral sequence collapses and we have

$$Tor_i^{\mathbb{Z}G_{1,\chi}}(\mathbb{Z}, \prod \mathbb{Z}G_{1,\chi}) \simeq Tor_i^{\mathbb{Z}Q_{1,\chi}}(\mathbb{Z}, (\prod \mathbb{Z}G_{1,\chi})_{A_1}) \text{ for all } i.$$

We note that any maximal regular sequence in the commutative ring $\mathbb{Z}Q_{1,\chi}$ gives rize to a projective resolution of Koszul type of the trivial $\mathbb{Z}Q_{1,\chi}$–module $\mathbb{Z}$. Using this projective resolution for a regular sequence $\{u-1 \in \mathbb{Z}Q_{1,\chi} \mid u \in U\}$, where $U = \{$ a basis of $Q'\}\cup\{$ a basis of $H\}\cup\{q^*\}$, we see that $Ext^0_{\mathbb{Z}Q'}(\mathbb{Z}, P) \simeq Tor^{\mathbb{Z}Q'}_{m-1}(\mathbb{Z}, P)$, $Ext^{n-m+2}_{\mathbb{Z}Q''}(\mathbb{Z}, P) \simeq Tor_0^{\mathbb{Z}Q''}(\mathbb{Z}, P)$, and that $Tor^{\mathbb{Z}Q_{1,\chi}}_{m-1}(\mathbb{Z}, P) \simeq Ext^{n-m+2}_{\mathbb{Z}Q_{1,\chi}}(\mathbb{Z}, P)$ for every $\mathbb{Z}Q_{1,\chi}$–module P. Finally we get that the canonical map $E^{Q'} \to E_{Q''}$ can be factored through $Tor^{\mathbb{Z}Q_{1,\chi}}_{m-1}(\mathbb{Z}, E)$, where $E = (\prod \mathbb{Z}G_{1,\chi})_A$, i.e.

$$E^{Q'} = Ext^0_{\mathbb{Z}Q'}(\mathbb{Z}, E) \simeq Tor^{\mathbb{Z}Q'}_{m-1}(\mathbb{Z}, E) \to Tor^{\mathbb{Z}Q_{1,\chi}}_{m-1}(\mathbb{Z}, E)$$

$$\simeq Ext^{n-m+2}_{\mathbb{Z}Q_{1,\chi}}(\mathbb{Z}, E) \to Ext^{n-m+2}_{\mathbb{Z}Q''}(\mathbb{Z}, E) \simeq Tor_0^{\mathbb{Z}Q''}(\mathbb{Z}, E) = E_{Q''}.$$

Using [B], Thm 1.3 and the fact that $G_{1,\chi}$ is of type FP_m we see that

$$Tor_i^{\mathbb{Z}G_{1,\chi}}(\mathbb{Z}, \prod \mathbb{Z}G_{1,\chi}) = 0$$

for all $1 \le i \le m-1$. In the rest of the paper we aim to show that in our assumptions the canonical map $E^{Q'} \to E_{Q''}$ is non-trivial and hence $0 \neq Tor^{\mathbb{Z}Q_{1,\chi}}_{m-1}(\mathbb{Z}, E) \simeq Tor^{\mathbb{Z}G_{1,\chi}}_{m-1}(\mathbb{Z}, \prod \mathbb{Z}G_{1,\chi})$, a contradiction.

3.2 A criterion for non-vanishing in $E_{Q''}$

Suppose $\{F^d_*\}_{d\in\mathbb{Z}}$ is a decreasing filtration of $\mathbb{Z}Q''$–submodules of A_1. Then as in [Å] we define an order function of the elements of the augmentation ideal of $\mathbb{Z}A_1$ given by

$$o(f) = sup\{d \mid f \equiv 0 \text{ in } \mathbb{Z}[A_1/F^d_*]\} \in \mathbb{Z}_\infty.$$

The following proposition is a modification of [K 1], Prop. 3.1, with the only difference that Q'' in our case is only a monoid, not a group.

Proposition. *Let $\pi : Q_1 \to G_1$ be a lifting of the projection $G_1 \to Q_1$, $\{F^d_*\}_{d\in\mathbb{Z}}$ be a decreasing filtration of (left) $\mathbb{Z}Q''$-submodules of A_1 such that $\cup_{d\in\mathbb{Z}} F^d_* = A_1$. Suppose $\alpha = (\alpha_j)_{j\in\mathbb{N}}$ is an element of $(\mathbb{Z}G_1)^{\mathbb{N}}$ with $\alpha_j = \sum_{q\in Q'} \alpha_{j,q}\pi(q), \alpha_{j,q} \in \mathbb{Z}A_1$. If α vanishes in $((\mathbb{Z}G_1)^{\mathbb{N}})_{G''}$, where G'' is the submonoid of G_1 generated by A_1 and $\pi(Q'')$, then all $\alpha_{j,q}$ are elements of the augmentation ideal of $\mathbb{Z}A_1$ and $inf\{o(\alpha_{j,q}) : j, q\} > -\infty$.*

3.3 The filtration $\{F^d_*\}_{d\in\mathbb{N}}$ of A_1

Let $P^*_1, \ldots, P^*_s$ be the minimal associated primes for the left $\mathbb{Z}Q_1$–module A_1. Then $N_1 := \mathbb{Z}Q_1/P^*_1 \oplus \ldots \oplus \mathbb{Z}Q_1/P^*_s$ embeds in A_1 and $\Sigma^1(G_1, \mathbb{Z})^c = \Sigma^c_{A_1}(Q_1) = \Sigma^c_{N_1}(Q_1) = \cup_{1\le i\le s}\Sigma^c_{\mathbb{Z}Q_1/P^*_i}(Q_1)$. Since $[\chi^*_i] \in \Sigma^1(G_1, \mathbb{Z})^c$, for some $j(i)$ we have

$[\chi_i^*] \in \Sigma^c_{\mathbb{Z}Q_1/P^*_{j(i)}}(Q_1)$. Then by [B-G 2], Thm 8.1, there exists a real valuation v_i^* (in Bourbaki sense) of $\mathbb{Z}Q_1/P^*_{j(i)}$ such that $v_i^*(\mathbb{Z}) \geq 0$ and χ_i^* is the restriction of v_i^* to Q_1. We extend v_i^* to a map $w_i^* : N_1 \to \mathbb{R}_\infty$ by $w_i^*(\lambda_1 \oplus \ldots \oplus \lambda_s) = v_i^*(\lambda_{j(i)})$. Then N_1 is equipped with a decreasing filtration $\{F_i^d\}_{d \in \mathbb{Z}}$ of $\mathbb{Z}Q''$–submodules given by $F_i^d = (w_i^*)^{-1}[d.\chi_i(q_{\chi_i}), \infty)$.

Let $\{F_*^d\}_{d\in\mathbb{Z}}$ be the decreasing filtration of $\mathbb{Z}Q''$–submodules of N_1 given by $F_*^d = \cap_{1\leq i\leq m} F_i^d$. Using [K 1], Lemma 3.2 for $R = \mathbb{Z}Q''$, we extend $\{F_*^d\}_{d\in\mathbb{Z}}$ to a decreasing filtration of $\mathbb{Z}Q''$–submodules of A_1. We call the new filtration $\{F_*^d\}_{d\in\mathbb{Z}}$ as well.

Let N be the direct sum $\mathbb{Z}Q/P_1 \oplus \ldots \oplus \mathbb{Z}Q/P_s \subseteq N_1$, where $P_i = P_i^* \cap \mathbb{Z}Q$, G be the subgroup of G_1 generated by N and $\pi(Q)$, and A be the abelian normal subgroup of G generated by N and $[\pi(Q), \pi(Q)]$. We define a decreasing filtration $\{F^d\}_{d\in\mathbb{Z}}$ of $\mathbb{Z}H$–submodules of A given by $F^d = F_*^d \cap A$ and denote $F^d_{\chi_i}$ the restriction of F_i^d to A.

3.4 The choice of the special element in E

We want to apply the main constructions from [K 1] to the finitely generated group G. As in [K 1] we have a free abelian group $Q \simeq \mathbb{Z}^n$ and some elements $q_{\chi_1}, \ldots, q_{\chi_m}$ from Q' that form a $(m-1)$–dimensional simplex in $\mathbb{R}^n$. We consider the subsets $Q'_k, S_k, k \in \mathbb{N}$ of Q', given by

$$Q'_k = \{q \in Q' : \chi_i(q) \geq -k\chi_i(q_{\chi_i}), 1 \leq i \leq m\}, \quad S_k = Q'_k \setminus Q'_{k-1}.$$

As in [K 1], Section 3.2, for every $q \in (Q'_k)^{-1}, k \in \mathbb{N}$, there is a special subset $W_{k,q}$ of G which in the assumptions of Theorem B is finite. Furthermore if G is a split extension of A by Q then $W_{k,q} = \{q\}$. All details about $W_{k,q}$ can be found in [K 1]. Later on we will need that $W_{k,q}\pi(q)^{-1} \subset A$, where π is the lifting used in the Proposition in Section 3.2.

Now we define some rings $B_i, 1 \leq i \leq m$. We note that the definition of the ring B_i is different from the one used in [K 1], section 3.3. The reason is that the filtrations defined in the previous section come from valuations, not from pseudovaluations. By the choice of the characters $\chi_1, \ldots, \chi_m$ for every $1 \leq i \leq m$ there exists an integer $j(i)$ such that the ring

$$B_i = (\mathbb{Z}Q_{\chi_i} + P_{j(i)}/P_{j(i)})/(v_i^{-1}[\chi_i(q_{\chi_i}), +\infty) \cap (\mathbb{Z}Q_{\chi_i} + P_{j(i)}/P_{j(i)}))$$

is non-trivial, where v_i is the restriction of the valuation v_i^* to $\mathbb{Z}Q/P_{j(i)}$. B_i is a non-trivial commutative ring of affine type and we fix a maximal ideal J_i in B_i. Then B_i/J_i is a finite field and there exists a positive integer r_i such that

$$(Ker(\chi_i))^{r_i} = 1 \text{ in } B_i/J_i.$$

Then we set

$$U_{i,k} = \{q \in S_k \mid qq^k_{\chi_i} \in (Ker\ \chi_i)^{r_i}, \chi_j(q) \geq -(k-1)\chi_j(q_{\chi_j}), 1 \leq j \neq i \leq m\},$$

$$U_k = \cup_{i=1}^m U_{i,k}.$$

Finally we define the special element $\alpha = (\alpha_k)_{k \geq k_1} \in \mathbb{Z}G^{\mathbb{N}} \subset (\mathbb{Z}G_1)^{\mathbb{N}}$ by

$$\alpha_k = \sum_{q \in (Q'_k)^{-1}} \alpha_{k,q} \prod_{u \in U_{k+t_0}} (T^u - 1) \in \mathbb{Z}G$$

where we write the elements of $\mathbb{Z}A$ as $\mathbb{Z}$–linear combinations of T^a, $a \in A$; $\alpha_{k,q}$ is the sum of the elements of $W_{k,q} \subset G$ in the algebra $\mathbb{Z}G$, $U_{k+t_0} \subset Q$ embeds in $N = \mathbb{Z}Q/P_1 \oplus \ldots \oplus \mathbb{Z}Q/P_s \subset A$ via the diagonal map, t_0 is a sufficiently large positive integer such that $W_{k,1_Q} \subseteq F_*^{-k-t_0+1}$ for all $k \in \mathbb{N}$ (compare with [K 1], Lemma 3.8). At the beginning of next section we will make some remarks about the choice of the positive integer k_1.

3.5 Some important properties of α

We claim that the image of α in $E = ((\mathbb{Z}G_1)^{\mathbb{N}})_{A_1}$ is Q'–invariant. Indeed α is defined copying the recipe of the special element in homology from [K 1] with only difference that the filtrations defined in section 3.3 are given by valuations, not by pseudovaluations. The argument of [K 1], Lemma 3.4, gives that if k_1 is a sufficiently large positive integer the image of α in $(\mathbb{Z}G^{\mathbb{N}})_A$ is Q'–invariant, so furthermore the image β of α in $E = ((\mathbb{Z}G_1)^{\mathbb{N}})_{A_1}$ is Q'–invariant.

We claim as well that the image of β in $E_{Q''}$ is non-trivial. The proof is quite technical and is a modification of the proof of [K 1], Thm 3.10. We outline the main steps in the proof. We assume the image of β in $E_{Q''}$ is trivial. Then by Proposition there exists $d \in \mathbb{Z}$ such that for all $k \geq k_1$

$$\alpha_{k,1_Q} \prod_{u \in U_{k+t_0}} (T^u - 1) = 0 \text{ in } \mathbb{Z}[A_1/F_*^d].$$

Since $W_{k,1_Q} \subseteq F_*^{-k-t_0+1}$ for all $k \in \mathbb{N}$ we deduce that for sufficiently big k, say $k \geq k^*$,

$$\prod_{u \in U_{k+t_0}} (T^u - 1) = 0 \text{ in } \mathbb{Z}[A_1/F_*^{-k-t_0+1}]$$

where the elements of $U_{k+t_0} \subset Q$ embed in $N = \mathbb{Z}Q/P_1 \oplus \ldots \oplus \mathbb{Z}Q/P_s \subset N_1 \subseteq A_1$ via the diagonal map. We remind the reader that $\{F^d\}_{d \in \mathbb{Z}}$ is the decreasing filtration of $\mathbb{Z}H$–submodules of A obtained by restricting the filtration $\{F_*^d\}_{d \in \mathbb{Z}}$ to A i.e. $F^d \cap N = \{\lambda_1 \oplus \ldots \oplus \lambda_s \in \mathbb{Z}Q/P_1 \oplus \ldots \oplus \mathbb{Z}Q/P_s \mid v_i(\lambda_{j(i)}) \geq d\chi_i(q_{\chi_i}) \text{ for all } 1 \leq i \leq m\}$, where v_i is the restriction of the valuation v_i^* to $\mathbb{Z}Q/P_{j(i)}$. Then

$$\prod_{u \in U_{k+t_0}} (T^u - 1) = 0 \text{ in } \mathbb{Z}[A/F^{-k-t_0+1}] \text{ for } k \geq k^*. \tag{1}$$

Let $V'_{i,k+t_0}$ be the abelian subgroup of $A/F_{\chi_i}^{-k-t_0+1}$ generated by the image of $U_{i,k+t_0}$ in N. As in [K 1] we deduce from (1) that for some $1 \leq i_0 \leq m$

$$\prod_{u \in U_{i_0,k+t_0}} (T^u - 1) = 0 \text{ in } \mathbb{Z}[V'_{i_0,k+t_0}] \text{ for } k = k^*. \tag{2}$$

We note that $V'_{i_0,k+t_0}$ embeds in $(\oplus_{1\le j\ne j(i_0)\le s}\mathbb{Z}Q/P_j)\oplus B'_{i_0}$ where

$$B'_{i_0} := \frac{q_{\chi_{i_0}}^{-k-t_0}\mathbb{Z}Q_{\chi_0}+P_{j(i_0)}/P_{j(i_0)}}{((q_{\chi_{i_0}}^{-k-t_0}\mathbb{Z}Q_{\chi_0}+P_{j(i_0)}/P_{j(i_0)})\cap v_{i_0}^{-1}[-(k+t_0-1)\chi_{i_0}(q_{\chi_{i_0}}),+\infty))}.$$

Obviously B'_{i_0} is isomorphic to B_{i_0} as a $\mathbb{Z}(Ker\chi_{i_0})$–module with isomorphism given by multiplication with $q_{\chi_0}^{k+t_0}$. By (2) we have

$$\prod_{u\in U_{i_0,k+t_0}}(T^u-1)=0 \text{ in } \mathbb{Z}[B'_{i_0}] \text{ for } k=k^* \text{ and so}$$

$$\prod_{u\in q_{\chi_{i_0}}^{k+t_0}U_{i_0,k+t_0}}(T^u-1)=0 \text{ in } \mathbb{Z}[B_{i_0}] \text{ and hence in } \mathbb{Z}[B_{i_0}/J_{i_0}] \text{ for } k=k^*,$$

where J_{i_0} is the ideal of B_{i_0} defined in section 3.4. By the definition of r_{i_0} all the elements of $(Ker\chi_{i_0})^{r_{i_0}}$ represent $1+J_{i_0}$ in B_{i_0}/J_{i_0} and $q_{\chi_{i_0}}^{k+t_0}U_{i_0,k+t_0}\subseteq(Ker\chi_{i_0})^{r_{i_0}}$, so

$$(T^{1+J_{i_0}}-1)^{|U_{i_0,k+t_0}|}=0 \text{ in } \mathbb{Z}[B_{i_0}/J_{i_0}] \text{ for } k=k^*.$$

This contradicts the fact that $1+J_{i_0}$ is a non-trivial element of the finite field B_{i_0}/J_{i_0} and $\mathbb{Z}[B_{i_0}/J_{i_0}]$ does not have non-trivial nilpotent elements. Therefore the image of β in $E_{Q''}$ is non-trivial.

We note that the described properties of α show that the canonical map

$$H^0(Q',((\mathbb{Z}G_1)^{\mathbb{N}})_{A_1})\to H_0(Q'',((\mathbb{Z}G_1)^{\mathbb{N}})_{A_1})$$

is non-trivial. Then $\mathbb{Z}G_{1,\chi}$ is not of type FP_m, a contradiction with the assumption that $[\chi]\in\Sigma^m(G_1,\mathbb{Z})$. This completes the proof of Theorem B.

References

[Å] H. Åberg, *Bieri–Strebel valuations (of finite rank)*, Proc. London Math. Soc. (3) **52** (1986), 269–304.

[B] R. Bieri, *Homological dimension of discrete groups*, Queen Mary College Mathematics Notes, London, 2nd ed. 1981.

[B-G 1] R. Bieri, J.R.J. Groves, *Metabelian groups of type $(FP)_\infty$ are virtually of type (FP)*, Proc. London Math. Soc. (3) **45** (1982), 365–384.

[B-G 2] R. Bieri, J.R.J. Groves, *The geometry of the set of characters induced by valuations*, J. Reine Angew. Math. **347** (1984), 168–195.

[B-N-S] R.Bieri. W.D. Neumann, R. Strebel, *A geometric invariant of discrete groups*, Invent. Math. **90** (1987), 451–477.

[B-R] R.Bieri, B.Renz, *Valuations on free resolutions and higher geometric invariants of groups*, Comment. Math. Helv. **63** (1988), 464–497.

[B-S 1] R. Bieri, R. Strebel, *Valuations and finitely presented metabelian groups*, Proc. London Math. Soc. (3) **41** (1980), 439–464.

[B-S 2] R. Bieri, R. Strebel, *Geometric invariants for discrete groups*, a book in preparation.

[Ge] R. Gehrke, *The higher geometric invariants for groups with sufficient commutators*, Comm. Algebra, to appear.

[K 1] D.H. Kochloukova, *The FP_m–Conjecture for a class of metabelian groups*, J. Algebra, 184(1996), 1175-1204.

[K 2] D.H. Kochloukova, *The Σ^2–Conjecture for metabelian groups: the split extension case*, preprint.

[M 1] H. Meinert, *The homological invariants for metabelian groups of finite Prufer rank: a proof of the Σ^m–Conjecture*, Proc. London Math. Soc. (3) **72** (1996), 385–424.

[M 2] H. Meinert, *Actions on 2–Complexes and the Homotopical Invariant Σ^2 of a Group*, J. Pure Appl. Algebra, to appear.

[N] G.A. Noskov, Bieri-Strebel invariant and homological finiteness properties of metabelian groups, SFB-Preprint 93-028, Universität. Bielefeld 1993.

[Re] B. Renz, *Geometrische Invarianten und Endlichkeitseigenschaften von Gruppen.* Dissertation. Universität Frankfurt a.M. (1988).

[Rot] J.J. Rotman, *An introduction to homological algebra*, Academic press, New York, 1979.

RINGS WITH PERIODIC GROUPS OF UNITS II

JAN KREMPA

Institute of Mathematics, Warsaw University, ul. Banacha 2, 02-097 Warszawa, Poland

Abstract

In this note we will survey and extend some results on periodicity of unit groups of associative rings. Special attention will be paid to group rings.

1 Preliminaries

In this paper we assume that rings are associative, in general with $1 \neq 0$. If R is a ring then $\mathrm{U}(R)$ will denote the unit group of the ring R, R^+ the additive group of R, R_U the subring of R generated by $\mathrm{U}(R)$ and $\mathrm{J}(R)$ the (Jacobson) radical of the ring R. By an order we mean here a $\mathbb{Z}$-order. For other notions and results of ring theory one can consult for example [17].

We will apply rather standard notation and terminology on groups. For example, C_n will denote the cyclic group of order n and Q_8 the quaternion group of order 8. For further information about groups see for example [18, 23].

Various finiteness conditions for groups of units of associative rings, in particular of group rings are studied in the literature (see for example [25, 20, 13, 26, 15]). In this paper we are going to concentrate on periodicity of groups of units.

Many years ago Graham Higman (see [10], or for example [13, 25, 12]), described all integral group rings with finite and with periodic groups of units by the following result:

Theorem 1.1 (Higman) *Let G be a periodic (finite) group. Then the following conditions are equivalent:*

1. $\mathrm{U}(\mathbb{Z}G)$ *is periodic (finite);*
2. *Either G is Abelian of exponent* $2, 3, 4$ *or* 6, *or G is a Hamiltonian 2-group;*
3. $\mathrm{U}(\mathbb{Z}G) = \pm G$.

Several authors, (see [3, 4, 5, 21] and [25, 13]), generalized the above theorem to periodic normal subgroups in $\mathbb{Z}G$. In Section 2 we give another generalization. Using some results from [16], we are going to extend the theorem of Higman to group rings with bigger class of coefficients. These coefficient rings will however still have torsion free additive groups.

Now we will present some simple results in which no assumption on the additive structure of the ring is needed. The first observation can be found for example in [7, 16]

Lemma 1.2 *Let R be a ring and let $J \subseteq \mathrm{J}(R)$ be an ideal. Then:*

(i) $1 + J \lhd \mathrm{U}(R)$ *and* $\mathrm{U}(R)/(1 + J) \cong \mathrm{U}(R/J)$;

(ii) *The group* $1+J$ *is periodic if and only if* J *is a nil ring with* J^+ *periodic.*

As a generalization of a standard lemma to nonnecessarily commutative coefficients we have:

Lemma 1.3 *Let* R *be a ring and let* $C_2 = \langle g\rangle$. *If* $a = r+sg \in RC_2$ *then* $a \in \mathrm{U}(RC_2)$ *if and only if* $r+s$ *and* $r-s \in \mathrm{U}(R)$.

Proof $\Leftarrow$ Let $r+s, r-s \in \mathrm{U}(R)$ and let $q = (r+s)^{-1}$. Then clearly $q \in \mathrm{U}(R)$ and qr commutes with $qs = 1-qr$. Thus

$$(qrq - qsqg)a = (qr - qsg)(qr + qsg) = (qr)^2 - (qs)^2.$$

Hence we have:

$$(qrq - qsqg)a = (qr)^2 - (qs)^2 = q(r-s)q(r+s) \in \mathrm{U}(R) \subseteq \mathrm{U}(RC_2).$$

It means that a is left invertible in RC_2. Similarly one can see that a is right invertible, and consequently, $a \in \mathrm{U}(RC_2)$.

$\Rightarrow$ Now let $a \in \mathrm{U}(RC_2)$. If we substitute $g = 1$ or $g = -1$ then we obtain that $r+s$ and $r-s \in \mathrm{U}(R)$, respectively. □

Theorem 1.4 *Let* R *be a ring and let* G *be an elementary Abelian 2-group. Then* $\mathrm{U}(RG)$ *is periodic if and only if* $\mathrm{U}(R)$ *is periodic.*

Proof $\Leftarrow$ Let $\mathrm{U}(R)$ be periodic and G be an elementary Abelian 2-group. In particular G is locally finite. Hence, without loss of generality, we can assume that $|G| = 2^n < \infty$. If $n > 1$ then $G = H \times C_2$ where $|H| = 2^{n-1}$. Thus it's enough to consider the case $G = \langle g\rangle = C_2$.

Let $\phi : RC_2 \longrightarrow R \oplus R$ be a homomorphism given by the formula $\phi(r+sg) = (r+s, r-s)$ for $r, s \in R$, and let I denote the kernel of ϕ. It is easy to see that $2I = 0 = I^2$, hence $1 + I \subseteq \mathrm{U}(RC_2)$ is a normal subgroup.

If we consider ϕ as a homomorphism from $\mathrm{U}(RC_2)$ into $\mathrm{U}(R \oplus R) = (\mathrm{U}(R)) \times \mathrm{U}(R)$ then, by assumption, the image of ϕ is periodic and the kernel, equal to $1 + I$ is an elementary Abelian 2-group, thus it is periodic too. Hence $\mathrm{U}(RC_2)$ is periodic.

$\Rightarrow$ This implication is obvious, because $\mathrm{U}(R) \subset \mathrm{U}(RC_2)$. □

In Section 3 we are going to show that if R^+ is torsion then periodicity of $\mathrm{U}(R)$ is strongly related to algebraicity of R. This is of special interest in the case of algebras over fields of positive characteristic.

Now we will show that, even if our ring R under consideration is semisimple, but R^+ is mixed, then a reduction to positive characteristic case and to torsion free case is in general impossible. Let us consider the following example:

Example 1.5 Let $A = \mathbb{Z}[x_1, y_1, x_2, y_2]$ be the integral polynomial ring in noncommutative indeterminates and let $J \subset A$ denote the ideal generated by the elements $x_1y_1 - 1$, $y_1x_1 - 1$, $x_2y_2 - 1$, $y_2x_2 - 1$. For a prime number $p \in \mathbb{N}$ set $R = A/pJ$

and let T stand for the torsion part of R^+. Because $pJ = pA \cap J$, one can compute that R is semisimple. Moreover, $\mathrm{U}(R) = C_2$, $T = J/pJ$. Thus $R/T = A/J$ is isomorphic to the group ring $\mathbb{Z}F$ where F is the free group of rank 2. In particular $\mathrm{U}(R/T)$ is nonperiodic and satisfies no nontrivial group identity.

If in the above example we would considered A as a polynomial ring in commuting indeterminates then $\mathrm{U}(R/T)$ would be nonperiodic Abelian while still $\mathrm{U}(R)$ would be equal to C_2. If in either commutative or noncommutative version of this example we would apply an infinite set of indeterminates, then $\mathrm{U}(R/T)$ would not be finitely generated, while still $\mathrm{U}(R) = C_2$.

2 Case of characteristic zero

For studying torsion free Abelian groups with finite groups of automorphisms Hallett and Hirsch, (see [7, 16]), distinguished in [9] six small finite groups, which we will call here *acceptable*. They are as follows: C_2, C_4, C_6, Q_8, $DC_{12} = \langle a, b \mid a^3 = b^2 = (ab)^2 = 1\rangle$ (of order 12), and $BT_{24} = \langle a, b \mid a^3 = b^3 = (ab)^2 = 1\rangle$ (of order 24).

To realize acceptable groups as units of rings Hirsch and Zassenhaus distinguished in [11] (see also [16]) the following six orders: $\mathbb{Z}$, $\mathbb{Z}[i]$, $\mathbb{Z}[\zeta_3]$, $\mathbb{Z}[i,j]$, $\mathbb{Z}[\zeta_3, j]$, and $\mathbb{Z}[i,j,l]$, where $l = \frac{1+i+j+k}{2}$. Those orders, presented here with the use of rather standard notation for quaternion algebras, will also be called *acceptable*. From [11, 16] and the choice of acceptable rings we have:

Theorem 2.1 *Let R be a prime order such that $R = R_U$. Then*

(i) *R is acceptable if and only if $\mathrm{U}(R)$ is acceptable;*

(ii) *If R is acceptable, then $\mathrm{U}(R) \cap (1 + 2R) = \{1, -1\}$;*

(iii) *If R is not acceptable then $\mathrm{U}(R)$ is not periodic.*

Further, if P is a minimal prime ideal of a ring R then the factor ring R/P will be called a *main factor (of R)*.

Now, as in [16] let R be an order and let A be its ring of quotients. Then $A = \mathbb{Q}R$ and it is a semisimple finite dimensional $\mathbb{Q}$-algebra. By Wedderburn Theorem we have a unique decomposition: $A = \sum_{i=1}^{n} Ae_i$ where e_i is a minimal central idempotent in A for any $1 \leq i \leq n$. Let us put $\tilde{R} = \sum_{i=1}^{n} Re_i$. Using results from [1] one can prove, as in [16] the following facts:

Proposition 2.2 *Under the above notation we have:*

(i) *The orders R and $\tilde{R}$ have the same main factors Re_i for $i = 1, \ldots, n$;*

(ii) *If $P \subset R$ is a prime ideal such that $(R/P)^+$ is torsion free then P is a minimal prime ideal of R;*

(iii) *$\mathrm{U}(R)$ is of finite index in $\mathrm{U}(\tilde{R})$;*

(iv) *$\mathrm{U}(\tilde{R}) = \prod_{i=1}^{n} \mathrm{U}(Re_i)$.*

As another result from [16] we have:

Theorem 2.3 *Let R be a ring such that $R = R_U$. If R^+ is torsion free then the following conditions are equivalent:*

1. *The group* $\mathrm{U}(R)$ *is periodic;*
2. *Every finitely generated subring of* R *is an order with a finite group of units;*
3. R *is reduced and all its main factors are acceptable rings;*
4. $\mathrm{U}(R)$ *is a subdirect product of acceptable groups;*
5. *The group* $\mathrm{U}(R)$ *is locally finite.*

Corollary 2.4 *Let R satisfies any condition of the above theorem. Then no rational prime is invertible in R and $\mathrm{U}(R) \cap (1+2R) = \{1, -1\}$. Moreover, any factor of R with torsion free additive group again satisfies conditions of the theorem above.*

If R fulfills the conditions of the above theorem then from the description of acceptable rings we have that R satisfies many polynomial identities. From the description of acceptable groups we see that $\mathrm{U}(R)$ is a solvable group of class at most 3 and of exponent dividing 12.

Now we are going to apply the above results to group rings with periodic groups of units.

Lemma 2.5 *Let R be an acceptable ring. Then for some small finite groups G we have the following table showing when $\mathrm{U}(RG)$ is periodic:*

$R =$	$\mathbb{Z}$	$\mathbb{Z}[i]$	$\mathbb{Z}[\zeta_3]$	$\mathbb{Z}[i,j]$	$\mathbb{Z}[\zeta_3,j]$	$\mathbb{Z}[i,j,l]$
$G = C_3$	*yes*	*no*	*yes*	*no*	*no*	*no*
$G = C_4$	*yes*	*yes*	*no*	*no*	*no*	*no*
$G = Q_8$	*yes*	*no*	*no*	*no*	*no*	*no*

Proof In any case RG is an order in $\mathbb{Q}RG = \mathbb{Q}R \otimes_{\mathbb{Q}} \mathbb{Q}G$. Now it is enough to look for main factors of this order and compare them with the definition of acceptable orders and with Theorem 2.1. This will give the proof in any desired case. □

Using the above facts and Theorem 2.3 we can obtain the following partial generalization of the theorem of Higman:

Theorem 2.6 *Let R be a ring such that $R = R_U$ and R^+ is torsion free. Let G denote a nontrivial group. Then*

1. *If G is an elementary Abelian 2-group then $\mathrm{U}(RG)$ is periodic if and only if $\mathrm{U}(R)$ is periodic;*
2. *If G is Abelian of exponent 3 or 6 then $\mathrm{U}(RG)$ is periodic if and only if R is reduced and any of its main factors is isomorphic either to $\mathbb{Z}$ or to $\mathbb{Z}[\zeta_3]$;*
3. *If G is Abelian of exponent 4 then $\mathrm{U}(RG)$ is periodic if and only if R is reduced and any of its main factors is isomorphic either to $\mathbb{Z}$ or to $\mathbb{Z}[i]$;*
4. *If G is a Hamiltonian 2-group then $\mathrm{U}(RG)$ is periodic if and only if R is reduced and any of its main factors is isomorphic to $\mathbb{Z}$.*
5. *In any other case $\mathrm{U}(RG)$ is not periodic.*

Proof Let $\mathrm{U}(RG)$ be periodic. Then certainly $\mathrm{U}(R)$ is periodic. Moreover, by assumption, $\mathbb{Z} \subseteq R$. Thus, by the theorem of Higman, G is locally finite of special type. On the other hand groups listed in statements 1–4 are locally finite too. Hence, without loss of generality we can assume that G is finite.

The statement 1 is an immediate consequence of Theorem 1.4.

Let G be Abelian of exponent 3 and of order 3^n. If $G = C_3$ then from Lemma 2.5, Theorem 2.3 and Corollary 2.4 the desired form of prime factors of R follows. Now one can finish the proof by induction on n.

The statement 3 follows easily, and the proof is similar to that of statements 1 and 2.

If G is a Hamiltonian 2-group then, using statement 1, one can reduce considerations to the case when $G = Q_8$. Now it's enough to procede as in the case of $G = C_3$.

Now from the theorem of Higman, and the previous part of the proof, statement 5 follows immediately. □

For convenience of further notation let $\mathrm{T}(RG)$ be the subgroup of $\mathrm{U}(RG)$ generated by all elements of the form $eg + 1 - e$ where $e \in R$ is a central idempotent and $g \in G$. Below we give more convenient description of elements of $\mathrm{T}(RG)$.

Lemma 2.7 (cf. [14, 13]) *Let $v \in RG$. Then $v \in \mathrm{T}(RG)$ if and only if there exists $n \geq 1$ such that $v = \sum_{i=1}^{n} e_i g_i$ where $1 = \sum_{i=1}^{n} e_i$ is a decomposition of $1 \in R$ into a sum of orthogonal central idempotents and all $g_i \in G$.*

Now we have the following extension of another part of the theorem of Higman:

Theorem 2.8 *Let R be a ring such that $R = R_U$ and R^+ is torsion free. Moreover let G be a nontrivial group. Then $\mathrm{U}(RG)$ is periodic if and only if $\mathrm{U}(R)$ and G are periodic and $\mathrm{U}(RG) = \mathrm{U}(R)\mathrm{T}(RG)$.*

Proof $\Rightarrow$ Let $\mathrm{U}(RG)$ be periodic. Then certainly $\mathrm{U}(R)$ and G are periodic and G is locally finite by the theorem of Higman.

Now let $u \in \mathrm{U}(RG)$ be a normalized unit. If R is a commutative domain then $u \in G$ by a result of Saksonov and Bass, ([13, 25]) and Corollary 2.4. If R is a noncommutative domain, then from the above theorem G is an elementary Abelian 2-group, hence by Theorem 1.4 and Corollary 2.4 $u \in G$ too.

Let R be an order. By Proposition 2.2 we can assume that R is a finite direct product of orders which are domains. Hence in this case $u \in \mathrm{T}(RG)$ by the first step of the proof.

Now, by Theorem 2.3, every finitely generated subring of R is an order, and the implication follows in full generality.

The converse implication is evident. □

3 Case of positive characteristic

In this section we are going to consider algebras with periodic groups of units over a field of positive characteristic. Certainly we can restrict to algebraic extensions of prime fields, because they are characterized by the periodicity of their groups of their units. So let us fix a prime number p, let F, be the prime field with p elements and K – an algebraic extension of $\mathbb{F}$. Let us begin with the following simple observation:

Lemma 3.1 *Let A be a K-algebra.*

1. *A is algebraic over K if and only if A is algebraic over $\mathbb{F}$;*
2. *If A is algebraic over K then $\mathrm{U}(A)$ is periodic;*
3. *If $A = A_U$ then A is a locally finite K-algebra if and only if the group $\mathrm{U}(A)$ is locally finite.*

Theorem 2.3 suggests the following question: *Let A be a K-algebra such that $\mathrm{U}(A)$ is periodic. Is A_U algebraic over K?*

It appears that an answer to the above question seems to be difficult, even partial results need nontrivial facts.

Theorem 3.2 *Let A be a K-algebra satisfying a nontrivial polynomial identity. If $\mathrm{U}(A)$ is periodic then A_U is locally finite. Hence $\mathrm{U}(A)$ is locally finite.*

Proof From the assumption we have that A_U has an algebraic basis, which is a subset of $\mathrm{U}(A)$. By the theorem of Shirshov, Procesi and others, (see [22, Theorem 6.2.5] or [24, Theorem 4.2.8]), we obtain immediately that A_U is locally finite. Hence the group $\mathrm{U}(A)$ is locally finite. □

Example 3.3 Let L and M be nontrivial algebraic extensions of K and A denote the free coproduct of these extensions over K. From results of G.M. Bergman and P.M. Cohn, (see [6]), we have that A is a domain, $\mathrm{U}(A)$ is generated by the set T of periodic elements of $\mathrm{U}(A)$ and $T \neq \mathrm{U}(A)$. Moreover, $A = A_U$ is not algebraic over K.

Now we are going to answer the above question on "stable" level. For this we will use the following lemma of Macbeth (see [18])

Lemma 3.4 *Let $G = \langle G_1, G_2 \rangle$ be a group of permutations of a set Ω, where G_1, G_2 are nontrivial subgroups of G and at least one of them is different from C_2. Let Ω_1 and Ω_2 be nonempty disjoined subsets of Ω such that $(G_1 \setminus 1) \cdot \Omega_1 \subseteq \Omega_2$ and $(G_2 \setminus 1) \cdot \Omega_2 \subseteq \Omega_1$. Then G is the free product of G_1 and G_2.*

The result below can be considered as a consequence of Macbeth's Lemma. We present here its complete proof, which was kindly communicated by A. Salwa.

Lemma 3.5 *Let $R = A[x]$ where A is any ring. Then the group $\mathrm{GL}_2(R)$ contains a nontrivial free subgroup.*

Proof Let $u = \begin{bmatrix} 1 & x \\ 0 & 1 \end{bmatrix}$ and $v = \begin{bmatrix} -1 & 1 \\ -1 & 0 \end{bmatrix}$. Further let $G_1 = \langle u \rangle$, $G_2 = \langle v \rangle$ and $G = \langle G_1, G_2 \rangle = \langle u, v \rangle \subset \mathrm{GL}_2(R)$. Clearly any element of G_1 is of the form

$\begin{bmatrix} 1 & ax \\ 0 & 1 \end{bmatrix}$ for some $a = n \cdot 1, n \in \mathbb{Z}$ Moreover, $v^2 = \begin{bmatrix} 0 & -1 \\ 1 & -1 \end{bmatrix}$ and $v^3 = 1$, hence $|G_2| = 3$.

We will show that the commutators $[u, v]$ and $[u, v^2]$ generate a free subgroup of G.

Let first A be a domain. Put $\Omega = R \oplus R \setminus (0,0)$, $\Omega_1 = \{(f,g) : \deg(f) \leq \deg(g)$, and $\Omega_2 = \{(f,g) : \deg(f) > \deg(g)$. Immediate calculation shows that under this notation, the assumptions of Macteth's lemma are satisfied, hence G is the free product of G_1 and G_2. It means that the commutators $[u,v], [u,v^2]$ generate a nontrivial free subgroup of G.

Now let A be arbitrary. Because of the definition of the matrices considered, we can assume that A is a homomorphic image of $\mathbb{Z}$. By passing to any of its prime images, one can obtain the result from the first step of the proof. □

Remark If A is a domain of characteristic different from 2 then the above result is in fact well known (see [27, Proposition 2.8]) with the use of the matrix $u^* = \begin{bmatrix} 1 & 0 \\ x & 1 \end{bmatrix}$ instead of v. When A is a field of characteristic 2 the group $\langle u, u^* \rangle$ is isomorphic to the infinite dihedral group, so it does not contain a nontrivial free subgroup.

As in [2, Definition 10] let us agree that an K-algebra A is *matrix algebraic* if for any $n \in \mathbb{N}$ the matrix ring $M_n(A)$ is algebraic over K. This concept is connected with our considerations because of the following result:

Theorem 3.6 *Let K be an absolute field and let A be a K-algebra. Then the following conditions are equivalent:*

1. *The group $GL_n(A)$ is periodic for any $n \in \mathbb{N}$;*
2. *The group $GL_n(A)$ does not contain a noncyclic free group for any $n \in \mathbb{N}$;*
3. *A is matrix algebraic.*

Proof Clearly $1 \Rightarrow 2$. Now let us assume that A is not matrix algebraic. Then for some $n \in \mathbb{N}$ the matrix ring $M_n(A)$ contains an element c transcendental over K. From Lemma 3.5 the unit group of the algebra $M_2(M_n(A)) \cong M_{2n}(A)$ contains a noncyclic free subgroup. In this way we proved that $2 \Rightarrow 3$.

$3 \Rightarrow 1$. Let A be matrix algebraic, and $n \in \mathbb{N}$. Then, by assumption, the ring $M_n(A)$ is algebraic over K. Hence, from Proposition 3.1, we have that the group $GL_n(A) = U(M_n(A))$ is periodic. This completes the proof. □

Several characterizations of matrix algebraic algebras in terms of polynomial algebras and their central localizations can be found, for example in [2, Proposition 13].

Example 3.7 Let B be any unital algebra of Golod type. It was noticed by Amitsur and others that B is matrix algebraic and, in fact, satisfies some stronger properties. However B is not locally finite and $B = K + J(B)$. This shows that an algebra with "stably periodic" group of units need not be locally finite when the radical is very large.

Following ideas of Amitsur, (see [2, page 161]), we show that for semisimple algebras the situation is similar. In this way we also answer Question 4.6 from [16] in the negative.

Theorem 3.8 *Let K be a field of positive characteristic which is algebraic over its prime subfield. Then there exists a semisimple, (and even left and right primitive), K-algebra $A = A_U$ such that* $\mathrm{U}(A)$ *is stably periodic but not locally finite. Moreover* $\mathrm{U}(A)$ *satisfies no group identity, but A satisfies a generalized polynomial identity.*

Proof Let B be any K-algebra of Golod type. Taking the regular representation, B can be viewed as an algebra of linear transformations of the K-vector space B. Let $A \subset \mathrm{End}_K(B)$ be a subalgebra generated by B and by the ideal I of all linear transformations of finite rank. Because I is an locally finite ideal in A and A/I is a homomorphic image of B then, as pointed in [2], A is matrix algebraic, hence $\mathrm{U}(A)$ is stably periodic. On the other hand $\mathrm{U}(A)$ contains a group $\mathrm{U}(B)$ which is not locally finite.

The group $\mathrm{U}(A)$ satisfies no group identity because it contains $\mathrm{GL}_n(K)$ for any $n \in \mathbb{N}$. However, A satisfies a generalized polynomial identity because it is prime with an idempotent e such that $eAe \cong K$. □

For group algebras we have the following result:

Proposition 3.9 *Let A be an algebra over an absolute field K. Then A is matrix algebraic if and only if* $\mathrm{U}(AG)$ *is periodic for any finite group G.*

Proof For any n there exists a finite group G such that KG contains $\mathrm{M}_n(K)$ as a direct summand. On the other hand for any finite G the ring AG is contained in $\mathrm{M}_n(K)$ for enough large n. Hence the result follows. □

If a group G is not locally finite then situation can be complicated even for the case $A = K$, as observed by A.K. Lichtman (see [20, §10.1]). However, if $\mathrm{U}(KG)$ is periodic and satisfies a nontrivial group identity, then in many cases we know, by results from [8, 19] that G has to be locally finite.

References

[1] D.M. Arnold, *"Finite rank torsion free Abelian groups and rings"*, LNM 931, Springer-Verlag, Berlin 1982.

[2] G.M. Bergman, *Radicals, tensor products, and algebraicity,* Israel Math. Conf. Proc. vol. 1, The Weizmann Science Press of Israel (1989), pp 150-192.

[3] A.A. Bovdi, *The periodic normal divisors of the multiplicative group of a group ring I,* Sibirsk. Mat. Zh. 9(1968), 495-498.

[4] A.A. Bovdi, *The periodic normal divisors of the multiplicative groups of a group ring II,* Sibirsk. Mat. Zh. 11(1970), 492-511.

[5] S.P. Coelho & C. Polcino Milies, *Group rings whose torsion units form a subgroup,* Proc. Edinburgh Math. Soc. 37(1994), 201-205.

[6] P.M. Cohn, *"Skew fields"*, Encyclopedia of Math. Appl. vol. 57, Cambridge 1995.
[7] L. Fuchs, *"Infinite Abelian groups"*, vol. 2, Academic Press, New York 1973.
[8] A. Giambruno, S.K. Sehgal & A. Valenti, *Group algebras whose units satisfy a group identity,* Proc. Amer. Math. Soc., (to appear).
[9] J.T. Hallett, & K.A. Hirsch, *Torsion-free groups having finite automorphism groups,* J. Algebra 2(1965), 287-298.
[10] G. Higman, *The units of group rings,* Proc. London Math. Soc. 46(1940), 231-248.
[11] K.A. Hirsch, & H. Zassenhaus, *Finite automorphism groups of torsion-free groups,* J. London Math. Soc. 41(1966), 545-549.
[12] E. Jespers, M.M. Parmenter, & P.F. Smith, *Revisiting a theorem of Higman,* in: *"Groups '93 Galway/St Andrews"*, vol. 1, London Math. Soc. Lecture Note 211, Cambridge University Press, Cambridge 1995, pp. 269-273.
[13] G. Karpilovsky, *"Unit groups of group rings"*, Longman, Essex 1989.
[14] J. Krempa, *Homomorphisms of group rings,* in: Banach Center Publication vol. 9, PWN Warsaw 1982, 233-255.
[15] J. Krempa, *On finite generation of unit group for group rings,* in: "Groups '93 Galway/St Andrews", vol. 2, London Math. Soc. Lecture Note 212, Cambridge University Press, Cambridge 1995, pp. 352-367.
[16] J. Krempa, *Rings with periodic unit groups,* in: "Abelian groups and modules", A. Facchini & C. Menini (eds.), Kluwer Academic Publishers, Dordrecht 1995, pp. 313-321.
[17] T.Y. Lam, *"A first course in noncommutative rings"*, Springer-Verlag, Berlin 1991.
[18] R.C. Lyndon & P.E. Schupp, *"Combinatorial group theory"*, Springer-Verlag, Berlin 1977.
[19] D.S. Passman, *Group algebras whose units satisfy a group identity II,* Proc. Amer. Math. Soc. (to appear).
[20] D.S. Passman, *"The algebraic structure of group rings"*, Wiley-Interscience Publications, New York 1977.
[21] C. Polcino Milies, *Group rings whose torsion units form a group II,* Comm. Algebra 9(1981), 699-712.
[22] C. Procesi, *"Rings with polynomial identities"*, Marcel Dekker Inc., New York 1973.
[23] D.J.S. Robinson, *"A course in the theory of groups,* Springer-Verlag, Berlin 1982.
[24] L.H. Rowen, *"Polynomial identities in ring theory"*, Academic Press, New York 1980.
[25] S.K. Sehgal, *"Topics in group rings"*, Marcel Dekker Inc., New York 1978.
[26] S.K. Sehgal, *"Units in integral group rings"*, Longman, Essex 1993.
[27] B.A.F. Wehrfritz, *"Infinite linear groups"*, Springer-Verlag, Berlin 1973.

SOME FREE-BY-CYCLIC GROUPS

IAN J. LEARY*[1], GRAHAM A. NIBLO*[2]and DANIEL T. WISE†[3]

*Faculty of Mathematical Studies, University of Southampton, Southampton SO17 1BJ, England

†Department of Mathematics, Cornell University, Ithaca, NY 14853, U.S.A.

A group is said to be locally free if every finitely generated subgroup of it is free. One example is the additive group of the rationals. We exhibit a finitely generated group G that is free-by-cyclic and contains a non-free, locally free subgroup. The smallest such example that we have found is of the form $G \cong F_n \rtimes \mathbb{Z}$ for $n = 3$. We also construct word-hyperbolic examples for larger values of n, and show that the groups are not subgroup separable. We used Bestvina and Brady's 'Morse theory for cube complexes' in the construction of these groups. The authors thank Jim Anderson, who posed a question concerning 3-manifolds that led to these examples, and the referee, whose comments were very helpful. This work was started at a conference at Southampton, immediately before Groups St Andrews, which was funded by EPSRC visitor grants GR/L06928 and GR/L31135, and by a grant from the LMS.

Throughout this note, F_n denotes a free group of rank n, $\bar{x}$ denotes x^{-1}, and $x^y = \bar{y}xy$.

Proposition 1 *The group G given by the presentation*

$$G = \langle a, b, t : a^t = b,\ b^t = a b \bar{a} \rangle$$

contains a non-free, locally free subgroup, and is isomorphic to a split extension $F_3 \rtimes \mathbb{Z}$.

Proof The given presentation expresses G as an ascending HNN extension, with base group freely generated by a and b and with stable letter t. Define $\phi: G \to \mathbb{Z}$ by $\phi(t) = 1$, $\phi(a) = \phi(b) = 0$, and let K be the kernel of ϕ. Then K is a strictly ascending union of 2-generator free groups

$$\langle a, b \rangle \subseteq \langle a, b \rangle^{\bar{t}} \subseteq \langle a, b \rangle^{\bar{t}^2} \subseteq \cdots \subseteq K.$$

Any such group is locally free and not free, since it is not finitely generated and the rank of its abelianization is at most two. (In fact, the abelianization of K is infinite cyclic.)

It remains to show that G is free-by-cyclic. Define $\psi: G \to \mathbb{Z}$ by $\psi(t) = \psi(a) = \psi(b) = 1$. It will be shown that the kernel of ψ is free of rank 3. A presentation 2-complex Y for G may be constructed by attaching two 2-cells to a rose with edges a, b, and t according to the maps given in Figure 1. Since Y is obtained from the

[1]Partially supported by EPSRC grant no. GR/L69398

[2]Partially supported by EPSRC grant no. GR/K25618

[3]Supported by NSF grant no. DMS-9627506

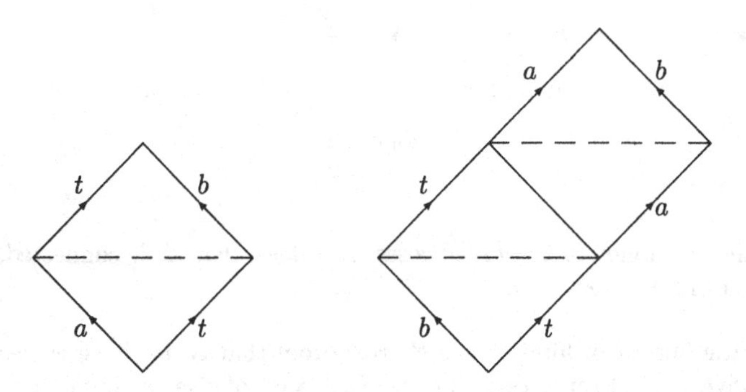

Figure 1.

presentation of G as an HNN extension with free base group, it follows that Y is an Eilenberg-MacLane space for G (see proposition 3.6 of [3]). Represent the three 1-cells of Y as unit intervals, and represent the two 2-cells of Y as a unit square and a 2×1 rectangle, as indicated in Figure 1. This makes Y into an affine cell complex in the sense of Bestvina and Brady ([1], Def. 2.1). Now take $S^1 = \mathbb{R}/\mathbb{Z}$, viewed as a cell complex with one vertex and one edge of length 1, as an Eilenberg-MacLane space for the integers. A cellular map $g: Y \to S^1$ may be defined that induces the homomorphism $\psi: G \to \mathbb{Z}$ on fundamental groups and is affine on each cell. In Figure 1 this map is represented by 'height modulo one', where the length of each edge is chosen so that its height is one. The inverse image of the vertex v of S^1 is a rose consisting of one vertex and three 1-cells (the dotted lines on figure 1). Now let X be the cover of Y corresponding to the subgroup $H = \ker(\psi)$. The map g lifts to a map $f: X \to \mathbb{R}$. X is an affine cell complex, and f is a Morse function in the sense of [1], Def. 2.2. By construction, X is an Eilenberg-MacLane space for H, and for any integer t, $X_t = f^{-1}(t)$ consists of a disjoint union of copies of a 3-petalled rose. (Clearly, X_t is a disjoint union of connected covers of $g^{-1}(v)$, but since any loop in $g^{-1}(v)$ represents an element of $G = \pi_1(Y)$ in the kernel of ψ, every lift in X_t of such a loop is itself a loop, and hence X_t is a disjoint union of 1-fold covers.)

Bestvina and Brady's Morse theory allows one to compare X and X_t: by Lemma 2.5 of [1], a space homotopy equivalent to X may be obtained from X_t by coning off a subspace homeomorphic to a copy of the descending link (resp. ascending link) at v for each vertex v of X such that $f(v) > t$ (resp. $f(v) < t$). (Ascending and descending links are defined in Section 2 of [1].) All vertices of X have isomorphic links, since Y has only one vertex, and the ascending and descending links at each vertex are as shown in Figure 2. Both the ascending and descending link are contractible. Since coning off a contractible subspace does not change the homotopy type of a space, it follows that X is homotopy equivalent to X_t. But it is already known that X is an Eilenberg-MacLane space for H, and that

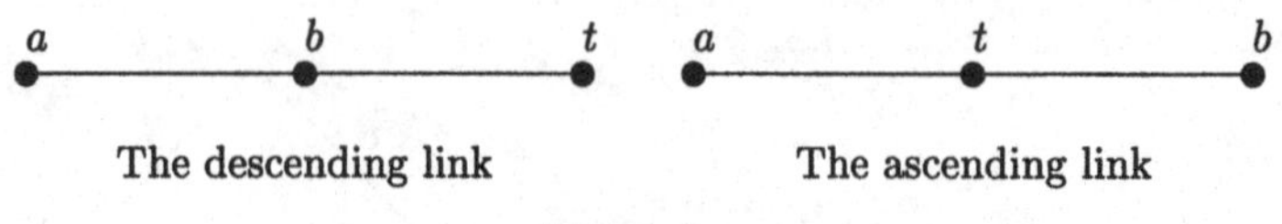

Figure 2.

X_t is a disjoint union of 3-petalled roses. It follows that X_t is connected, and that H is free of rank three. □

With the benefit of hindsight, a shorter proof that G as above is free-by-cyclic may be given—see Proposition 2 below. Such a proof gives no indication as to how G was discovered however. Moreover, the techniques of Proposition 1 generalize easily to more complicated presentations such as those given in Proposition 3.

Proposition 2 *Let H' be freely generated by x, y and z, and define an automorphism θ of H' by*

$$\theta(x) = y, \quad \theta(y) = z, \quad \theta(z) = y^2\bar{x}.$$

The group G of Proposition 1 is isomorphic to $H' \rtimes \langle t \rangle$, where the conjugation action of t on H' is given by θ.

Proof First, check that the endomorphism θ is an automorphism of H' by exhibiting an inverse:

$$\theta^{-1}(z) = y, \quad \theta^{-1}(y) = x, \quad \theta^{-1}(x) = \bar{z}x^2.$$

Now, eliminate b from the given presentation for G to obtain

$$G = \langle a, t : \bar{t}\bar{t}attat\bar{a}\bar{t}\bar{a} \rangle.$$

Substitute $a = xt$, and eliminate a, obtaining

$$G = \langle x, t : \bar{t}\bar{t}(xt)tt(xt)\bar{t}(\bar{t}\bar{x})t(\bar{t}\bar{x}) \rangle = \langle x, t : \bar{t}^2xt^3x\bar{t}\bar{x}^2 \rangle.$$

Add new generators $y = x^t$ and $z = x^{t^2}$, obtaining

$$\begin{aligned} G &= \langle x, y, z, t : x^t = y,\ y^t = z,\ ztx\bar{y}^2\bar{t} \rangle \\ &= \langle x, y, z, t : x^t = y,\ y^t = z,\ z^t = y^2\bar{x} \rangle. \end{aligned}$$

Thus G is seen to be isomorphic to $H' \rtimes \langle t \rangle$ as claimed. □

Next, we show how to construct a word-hyperbolic group having similar properties to the group G.

Proposition 3 *For $s \geq 3$ define a word $W(x, y)$ by*

$$W(x, y) = xy^4xy^5x \cdots xy^{4+s}x.$$

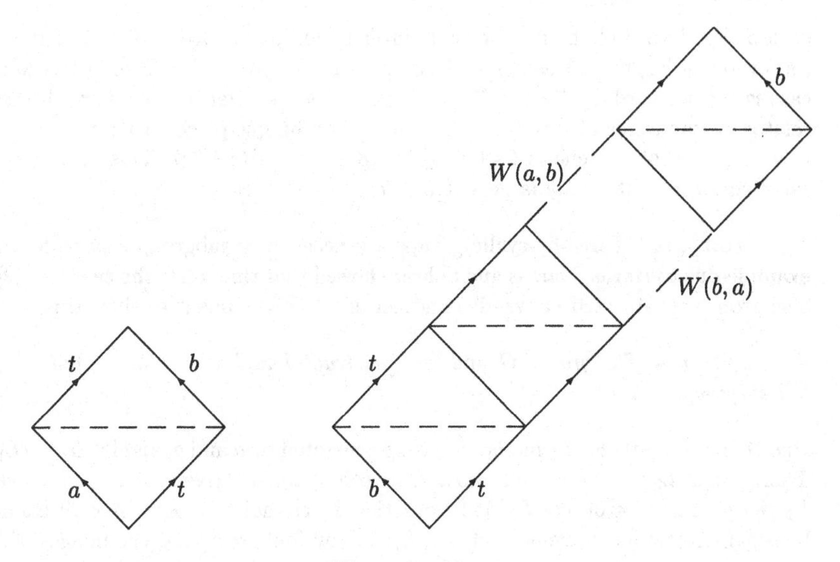

Figure 3.

The group G_s with presentation

$$G_s = \langle a, b, t : a^t = b,\ b^t = W(b,a)b(W(a,b))^{-1}\rangle$$

is free-by-cyclic and contains a non-free, locally free subgroup. For s sufficiently large, G_s is word-hyperbolic.

Proof As in Proposition 1, G_s is a strictly ascending HNN-extension with base group freely generated by a and b, so contains a non-free, locally free subgroup. As in Proposition 1, an Eilenberg-MacLane space for G_s with an affine cell structure can be made by attaching a unit square and an $m \times 1$ rectangle to a rose with three petals of length 1. (Here m is one more than the length of the word W, as shown in figure 3.) The argument given in Proposition 1 shows that G_s is expressible as $F_n \rtimes \mathbb{Z}$, where n is the total area of the two 2-cells in figure 3, i.e., $n = m+1 = s+4+(8+s)(s+1)/2$. It remains to show that G_s is word-hyperbolic for s sufficiently large. For this, it suffices to show that some presentation for G_s satisfies the $C'(1/7)$ small cancellation condition (see [4]). Eliminate b from the presentation for G_s. This leaves a 1-relator group, with relator

$$btab^4ab^5a\cdots ab^{4+s}a\bar{b}^2\bar{a}^{4+s}\bar{b}\cdots\bar{b}\bar{a}^5\bar{b}\bar{a}^4\bar{b}\bar{t}$$

$$= \bar{t}at^2a(\bar{t}a^4ta)(\bar{t}a^5ta)\cdots(\bar{t}a^{4+s}ta)\bar{t}\bar{a}^2(t\bar{a}^{4+s}\bar{t}\bar{a})(t\bar{a}^{3+s}\bar{t}\bar{a})\cdots(t\bar{a}^5\bar{t}\bar{a})(t\bar{a}^4\bar{t}\bar{a}).$$

The total length of this relator as a cyclic word in a and t may be seen to be $8+(14+s)(s+1)$. (The bracketing of the word is intended to facilitate this check.) For any $4 \le r \le 4+s$, the four words $(\bar{t}a^rt)^{\pm1}$ and $(ta^r\bar{t})^{\pm1}$ occur exactly once each as a subword of the relator or its inverse, and any subword of the relator or its inverse of length at least $2s + 15$ contains a subword of this form. (The worst case is the subword $a^{4+s}ta\bar{t}\bar{a}^2t\bar{a}^{4+s}$, of length $2s + 14$.) Hence any subword of the relator or its inverse of length $2s + 15$ occurs in a unique place. It follows that G_s is word-hyperbolic whenever $2s+15 \leq 1/7\,(8 + (14 + s)(s + 1))$. This inequality is satisfied for all sufficiently large s. (In fact, $s \geq 9$ suffices.) □

P. Scott asked if free-by-cyclic groups are necessarily subgroup separable. An example due to Burns, Karass and Solitar showed that this is not the case (see [2]). The groups constructed above give another, simpler, argument to show this.

Proposition 4 *The groups G and G_s, constructed in Propositions 1 and 3, are not subgroup separable.*

Proof In each case, let L_1 be the subgroup generated by a and b, and let $L_2 = tL_1\bar{t}$. Then L_1 and L_2 are free of rank two and are conjugate in G (resp. in G_s). Moreover, L_1 is a proper subgroup of L_2. It follows that L_1 cannot be closed, since it cannot be separated from any element of $L_2 \setminus L_1$: in any finite quotient, the images of L_1 and L_2 have the same order, since they are conjugate, and so must be equal since the image of L_1 is a subgroup of the image of L_2. □

References

[1] M. Bestvina and N. Brady, Morse theory and finiteness properties of groups, to appear in *Inventiones Math.*

[2] R. G. Burns, A. Karrass,and D. Solitar, A note on groups with separable finitely generated subgroups, *Bull. Austral. Math. Soc.* 36 (1987), 153–160.

[3] P. Scott and C. T. C. Wall, Topological methods in group theory, in *Homological Group Theory*, London Math. Soc. Lecture Notes 36 (ed. by C. T. C. Wall), Cambridge Univ. Press, Cambridge 1979.

[4] R. Strebel, Small cancellation groups, Appendix to: *Sur les groupes hyperboliques d'aprés Mikhael Gromov*, 227–273, Progr. Math., 83, Birkhäuser Boston, Boston, MA, 1990.

THE RESIDUALLY WEAKLY PRIMITIVE GEOMETRIES OF THE SUZUKI SIMPLE GROUP $Sz(8)$

DIMITRI LEEMANS[1]

Université Libre de Bruxelles, Département de Mathématiques - C.P.216, Boulevard du Triomphe, B-1050 Bruxelles, Belgium

Abstract

We determine all firm and residually connected geometries on which the group $Sz(8)$ acts flag-transitively and fulfills the primitivity condition RWPRI, requiring that the stabilizer of each flag $\mathcal{F}$ acts primitively on the elements of some type in the residue $\Gamma_{\mathcal{F}}$. This work was the starting point of a more ambitious work: the classification of all geometries of a Suzuki simple group $Sz(q)$. The case $q = 8$ which is solved here, is the smallest case and the only one that is currently possible to analyse completely using the computer algebra package MAGMA. The rank 2 case was classified for all q (see Theorem 7.1 in [17]). The results obtained here rely partially on computer algebra.

1 Introduction

The present paper gives a complete classification of the firm and residually connected geometries on which the group $Sz(8)$ acts flag-transitively and residually weakly primitively (see Section 2 for the definitions).

This work continues a systematic investigation of groups that has started some years ago (see [6, 8, 7, 12, 13, 14, 15, 16]).

Here we study the smallest Suzuki simple group, namely $Sz(8)$. One of the reasons for this choice is that the infinite class of Suzuki groups $Sz(q)$, with $q = 2^{2e+1}$ and $e \geq 1$, looks particularly attractive for a general study and that a previous treatment of the smallest case may help in guessing the way to follow in general. Such a work is already accomplished for the rank 2 geometries (see [17]) and the ranks ≥ 3 are currently under investigation.

The paper is organized as follows. In Section 2, we recall definitions and notation for incidence geometry. In Section 3, we give the diagrams of the geometries obtained for $Sz(8)$ which have been determined with help of MAGMA [1].

At the end of the paper, we give the subgroup pattern of $Sz(8)$. This pattern has been computed by Michel Dehon in August 1994, using the help of the computer algebra package CAYLEY [9, 10].

For an introduction on the Suzuki groups, we refer to [18]. Facts about $Sz(8)$ can also be found in the Atlas of Finite Groups [11]. For a historical survey of the Suzuki groups, we refer to [5].

---

[1]We gratefully acknowledge support from the British Council and the "Fonds National de la Recherche Scientifique de Belgique".

2 Definitions and notation

The basic concepts about geometries constructed from a group and some of its subgroups are due to Tits [19] (see also [4], Chapter 3).

Let G be a group together with a finite family of subgroups $(G_i)_{i\in I}$. We define the *pre-geometry* $\Gamma = \Gamma(G, (G_i)_{i\in I})$ as follows. The set X of *elements* of Γ consists of all cosets gG_i, $g \in G$, $i \in I$. We define an *incidence relation* * on X by :

$$g_1G_i * g_2G_j \text{ iff } g_1G_i \cap g_2G_j \text{ is non-empty in } G.$$

The *type function* t on Γ is defined by $t(gG_i) = i$. The *type* of a subset Y of X is the set $t(Y)$; its *rank* is the cardinality of $t(Y)$ and we call $\mid t(X) \mid$ the *rank* of Γ. The *Borel subgroup* of the pre-geometry is the subgroup $B = \cap_{i\in I}G_i$. A *flag* is a set of pairwise incident elements of X and a *chamber* of Γ is a flag of type I. An element of type i is also called an *i-element.*

The group G acts on Γ as an automorphism group, by left translation, preserving the type of each element.

As in [12], we call Γ a *geometry* provided that every flag of Γ is contained in some chamber and we call Γ *flag-transitive* (FT) provided that G acts transitively on all chambers of Γ, hence also on all flags of any type J, where J is a subset of I. Assuming that Γ is a flag-transitive geometry and that F is a flag of Γ, the *residue* of F is the pre-geometry

$$\Gamma_F = \Gamma(\cap_{j\in t(F)}G_j, (G_i \cap (\cap_{j\in t(F)}G_j))_{i\in I\setminus t(F)})$$

and we readily see that Γ_F is a flag-transitive geometry.

Let J be a subset of I. The *J-truncation* of Γ is the geometry consisting of the elements of type $j \in J$, together with the restricted type-function and induced incidence relation. In group-geometry terms, the J-truncation of $\Gamma(G, (G_i)_{i\in I})$ is the geometry $\Gamma(G, (G_j)_{j\in J})$.

We call Γ *firm* (F) (resp. *thick, thin*) provided that every flag of rank $\mid I \mid -1$ is contained in at least two (resp. three, exactly two) chambers. We call Γ *residually connected* (RC) provided that the incidence graph of each residue of rank ≥ 2 is a connected graph. We call Γ *primitive* (PRI) provided that G acts primitively on the set of i-elements of Γ, for each $i \in I$.

As in [8], we call Γ *residually primitive* (RPRI) if each residue Γ_F of a flag F is primitive for the group induced on Γ_F by the stabilizer G_F of F.

We call Γ *weakly primitive* (WPRI) provided there exists some $i \in I$ such that G acts primitively on the set of i-elements of Γ and we call Γ *residually weakly primitive* (RWPRI) provided that each residue Γ_F of a flag F is weakly primitive for the group induced on Γ_F by the stabilizer G_F of F.

A subgroup H of a group G is called *quasi-maximal* if there exists a unique chain of intermediate subgroups from G to H.

We call Γ *quasi-primitive* (QPRI) provided that each subgroup G_i forming the geometry is a quasi-maximal subgroup of G.

If Γ is a geometry of rank 2 with $I = \{0, 1\}$ such that each of its 0-elements is incident with each of its 1-elements, then we call Γ a *generalized digon.*

We say that Γ satisfies the *rank two intersection property* $(IP)_2$ if in every rank 2 residue of Γ other than a generalized digon, any two elements of the same type are incident with at most one element of the other type.

We call Γ *locally 2-transitive* and we write $(2T)_1$ for this, provided that the stabilizer G_F of any flag F of rank $|I|-1$ acts 2-transitively on the residue Γ_F.

Following [2] and [3], the *diagram* of a firm, RC, FT geometry Γ is a graph together with additional structure, whose vertices are the elements of I, which is further described as follows. To each vertex $i \in I$, we attach the *order* s_i which is $|\Gamma_F|-1$, where F is any flag of type $I\backslash\{i\}$, the *number* n_i *of varieties of type* i, which is the index of G_i in G, and the subgroup G_i. Elements i, j of I are not joined by an edge of the diagram provided that a residue Γ_F of type $\{i,j\}$ is a generalized digon. Otherwise, i and j are joined by an edge endowed with three positive integers d_{ij}, g_{ij}, d_{ji} where g_{ij} (the *gonality*) is equal to half the girth of the incidence graph of a residue Γ_F of type $\{i,j\}$ and d_{ij} (resp. d_{ji}), the *i-diameter* (resp. *j-diameter*) is the greatest distance from some fixed i-element (resp. j-element) to any other element in the incidence graph of Γ_F.

On a picture of the diagram, this structure will often be depicted as follows.

$$\overset{d_{ij}g_{ij}d_{ji}}{\underset{\substack{s_i\\ n_i\\ G_i}}{\circ}\!\!-\!\!-\!\!-\!\!-\!\!\underset{\substack{s_j\\ n_j\\ G_j}}{\circ}}$$

If $g_{ij} = d_{ij} = d_{ji} = n$, then Γ_F is called a *generalized n-gon* and on a picture, we do not write d_{ij} and d_{ji}. The ordered pairs (Γ,G) and (Γ',G) are *isomorphic* (resp. *conjugate*) if there exists an automorphism (resp. internal automorphism) of G mapping Γ onto Γ'. The group *Cor*(Γ,G) (resp. *Aut*(Γ,G)) is the group of automorphisms (resp. type-preserving automorphisms) of the pair (Γ,G).

As to notation for groups, we follow the conventions of the Atlas [11] up to slight variations. The symbol ":" stands for split extensions, the "hat" symbol "$\hat{\cdot}$" stands for non split extensions and the symbol $\times$ stands for direct products.

3 The geometries of the Suzuki group $Sz(8)$

In this section we mention all the F, RC, FT, RWPRI geometries that can be constructed from *Sz*(8).

Besides the numbering of each geometry, there is a number between parenthesis giving the number of non-conjugate geometries that are fused under the action of *Aut*(G). We mention when a geometry satisfies the conditions PRI, QPRI, RPRI, $(2T)_1$. We also give diagrams of the geometries that do not satisfy the intersection property $(IP)_2$ as there are only 5 such geometries.

Up to isomorphism, there are 29 geometries of rank 2 and 151 of rank 3. There is no geometry of rank ≥ 4.

3.1 Rank 2 geometries

The rank 2 geometries have been determined using the classification theorem stated in [17]. Another way to obtain these results is the following. We look at the subgroup pattern of $Sz(8)$. We choose a subgroup G_0 maximal in G. Then we find all its maximal subgroups. Finally, all the subgroups containing a maximal subgroup of G_0 as a maximal subgroup are taken as G_1 subgroups. These couples of subgroups give us the rank 2 geometries of G. Then we still have to compute how many of them are pairwise non-isomorphic, which is a quite long (but not hard) task. And when we have all the geometries, we can compute their diagrams using the computer algebra package MAGMA. Table 1 gives for each rank 2 geometry Γ, the rank 3 geometries containing it as a truncation, $Aut(\Gamma, G)$ and $Cor(\Gamma, G)$. The first entry is the number of the rank 2 geometry that is analysed. The second entry is a list of numbers corresponding to the rank 3 geometries that contain the corresponding rank 2 geometry as a truncation. When G_0 is not isomorphic to G_1, it is not possible to find non-type preserving automorphisms. Thus in these cases, we write nothing in the $Cor(\Gamma, G)$ column.

Theorem 3.1 *Up to isomorphism, there are* 29 *firm, residually connected geometries of rank* 2 *on which* $Sz(8)$ *acts flag-transitively and residually weakly primitively. Their diagrams are given in Figure 1.*

Proof This is a direct application of the theorem stated in [17]. □

3.2 Rank 3 geometries

For the rank 3 case, we use the fact that all the rank 2 geometries of the subgroups of $Sz(8)$ can easily be constructed using MAGMA.

Theorem 3.2 *Up to isomorphism, there are* 151 *firm, residually connected geometries of rank* 3 *on which* $Sz(8)$ *acts flag-transitively and residually weakly primitively. Their diagrams are given below.*

Sketch of the proof We do not give a complete proof here since it is quite lengthy. The idea consists in looking at the rank 2 geometries of the maximal subgroups of $Sz(8)$ which are respectively $2^{3+3}:7$, $13:4$, $5:4$, $D_{14} \cong 7:2$ (see picture of the subgroup pattern at the end of the paper).

It is easy to see that if we take $2^{3+3}:7$ as one of the subgroups, say G_0, forming the geometry, we are not able to construct a geometry satisfying the RWPRI condition.

If we take $13:4$ (resp. $5:4$) as G_0, then there must be one of the G_{0i} isomorphic to $13:2 \cong D_{26}$ (resp. $5:2 \cong D_{10}$). By looking at the subgroup pattern we readily see that there is no possibility to find a subgroup G_i that does not contradict the RWPRI condition.

Thus the only maximal subgroup that can be taken as G_0 is $7:2 \cong D_{14}$. And the geometries of rank two of this group are known (see [8]). We then start from these geometries to construct configurations satisfying the RWPRI condition. And then, by using MAGMA, we test the firmness, the residual connectedness and the flag-transitivity on our configurations. The diagrams are computed with MAGMA while they could be computed by hand. □

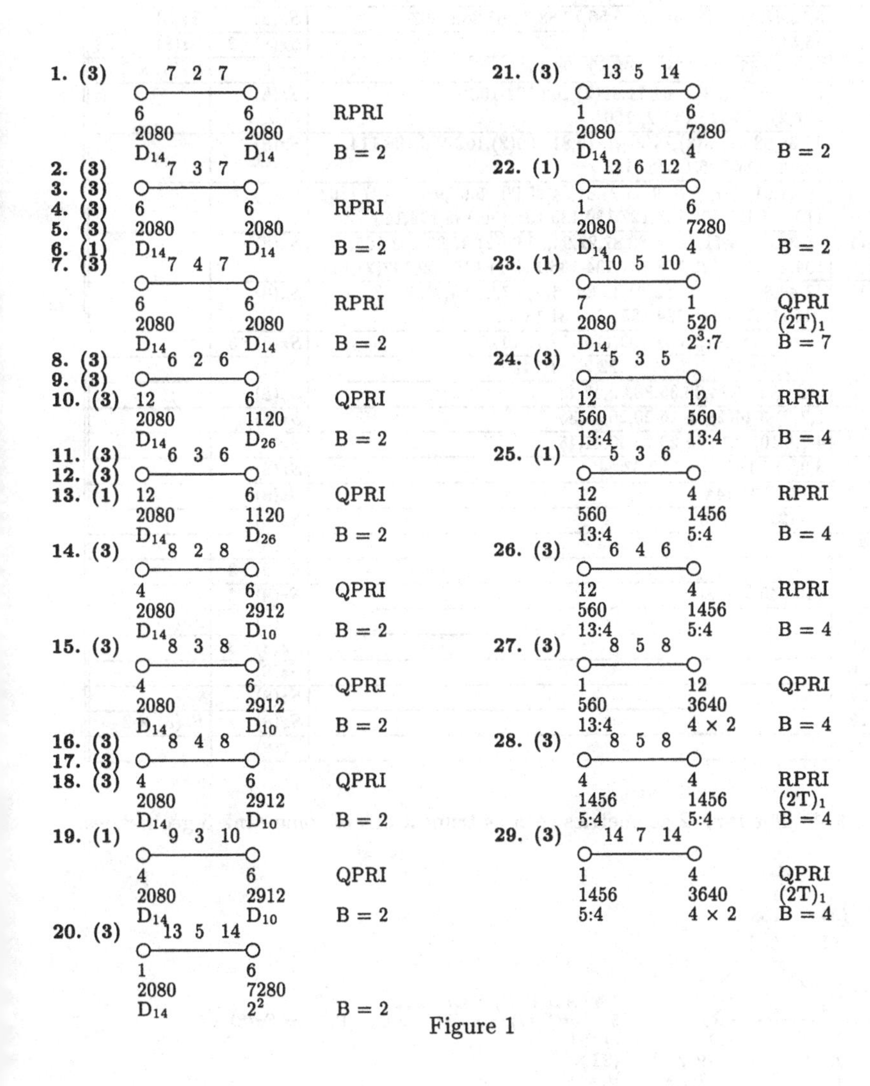
1. (3)
7 2 7
6 6 RPRI
2080 2080
D14 D14 B = 2
2. (3)
3. (3)
4. (3)
5. (3)
6. (1)
7. (3)
7 3 7
6 6 RPRI
2080 2080
D14 D14 B = 2
7 4 7
6 6 RPRI
2080 2080
D14 D14 B = 2
8. (3)
9. (3)
10. (3)
6 2 6
12 6 QPRI
2080 1120
D14 D26 B = 2
11. (3)
12. (3)
13. (1)
6 3 6
12 6 QPRI
2080 1120
D14 D26 B = 2
14. (3)
8 2 8
4 6 QPRI
2080 2912
D14 D10 B = 2
15. (3)
8 3 8
4 6 QPRI
2080 2912
D14 D10 B = 2
16. (3)
17. (3)
18. (3)
8 4 8
4 6 QPRI
2080 2912
D14 D10 B = 2
19. (1)
9 3 10
4 6 QPRI
2080 2912
D14 D10 B = 2
20. (3)
13 5 14
1 6
2080 7280
D14 2^2 B = 2
21. (3)
13 5 14
1 6
2080 7280
D14 4 B = 2
22. (1)
12 6 12
1 6
2080 7280
D14 4 B = 2
23. (1)
10 5 10
7 1 QPRI
2080 520 (2T)1
D14 2^3:7 B = 7
24. (3)
5 3 5
12 12 RPRI
560 560
13:4 13:4 B = 4
25. (1)
5 3 6
12 4 RPRI
560 1456
13:4 5:4 B = 4
26. (3)
6 4 6
12 4 RPRI
560 1456
13:4 5:4 B = 4
27. (3)
8 5 8
1 12 QPRI
560 3640
13:4 4 × 2 B = 4
28. (3)
8 5 8
4 4 RPRI
1456 1456 (2T)1
5:4 5:4 B = 4
29. (3)
14 7 14
1 4 QPRI
1456 3640 (2T)1
5:4 4 × 2 B = 4

Figure 1

No.	Rank 3 geometries containing it as a truncation	$Aut(\Gamma, G)$	$Cor(\Gamma, G)$
1	10,11,18,48,52,54,55,57(2),58(2),61,63,68,69,74,81,84,86,146	$Sz(8)$	$Sz(8)$
2	13-15,54,55,56(2),85,94	$Sz(8)$	$Sz(8) \times 2$
3	6,46-50,59(3),60,62,64,72,73,75,80,82,83,87,89	$Sz(8)$	$Sz(8)$
4	7,17,46,47,50(2),51,52(2),53,65,66,71,76,78,79,91,98,143	$Sz(8)$	$Sz(8)$
5	8,12,47,48,49(2),51,53,55,56,77,88,90,92,93,95,97	$Sz(8)$	$Sz(8)$
6	46,51,67	$Sz(8):3$	$Sz(8):3 \times 2$
7	9,16,19,53,54,57,58,70,96,99,100,101	$Sz(8)$	$Sz(8) \times 2$
8	20-27,60,62,63(2),64-71,99(2),102,104-106, 107(2),108-113,140(2),150	$Sz(8)$	
9	35-38,68,72,73(2),74(2),75-79,81,100(2),102,106,108,114, 116(2),117-126,141(2),148	$Sz(8)$	
10	39-41,61,67,75,77-79,80(2),81,84,85,90,95,97,98,101(2),110, 113,115,121,122,126,127,130,133,135,136,138,139,149	$Sz(8)$	
11	28-34,65,69,71,72,76,83,87,92,94,95,96(2),97,98,103,105, 109,111,119,123,128,131,134-136,137(2),138,139,142(2),145	$Sz(8)$	
12	42-44,60,64,66,70,82(2),83-85,86(2),87,88,89(2),90-93, 104,112,114,118,124,125,127-131,132(2),133,134	$Sz(8)$	
13	45,61,62,88,91,93,94,103,115,117,120,129	$Sz(8):3$	
14	4,5(2),11,17,18(2),19(2),22,25,31,147	$Sz(8)$	
15	6(2),7,8,21,24,33,35,38,39,42,43	$Sz(8)$	
16	1,2(2),3,14(2),15,16,20,26,32,36	$Sz(8)$	
17	4,7,8,10,12,15,16,28,29,40,41,45	$Sz(8)$	
18	3,9(2),10-13,23,27,30,37,44	$Sz(8)$	
19	1,13,17,34,144	$Sz(8):3$	
20	146(2),147-150	$Sz(8)$	
21		$Sz(8)$	
22		$Sz(8):3$	
23	143-145,151(2)	$Sz(8):3$	
24		$Sz(8)$	$Sz(8) \times 2$
25		$Sz(8):3$	
26		$Sz(8)$	
27		$Sz(8)$	
28		$Sz(8)$	$Sz(8) \times 2$
29		$Sz(8)$	

Table 1. The rank 2 geometries seen as truncations of some rank 3 geometries.

1. to
5. (3)

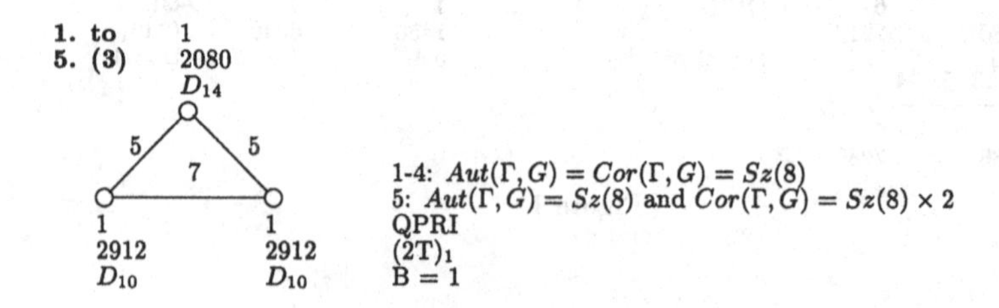

1-4: $Aut(\Gamma, G) = Cor(\Gamma, G) = Sz(8)$
5: $Aut(\Gamma, G) = Sz(8)$ and $Cor(\Gamma, G) = Sz(8) \times 2$
QPRI
$(2T)_1$
B = 1

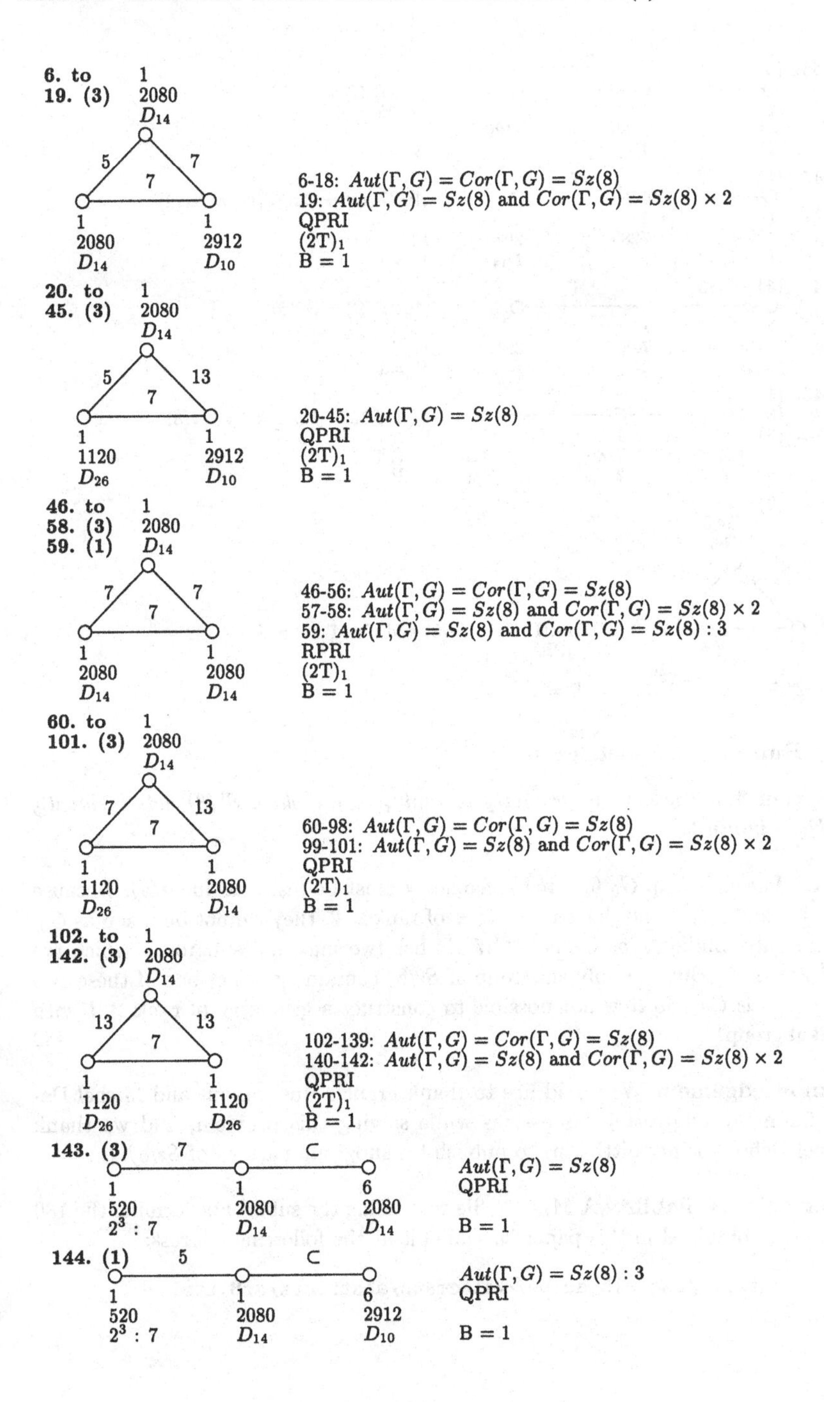

6. to
19. (3)

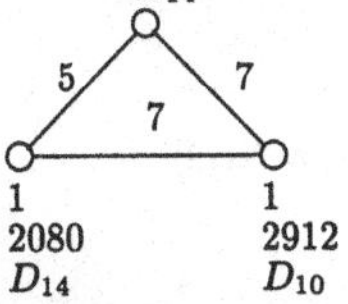
1
2080
D_{14}
5
7
7
1
2080
D_{14}
1
2912
D_{10}

6-18: $Aut(\Gamma, G) = Cor(\Gamma, G) = Sz(8)$
19: $Aut(\Gamma, G) = Sz(8)$ and $Cor(\Gamma, G) = Sz(8) \times 2$
QPRI
$(2T)_1$
B = 1

20. to
45. (3)

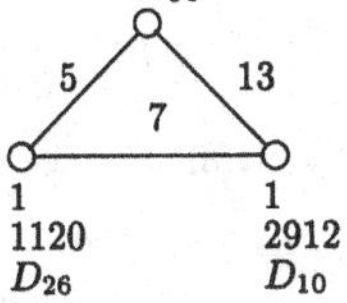
1
2080
D_{14}
5
13
7
1
1120
D_{26}
1
2912
D_{10}

20-45: $Aut(\Gamma, G) = Sz(8)$
QPRI
$(2T)_1$
B = 1

46. to
58. (3)
59. (1)

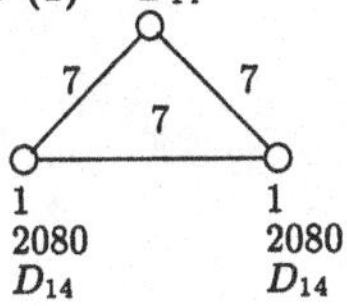
1
2080
D_{14}
7
7
7
1
2080
D_{14}
1
2080
D_{14}

46-56: $Aut(\Gamma, G) = Cor(\Gamma, G) = Sz(8)$
57-58: $Aut(\Gamma, G) = Sz(8)$ and $Cor(\Gamma, G) = Sz(8) \times 2$
59: $Aut(\Gamma, G) = Sz(8)$ and $Cor(\Gamma, G) = Sz(8) : 3$
RPRI
$(2T)_1$
B = 1

60. to
101. (3)

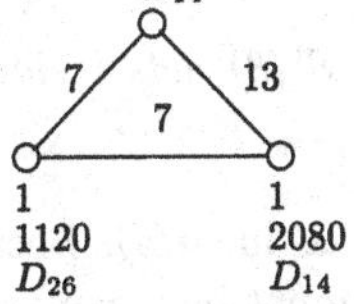
1
2080
D_{14}
7
13
7
1
1120
D_{26}
1
2080
D_{14}

60-98: $Aut(\Gamma, G) = Cor(\Gamma, G) = Sz(8)$
99-101: $Aut(\Gamma, G) = Sz(8)$ and $Cor(\Gamma, G) = Sz(8) \times 2$
QPRI
$(2T)_1$
B = 1

102. to
142. (3)

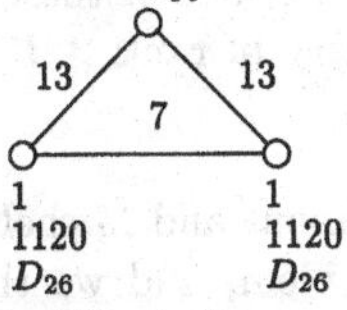
1
2080
D_{14}
13
13
7
1
1120
D_{26}
1
1120
D_{26}

102-139: $Aut(\Gamma, G) = Cor(\Gamma, G) = Sz(8)$
140-142: $Aut(\Gamma, G) = Sz(8)$ and $Cor(\Gamma, G) = Sz(8) \times 2$
QPRI
$(2T)_1$
B = 1

143. (3)

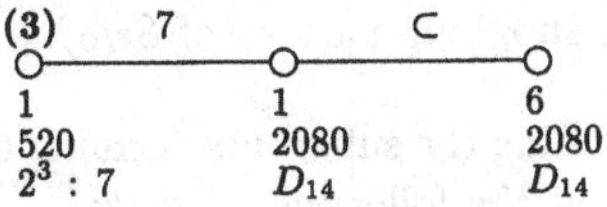
7
⊂
1
520
$2^3 : 7$
1
2080
D_{14}
6
2080
D_{14}

$Aut(\Gamma, G) = Sz(8)$
QPRI

B = 1

144. (1)

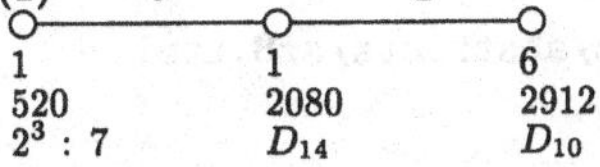
5
⊂
1
520
$2^3 : 7$
1
2080
D_{14}
6
2912
D_{10}

$Aut(\Gamma, G) = Sz(8) : 3$
QPRI

B = 1

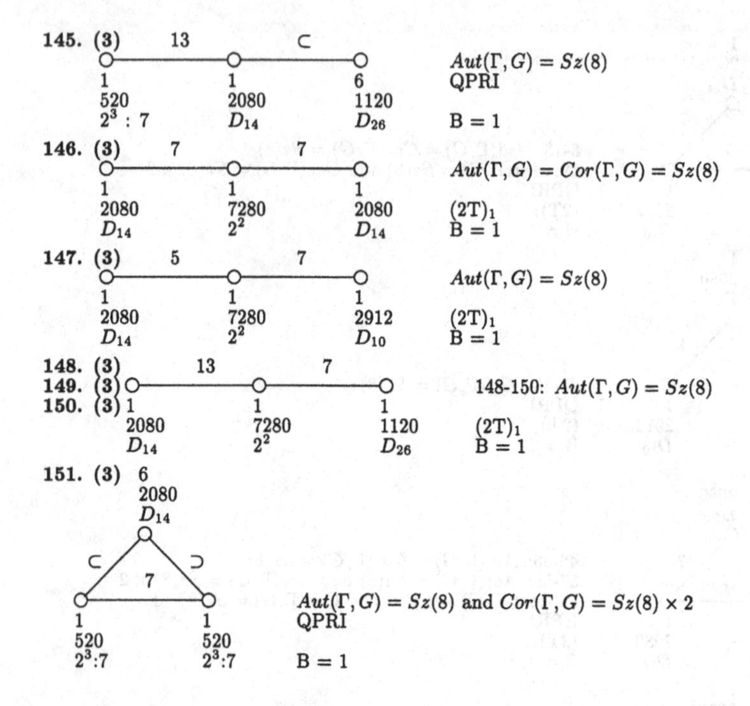

3.3 Rank $\geq$ 4 geometries

Theorem 3.3 *There is no geometry of rank ≥ 4 on which Sz(8) acts residually weakly primitively.*

Proof The subgroup G_0 forming a geometry must be maximal in *Sz(8)*. Because 13:4, 5:4 and D_{14} do not have geometries of rank > 2, they cannot be taken as G_0. So the only candidate as G_0 is 2^{3+3}:7. It has two maximal subgroups which are 2^{3+3} and 2^3:7. But the only subgroup of *Sz(8)* containing either one of these two subgroups is G_0. So it is not possible to construct a geometry of rank > 1 with this subgroup. □

Acknowledgement We would like to thank Francis Buekenhout and Michel Dehon for many interesting discussions while solving this problem, and we thank Michel Dehon for permitting us to publish his subgroup pattern of *Sz(8)*.

Electronic Availability A MAGMA file containing the subgroups forming the 180 geometries obtained in this paper, is available at the following address:

`http://cso.ulb.ac.be/~dleemans/abstracts/sz8.html`

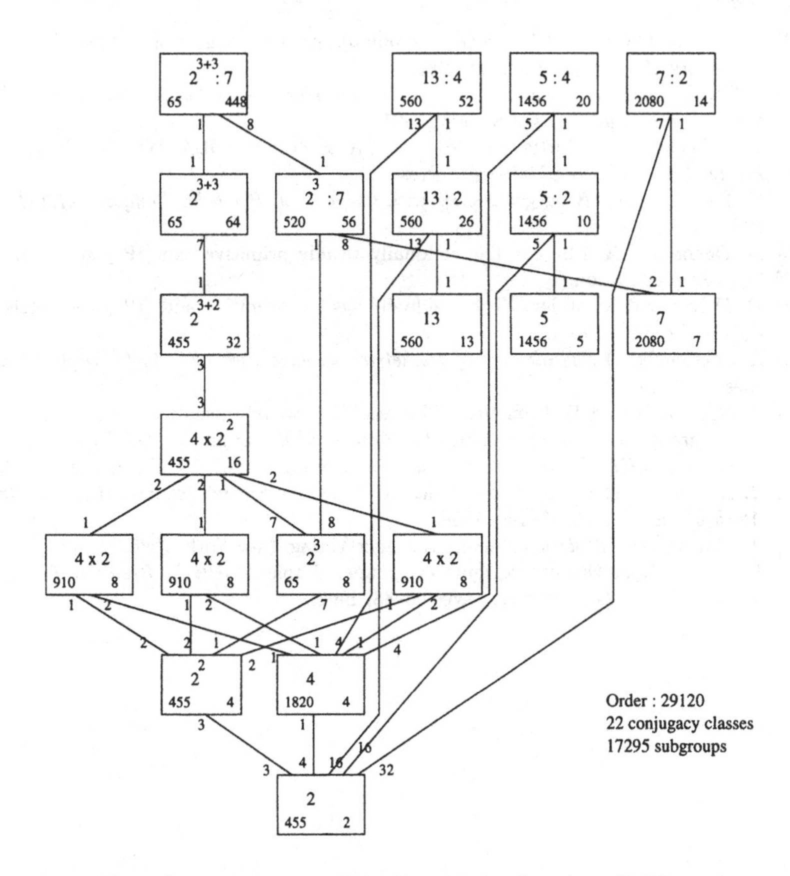

The subgroup pattern of the Suzuki simple group $Sz(8)$

References

[1] W. Bosma, J. Cannon, and C. Playoust. The Magma Algebra System I: the user language. *J. Symbolic Comput.*, (3/4):235–265, 1997.

[2] F. Buekenhout. Diagrams for geometries and groups. *J. Combin. Theory Ser. A*, 27:121–151, 1979.

[3] F. Buekenhout. (g, d, d*)-gons. In Johnson N.L., Kallaher M.J., and Long C.T., editors, *Finite Geometries*, pages 93–102, Marcel Dekker, New York, 1983.

[4] F. Buekenhout, editor. *Handbook of Incidence Geometry.* Elsevier, Amsterdam, 1995.

[5] F. Buekenhout. About the history of the Suzuki Groups. Preprint, 1996.

[6] F. Buekenhout, P. Cara, and M. Dehon. Geometries of small almost simple groups based on maximal subgroups. *Bull. Belg. Math. Soc. - Simon Stevin Suppl.*, 1998.

[7] F. Buekenhout, M. Dehon, and D. Leemans. All geometries of the Mathieu group M_{11} based on maximal subgroups. *Experiment. Math.*, 5:101–110, 1996.

[8] F. Buekenhout, M. Dehon, and D. Leemans. An Atlas of residually weakly primitive geometries for small groups. *Mém. Acad. Royale Belg., Classe des Sciences*, 1996. To appear.

[9] J. Cannon. CAYLEY: *A language for group theory.* Department of Pure Mathematics, University of Sydney, December 1982.

[10] J. Cannon and W. Bosma. CAYLEY: *quick reference guide.* Department of Pure Mathematics, University of Sydney, 1991.

[11] J.H. Conway, R.T. Curtis, S.P. Norton, R.A. Parker, and R.A. Wilson. *An Atlas of Finite Groups.* Oxford University Press, 1985.

[12] M. Dehon. Classifying geometries with Cayley. *J. Symbolic Comput.*, 17:259–276, 1994.

[13] M. Dehon and X. Miller. The residually weakly primitive and $(IP)_2$ geometries of M_{11}. In preparation.

[14] M. Dehon and X. Miller. The residually weakly primitive and $(IP)_2$ geometries of U(4,2). In preparation.

[15] H. Gottschalk. *A classification of geometries associated with PSL(3,4).* Diplomarbeit, Giessen, 1995.

[16] H. Gottschalk and D. Leemans. The residually weakly primitive geometries of the Janko group J_1. In Di Martino L., Kantor W.E., Lunardon G., Pasini A., and Tamburini M.C., editors, *Groups and Geometries*, pages 65–79. Birkhäuser, 1998.

[17] D. Leemans. The rank 2 geometries of the simple Suzuki groups Sz(q). *Beiträge Algebra Geom.*, 39(1):97–120, 1998.

[18] H. Lüneburg. *Translation Planes.* Springer-Verlag, New York, 1980.

[19] J. Tits. Géométries polyédriques et groupes simples. *Atti 2a Riunione Groupem. Math. Express. Lat. Firenze*, pages 66–88, 1962.

SEMIGROUP IDENTITIES AND ENGEL GROUPS

PATRIZIA LONGOBARDI* and MERCEDE MAJ†

*Dipartimento di Matematica e Applicazioni "R. Caccioppoli", Via Cintia, Monte S. Angelo, 80126 Napoli, Italy
†Dipartimento di Ingegneria dell'Informazione Matematica Applicata, via Salvator Allende, 84081 Baronissi, Salerno, Italy

Abstract

In this paper we review some results about groups that satisfy a nontrivial semigroup law, and, more generally, about groups with no free subsemigroups on two letters. We show that 4-Engel torsion-free groups satisfy a semigroup law, and we deduce that 4-Engel right-ordered groups are nilpotent. Moreover we show that 4-Engel group has no free subsemigroups on two letters.

1 Introduction

Let $F = F(x_1, x_2, \dots, x_n)$ be the free group on the letters $x_1, x_2, \dots, x_n$.

A word $u = u(x_1, x_2, \dots, x_n) \in F$ is called a *positive word* if it does not involve ${x_i}^{-1}$ for any $i \in \{1, 2, \dots, n\}$. If $u = u(x_1, x_2, \dots, x_n), \nu(x_1, x_2, \dots, x_n) \in F$, a law $u = \nu$ is a *semigroup law* if u, ν are positive words.

Obviously every group of finite exponent satisfies a nontrivial positive law, and every abelian group satisfies the positive law $xy = yx$.

Mal'cev proved in [4] (see also [5]) that every nilpotent group of class at most c satisfies the positive law $u_c(x, y) = \nu_c(x, y)$, where the words $u_n(x, y)$ and $\nu_n(x, y)$ are defined inductively by putting

$$u_0(x, y) = x, \nu_0(x, y) = y,$$
$$u_{i+1}(x, y) = u_i(x, y)\nu_i(x, y), \nu_{i+1}(x, y) = \nu_i(x, y)u_i(x, y).$$

Therefore a group which is a finite exponent extension of a nilpotent group satisfies a nontrivial semigroup law.

It is easy to prove that if a group G satisfies a nontrivial semigroup law, then G satisfies a nontrivial semigroup law in two letters. Hence groups that satisfy a nontrivial semigroup law belong to the class of groups which contain no free subsemigroups on two generators.

Groups with a nontrivial semigroup law and, more generally, groups with no free subsemigroups have been studied by many authors. By Mal'cev' result every group G which is a periodic extension of a locally nilpotent group has no free non-abelian subsemigroup. In [6] Rosenblatt proved that conversely a finitely generated soluble group with no free nonabelian subsemigroups is nilpotent-by-finite. The same result for finitely generated soluble groups that satisfy a nontrivial semigroup law had been proved in [2] by J.A. Lewin and T. Lewin. In [8] Shalev proved that every residually finite group that satisfies a nontrivial semigroup law is nilpotent-by-finite. More generally Y. Kim and A.H. Rhemtulla proved in [1] that if G is locally graded and there is a bound N such that for all ordered pairs of elements of G there is a relation

$$a^{r_1}b^{s_1}\dots a^{r_j}b^{s_j} = b^{m_1}a^{n_1}\dots b^{m_k}a^{m_k},$$

where r_i, s_i, m_i and n_i are all non-negative, r_1 and m_1 are positive integers and $r_1 + s_1 + \dots + r_j + s_j + m_1 + n_1 + \dots + m_k + n_k$ is at most N, then G is locally (nilpotent-by-finite). Notice that in [7] Ol'shanskii and Storozhev proved that a group with no free nonabelian subsemigroup need not be a periodic extension of a locally soluble group.

In this paper we are concerned with semigroup identities in some classes of generalized nilpotent groups. We consider the class of n-Engel groups. If n is a positive integer, a group G is said to be an *n-Engel group* if, for all a, b in G, $[a, {}_n b] = 1$, where $[a, {}_0 b] = a, [a, {}_1 b] = [a, b]$ and, by induction, for $i > 1$, $[a, {}_{i+1} b] = [[a, {}_i b], b]$.

Obviously 1-Engel groups are the abelian ones and therefore satisfy a nontrivial semigroup law. In [9] Shirshov proved that 2-Engel groups and 3-Engel groups satisfy a semigroup law. He proved that a group G is 2-Engel if and only if G satisfies the law $xy^2x = yx^2y$ (i.e. the identity $u_2(x,y) = \nu_2(x,y)$), and that G is a 3-Engel group if and only if G satisfies the laws (i) $u_3(x,y) = \nu_3(x,y)$ and (ii) $xy^2xyxyx^2y = yx^2y^2x^2y^2x$.

We study 4-Engel groups and we prove the following results:

Theorem A *Let G be a torsion-free 4-Engel group. Then G satisfies the positive identity $u_5(x,y) = \nu_5(x,y)$.*

Theorem B *Let G be a 4-Engel group. Then G has no free nonabelian subsemigroups.*

We have been unable to decide whether every torsion-free 4-Engel group is nilpotent. By a result of Traustason (see [10]) all 4-Engel groups are locally nilpotent if torsion-free 4-Engel groups are nilpotent and 4-Engel p-groups are locally finite for any prime p.

It would be also nice to get a characterization of torsion-free 4-Engel groups in terms of semigroups law, as Shirshov did in [9] for 2-Engel and 3-Engel groups.

Theorem A has application when G is a right ordered group. We recall that G is right orderable if there exists a total order relation $\geq$ on G such that for all a, b, g in G, $a \geq b$ implies $ag \geq bg$. A right ordered group is obviously torsion-free, hence, by Theorem A, a right ordered 4-Engel group G satisfies a nontrivial semigroup law. Then G is locally graded, by a result of [3], and G is locally nilpotent by a result of [1]. But a torsion-free locally nilpotent n-Engel group is nilpotent by a result of Zelmanov (see [11]). Hence we have:

Theorem C *A 4-Engel right ordered group is nilpotent.*

Arguing exactly as in the Proof of Corollary C we get that a right ordered n-Engel group is nilpotent if it satisfies a nontrivial semigroup law. Hence the well-known open question, whether every right ordered n-Engel group is nilpotent, may be reduced to the following:

Question Does every right ordered n-Engel group satisfy a non trivial semigroup law?

We would like to thank professor Akbar Rhemtulla for helpful discussion.

2 Proofs

We begin with a couple of lemmas that are probably well-known.

Lemma 1 *Let G be a torsion-free n-Engel group, $a, b \in G$. If $[a^s, b] = 1$, for some integer $s \geq 1$, then $[a, b] = 1$.*

Proof Let $i \geq 1$ be minimum such that $[b, {}_i a] = 1$, and assume $i > 1$. Then we have $[[b, {}_{i-1}a], a] = 1$ and $[[b, {}_{i-2}a], a^s] = 1$. Therefore $[[b, {}_{i-2}a], a]^s = 1$, and $[b, {}_{i-1}a] = 1$ since G is torsion-free, a contradiction. □

Lemma 2 *Let G be a torsion-free n-Engel group, $a, b \in G$. If $< a^s, b >$ is nilpotent of class k, $< a, b >$ is also nilpotent of class k.*

Proof From Lemma 1 it easily follows that

$$Z(< a^s, b >) = Z(< a, b >) \cap < a^s, b > .$$

Assume $k > 1$, and argue by induction on k.

The factor group $< a, b > /Z(< a, b >)$ is still a torsion-free n-Engel group, and $< a^s, b > Z(< a, b >)/Z(< a, b >) \simeq < a^s, b > /Z(< a^s, b >)$ is nilpotent of class $k-1$, hence, by induction, $< a, b > /Z(< a, b >)$ is nilpotent of class k - 1, and $< a, b >$ is nilpotent of class k. □

Now we assume that G is a 4-Engel group. First we have:

Lemma 3 *Let G be a 4-Engel group. If $x, y \in G$, and $a = x^{-1}yx$, then $< a, a^y >$ is nilpotent of class at most 2.*

Proof From $[x^{-1}, y, y, y, y] = 1$, we get $[[y, x]^{x^{-1}}, y, y, y] = 1$, and $[y, x, y^x, y^x, y^x] = 1$.

Then $1 = [y^{-1}y^x, y^x, y^x, y^x] = [y^{-1}, y^x, y^x, y^x]^{y^x}$ and $[y^{-1}, a, a, a] = 1$, *i.e.* $[[a, y]^{y^{-1}}, a, a] = 1$, and $1 = [a, y, a^y, a^y] = [a^{-1}a^y, a^y, a^y]$.

Hence $[a^{-1}, a^y, a^y] = 1$.

Arguing similarly with y^{-1} instead of y and a^{-1} instead of a, we get that $[a, (a^{-1})^{y^{-1}}]$ commutes with $a^{y^{-1}}$, *i.e.* $[a^y, a^{-1}]$ commutes with a. Hence $[a^y, a^{-1}] = [a, a^y]^{a^{-1}}$ commutes with a and a^y and $[a, a^y] \in Z(< a, a^y >)$, *i.e.* $< a, a^y >$ is nilpotent of class at most 2. □

From Lemmas 1, 2 and 3 it easily follows

Lemma 4 *Let G be a torsion free 4-Engel group. If a, y are conjugate elements of G, then the groups*

$$< a, a^y >, < y, y^a >, < y, y^{a^3} >, < a, a^{y^3} >, < a, a^{y^2} >$$

are all nilpotent of class ≤ 2.

Proof The nilpotence of $< a, a^y >$ and of $< y, y^a >$ follows from Lemma 3. Similarly $< y^3, (y^3)^{a^3} >$ is nilpotent of class ≤ 2, and, by Lemma 2, $< y, y^{a^3} >$ is nilpotent of class ≤ 2. We can argue similarly in the other cases. □

Now we are able to prove Theorem A.

Proof of Theorem A Let G be a torsion-free 4-Engel group. We will show that, if a, y are conjugate elements of G, then $< a, y >$ is nilpotent of class ≤ 4. From that it follows that, for any $g, h \in G, < gh, hg > = < gh, g^{-1}(gh)g >$ is nilpotent of class ≤ 4. Then $u_4(gh, hg) = \nu_4(gh, hg)$, by a result of Mal'cev (see [4]), and $u_5(g, h) = \nu_5(g, h)$, as required.

So assume that a, y are conjugate elements of G. Then $< [a^3, y], y >$ is nilpotent of class ≤ 2, by Lemma 4. But $[a^3, y] = [a^2, y]^a[a, y] = [a, y]^{a^2}[a, y]^a[a, y] = [a, y][a, y, a]^2[a, y][a, y, a][a, y] = [a, y]^3[a, y, a]$ since $< [a, y], a >$ is nilpotent of class ≤ 2, again by Lemma 4.

Hence $[a^3, y] = [a, y]^3[a, y]^{-3}([a, y]^a)^3 = ([a, y]^a)^3$.

Therefore $< ([a, y]^a)^3, y >$ is nilpotent of class ≤ 2, and $< [a, y]^a, y >$ is nilpotent of class ≤ 2, by Lemma 2.

It follows that $[a, y, y^{a^{-1}}]$ commutes with $y^{a^{-1}}$, i.e. that $[[a, y], y[y, a^{-1}]] = [[a, y], [y, a^{-1}]][a, y, y]^{[y, a^{-1}]}$ commutes with $y^{a^{-1}}$.

But $[[a, y], [y, a^{-1}]] = [[a, y], [a, y]^{a^{-1}}] = [[a, y], [a, y][a, y, a^{-1}]] = 1$ since $< [a, y], a >$ is nilpotent of class ≤ 2. Hence $[a, y, y]^{y^{-1}aya^{-1}}$ commutes with y^{a-1}, and $[a, y, y]^{y^{a^{-1}}}$ commutes with y.

Also $[a, y, y, y^{-1}] = 1$, since $< y, y^a >$ is nilpotent of class ≤ 2 and we have that $[a, y, y]^{y^{a^{-1}}} = [a, y, y]^a$ commutes with y, and $[a, y, y, a]$ commutes with y. But $[a, y, y, a] = [a, y, y]^{-1}[a, y, y]^a$ also commutes with y^a, since $[a, y, y]$ does and $[a, y, y, y] = 1$. Therefore $[a, y, y, a]$ commutes with y and $[a, y]$.

From $< [a, y^2], a >$ nilpotent of class ≤ 2, by Lemma 4, we also get that $[[a, y]^2[a, y, y], a]$ commutes with a. Hence $[[a, y, y][a, y]^2, a] = [a, y, y, a]^{[a,y]^2}$

$[[a, y]^2, a]$ commutes with a, from which $[a, y, y, a]$ commutes with a, since $[[a, y]^2, a] = [a, y, a]^2$ commutes with a, because $< [a, y], a >$ is nilpotent of class ≤ 2. Hence $[a, y, y, a] \in Z(< a, y >)$. From $[a, y, y, y] = 1$ we get that $[a, y, y] \in Z_2(< a, y >)$.

Arguing similarly $[y, a, a] \in Z_2(< a, y >)$. Hence $[y, a] \in Z_3(< a, y >)$ and $< a, y >$ is nilpotent of class ≤ 4, as required. □

The following Lemma will be used in the Proof of Theorem B.

Lemma 5 *Let G be a 4-Engel group with periodic derived subgroup. Then, for any $x, y \in G$, there exists a positive integer $n = n(x, y)$ such that $(xy)^n yx = yx(xy)^n$.*

Proof Let b and $a = b^c$ be conjugate elements of G. Then $< a, a^b >$ is nilpotent of class ≤ 2 by Lemma 3. Since G' is periodic,$[a, a^b]^s = 1$, for a suitable $s > 0$, hence $[a^s, a^b] = [a, a^b]^s = 1$. Therefore $[a^s, (a^s)^b] = 1$, and $[[a^s, b], a^s] = 1$. Moreover $[a^s, b]^t = 1$, for a suitable $t > 0$, whence $[a^{st}, b] = 1$. Therefore if b and a are conjugate elements of G, a suitable power of a commutes with b. The result follows with $b = yx$ and $a = xy = (yx)^y$. □

Now we can prove Theorem B.

Proof of Theorem B Let G be a 4-Engel group, and denote by T the set of all torsion elements of G. Then T is a subgroup of G, by Theorem 1 of [10].

Moreover, by Theorem A, for any $x, y \in G$,

$$u_5(x, y)T = \nu_5(x, y)T.$$

Write $a = u_5(x, y), b = \nu_5(x, y)$. Then $a = bd$, where $d \in T$.

Write $H =< b, d >$. Then $a \in H$. Moreover $H \leq T < b >$, and $H/H \cap T$ is abelian. Hence $H' \leq T$ is periodic, and by Lemma 5, $(ab)^n ba = ba(ab)^n$, for some $n > 0$. Therefore we have $(u_5(x, y)\nu_5(x, y))^n \nu_5(x, y)u_5(x, y) = \nu_5(x, y)u_5(x, y)$ $(u_5(x, y)\nu_5(x, y))^n$, and the subsemigroup generated by x and y is not free non abelian, as required. □

References

[1] Y. Kim and A. H. Rhemtulla, *"On Locally Graded Groups"*, Proc. Groups Korea '94, de Gruyter, 1995, 189–197.

[2] J. A. Lewin and T. Lewin, *"Semigroup laws in varieties of solvable groups"*, Math. Proc. Cambridge Philos. Soc. **65** (1969), 1–9.

[3] P. Longobardi, M. Maj and A. H. Rhemtulla, *"Groups with no free subsemigroups"*, Trans. Amer. Math. Soc. **347** (1995), 1419–1427.

[4] A. I. Mal'cev, *"Nilpotent subsemigroups"*, Ivanov. Gos. Ped. Inst. Ucen. Zap. Fiz.-Mat. Nauki **8** (1958), 49–60.

[5] B. H. Neumann and T. Taylor, *"Subsemigroups of nilpotent groups"*, Proc. Roy. Soc. London **A 274** (1963), 1–4.

[6] J. M. Rosenblatt, *"Invariant measures and growth conditions"*, Trans. Amer. Math. Soc. **197** (1974), 33–53.

[7] A. Yu. Ol'shanskii and A. Storozhev *"A group variety defined by a semigroup law"*, J. Austral. Math. Soc. (Series A) **60** (1996), 255–259.

[8] A. Shalev, *"Combinatorial conditions in residually finite groups II"*, J. Algebra **157** (1993), 51–62.

[9] A. I. Shirshov, *"Certain Almost-Engel Groups"*, Algebra i Logika, **2** (1963), 5–18.

[10] G. Traustason, *"On 4-Engel groups"*, J. Algebra **178** (1995), 414–429.

[11] E. I. Zelmanov, *"On some problems of group theory and Lie algebras"*, Math. USSR-Sb. **66** (1990), 159–167.

GROUPS WHOSE ELEMENTS HAVE GIVEN ORDERS

V.D. MAZUROV* and W.J. SHI†

*Institute of Mathematics, Novosibirsk, 630080 Russia
†Department of Mathematics, Southwest-China Teachers University, Beibei, Chongqing, Sichuan, People's Republic of China

"Element orders" is one of the most fundamental concepts in group theory. It plays an important role in research in group theory, which can be seen from the famous Burnside problem. Some well-known group theory specialists, such as B.H. Neumann, G. Higman, M. Suzuki and others, have studied the groups whose element orders are of the special values (see [1, 2, 3]). In 1981 Shi investigated the finite groups all of whose elements are of prime order except the identity element, and got the interesting result: The alternating group A_5 can be characterized only by its element orders (see [4]). The above work was repeated in [5] since [4] was published in Chinese and not reviewed in "Mathematical Reviews".

Let G be a group. Denote by $\pi_e(G)$ the set of all orders of elements in G. Obviously, $\pi_e(G)$ is a subset of the set Z^+ of positive integers, and it is closed and partially ordered under divisibility. The converse problem, which divisibility closed subsets of Z^+ can be the sets of element orders of groups, is more difficult. Let Γ be a subset of Z^+ and $h(\Gamma)$ be the number of isomorphism classes of groups G such that $\pi_e(G) = \Gamma$. For a given Γ, groups G such that $\pi_e(G) = \Gamma$ do not necessarily exist. However, for a given group G, we have $h(\pi_e(G)) \geq 1$. Using this function h we give the following definitions:

A group G is called *characterizable* if $h(\pi_e(G)) = 1$.

A group G is called *recognizable* if $h(\pi_e(G)) < \infty$.

A group G is called *irrecognizable* if $h(\pi_e(G)) = \infty$.

Thus we may divide all groups into the following three classes: *characterizable groups*, *irrecognizable groups* , and recognizable but not characterizable groups (in short, *recognizable groups*).

1 Characterizable groups

For finite characterizable groups Shi and his collaborators, and also Mazurov have obtained many results, which can be summed up as follows (see [6-25], there is a printer's error $U_5(2)$ in [25, Theorem 4.3]):

Theorem 1.1 *The following groups are finite characterizable:*

(1) *Alternating groups* A_n, $n = 5, 7, 8, 9, 11, 12, 13$.

(2) *Symmetric groups* S_n, $n = 7, 9, 11, 12, 13$.

(3) *All sporadic groups except* J_2.

(4) *Simple groups of Lie type* $L_2(q)$, $q \neq 9$, $Sz(2^{2m+1})$, $m \geq 1$, $R(3^{2m+1})$, $m \geq 1$, $L_3(4)$, $L_3(7)$, $L_3(8)$, $L_4(3)$, $S_4(7)$, $U_3(4)$, $U_4(3)$, $U_6(2)$, $G_2(3)$, ${}^2F_4(2)'$, $O_8^-(2)$, *and* $O_{10}^-(2)$.

(5) *Almost simple groups* M_{10}, $L_3(4).2_1$.

The proofs of most above-mentioned characterizable groups used the classification of finite simple groups directly or indirectly. In [26] Shi proved that A_5 is a finite characterizable group using an elementary method independent of the classification theorem.

Problem 1.2 Does there exist an infinite characterizable group?

From Theorem 1.1 we see that $L_3(2)$, $L_3(4)$ and $L_3(8)$ are all finite characterizable. Because of this, we have:

Problem 1.3 For $n \geq 4$, are $L_3(2^n)$ all finite characterizable groups ?

Similarly, we know that the Suzuki-Ree groups consist of the three families of infinite series $Sz(q)$, $R(q)$ and ${}^2F_4(q)$, and we proved that $Sz(q)$, $R(q)$ are all finite characterizable. Are ${}^2F_4(q)$ finite characterizable?

Problem 1.4 Are ${}^2F_4(2^{2m+1})$ $(m \geq 1)$ finite characterizable groups?

In the proof of Theorem 1.1 we often reduce the existence of some orders of elements to the action of a group on another group, and to the Brauer characters in some cases (see [8-10, 13]). However, only some sufficient conditions for the existence of some orders are established. We did not find necessary and sufficient condition for the existence of some orders.

As above-mentioned it is a difficult problem which subset of Z^+ can become the set of element orders of a group. In [6] R. Brandl and Shi studied the finite groups whose element orders are consecutive integers and got the following result.

Theorem 1.5 *Let G be a finite group whose element orders are consecutive integers. Then the maximal order of the element orders in G is* ≤ 8.

Moreover, we classified all such finite groups satisfying the assumption of Theorem 1.5.

Problem 1.6 Let G be an infinite group and $\pi_e(G) = \{1, 2, ..., n\}$, $n \in Z^+$. What is the maximum number n? The group G is obviously periodic. Is it locally finite? For $n \leq 4$, the last question has an affirmative answer.

Motivated by of [6], the author of [27] discussed those finite groups whose element orders are consecutive integers except primes, and got the complete classification of such groups. If we divide the set of element orders $\pi_e(G)$ into $\{1\}$, the set $\pi_e'(G)$ consisting of primes and the set $\pi_e''(G)$ consisting of composite numbers, then we may give a criterion for the simplicity of finite groups using the number of elements among these sets (see [28]).

Theorem 1.7 *Let G be a finite group. Then* $|\pi_e'(G)| \leq |\pi_e''(G)| + 3$, *and if the equality holds, then G is simple. Moreover these simple groups are all characterizable.*

Other than the significance of element orders in group theory, they also have a clear meaning in the study of graph automorphisms. The authors of [29] constructed a special graph on 30 vertices whose automorphism group is A_5 using the characteristic property of A_5. In other branches of mathematics, even in physics and chemistry the "element orders" also has a clear meaning, whether or not there is some relative application from the characterizable groups. The authors believe that there exist broad prospects in these fields.

2 Irrecognizable groups

Two groups G_1 and G_2 are called *conformal* if $\pi_e(G_1) = \pi_e(G_2)$, and for all $n \in \pi_e(G_1)$, the number of elements of order n in G_1 and G_2 are equal. The conformality of groups is also called *the Grassmann equivalence*. An important problem raised by J.G. Thompson is the following:

Problem 2.1 Is it true that a group which is conformal to some finite soluble group is also soluble?

Thompson has given an example of conformal groups which are not isomorphic: $G_1 = 2^4{:}A_7$ and $G_2 = L_3(4){:}2_2$, both are maximal subgroups of M_{23}. The significance of the study of conformal groups which are not isomorphic stems from the needs of research in algebraic number theory and differential geometry. Obviously, conformal groups which are not isomorphic are, in particular, non-characterizable groups. So it is necessary to study the irrecognizable and recognizable groups. The following theorem gives a sufficient condition for the irrecognizable groups (see [15]).

Theorem 2.2 *If the minimal normal subgroup of a group G is an elementary abelian p-group, then G is irrecognizable. In the other words, every finite recognizable group is an extension of the direct product P of non-abelian simple groups by an outer automorphism group of H. In particular, the solvable groups are irrecognizable.*

Chen and Shi had believed that if the minimal normal subgroup N of a group G contains an element of order $exp(N)$, where $exp(N)$ is the exponent of N, then G is irrecognizable. In [30] Mazurov displayed a counterexample to the above statement.

Problem 2.3 What is the characteristic property of irrecognizable groups?

The list of known almost simple irrecognizable groups includes S_5, A_6, S_6, S_8, A_{10}, $U_3(3), U_4(2), U_5(2)$, $U_3(5), U_3(7), J_2$, $L_3(4) : 2_2$ and infinitely many groups $PGL_p(r^s)$ (if p, r are odd primes, $r-1$ is divisible by p but not by p^2 and s is a natural number non-divisible by p).

Problem 2.4 Are there infinitely many simple irrecognizable groups?

3 Recognizable groups

Does there exist any recognizable group? Shi posed this problem at the Third International Conference on the Theory of Groups and Related Topics, which was held in Canberra, 1989. It appeared in [8] as a conjecture in the negative form. This problem was also collected in "Unsolved Problems in Group Theory", Novosibirsk, 1992 (see [31]). Recently, the authors and their co -authors discovered some finite recognizable groups (see [32, 33, 20]).

Theorem 3.1 *The following almost simple groups G are finite recognizable groups with $h(\pi_e(G)) = 2$.*

(1) $L_3(5)$, $L_3(5).2$ ($\simeq Aut(L_3(5))$).
(2) $L_3(9)$, $L_3(9).2_1$.
(3) $S_6(2)$, $O_8^+(2)$.
(4) $O_7(3)$, $O_8^+(3)$.

Considering the above four pairs of recognizable groups we may easily find that one of the groups in each pair is a *simple section* of another group. So we have the following problem.

Problem 3.2 Does there exist a pair of finite recognizable groups which are section-free?

Problem 3.3 Does there exist a finite recognizable group G with $h(\pi_e(G)) \geq 3$?

Recently, Lipschutz and Shi proved:

Theorem 3.4 *Let $\Gamma \subseteq \{1, 2, ..., 20\}$. Then $h(\pi_e(G)) \in \{0, 1, 2, \infty\}$.*

Problem 3.5 Does there exist an infinite recognizable group?

According to Theorem 2.2, every finite recognizable group G satisfies the following

Condition (*) If $G \simeq H/N$ and $\pi_e(G) = \pi_e(H)$ then $N = 1$.

Problem 3.6 Which (almost) simple finite groups satisfy the condition (*)?

It was proved by Mazurov and A. Zavarnicyn (to appear) that A_n and S_n satisfy this condition for all $n \geq 5$.

The Problems 1.6, 2.1 and 3.3 in this talk have been published in the 13th edition of [31] as Problems 12.37, 13.64 and 13.63 respectively.

References

[1] B.H. Neumann, Groups whose elements have bounded orders, *J. London Math. Soc.* **12**(1937), 195-198.

[2] G. Higman, Finite groups in which every element has prime power order, *J. London Math. Soc.* **32**(1957), 335-342.

[3] M. Suzuki, On a class of doubly transitive groups, *Ann. Math.* **75**(1962), 105-145.
[4] W.J. Shi and W.Z. Yang, A new characterization of A_5 and the finite groups in which every non-identity element has prime order (in Chinese), *J. Southwest-China Teachers College*, **9**(1984),36-40.
[5] M. Deaconescu, Classification of finite groups with all elements of prime order, *Proc. Amer. Math. Soc.* **106**(1989), 625-629; K.N. Cheng, M. Deaconecus, M.L. Lang and W.J. Shi, Corrigendum and addendum to "Classification of finite groups with all elements of prime order", *Proc. Amer. Math. Soc.* **117**(1993), 1205-1207.
[6] R. Brandl and W.J. Shi, Finite groups whose element orders are consecutive integers, *J. Algebra* **143**(1991), 388-400.
[7] W.J. Shi, A characteristic property of A_8, *Acta Math. Sinica (N.S.)* **3**(1987), 92-96.
[8] C.E. Praeger and W.J. Shi, A characterization of some alternating and symmetric groups, *Comm. Algebra* **22**(1994), 1507-1530.
[9] V.D. Mazurov, Characterizations of finite groups by the sets of orders of their elements, *Algebra and Logic* **36**(1997), 32-44.
[10] W.J. Shi, The characterization of the sporadic simple groups by their element orders, *Algebra Colloq.* **1**(1994), 159-166.
[11] R. Brandl and W.J. Shi, The characterization of $PSL(2,q)$ by its element orders, *J. Algebra* **163**(1994), 109-114.
[12] W.J. Shi, A characterization of Suzuki's simple groups, *Proc. Amer. Math. Soc.* **114**(1992), 589-591.
[13] R. Brandl and W.J. Shi, A characterization of finite simple groups with abelian Sylow 2-subgroups, *Ricerche Mat.* **42**(1993), 193-198.
[14] W.J. Shi, A characterization of some projective special linear groups, *J. Math. (PRC)* **5**(1985), 191-200.
[15] W.J. Shi, A characteristic property of Mathieu groups, (in Chinese), *J. of Contemporary Math.* **9**(1988), 317-326.
[16] F.J. Liu, A characteristic property of the projective special linear group $L_3(8)$ (in Chinese), *J. Southwest-China Teachers Univ.* **22**(1997), 131-134.
[17] F. Zhou, A characteristic property of some unitary group (in Chinese), *J. Shanghai Jiaotong Univ.* **28**(1994), 106-109.
[18] W.J. Shi, A characterization of the finite simple group $U_4(3)$, *An. Univ. Timisoara* **30**(1992), 319-323.
[19] W.J. Shi and H.L. Li, A characteristic property of M_{12} and $PSU(6,2)$ (in Chinese), *Acta Math. Sinica* **32**(1989),758-764.
[20] W.J. Shi and C.Y. Tang, A characterization of some orthogonal groups, *Progress in Natural Science* **7**(1997), 155-162.
[21] H.L. Li and W.J. Shi, A characteristic property of some sporadic simple groups (in Chinese), *Chin. Ann. Math.* **14A**(1993), 144-151.
[22] W.J. Shi, A characterization of the Higman-Sims group, *Houston J. Math.* **16**(1990), 597-602.
[23] W.J. Shi, A characteristic property of $PSL_2(7)$, *J. Austral. Math. Soc. (Ser.A)* **36**(1984), 354-356.
[24] W.J. Shi, A characteristic property of J_1 and $PSL_2(2^n)$ (in Chinese), *Adv. in Math.* **16**(1987), 397-401.
[25] W.J. Shi, The quantitative structure of groups and related topics, *Group Theory in China, Math. and Its Appl. (China Ser.)* Sci. Press, New York/Beijing (1996), 163-181.
[26] W.J. Shi, A characteristic property of A_5 (in Chinese), *J. Southwest-China Teachers Univ.* **11**(1986),11-14.
[27] M.C. Xu, Finite groups whose element orders are consecutive integers except some

primes (in Chinese), *J. Southwest-China Teachers Univ.* **19**(1994), 116-122.

[28] H.W. Deng and W.J. Shi, A simplicity criterion for finite groups, *J. Algebra* **171**(1997), 371-381.

[29] J.S. Zhang and W.J. Shi, An application of the quantitative characterization of A_5 in graph theory (in Chinese), *J. Southwest-China Teachers Univ.* **16**(1991), 399-402.

[30] V.D. Mazurov, Recognition of finite non-simple groups by the set of orders of their elements, *Algebra and Logic* **36**(1997), 262-278.

[31] V.D. Mazurov, E.I. Khukhro, editors. *Unsolved Problems in Group Theory, The Kourovka Notebook* 13th edition, Novosibirsk, 1995.

[32] V.D. Mazurov, On the set of element orders of finite groups, *Algebra and Logic* **33**(1994), 49-55.

[33] N. Chigira and W.J. Shi, More on the set of element orders of finite groups, *Northeast. Math. J.* **12**(1996), 257-260.

THE BURNSIDE GROUPS AND SMALL CANCELLATION THEORY

JONATHAN P. McCAMMOND

Department of Mathematics, Texas A & M University, College Station, TX 77843, U.S.A.

Abstract

In a pair of recent articles[1], the author develops a general version of small cancellation theory applicable in higher dimensions ([5]), and then applies this theory to the Burnside groups of sufficiently large exponent ([6]). More specifically, these articles prove that the free Burnside groups of exponent $n \geq 1260$ are infinite groups which have a decidable word problem. The structure of the finite subgroups of the free Burnside groups is calculated from the automorphism groups of the general relators used in the presentation. The present article gives a brief introduction to the methods and techniques involved in the proof. Many of the ideas originate with the recent work of A. Yu. Ol'shanskii and S. V. Ivanov. Some familiarity with their work on the Burnside problem will be assumed.

1 Introduction

Ninety-five years ago William Burnside asked whether every finitely generated group of finite exponent must be finite [2]. Since then, significant progress has been made in the study of the free Burnside groups, much of it coming in recent years. In this first section the recent history of their study will be briefly reviewed so that the results announced below can be viewed in their proper context. The second section introduces many of the concepts from which the general theory is constructed. The third and the fourth sections will describe the results of general small cancellation theory and the results on the Burnside groups in greater detail.

1.1 Historical context

The existence of exponents for which Burnside's conjecture is false was first demonstrated by Novikov and Adian in a series of articles published in 1968 [7]. More specifically, for $n \geq 4381$ and odd, they proved that the free Burnside group of exponent n is infinite, that these groups have decidable word problems, and that their only finite subgroups are cyclic. In a book-length reworking of the results [1], Adian was able to extend coverage of the proof to include all cases with $n \geq 665$ and odd. Both versions of the proof are several hundred pages in length.

In 1982, A. Yu. Ol'shanskii published an alternative proof that the Burnside groups of exponent $n > 10^{10}$ and odd are infinite, that they have a decidable word

[1]As of 1 Dec 1997, one article has been accepted for publication and the second will soon be available as a preprint. Until the second article has been distributed, all of the results which refer to the Burnside groups in particular should be considered preliminary announcements only. The exponent of 1260 is especially provisional.

problem, and that their only finite subgroups are cyclic [8]. Ol'shanskii used small cancellation theory and especially van Kampen diagrams to create a significantly shorter proof of Adian and Novikov's results. By introducing more geometry, the proof was reduced to a mere thirty-page article.

In 1994, S. V. Ivanov published a substantial article on the Burnside groups in which he showed that when n is at least 2^{48} and either odd or divisible by 2^9, then the free Burnside group of exponent n is infinite [3]. In addition, he showed that these groups have a decidable word problem and he effectively described their finite subgroups. Since every Burnside group with exponent at least 2^{57} has one of these groups as a homomorphic image, Ivanov's proof shows that Burnside's original conjecture can be answered in the negative for almost all exponents n.

Finally, in 1996 I. Lysionok published a proof that the Burnside groups of exponent $n = 16k \geq 8000$ are infinite [4]. Combined with the earlier proofs regarding Burnside groups of large odd exponent, this work also shows that almost all of the Burnside groups are infinite.

As can be seen from the above review, the study of an arbitrary Burnside group of large exponent has proceeded indirectly through the study of the homomorphic images onto Burnside groups whose exponent is either large and odd or divisible by a large power of 2. In particular, the restriction that n be either large and odd or divisible by a large power of 2 has always been needed to show the decidability of the word problem, the structure of the finite subgroups, and other detailed structural information. The results described below remove these restrictions.

1.2 Announced results

The author is currently in the process of publishing and preparing to publish a pair of articles which first develop a general version of small cancellation theory applicable in higher dimensions ([5]), and then apply this theory to the Burnside groups of sufficiently large exponent ([6]). In combination these articles prove that the free Burnside groups of exponent $n \geq 1260$ are infinite groups which have a decidable word problem. The structure of the finite subgroups of these groups is calculated from the automorphism groups of the general relators used in the presentation. The present article gives a brief introduction to the methods and techniques involved in the proof. The definition of a general relator, the axioms of general small cancellation theory, and a description of the inductive construction involved for the Burnside groups are included. Many of the ideas involved originate with the recent work of A. Yu. Ol'shanskii and S. V. Ivanov. Some familiarity with their work on the Burnside problem will be assumed.

The theory developed by the author in [5] is a generalized version of small cancellation theory which is applicable to specific types of high-dimensional simplicial complexes. The usual results on small cancellation groups are shown to hold in this new setting with only slight modifications. The major results of this theory are summarized below in Theorem A. The definitions of the unfamiliar terms will be given later in the article.

Theorem A *If $G = \langle A|\mathcal{R}\rangle$ is a general small cancellation presentation with $\alpha \leq \frac{1}{12}$, then the word and conjugacy problems for G are decidable, the Cayley graph is constructible, the Cayley category of the presentation is contractible, and G is the direct limit of hyperbolic groups. If the presentation satisfies a few additional hypotheses, then every finite subgroup of G is a subgroup of the automorphism group of some general relator in $\mathcal{R}$.*

The generalized small cancellation theory developed in [5] overlaps significantly with the general theories developed by Rips ([10]) and Ol'shanskii ([8]). In fact, the theory described below attempts to provide an underlying geometric object whose existence explains the success of these earlier constructions.

The main results on the Burnside groups which are proved in [6] are summarized below in Theorem B. The notation $\mathcal{B}(m, n)$ refers to the m-generated free Burnside group of exponent n.

Theorem B *For any $m > 1$, and any $n \geq 1260$, the Burnside group $\mathcal{B}(m, n)$ possesses a general small cancellation presentation which satisfies all of the hypotheses and conclusions of Theorem A. In addition, every finite subgroup of the group $\mathcal{B}(m, n)$ is contained in a direct product of a dihedral of order $2n$ with a finite number of dihedral 2-groups whose exponent divides n. As a consequence, all of the groups $\mathcal{B}(m, n)$ are infinite.*

The approach used in these articles has succeeded in significantly lowering the previous bounds beyond which all of the Burnside groups are known to be infinite, and also in proving for the first time that the word problem for all Burnside groups of sufficiently large exponent is decidable. Finally, in contrast with earlier work on the Burnside groups, no distinction is made between Burnside groups with even or odd exponents.

2 Preliminaries

The generalization of small cancellation theory alluded to above will depend on an expansion of the usual definition of a relator. To help motivate the revised definition, we will begin with a few familiar examples, followed by the technical definitions.

2.1 First examples

The following examples are groups with familiar presentations and well-known properties which can be subsumed under the rubric of general small cancellation theory. The explanation of the exact definitions will be postponed until after the examples have been described.

Example 2.1 **[The modular group]** Let

$$S = \begin{pmatrix} 0 & -1 \\ 1 & 0 \end{pmatrix} \text{ and } T = \begin{pmatrix} 1 & 1 \\ 0 & 1 \end{pmatrix}$$

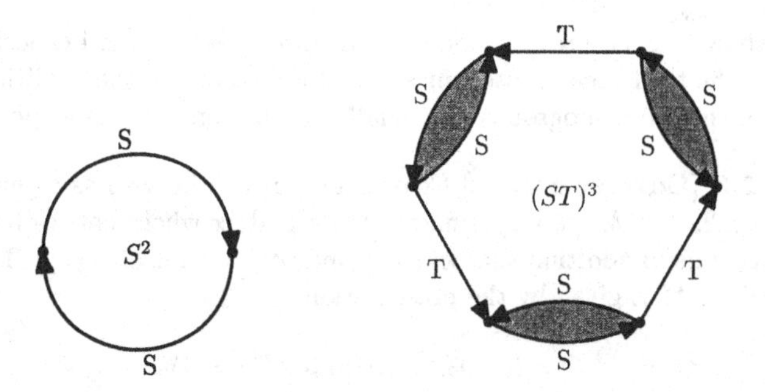

Figure 1. General relators for the modular group

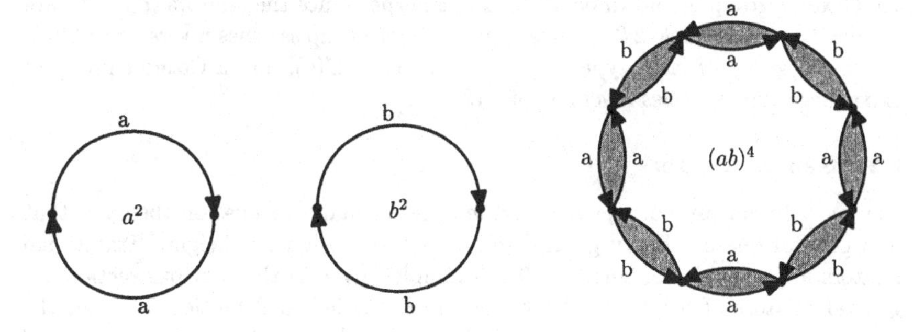

Figure 2. General relators for the dihedral group of order 8

Then $S^2 = -I, (ST)^3 = -I$. The modular group $\mathrm{PSL}_2(\mathbb{Z}) = \mathrm{SL}_2(\mathbb{Z})/\pm 1$ is the free product of the cyclic groups of order 2 and 3 which are generated by S and ST. In particular, $\mathrm{PSL}_2(\mathbb{Z})$ is generated by S and T, and it has the following presentation:

$$\mathrm{PSL}_2(\mathbb{Z}) = \langle S, T | S^2 = (ST)^3 = 1 \rangle.$$

This particular presentation does not satisfy any of the usual small cancellation conditions such as $C(6)$ or $C'(\frac{1}{6})$ since the letter S is a piece and it represents one-half of the boundary of the first relator. The general small cancellation theory under discussion would replace this presentation with a 'general presentation' which still has S and T as generators, but the relators will be altered to appear as in Figure 1.

Example 2.2 [Dihedral groups] Consider the standard presentation of the dihedral group of order $2n$:

$$\mathrm{D}_{2n} = \langle a, b | a^2 = b^2 = (ab)^n = 1 \rangle$$

This presentation also has a modification which satisfies some general small cancellation conditions. The resulting general relators are given in Figure 2 Moreover, it is easy to show that no presentation of a dihedral is ever a small cancellation presentation in the traditional sense, since one of the results of the traditional theory is that all of the finite subgroups of a small cancellation group are cyclic.

Example 2.3 [Coxeter groups] Coxeter groups can be seen as a generalization of Example 2.2. Let M be a symmetric $n \times n$ matrix whose entries lie in the set $\mathbb{Z}^+ \cup \infty$. Assume in addition that $m_{ii} = 1$ and $m_{ij} > 1$ for all $i \neq j$. The Coxeter group based on M is given by the presentation

$$G = \langle a_1, a_2, \ldots, a_n | (a_i a_j)^{m_{ij}} = 1 \rangle$$

where no relation is added in the case $m_{ij} = \infty$. If m_{ij} is never equal to 2, then the Coxeter group is said to be of large type. If m_{ij} is never equal to 2 or 3, then the Coxeter group is said to be of extra-large type. Once the relators $(a_i a_j)^{m_{ij}}$ are modified as in Example 2.2, the resulting Coxeter group satisfies a version of $C(4)$, a Coxeter group of large type satisfies a version of $C(6)$, and a Coxeter group of extra-large type satisfies a version of $C(8)$.

2.2 General relators

The most important concept involved in general small cancellation theory is that of a general relator. Every general relator will be assigned a height. Traditional relators are examples of height 2. The examples given in the previous section are general relators of height 3. Before the precise technical definition is given, the examples from the previous section will be discussed in greater detail, and we will present a single example of a general relator of height 4.

Definition 2.4 [Boundaries of general relators] In the same way that a relator in the traditional sense is often viewed as a labeled disk rather than a labeled circle, the structures shown in Figure 1 and Figure 2 are, technically speaking, merely the boundaries of general relators and not the general relators themselves. A general relator will be the topological cone over its roughly circular boundary. Thus, the general relator corresponding to $(ab)^4$ (whose boundary is shown in Figure 2) will be composed of eight solid circular cones which are adjoined along their lateral sides. This particular general relator is shown schematically in Figure 3. Similarly the general relator corresponding to $(ST)^3$ (whose boundary is shown in Figure 1) will consist of three triangles and three solid circular cones.

Definition 2.5 [Heights] In a traditional relator, the vertices, the open edges and the interior of the disk can be thought of as being represented by their barycenters. Moreover, these barycenters can be assigned heights based on whether they come from a vertex (height 0), an edge (height 1), or a disk (height 2). In a similar way, the apex of the cone over the dihedral structure in Figure 3 can be assigned a height of 3 since it contains a disk of height 2 in its boundary. Once these barycenters have been assigned heights, there is a notion of a 1-, 2-, and 3-skeleton which is the union of all of the pieces assigned a height of at most 1, at most 2, or at most

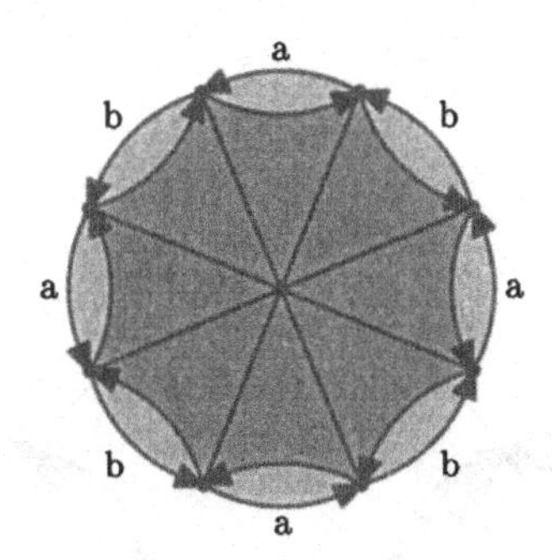

Figure 3. A more precise picture of a general relator

3. The rightmost picture in Figure 2 is thus the 2-skeleton of the general relator, as well as its boundary.

Example 2.6 [A general relator of height 4] The following explicit example will be used to illustrate several key concepts which are precisely defined below. Let $G = \langle a, b, c | a^2, b^2, (ab)^4, (abcb)^2 \rangle$. A modification of this presentation which uses general relators begins with the three (general) relators whose boundaries are shown in Figure 2. The fourth relator, $(abcb)^2$, is modified as shown in Figure 4. It is obtained from the cycle labeled $(abcb)^2$ by attaching the general relator for $(ab)^4$ to each of the paths labeled bab. The top figure shows the 2-skeleton of this relator, while the bottom figure shows its 3-skeleton. As described above, the general relator itself will be the topological cone over this roughly circular boundary, and the apex of the cone will be assigned a height of 4 since it contains a point of height 3 in its boundary.

Examples of general relators of higher height can be obtained as follows: Given a general relator R of height k, introduce a new generator d and attach an arrow labeled d to any two distinct vertices in R. The resulting structure qualifies as the boundary of a general relator, and since it contains R in its boundary it has height $k + 1$. Variations on this theme can create presentations which contain general relators of arbitrary height.

As can be seen from these examples, a general relator is no longer simply a word or a cycle, but rather a particular type of simplicial complex with a labeling on its 1-skeleton. The detailed structure of a general relator is best captured by the partially ordered set of its barycenters. In each of the examples given so far we can associate a partially ordered set to each general relator. The elements are the barycenters, and the ordering is defined by setting $p < q$ if and only if p is contained in the boundary of q. Perhaps more surprising, this partially ordered set, or poset, contains all of the information needed to reconstruct the complex. (The procedure for creating a simplicial complex from a poset is due to D. Quillen [9].) For an arbitrary partially ordered set the corresponding simplicial complex is derived by taking the finite chains as the simplices. For a traditional relator, the poset has

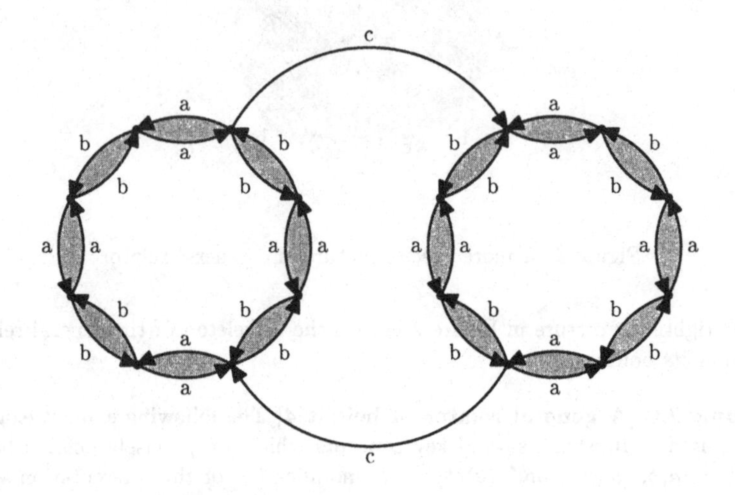

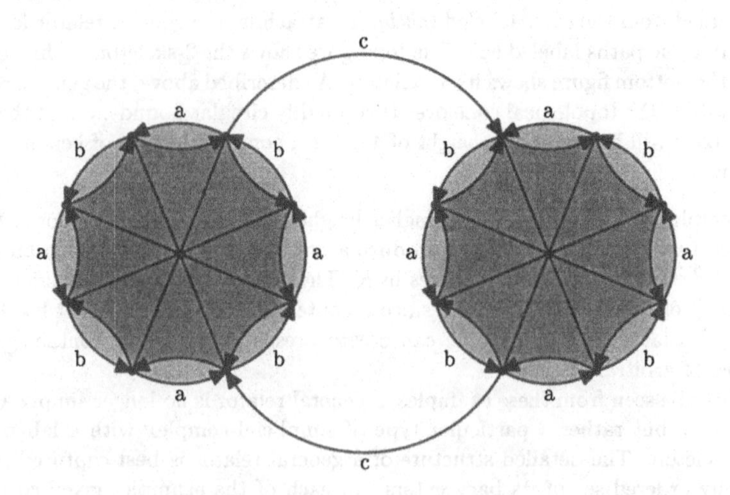

Figure 4. The 2- and 3-skeletons of a general relator of height 4

only 3 levels: the vertices are the elements of height 0, the open edges are the elements of height 1, and the single open 2-cell is the unique element of height 2. The simplicial complex corresponding to this poset is a simplicial subdivision of the original 2-cell. The k-skeleton of such a poset will refer to the geometric realization of the sub-poset consisting only of those elements of height k or less. Thus in the traditional case, the 1-skeleton is the cyclic boundary and the 0-skeleton consists of the vertices alone. For the general relator shown in Figure 4, the poset has one element of height 4, two elements of height 3, sixteen elements of height 2, and sixteen elements of height 1.

We will now precisely define a general relator using these partially ordered sets.

Definition 2.7 **[General relators]** A general relator R is (the simplicial complex corresponding to) a finite poset with a unique maximum element subject to certain additional restrictions. First, the 1-skeleton of the poset must be a simplicial subdivision of a graph, and the unsubdivided graph needs to be deterministically labeled by a set of generators. Next, for all elements p of height at least 2, the geometric realization of the ideal of all elements strictly below p must be homotopically equivalent to a circle. In other words, the simplicial complex corresponding to this ideal should be homotopically equivalent to the space S^1. If the unique maximum element has height k then the general relator R is said to be of height k, and the $(k-1)$-skeleton of R is called its boundary. General relators were introduced by the author in [5].

Notice that the boundary of a general relator is allowed to be fairly complicated as long as it is homotopically equivalent to the unit circle. Although a general relator can contain a 1-skeleton which can be significantly more complicated than is allowed in the traditional theory, the local structure is often less important than its global topology. One example of a global topological concept which plays a key role in the theory is that of the winding number of a loop. A winding number is definable in this context because of the homotopic equivalence to the unit circle. A loop in the boundary of a general relator with winding number 1 will be called a representative of the general relator. Using representatives, it is possible to define more or less traditional van Kampen diagrams over collections of general relators by requiring that the label of every 2-cell in the planar van Kampen diagram be the label of a representative loop in the boundary of some general relator.

Definition 2.8 **[Representatives]** A representative of a general relator R is any path in R which forms a loop with winding number ± 1. There are no other restrictions on representatives: they can complete five clockwise rotations followed by four counterclockwise ones, they can self-intersect, they can even be unreduced in the free group. The main use to which these representatives will be put is to define a van Kampen diagram over a set of general relators.

Definition 2.9 **[Van Kampen diagrams]** A van Kampen diagram over a set $\mathcal{R}$ of general relators is exactly like a van Kampen diagram over a set of traditional relators except that the boundary cycle of each 2-cell in the diagram is a path

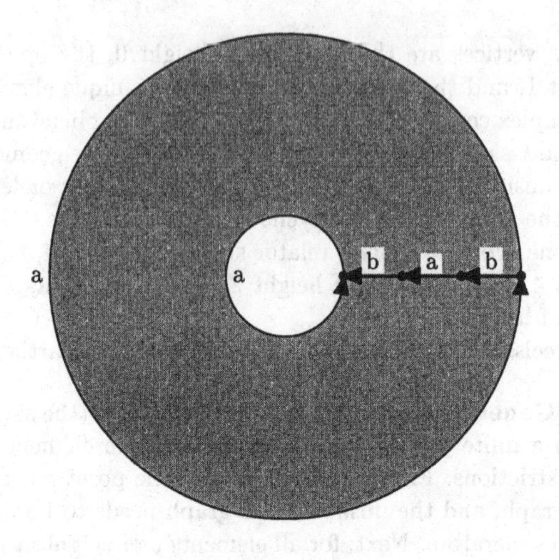

Figure 5. A 2-cell with a large overlap

which maps onto a representative path in one of the general relators in $\mathcal{R}$, instead of being a relator itself.

In traditional small cancellation theory van Kampen diagrams are reduced and cancellable pairs are removed until in the reduced diagram a curvature condition forces a large portion of a relator to exist on the boundary. In the general theory described here a similar strategy is pursued, but one situation arises which does not need to be considered in the traditional case.

Example 2.10 [Self-bordering cells] The annular diagram shown in Figure 5 is a van Kampen diagram over the presentation given in Figure 2. More explicitly, starting at the rightmost vertex, and reading counterclockwise around the outer circle, the boundary of the 2-cell in the diagram reads the word $W = ababa^{-1}b^{-1}a^{-1}b^{-1}$, and this is a path which is a representative of the rightmost relator in Figure 2. The difficulty that this diagram causes is that the word *bab* is long relative to the length of the boundary. In the traditional theory, an overlap longer than a piece is prohibited by the fact that no (nontrivial) word in the free group is conjugate to its own inverse. The word W, however, is conjugate to its own inverse, as Figure 2 demonstrates.

When diagrams such as this are encountered as subdiagrams during the reduction of a van Kampen diagram, there must be some way to remove them since the goal of the reductions is to produce a reduced diagram with no long internal arcs. For Example 2.10 the reduction is easy since the entire annular subdiagram can be removed and the inner and the outer loops labeled a can be identified. (The general situation is more complicated but those details will be glossed over here.)

There is another way to explain the existence of self-bordering cells in the general case. Notice that corresponding to a self-bordering 2-cell which represents a general relator R there is a label-preserving automorphism of the boundary of R which reverses the orientation of its boundary. In particular, the self-bordering 2-cell in Figure 5 corresponds to the automorphism of Figure 3 which rotates the left- and right-hand a^2 loops by 180 degrees. This automorphism will preserve the labels on the edges but it will reverse the orientation of the circular boundary. Traditional relators cannot have automorphisms of this type.

Although general relators can have automorphisms which are more complicated than traditional relators, the structure of the possible automorphism groups is fairly restricted by group theoretic standards. Before the relevant theorem is quoted, however, it is necessary to make plain the meaning of two of the expressions used in its statement. First, the phrase 'closed under subcones' is just another way of stipulating that a general relator which is contained in the boundary of a general relator in the set $\mathcal{R}$ must also be contained in $\mathcal{R}$. Thus a set of general relators which contains the one whose boundary is shown in Figure 4 must also contain the general relator shown in Figure 3.

Next, a crucial cone in the boundary of a general relator R is, loosely speaking, an edge or a general relator whose removal from the boundary of R makes it impossible to complete a representative. In a traditional relator all of its edges are crucial. In the general case, there are some additional restrictions on the way in which the crucial cone is situated in the boundary. See [5] for details.

Theorem 2.11 ([5]) *Let $\mathcal{R}$ be a set of general relators closed under subcones and suppose that all general relators in $\mathcal{R}$ have at least one crucial cone in their boundary. If H is a subgroup of the automorphism group of a general relator $R \in \mathcal{R}$, then for some r there is a group homomorphism from H to D_{2r} whose kernel is a 2-group. In particular, H is isomorphic to a cyclic or a dihedral group extended by a 2-group.*

The main advantage which general relators present over the traditional type is that certain symmetries which exist in the group can be captured by the relations. This obviates the need to break these symmetries by making arbitrary choices, such as selecting one representative loop of length 8 in the final relator in Figure 2 and not any of the others. When these symmetries are incorporated into the general relators themselves, the structure of the group beyond the relator level becomes more visible.

2.3 Length and width

Because general relators are allowed to have boundaries which are 'thick', the portion of the boundary traversed by a path requires a bit of exposition. Along the way, a length and a width will be prescribed for each general relator. A single example will be sufficient to illustrate the nuances of these definitions.

Example 2.12 **[Twisted boundaries]** Consider the presentation

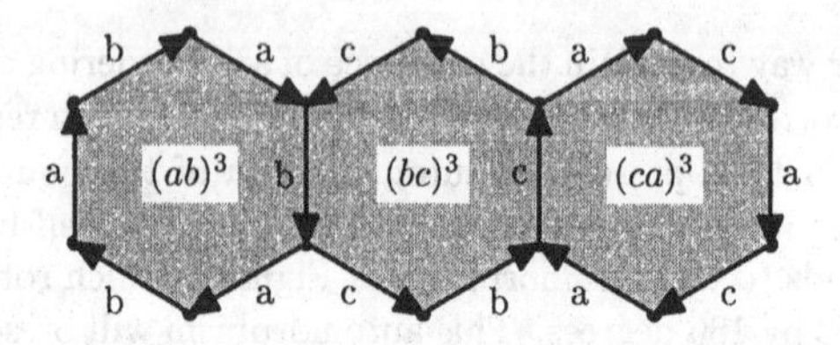

Figure 6. A construction based on the word W

$$G = \langle a, b, c | (ab)^3 = (bc)^3 = (ca)^3 = W = 1 \rangle$$

where $W = babcba^{-1}c^{-1}$. If the relator W had been excluded, the presentation would satisfy $C(6)$. The relator W can be altered to form a general relator by attaching a relator to W whenever the cycle W contains a sizable portion of this other relator. This process quickly stops at a structure which is a Möbius strip. The result is shown in Figure 6 except that the arrow on the far left labeled a must be glued (in an orientation-preserving way) to the arrow on the far right labeled a. The word W can be traced in the diagram from the upper left to the lower righthand corner. Notice also that this path becomes a loop under the gluing operation.

The resulting structure satisfies all of the conditions necessary to qualify as the boundary of a general relator. Let R be the name of the general relator which results. The general relator R is technically the topological cone over the Möbius strip, the Möbius strip being merely the boundary of R. The twisting of the boundary which occurs in this example occurs in a great many general small cancellation presentations, including those constructed for the Burnside groups.

The general relator R decribed above will now be used to illustrate several key definitions. The first such concept is that of length of a general relator. For a traditional relator, the length of its boundary is simply the length of the unique reduced loop of winding number 1. In the general case the situation is more subtle. In Example 2.12 the loop W is the shortest representative of general relator R and it has length 7. But among those loops with winding number 2, some have length 12. For loops with extremely large winding number the geodesic length of the path is more closely approximated by six times the winding number rather than by seven. In order to accurately estimate large geodesic distances, the length of a general relator must be based on the smallest average length per winding number. Thus the length of the general relator R will be defined to be 6 even though there is no representative loop with that length.

Definition 2.13 [Length of a general relator] The length of a general relator R is defined to be the smallest value of the length of a loop in R divided by its nonzero winding number. It is easy to show that such a minimum value must exist, and to create examples where the length of the relator is a fraction instead of an integer. The length of R will be denoted $|R|$. Finally, notice that for traditional relators, this procedure gives the traditional value for the length.

Now that the length of the relator itself has been clarified, it is possible to define the relative length of a path in the boundary of a general relator.

Definition 2.14 [Graph metric] The (normalized) graph metric for a general relator R is a function which assigns a nonnegative length to every path in the boundary of R. In particular, if U is a path in R whose lift to the universal cover of the boundary of R is a geodesic path, then the length of U relative to R, denoted $|U|_R$, is the length of U divided by the length of R. If the path U is not a geodesic when it is lifted to the universal cover of the boundary of R, then let V be a geodesic path in the universal cover between the same vertices as U and set $|U|_R$ equal to $|V|/|R|$. Equivalently, the normalized length of U is the length of the shortest path which is homotopic to U in the boundary of R relative to its endpoints, divided by the length of R defined above.

Consider the path W in the relator R from Example 2.12. Since the length of the general relator R is 6, since the length of the word W is 7, and since W is a geodesic when lifted to the universal cover of the boundary of R, $|W|_R = \frac{7}{6}$. A metric similar (and possibly identical) to the one just described will be used to test whether a set of general relators satisfies a general small cancelation condition.

An additional concept to be illustrated by Example 2.12 is that of width. Let R^∞ represent the universal cover of the boundary of a general relator R. Since the boundary of R is by definition topologically equivalent to a circle, it seems clear that the universal cover should look topologically like the real line. Thus R^∞ should have two infinite 'ends', one going in the positive direction and the other going in the negative. These intuitions are true and they can be made precise but we will not do so here. See [5] for further details.

The width of a general relator is the smallest integer needed to guarantee that the infinite ends of R^∞ will be disconnected from each other upon the removal of an open ball, regardless of the vertex used for the center.

Definition 2.15 [Width of a general relator] The width of a general relator R, denoted ω_R, is the unique smallest integer needed to guarantee that the removal of an open ball of radius ω_R centered at an arbitrary vertex in R^∞ will always disconnect the two infinite ends. Technically, only the 1-skeleton must be disconnected; the higher-dimensional skeleta are ignored. Also, for convenience, an open ball will be defined so that it always contains the vertex used as its center. As a result, the width of a traditional relator is 0.

In Example 2.12, the width of the general relator R is 3. To see this, consider the vertex directly below the label $(bc)^3$ in Figure 6, and call this vertex u. Assume in addition that the figure is a portion of the universal cover, so that the row of adjacent hexagons extends infinitely in both directions. An open ball of radius 1 centered at u will remove only the vertex u and the two open edges incident with it. An open ball of radius 2 will remove a total of six edges and three vertices, but a path running along the top edge of the structure still succeeds in connecting the ends. With a radius of 3, the ends are disconnected, and the three connected components which result are the positive end, the negative end, and an isolated vertex which is directly above the vertex u.

2.4 General presentations

A general presentation is a set of generators A together with a collection of general relators $\mathcal{R}$ which are labeled by A. Such a presentation will be denoted $G = \langle A|\mathcal{R}\rangle$. The group assigned to a particular general presentation is based on the fundamental group of a topological space constructed from the generators and relators. For traditional presentations the standard construction of a 2-complex has the described group as its fundamental group. This complex is sometimes referred to as Poincaré's construction. For a general presentation there is a variation on this procedure. The resulting complex will be called the Poincaré construction of the presentation.

Definition 2.16 [Poincaré constructions] The Poincaré construction of a general presentation is formed by taking a single vertex, attaching an edge for each of the generators, and then attaching the general relators of successively higher height, in order. The attaching of the general relators of height 2 yields a structure that is precisely the standard 2-complex corresponding to these generators and relations. If the set of general relators, $\mathcal{R}$, is closed under subcones and the relators are attached in order of increasing height, then when a relator R is attached, all the edges and general relators used to construct the boundary of R have already been added to the Poincaré construction. In addition there exists a unique map from the boundary of R into the structure as constructed so far. The general relator R is then attached using the map from the boundary to the developing structure as the attaching map. After all the relators have been added, the result is the completed Poincaré construction.

Example 2.17 The idea behind this construction can be illustrated using Example 2.2. Of the three general relators in this presentation, two are traditional relators of height 2 and the final relator is of height 3. The Poincaré construction for this presentation begins by attaching the disks labeled a^2 and b^2 to a bouquet of two circles labeled a and b. The construction at this point is the standard 2-complex whose fundamental group is the infinite dihedral group. The final general relator is now attached along its boundary to the construction. Its boundary is illustrated in Figure 2. The relator itself is shown in Figure 3. Once attached the new complex will have the dihedral group of order 8 as its fundamental group.

Definition 2.18 [Cayley categories] The Cayley category of a presentation is based on the universal cover of the Poincaré construction, but the duplication which arises in the universal cover is removed. The details are as follows. The universal cover of the Poincaré construction contains a 1-skeleton which is the Cayley graph of this group. As in the traditional theory, the universal cover of the Poincaré construction may contain multiple copies of a general relator attached to this Cayley graph by functions which agree on the 1-skeletons in their boundaries. This occurs precisely when the general relator under consideration possesses nontrivial automorphisms. The number of such multiplicities is governed by the size of the automorphism group of the particular general relator involved. Once these multiple copies are eliminated through a suitable identification process, the resulting structure is called the Cayley category of the presentation. The map from the universal cover to the Poincaré construction factors through the Cayley category.

Example 2.19 If we again use the dihedral group of order 8 as an example, the universal cover of the Poincaré construction has the Cayley graph as its 1-skeleton, and has two disks labeled a^2 attached to each a^2 loop, and two disks labeled b^2 attached to each b^2 loop. If we forget the way they are mapped into the Poincaré construction, there is a natural way to identify two disks which are already attached along their boundaries. Once we perform these identifications, we next look at the 3-skeleton. We now have eight copies of Figure 3 attached to the 2-skeleton, one for each of the distinct label-preserving automorphisms of its boundary. Again, we ignore the eight distinct ways in which these are mapped to the Poincaré construction and simply identify the interiors. The result in this case is the structure shown in Figure 3. In general the Cayley category is infinite. This example is a special case since the relator contains the entire Cayley graph in its 1-skeleton.

The height of a general relator defines a filtration on any set of general relators that is sufficient for the above constructions, but it is a filtration which is often not fine enough for more involved operations. For this reason a rank function is allowed which provides an alternative way to order the general relators in a presentation.

Definition 2.20 [Rank of a relator] A rank function on a presentation $G = \langle A|\mathcal{R}\rangle$ is a function from $A \cup \mathcal{R}$ to the natural numbers which assigns each generator a rank of 1, and all of the general relators in $\mathcal{R}$ a rank of at least 2. The only restriction is that if R is a general relator which is contained in the boundary of the general relator S, then $\text{rank}(R) < \text{rank}(S)$.

3 General small cancellation theory

A very brief overview of general small cancellation theory will now be given. The reader is referred to [5] for further details and for the definition of any concepts which are not defined here. A general small cancellation presentation of a group G is a measured presentation $G = \langle A|\mathcal{R}\rangle$ which satisfies the axioms of general small cancellation theory. The presentation is measured in the sense that each of the general relators in the set $\mathcal{R}$ is required to possess a function which measures the length of paths in the boundary of the relator for the purposes of the small cancellation conditions. Following a discussion of these metrics, the axioms of general small cancellation theory will be discussed in detail. The section concludes with a statement of the results derived in [5].

3.1 Relator metrics

To generalize the small cancellation hypotheses to the context of general relators requires the introduction of functions which measure the extent to which a particular path wraps around the boundary of a general relator. The function on paths in R will be called the relator metric for R and it will be denoted d_R. The use of the term 'metric' is justified by the fact that the function d_R will induce a metric on the points in the universal cover of the boundary of R. The exact properties required of the function d_R are listed below.

Definition 3.1 [Relator metrics] A relator metric for a general relator R is a function d_R which assigns a nonnegative real number to every path in R. The function d_R must also satisfy the following six properties:

1. $d_R(U) = d_R(V)$ whenever UV^{-1} is a contractible loop in ∂R,
2. $d_R(U) \geq 0$ and $d_R(U) = 0$ iff U is a contractible loop in ∂R,
3. $d_R(U) = d_R(U^{-1})$,
4. $d_R(UV) \leq d_R(U) + d_R(V)$,
5. if U is a path which forms a loop in ∂R then $d_R(U) \geq$ wind.num.(U),
6. if U and V are paths in R which differ by an automorphism of R then $d_R(U) = d_R(V)$.

One example of a relator metric which is always available is the normalized graph metric defined in the previous section. Although the graph metric always satisfies the definition, the added flexibility of allowing other metrics will be retained since other possibilities prove to be useful in the construction of the general small cancellation presentation for the Burnside groups. A general relator together with a specified relator metric is said to be measured. Similarly, a general presentation $G = \langle A|\mathcal{R}\rangle$ over a set of measured relators is called a measured presentation.

Definition 3.2 [Reduced words] Corresponding to the various ways of measuring the length of a path, there are several ways of describing the 'tautness' of a word W. Most familiarly, a geodesic is a word which is not equivalent to any strictly shorter word in G. The word W will be called Dehn-reduced if there do not exist words U and V such that U is a subword of W, UV is readable as a loop in one of the general relators in $\mathcal{R}$, and the length of V is strictly less than the length of V. As in the traditional small cancellation theory, a geodesic is always Dehn-reduced but the converse is false. Finally, W is said to be μ-free if it is reduced in the free group and does not contain more than μ of a relator in $\mathcal{R}$ as measured by its relator metric. Specifically, there cannot exist a word U and a relator R such that U is a subword of W as well as a path in R and $d_R(U) > \mu$.

The multiplicity of metrics and the variety of corresponding reductions arise from various facets implicit in small cancellation theory. For instance, the curvature condition on van Kampen diagrams arises from the purely combinatorial and decidedly nonmetric condition $C(6)$, while the strict shortening of the length of a word is necessary in many cases to guarantee that a process will stop in a finite number of steps. As long as the metrics are suitably related to each other, different metrics can fulfill these two roles. In the present system these roles will be performed by the relator metrics and the (normalized) graph metrics, respectively.

3.2 The axioms

This particular version of general cancellation theory requires seven axioms involving five constants. While the numbers may seem large, the axioms are far from independent, as are the constants. The redundancy built into the system is intended to improve the overall conceptual clarity. After the constants and the axioms are described informally, the exact statements will be given.

The five constants ($\alpha, \beta, \gamma, \delta$, and ϵ) are functionally defined in the first five axioms. Of these, the constant α is the most analogous to the constant used in traditional small cancellation theory in that it measures the degree to which one relator can be contained in another without being subsumed. Specifically, the first axiom states that if U is a path which can be found in two distinct general relators (or in one general relator in two non-isomorphic ways) then either the measure of U in the relator metrics is small (less than α) or else the general relator R for which $d_R(U) \geq \alpha$ is contained inside the boundary of the other relator. This can be said more precisely using maps. The second axiom states that if a lower ranked relator overlaps with a higher ranked relator then any path contained in the overlap is small ($< \beta$) when measured by the relator metric for the higher ranked relator. The situation described in the second axiom is in some sense redundant since whenever the first axiom holds for some α, the second axiom holds for β equal to the same constant.

Axiom three specifies that the ratio of the width of a general relator to its length must be at most γ. The fourth states that the value given by the relator metric d_R never differs from the value given by the normalized graph metric $|\cdot|_R$ by more than δ. The fifth is probably the least used. If U is the shortest path from a vertex in R to a loop in R with a non-trivial winding number, then the measure of U using the relator metric is at most ϵ. The seventh axiom describes a few convenient relationships between the constants described above. The only constraints which have been included are those needed to derive the most basic results for general small cancellation theory. Other restrictions on the constants will be listed in the statements of the theorems as they are needed.

The only axiom which has not yet been described is Axiom 3.6. There is a situation which arises in the reduction of a van Kampen diagram to a reduced form in which a 2-cell in the diagram overlaps with itself to form an annular subdiagram. In traditional small cancellation theory the overlapping path can contain at most a subword of a piece since anything longer would imply that a word in the free group is conjugate in the free group to its own inverse, and this is a contradiction. In the case of a representative loop of a general relator, however, such a long overlap is possible, and the sixth axiom simply states that should this situation arise, there is always a way to reduce the diagram further. Figure 5 shows an example of how this can occur.

Definition 3.3 **[The axioms]** The axioms of general small cancellation theory are as follows:

Axiom 3.1 There is a constant α such that every general relator $R \in \mathcal{R}$ is α-closed with respect to $\mathcal{R}$. In particular, if U is a word readable in general relators R and S via $\mathcal{R}$-functors f and g respectively, and $d_R(U) \geq \alpha$, then, since S is α-closed, there exists a unique $\mathcal{R}$-functor $h : R \to S$ such that $hf = g$.

Axiom 3.2 There is a constant β such that whenever a word U is readable in general relators $R, S \in \mathcal{R}$ by $\mathcal{R}$-functors f and g respectively, and either rank$(R) <$ rank(S) or rank$(R) =$ rank(S) but there does not exist an $\mathcal{R}$-functor $h : R \to S$ with $hf = g$, then $d_S(U) < \beta$.

Axiom 3.3 There is a constant γ such that for all general relators $R \in \mathcal{R}$, $\omega_R \leq \gamma|R|$.

Axiom 3.4 There is a constant δ such that the length of a path U in a general relator $R \in \mathcal{R}$ in the relator metric d_R is within δ of its length in the normalized graph metric on the boundary of R. Specifically $|d_R(U) - |U|_R| \leq \delta$.

Axiom 3.5 There is a constant ϵ such that whenever U is the shortest possible path from a vertex in a general relator $R \in \mathcal{R}$ to a loop with nonzero winding number in R, the length of U in the relator metric is at most ϵ. That is, $d_R(U) \leq \epsilon$.

Axiom 3.6 If $W = XUYU^{-1}$ is a representative of a general relator R in which both instances of U are properly oriented with respect to W, and $d_R(U) \geq \alpha$, then there exists a word V such that the cycle $XVYV^{-1}$ is readable as a contractible loop in ∂R extending the reading of X given by W. The cycle thus bounds a connected and simply connected $\mathcal{R}$-diagram Δ with rank$(\Delta) <$ rank(R).

Axiom 3.7 The constants $\alpha, \beta, \gamma, \delta$, and ϵ satisfy the following constraints: $\beta \leq \alpha$, and $\gamma, \delta, \epsilon < \alpha$, and $2\gamma + \delta \leq \alpha \leq \frac{1}{6}$.

A general small cancellation presentation $G = \langle A|\mathcal{R}\rangle$ is a measured presentation which satisfies the axioms listed above. A group which possesses a general small cancellation presentation is called a general small cancellation group. The key example of an interesting general small cancellation group is, of course, the Burnside groups. In [6], a general small cancellation presentation is constructed for each of the free Burnside groups with exponent greater than or equal to 1260. The details of the construction of this general presentation are discussed in Section 4.

3.3 The results

Once the proper definitions are in place the proofs of the usual small cancellation results are slight variations on the traditional proofs. It thus follows quickly that subject to a few restrictions on the values of the constants, the word problem and the conjugacy problem are decidable.

Theorem 3.4 ([5]) *If $G = \langle A|\mathcal{R}\rangle$ is a general small cancellation presentation with $3\alpha + 2\gamma + \delta \leq \frac{1}{2}$, then the group G has a decidable word problem. In particular, these results are true when $\alpha \leq \frac{1}{6}$ and $\gamma = \delta = 0$ or whenever $\alpha \leq \frac{1}{8}$.*

Theorem 3.5 ([5]) *If $G = \langle A|\mathcal{R}\rangle$ is a general small cancellation presentation with $4\alpha+2\gamma+\delta \leq \frac{1}{2}$, then the group G has a decidable conjugacy problem. In particular, this is true when $\alpha \leq \frac{1}{8}$ and $\gamma = \delta = 0$ or whenever $\alpha \leq \frac{1}{10}$.*

Similarly, a finitely presented general small cancellation presentation is word-hyperbolic. The proof of this particular theorem uses small cancellation conditions to prove that these groups satisfy Rip's thin triangle conditon.

Theorem 3.6 ([5]) *If $G = \langle A|\mathcal{R}\rangle$ is a finitely presented general small cancellation presentation with $3\alpha + \delta \leq \frac{1}{3}$ and $\alpha \leq \frac{1}{10}$, then the group G is word hyperbolic. In particular, the result is true when $\alpha = \frac{1}{10}$ and $\gamma = \delta = 0$, or whenever $\alpha \leq \frac{1}{12}$.*

In the traditional theory, the Cayley complex of a small cancellation presentation is shown to be aspherical. The following is the general small cancellation version of this fact.

Theorem 3.7 ([5]) *If $G = \langle A|\mathcal{R}\rangle$ is a general small cancellation presentation, and C is the Cayley category of the presentation, then C is contractible.*

As discussed earlier, the difference between the universal cover of the Poincaré construction and the Cayley category of a general presentation is determined by the automorphism groups of the relators. As a result, presentations whose relators have no nontrivial automorphisms have additional properties.

Theorem 3.8 ([5]) *If $G = \langle A|\mathcal{R}\rangle$ is a finitely presented general small cancellation presentation with $\alpha \leq \frac{1}{10}$, then the following five conditions are equivalent:*

(1) *the group G is torsion-free,*

(2) *all of the general relators in $\mathcal{R}$ have no nontrivial automorphisms,*

(3) *the universal cover of the Poincaré construction is collapsed,*

(4) *the universal cover of the Poincaré construction is contractible,*

(5) *the Poincaré construction is a $K(G,1)$-space.*

The final result of general small cancellation theory concerns the relationship between the automorphism groups of the general relators in a presentation and the finite subgroups of the group described by the presentation. Under fairly mild restrictions the two lists are identical.

Theorem 3.9 ([5]) *If $G = \langle A|\mathcal{R}\rangle$ is a general small cancellation presentation with $2\beta + 2\gamma + \delta \leq \alpha \leq \frac{1}{12}$ in which $str(W)$ is finite and effectively constructible for all words $W \in A^*$, and such that all general relators in $\mathcal{R}$ have at least one crucial cone in their boundary, then every finite subgroup of G is a subgroup of the automorphism group of some general relator in $\mathcal{R}$.*

Since the automorphism groups of general relators possessing crucial cones were described in Theorem 2.11, the following corollary is immediate.

Corollary 3.10 ([5]) *If $G = \langle A|\mathcal{R}\rangle$ is a general small cancellation presentation with $2\beta + 2\gamma + \delta \leq \alpha \leq \frac{1}{12}$ in which str(W) is finite and effectively constructible for all words $W \in A^*$, and such that all general relators in $\mathcal{R}$ have at least one crucial cone in their boundary, then each finite subgroup of G can be embedded in the extension of a dihedral group by a 2-group. As a consequence, a general small cancellation group (satisfying the above conditions) which does not contain any elements of even order has only cyclic finite subgroups.*

4 Burnside groups

Before discussing the results concerning the Burnside groups in greater detail, the reader is reminded that the descriptions given in this final section are of a preliminary nature. In particular, the exponent 1260 is especially provisional.

In the tradition of Adian, Novikov, Ol'shanskii, and Ivanov, the general small cancellation presentation of the Burnside groups, $\mathcal{B}(m,n) = \langle A|\mathcal{R}\rangle$, is constructed inductively. In particular, most of the lemmas in [6] are proved by simultaneous induction on a single parameter k. The parameter k refers to the ranks of the general relators used in the presentation. The notations used are as follows: $\mathcal{R}_k$ represents the set of general relators of rank k, $\mathcal{R}(k)$ represents the set of general relators of rank at most k, and $\mathcal{R}$ is the union of the sets $\mathcal{R}_k$ over all positive integers k. There is a corresponding list of groups: $G(k) = \langle A|\mathcal{R}(k)\rangle$, and $G = \langle A|\mathcal{R}\rangle$. After a few of the details of the inductive construction are presented, the conclusions which result from the existence of a general small cancellation presentation for the Burnside groups will be listed. We begin with a simple illustration of the approach.

Example 4.1 [The main idea] Consider a presentation of the form

$$G = \langle A|X^n, Y^n\rangle$$

where n is large and X and Y are simple words. It is well-known that the only way that a high power of X can occur as a subword of Y^n is if the length of X is small compared to that of Y. More precisely, if X^i is a subword of Y^n then it is also a subword of a conjugate of Y^2. In this case, the construction described in outline below will attach disks with boundaries labeled X^n to each of the subwords X^i in Y^n. The result will be somewhat similar to the general relator shown in Figure 3 except that that general relator corresponds to the word $(abcb)^2$ instead of a higher power. In general, we will begin with a word Y^n and then we will attach previously constructed general relators whenever they have a high power of their defining word contained in the general relator under construction. Under appropriate restrictions, this process of attaching general relators stops, and the result has sufficiently nice properties to be of use in an inductive construction.

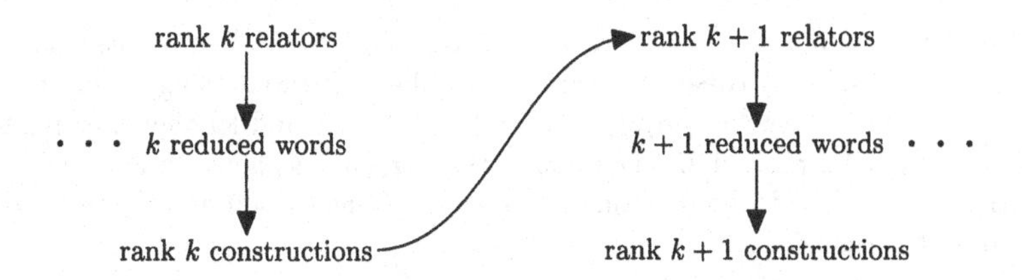

Figure 7. The order of definition

4.1 The inductive construction

The general relators used to present the Burnside groups are defined one rank at a time. The rank, k, indexes not only a set of general relators, but also a set of words, a set of cycles, and a set of constructions. The definitions of these concepts are intertwined and inductively defined. The cycle of definitions is as follows: reduced words and cycles in rank k are used to construct the rank k straightline and circular constructions. These constructions are then used to define the rank $k+1$ general relators, which are in turn used to define reduced words and cycles in rank $k+1$. See Figure 7.

Before describing the induction itself in more detail, a word should be said about the values of the constants used in the inductive construction and about the definitions in low cases. The values of the constants used are as follows:

$$\alpha = \frac{1}{12}, \beta = \frac{1}{210}, \gamma = \frac{1}{70}, \delta = \frac{1}{30}, \epsilon = \frac{1}{630}, n \geq 1260$$

These values are sufficient to satisfy all of the additional restrictions placed on the value of the constants in the statements of the results listed in Section 3.

The induction starts as follows. Since by definition all general relators have a rank of at least 2, there are no rank 1 relators and $\mathcal{R}_1 = \emptyset$. The cycle of definitions thus really begins with the notions of 1-reduced words and 1-reduced cycles. A word (cycle) is called 1-reduced iff it is reduced (cyclically reduced) in the free group. If W is a 1-reduced word, then $\text{str}_1(W)$ is simply the abstract path labeled by the word W. Similarly, if W is a 1-reduced cycle then $\text{cir}_1(W)$ is the abstract loop labeled by the cycle W. The definitions are then extended to arbitrary words (respectively, cycles) which are not 1-reduced by defining the rank 1 straightline (circular) construction on the word (cycle) to be the appropriate construction of its reduction in the free group. The rank 2 general relators are defined from the rank 1 constructions. In particular, if W is a simple word in the traditional sense (i.e. not equal to a proper power of a shorter word), and if the cycle of W does not contain βn powers of any simple word, then the cone over the construction $\text{cir}_1(W^n)$ is a rank 2 general relator. The set of all rank 2 general relators is called $\mathcal{R}_2$. Since these relators do not contain any βn powers of any simple word, they will not contain β of the boundary of any of the other rank 2 relators. Thus the rank 2 relators satisfy the traditional small cancellation hypothesis $C'(\beta)$. As a consequence, the group $G(2)$ is a traditional small cancellation group.

The rank 1 straightline and circular constructions have already been defined. The rank 2 versions of these constructions will be illustrated using Figure 6. Let $G = \langle a, b, c | (ab)^3 = (bc)^3 = (ca)^3 \rangle$, let $W = babcba^{-1}c^{-1}$, and let the three relators in G be assigned a rank of 2. The construction $\text{str}_2(W)$ is shown in Figure 6. The construction $\text{cir}_2(W)$ is the structure obtained by gluing the left and the right edges according to orientation.

Definition 4.2 [The straightline construction] The rank k straightline construction on a k-reduced word W is in some sense the smallest structure which contains W as a path and which is α-closed with respect to $\mathcal{R}(k)$. The latter condition means that given any path U in the structure and a general relator $R \in \mathcal{R}(k)$ such that $d_R(U) \geq \alpha$, the relator R is already attached to the structure along this path U.

Definition 4.3 [The circular construction] The circular construction is defined similarly. The rank k circular construction on a k-reduced cycle W is the smallest structure which contains W as a loop and is α-closed with respect to $\mathcal{R}(k)$.

If $G = \langle A | \mathcal{R} \rangle$ is any general small cancellation presentation with $\alpha \leq \frac{1}{8}$, it is shown in [5] that the rank k straightline and circular constructions on W exist and are well-defined. Under certain additional restrictions, the constructions which result are finite. The presentation constructed inductively for the Burnside groups satisfies these restrictions, and it is the rank k circular constructions which are used to define the boundaries of what will become the rank $k+1$ general relators. Once these relators have been constructed, the axioms of general small cancellation are shown to hold. At that point the results listed in the previous section become available, and the induction continues.

4.2 The results

The most important result on the Burnside groups of sufficiently large exponent that is established in [6] is that they possess a general small cancellation presentation.

Theorem 4.4 *For $n \geq 1260$ the Burnside group $\mathcal{B}(m, n)$ possesses a general small cancellation presentation.*

Most of the other results for these groups follow immediately from the existence of such a presentation, and from the general small cancellation theory described in the previous section. In particular, the following corollary is immediate from Theorem 3.4 and Theorem 3.5.

Corollary 4.5 *For all $n \geq 1260$, the Burnside group $\mathcal{B}(m, n)$ has a decidable word problem and a decidable conjugacy problem.*

As might be expected, the classification of the structure of the finite subgroups of the Burnside groups is more specific than the structure of the finite subgroups in a general small cancellation group in general.

Theorem 4.6 *Every finite subgroup of the group $\mathcal{B}(m,n)$ is contained in a direct product of a dihedral of order $2n$ with a finite number of dihedral 2-groups whose exponent divides n.*

It is from the classification of the finite subgroups that the last two major results are derived.

Corollary 4.7 *For all $n \geq 1260$ and all $m \geq 2$, the Burnside group $\mathcal{B}(m,n)$ is infinite.*

Proof If $\mathcal{B}(m,n)$ were finite, then it would itself be a finite subgroup. Thus, once it is shown that $\mathcal{B}(m,n)$ cannot be embedded in a finite product of dihedral groups, the proof will be complete. It is an easy matter to demonstrate this. □

Corollary 4.8 *For all $n \geq 1260$ and all $m \geq 2$, the Burnside group $\mathcal{B}(m,n)$ is not finitely presented.*

Proof If $\mathcal{B}(m,n)$ were finitely presented, then by Theorem 3.6 it would be a word hyperbolic group, and thus automatic. However, it is well known that finitely presented infinite torsion automatic groups do not exist. □

References

[1] S. Adian. *The Burnside Problem and Identities in Groups.* Springer-Verlag, New York, 1979.

[2] W. Burnside. *An unsettled question in the theory of discontinuous groups.* Quart. J. Pure Appl. Math., **33** (1902), 230–238.

[3] S. Ivanov. *The free Burnside groups of sufficiently large exponents.* Internat. J. Algebra and Comput., **4** (1994), 1–308.

[4] I. Lysionok. *Infinite Burnside groups of even period.* Izv. Ross. Akad. Nauk Ser. Mat. **60** (1996), 3–224.

[5] J. McCammond. *A general small cancellation theory* Internat. J. Algebra Comput., to appear.

[6] J. McCammond. *The Burnside groups via a general small cancellation theory.* In preparation.

[7] P. Novikov and S. Adian. *On infinite periodic groups, I, II, III.* Izv. Acad. Nauk. SSSR Ser. Matem., **32** (1968), 212–244, 251–524, 709–731.

[8] A. Yu. Ol'shanskii. *On the Novikov-Adian theorem.* Mat. Sb., **118** (1982), 203–235.

[9] D. Quillen. *Higher Algebraic K-Theory: I.* Springer-Verlag, New York, 1973.

[10] E. Rips. *Generalized small cancellation theory and applications I.* Israel J. Math. **41** (1982), 1–146.

SOLVABLE ENGEL GROUPS WITH NILPOTENT NORMAL CLOSURES

ROBERT FITZGERALD MORSE

Department of Mathematics and Computer Science, Eastern Nazarene College, Quincy, MA 02170, U.S.A.

Abstract

In this paper we investigate certain solvable $(n+1)$-Engel groups and bounded left Engel groups. We show that these $(n+1)$-Engel groups can be characterized as those groups in which the normal closure of each element in the group is nilpotent of class at most n. Similarly, the bounded left Engel groups investigated can be characterized as those groups in which the normal closure of each element is nilpotent.

1 Introduction

Given a group theoretic class $\mathfrak{X}$, we define the derived class of groups $L(\mathfrak{X})$ as the class of those groups in which the normal closure of each element in the group is an $\mathfrak{X}$-group. The property of being in the class $L(\mathfrak{X})$ is called the Levi-property generated by $\mathfrak{X}$. For example, the property of being a Dedekind group is a Levi-property. A group G is a Dedekind group if and only if the normal closure of each element in G is cyclic. Levi-properties were first introduced in [4] and are modeled after the groups first classified by F. W. Levi [7] where conjugates commute.

Let x, y be elements of a group G. The left n-Engel word is the left normed simple commutator $[x, {}_ny] = [x, y, \ldots, y]$ where the y is repeated n times. A group G is a bounded left Engel group if there exists some n depending only on y such that $[x, {}_ny] = 1$ for all $x, y \in G$. If for a fixed n, $[x, {}_ny] = 1$ for all x, y in G then the group is called an n-Engel group.

The Levi-property generated by nilpotency, denoted by $L(\mathfrak{N})$, implies the bounded left Engel property. It is straightforward to show that $L(\mathfrak{N})$ is exactly the class of Fitting groups: each element of a group G in $L(\mathfrak{N})$ is contained in a normal nilpotent subgroup, i.e. its normal closure, hence it is a Fitting group [9, 12.2.10]. Suppose G is a Fitting group then each element is contained in a normal nilpotent subgroup which contains the normal closure of the element. It follows the normal closure of each element is nilpotent and G is in $L(\mathfrak{N})$. Since Fitting groups are locally nilpotent, all groups in $L(\mathfrak{N})$ are locally nilpotent. However every Baer group, i.e. a group with every cyclic subgroup being subnormal, is a bounded left Engel group. Since the class of Baer groups strictly contains the Fitting groups [9, 12.2.10] in general we have $L(\mathfrak{N})$ is strictly contained in the class of bounded left Engel groups. For metabelian groups the property of being bounded left Engel is the Levi-property generated by nilpotency [6]. We will extend this result to a larger class of solvable groups.

The Levi-property generated by nilpotency of class n, denoted by $L(\mathfrak{N}_n)$, is the variety of groups defined by the commutator law

$$[x^{y_1}, \ldots, x^{y_n}, x] = 1.$$

This follows from the corollary below which in turn follows from a more general result dealing with outer commutator words [8, Theorem 4.3].

Corollary 1 *Let G be a group and let a be an elment of G. Then a^G is nilpotent of class at most n if and only if for all x_i in G, $i = 1, \ldots, n$,*

$$[a^{x_1}, \ldots, a^{x_n}, a] = 1.$$

The class $L(\mathfrak{N}_n)$ is contained in the class of $(n+1)$-Engel groups. For $n = 1$ we have $L(\mathfrak{N}_1) = \mathfrak{E}_2$ which was observed by Levi, and for $n = 2$, $L(\mathfrak{N}_2) = \mathfrak{E}_3$ [5] where $\mathfrak{E}_n$ denotes the class of n-Engel groups. Examples of nilpotent and solvable groups found in [2] show $L(\mathfrak{N}_n)$ is strictly contained in $\mathfrak{E}_{n+1}$ for $n = p + 1$ where p is a prime.

In this paper we will show that for nilpotent groups of class at most $n + 2$, the property $(n+1)$-Engel can be characterized as the Levi-property generated by nilpotent of class n. Moreover, this restriction on the nilpotency class is sharp. For certain solvable groups of derived length 4 and 5, we will show that bounded left Engel and $(n+1)$-Engel are Levi-properties generated by nilpotency and nilpotency of class n respectively. It is an open question whether all solvable groups of derived length 4 or 5 such that $(n + 1)$-Engel is the Levi-property generated by $\mathfrak{N}_n$ must be in this restricted class of solvable groups. A similar question arises for when bounded left Engel is the Levi-property generated by $\mathfrak{N}$.

2 Results

Gupta and Levin [2] construct a finite 4-Engel group of nilpotency class 6 in which there exists an element in the group whose normal closure is nilpotent of class exceeding 3. This example is best possible in the sense that there exist no nilpotent 4-Engel groups with class less than 6 which are not in $L(\mathfrak{N}_3)$. We generalize this situation in the following theorem for any $n \geq 3$:

Theorem 1 *Let $n \geq 3$ be a natural number and let G be a group of nilpotency class $n + 2$. Then G is $(n + 1)$-Engel if and only if $G \in L(\mathfrak{N}_n)$.*

For solvable groups the situation is less straightforward. For metabelian $(n+1)$-Engel groups it was shown in [6] that such groups are in $L(\mathfrak{N}_n)$. For the class of center-by-metabelian $(n+1)$-Engel groups the statement is also true [1]. However, for each prime $p \geq 3$ there exists a solvable group of derived length 3 such that the group is $(p + 2)$-Engel and the normal closure of an element in this group is not nilpotent of any class [2]. We give a brief description of these groups and show how the construction can be modified to include solvable groups of every derived length.

Let G be a free nilpotent of class k of exponent p group with countable rank. Let $\mathbb{Z}_pG$ be the group ring of G over the integers modulo p. Let

$$M_p = \left\{ \begin{pmatrix} g & 0 \\ r & 1 \end{pmatrix}, \quad g \in G, \quad r \in \mathbb{Z}_pG \right\}.$$

This set forms a group with the binary operation matrix multiplication. The group M_p is solvable, it has exponent p^2, and it is $(p+k)$-Engel. With an appropriate choice of p, M_p has an element whose normal closure in M_p is not nilpotent of any class. The original example given was for $k = 2$ and $p \geq 3$ in which case the derived length of M_p is 3. This example can be extended to include groups of derived length $k+1$ when p is chosen large enough.

A group is called nearly center-by-metabelian [3] if it satisfies the commutator law

$$[[[x_1, x_2, x_3], [x_4, x_5]], x_6] = 1. \tag{1}$$

We will show (Proposition 1 below) such groups are second-center-by-metabelian and hence are solvable of derived length 4. The following theorem shows the nearly center-by-metabelian property is a sufficient condition for an $(n+1)$-Engel group of derived length 4 to be in $L(\mathfrak{N}_n)$.

Theorem 2 *Let $n \geq 1$ be a natural number and let G be a nearly center-by-metabelian group. Then G is $(n+1)$-Engel if and only if G is in $L(\mathfrak{N}_n)$.*

We generalize Theorem 2 to solvable groups of derived length 5.

Theorem 3 *Let $n \geq 4$ be a natural number and let G be a group satisfying the commutator law*

$$[[[x_1, x_2, x_3, x_4], [x_5, x_6]], x_7] = 1. \tag{2}$$

Then G is $(n+1)$-Engel group if and only if G is in $L(\mathfrak{N}_n)$.

The restriction on n in Theorem 3 is required. The finite 4-Engel group constructed in [2] is nilpotent of class 6 and hence satisfies (2) but is not in $L(\mathfrak{N}_3)$.

The proofs of Theorems 2 and 3 are based on showing the left Engel lengths of individual elements impose a nilpotency restriction on their respective normal closures. Hence we are able to characterize groups satisfying (1) and (2) which are in $L(\mathfrak{N})$ as being exactly those that are bounded left Engel.

Theorem 4 *Let G be a group satisfying the commutator law (2). Then G is bounded left Engel if and only if G is in $L(\mathfrak{N})$.*

3 Proof of the results

The following lemma will facilitate the proof of Theorem 1.

Lemma 1 *Let $n \geq 2$ be a natural number, and G be a nilpotent of class $n+1$ group. Then the following identities hold for all $g, x, u_i \in G$.*

(i) $[g,_2 x, u_1, \ldots, u_{n-2}] = [x^g, x, u_1, \ldots, u_{n-2}]^{-1}$.
(ii) $[x^{g_1}, \ldots, x^{g_n}] = [x^{g_1},_{n-1} x]\,[x^{g_2},_{n-1} x]^{-1}$.

Proof We prove the lemma by induction on the nilpotency class of G.
(i) For $n = 2$ we have

$$[g,_2 x] = [x, [x, g]] = [x^g, x]^{-1}. \tag{3}$$

Suppose the statement is true for $n \geq 2$. If G is nilpotent of class $(n+1)+1$ then $G/Z(G)$ is nilpotent of class $n+1$. By the induction hypothesis we obtain

$$[g,_2 x, u_1, \ldots, u_{n-2}] = [x^g, x, u_1, \ldots, u_{n-2}]^{-1} z$$

for some $z \in Z(G)$. Commuting with u_{n-1} and the nilpotency restriction yield

$$\begin{aligned} [g,_2 x, u_1, \ldots, u_{n-1}] &= \left[[x^g, x, u_1, \ldots u_{n-2}]^{-1} z, u_{n-1}\right] \\ &= [x^g, x, u_1, \ldots, u_{n-2}, u_{n-1}]^{-1}. \end{aligned}$$

This completes the proof of *(i)*.
(ii) For $n = 2$ we obtain by expanding

$$[x^{g_1}, x^{g_2}] = [x^{g_1}, [x, g_2]]\,[x^{g_1}, x] = [x, [x, g_2]]\,[x^{g_1}, x].$$

This together with (3) and the nilpotency restriction yields

$$[x^{g_1}, x^{g_2}] = [x^{g_2}, x]^{-1}\,[x^{g_1}, x] = [x^{g_1}, x]\,[x^{g_2}, x]^{-1}.$$

Suppose the statement is true for $n \geq 2$. We will show that if G is nilpotent of class $(n+1)+1$, then $[x^{g_1}, \ldots, x^{g_{n+1}}] = [x^{g_1},_n x]\,[x^{g_2},_n x]^{-1}$. Since $G/Z(G)$ is nilpotent of class $n+1$. We obtain by the induction hypothesis

$$[x^{g_1}, \ldots, x^{g_n}] = [x^{g_1},_{n-1} x]\,[x^{g_2},_{n-1} x]^{-1} z.$$

for some $z \in Z(G)$. Commuting the above by $x^{g_{n+1}}$ yields

$$[x^{g_1}, \ldots, x^{g_{n+1}}] = \left[[x^{g_1},_{n-1} x]\,[x^{g_2},_{n-1} x]^{-1} z, x^{g_{n+1}}\right].$$

Because of the class restriction we can expand linearly and obtain

$$[x^{g_1}, \ldots, x^{g_{n+1}}] = [x^{g_1},_{n-1} x, x^{g_{n+1}}]\,[x^{g_2},_{n-1} x, x^{g_{n+1}}]^{-1}. \tag{4}$$

By *(i)* and the restriction on the class of G we observe

$$\left[x^g,_{n-2} x, x^h\right] = [g,_{n-1} x, x\,[x, h]]^{-1} = [x^g,_n x].$$

This together with (4) yields $[x^{g_1}, \ldots, x^{g_{n+1}}] = [x^{g_1},_n x]\,[x^{g_2},_n x]^{-1}$ as desired. This completes the proof of the lemma. □

Proof of Theorem 1 Let $n \geq 3$ and let G be a nilpotent of class $n+2$ group. Suppose G is $(n+1)$-Engel. To show the normal closure of each element of G is nilpotent of class at most n we will show G satisfies the identity

$$[x^{g_1}, \dots, x^{g_{n+1}}] = 1. \tag{5}$$

By *(ii)* of Lemma 1 we have $[x^{g_1}, \dots, x^{g_{n+1}}] = [x^{g_1},_n x]\,[x^{g_2},_n x]^{-1}$. Using *(i)* of Lemma 1 with $u_i = x$ for $i = 1, \dots, n-2$ and the assumption that G is $(n+1)$-Engel we have $1 = [g,_{n+1} x] = [x^g,_n x]^{-1}$. Therefore, G satisfies (5) as needed. □

Proposition 1 *Every nearly center-by-metabelian group is second-center-by-metabelian. Every group satisfying (2) also satisfies the commutator identities*

$$[x_1, x_2, x_3, [x_4, x_5], x_6, x_7] = 1 \quad \textit{and} \tag{6}$$
$$[x_1, x_2, [x_3, x_4], x_5, x_6, x_7] = 1. \tag{7}$$

Proof Let G be a nearly center-by-metabelian group. Let $H = G/Z(G)$. Then, $[\gamma_3(H), \gamma_2(H)] = 1$. Furthermore,

$$[\gamma_2(H), H, \gamma_2(H)] = 1 \qquad \text{and} \qquad [H, \gamma_2(H), \gamma_2(H)] = 1,$$

since $\gamma_3(H) = [\gamma_2(H), H] = [H, \gamma_2(H)]$. Therefore by the Three Subgroup Lemma (see [9, 5.1.10]) $[\gamma_2(H), \gamma_2(H), H] = 1$, and the first claim follows.

By a similar use of the Three Subgroup Lemma the second claim follows. □

The following lemmas provide the technical details to prove the remaining theorems. We prove Lemmas 2 and 3 for groups satisfying (2) with the appropriate restriction on the Engel length. Lemmas 2′ and 3′ state the case for nearly center-by-metabelian groups and are given without proof.

Lemma 2 *Let G be a group satisfying the commutator law (2). Let $u \in \gamma_3(G)$, $v \in G'$, and $j \geq 1$ an integer. Let $a_i \in G$, $i = 1, \dots, j$. Then there exists an element $w \in G'$ such that the following commutator identities hold:*

(i) $[uv, a_1, \dots, a_j] = [u, a_1, \dots, a_j]^w [v, a_1, \dots, a_j]$;

(ii) $[u^{-1}, a_1, \dots, a_j] = [u, a_1, \dots, a_j]^{-w}$.

Proof (i) We prove this identity by induction on j. For $j = 1$, we have $[uv, a_1] = [u, a_1]^w [v, a_1]$ where $v = w$. Assume that the identity holds for $j \geq 1$. Then

$$\begin{aligned}
[uv, a_1, \dots, a_j] &= [u, a_1, \dots, a_j]^w [v, a_1, \dots, a_j] \\
&= [u, a_1, \dots, a_j][u, a_1, \dots, a_j, w][v, a_1, \dots, a_j] \\
&= [u, a_1, \dots, a_j][v, a_1, \dots, a_j]z
\end{aligned}$$

where $z = [u, a_1, \dots, a_j, w] \in Z(G)$ by hypothesis. Thus

$$[uv, a_1, \dots, a_{j+1}] = [[u, a_1, \dots, a_j][v, a_1, \dots, a_j]z, a_{j+1}].$$

Normal commutator expansion yields

$$[uv, a_1, \dots, a_{j+1}] = [u, a_1, \dots, a_j]^w [v, a_1, \dots, a_{j+1}]$$

where $w = [v, a_1, \dots, a_j]$.

(ii) This is an immediate consequence of *(i)* by observing

$$1 = [uu^{-1}, a_1, \dots, a_j] = [u, a_1, \dots, a_j]^w [u^{-1}, a_1, \dots, a_j].$$

Hence $[u, a_1, \dots, a_j]^{-w} = [u^{-1}, a_1, \dots, a_j]$ as needed. □

Lemma 2′ *Let G be a nearly center-by-metabelian group. Let $u, v \in G'$, and $j \geq 1$ an integer. Let $a_i \in G$, $i = 1, \dots, j$. Then there exists an element $w \in G$ such that the following commutator identities hold:*

(i) $[uv, a_1, \dots, a_j] = [u, a_1, \dots, a_j]^w [v, a_1, \dots, a_j]$;

(ii) $[u^{-1}, a_1, \dots, a_j] = [u, a_1, \dots, a_j]^{-w}$.

We define the set of left n-Engel elements of a group G as

$$L_n(G) = \{ a \in G \mid [x,_n a] = 1 \text{ for all } x \in G \}.$$

Lemma 3 *Let G be group satisfying the commutator law (2). Let $a \in G$ and $n \geq 4$ an integer. Then the following statements are equivalent:*

(i) $a \in L_{n+1}(G)$;

(ii) a^G *is nilpotent of class at most n;*

(iii) a^G *is contained in $L_{n+1}(G)$.*

Proof Clearly *(iii)* implies *(i)*. Suppose $a \in L_{n+1}(G)$. To show that a^G is nilpotent of class at most n, we need to show $[a^{g_1}, \dots, a^{g_n}, a] = 1$ for all $g_i \in G$ [8]. We induct on k, the number of conjugates in $[a^{g_1}, \dots, a^{g_k},_{n-k+1} a]$. For $k = 1$ and $n \geq 4$ we obtain by using *(ii)* of Lemma 2

$$\begin{aligned} [a^{g_1},_n a] &= \left[[g_1, a]^{-1},_n a\right] \\ &= \left[[g_1, a, a]^{-1}\left[[g_1, a, a]^{-1}, [g_1, a]^{-1}\right],_{n-2} a\right] \\ &= [[g_1, a, a],_{n-2} a]^{-w_1}\left[[g_1, a, a]^{-1}, [g_1, a]^{-1},_{n-2} a\right] \\ &= \left[[g_1, a, a], [g_1, a]^{-1},_{n-2} a\right]^{-w_2} \\ &= 1 \end{aligned}$$

where w_1, w_2 are elements of G' and the last equality holds by (6). Suppose $[a^{g_1}, \dots, a^{g_k},_{n-k+2} a] = 1$. Let $u = [a^{g_1}, \dots, a^{g_{k-1}}]$. Then

$$\begin{aligned} [a^{g_1}, \dots, a^{g_k},_{n-k+1} a] &= [u, a^{g_k},_{n-k+1} a] \\ &= [u, [a, g_k],_{n-k+1} a]^{w_2} [u, a,_{n-k+1} a]^{w_1} [u, a, [a, g_k],_{n-k+1} a] \\ &= [u, [a, g_k],_{n-k+1} a]^{w_2} \end{aligned}$$

where $w_1, w_2 \in G'$. The last equality follows since $[u, a,_{n-k+1} a]^{w_1} = 1$ by the induction hypothesis and $[u, a, [a, g_k],_{n-k+1} a] = 1$ by (6) or (7): If $k = 2$ then $[u, a, [a, g_k],_{n-k+1} a] = 1$ by (7). If $k \geq 3$ then the equality holds by (6).

We complete the proof by showing

$$[u, [a, g_k],_{n-k+1} a] \tag{8}$$

is equal to the identity for all $k \geq 2$. For $k > 4$, it follows by (2) that (8) is equal the identity.

For $k = 2$ we expand (8) as follows:

$$\begin{aligned} & [u, [a, g_2],_{n-1} a] == [a^{g_1}, [a, g_2],_{n-1} a] \\ = \; & [a, [a, g_2],_{n-1} a]^{w_1} [a, [a, g_2], [a, g_1],_{n-1} a]^{w_2} [[a, g_1], [a, g_2],_{n-1} a] \end{aligned}$$

where $w_1, w_2 \in G'$. Each commutator in the product above is equal to the identity. The first term is equal to the identity as in the basis step:

$$[a, [a, g_2],_{n-1} a] \;\; = [[g_2, a]^{-1}, a,_{n-1} a]^{-w} = 1,$$

where $w \in G'$. The commutators $[a, [a, g_2], [a, g_1],_{n-1} a]$ and $[[a, g_1], [a, g_2],_{n-1} a]$ are equal to the identity since the commutator law (7) holds in G and $n - 1 \geq 3$. Hence (8) is equal to the identity for $k = 2$.

For $k = 3$ we expand (8) as follows:

$$\begin{aligned} [u, [a, g_3],_{n-2} a] &= [a^{g_1}, a^{g_2}, [a, g_3],_{n-2} a] \\ &= [a^{g_1}, [a, g_2], [a, g_3],_{n-2} a]^{w_2} \cdot [a^{g_1}, a, [a, g_3],_{n-2} a]^{w_1} \cdot \\ & \quad\; [a^{g_1}, a, [a, g_2], [a, g_3],_{n-2} a] \end{aligned}$$

where $w_1, w_2 \in G'$. Each term in the expansion above is equal to the identity by noting that the commutator identities (6), and (7) hold in each when $n - 2 \geq 2$. Therefore (8) equals the identity for $k = 3$.

By a similar argument (8) equals the identity for $k = 4$. This completes the proof that $[a^{g_1}, \ldots, a^{g_k},_{n-k+1} a] = 1$.

Let $k = n$. Then $[a^{g_1}, \ldots, a^{g_n}, a] = 1$ and by Corollary 1 a^G is nilpotent of class n. Hence *(i)* implies *(ii)*.

To complete the proof suppose a^G is nilpotent of class at most n. Let w be an arbitrary element of a^G and g an arbitrary element in G. Now $[g, w] \in a^G$ and since a^G is nilpotent of class n we have $[g, w,_n w] = 1$. Hence, a^G is contained in $L_{n+1}(G)$ and *(ii)* implies *(iii)*. This proves the lemma. □

Lemma 3′ *Let G be a nearly center-by-metabelian group, $a \in G$ and $n \geq 3$ an integer. Then the following statements are equivalent:*

(i) $a \in L_{n+1}(G)$;

(ii) a^G *is nilpotent of class at most n;*

(iii) a^G *is contained in* $L_{n+1}(G)$.

Proof of Theorems 2 and 3 For $n = 1$ and $n = 2$ the statement is true for all groups. Let G be a nearly center-by-metabelian group and $n \geq 3$. If G is $(n+1)$-Engel then $G = L_{n+1}(G)$. By Lemma 3′, we have that the normal closure of every element in G is nilpotent of class at most n. Hence $G \in L(\mathfrak{N}_n)$.

Suppose G satisfies (2). Then for $n \geq 4$ the result holds similarly by Lemma 3. □

Proof of Theorem 4 Let G be a group satisfying the commutator law (2). Suppose G is a bounded left Engel group and let $a \in G$. It follows from G being bounded left Engel that $a \in L_{n+1}(G)$ for some n. Choose m to be the maximum of $\{4, n\}$. Then $a \in L_m(G)$ as well. By Lemma 3 we have a^G is nilpotent which is all that is required. Hence $G \in L(\mathfrak{N})$. □

References

[1] M. A. Brodie and R. F. Morse, *Finite subnormal coverings of solvable groups*, Comm. Algebra, submitted.

[2] N. Gupta and F. Levin, *On soluble Engel groups and Lie algebras*, Arch. Math. **34** (1980), 289-295.

[3] N. D. Gupta and M. F. Newman, *Third Engel groups*, Bull. Austral. Math. Soc. **40** (1989), 215-230.

[4] L. C. Kappe, *On Levi-formations*, Arch. Math. **23** (1972), 561-572.

[5] L. C. Kappe and W. P. Kappe, *On 3-Engel groups*, Bull. Austral. Math. Soc. **7** (1972), 391-405.

[6] L. C. Kappe and R. F. Morse, *Levi-properties in metabelian groups*, Contemp. Math. **109** (1990), 59-72

[7] F. W. Levi, *Groups in which the commutator operation satisfies certain algebraic conditions*, J. Indian Math. Soc. **6** (1942), 87-97.

[8] R. F. Morse, *Levi-properties generated by varieties*, Contemp. Math. **169** (1994), 467-474.

[9] D. J. S. Robinson, "A course in the Theory of Groups", Springer-Verlag, Berlin, 1982.

NILPOTENT INJECTORS IN FINITE GROUPS

ANNI NEUMANN

Universität Tübingen, Mathematisches Institut, D-72074 Tübingen, Germany

Since Hall one of the main topics in the theory of finite solvable subgroups is to show the existence and conjugacy of subgroups which are maximal with respect to a group theoretic property. Fischer (1966) [4] and Fischer, Gaschütz, Hartley (1967) [5] investigated such objects and proved their existence and conjugacy, within the general framework of the theory of injectors of finite solvable groups.

In the following we concentrate on one special example of injectors. To this end we fix the following notation: G denotes a finite group and $\mathcal{N}$ the class of nilpotent groups.

Definition 1 A subgroup $U \leq G$ is an $\mathcal{N}$-injector of G, if for every subnormal subgroup S of G, $U \cap S$ is a maximal nilpotent subgroup of S.

$\mathcal{N}$-injectors for non-solvable groups have been introduced first by Mann (1971) [10]. He extented Fischer's results to $\mathcal{N}$-constrained groups i.e. to groups G, such that $C_G(F(G)) \subseteq F(G)$, where $F(G)$ denotes the Fitting subgroup of G. It is well known, that a solvable group is always $\mathcal{N}$-constrained.

Theorem 1 (Mann 1971, [10]) *Let S be a $\mathcal{N}$-constrained group.*

a) *$\mathcal{N}$-injectors exist and are conjugate in G.*

b) *The following are equivalent for $U \leq G$:*

  (1) *$U \in \mathcal{N}$ and whenever X is a nilpotent subgroup of G and $U \subseteq N(X)$ then $X \subseteq U$.*

  (2) *U is an $\mathcal{N}$-injector of G.*

  (3) *U is a maximal nilpotent subgroup of G and $F(G) \subseteq U$.*

Can parts of Theorem 1 be generalized to arbitrary finite groups?

Theorem 2 (Förster [7]; Iranzo, Pèrez-Monasor [9]) *$\mathcal{N}$-injectors exist in all finite groups.*

We give an example, showing that in general $\mathcal{N}$-injectors need not be conjugate. Let G be a quasi-simple group, then the $\mathcal{N}$-injectors are the maximal nilpotent subgroups. But the maximal nilpotent subgroups aren't conjugate.

Of course, considering the statements of Theorem 1b), (1) implies (3) and (2) implies (3) in every finite group G. However, it is easy to see that in general neither (2) nor (3) imply (1) and also (3) does not imply (2). So it remains the question whether (2) is a consequence of (1). This is answered by:

Theorem 3 *Let U be a nilpotent subgroup of G containing every U-invariant nilpotent subgroup X of G. Then U is an $\mathcal{N}$-injector of G.*

To describe some of the consequences of Theorem 3 we introduce a concept due to Bender, Bialostocki and Glauberman:

Definition 2 Now we define

a) $\mathcal{A}_2(G) := \{A \leq G \mid A$ has maximal order among all nilpotent subgroups of class $\leq 2\}$.

b) A subgroup U of G is called a B-$\mathcal{N}$-injector of G if U is a maximal nilpotent subgroup of G containing an element of $\mathcal{A}_2(G)$.

Our Definition 2 assures that any finite group G contains B-$\mathcal{N}$-injectors. By considering B-$\mathcal{N}$-injectors, we get the following result, which was proved originally by Bender.

Theorem 4 (Bender, cf. [6]) *Let U be a B-$\mathcal{N}$-injector of G. Then U contains every nilpotent subgroup of G which is normalized by U.*

Combining Theorems 3 and 4, we get the following:

Corollary 1 *B-$\mathcal{N}$-injectors are $\mathcal{N}$-injectors.*

This yields also a new proof for the existence of $\mathcal{N}$-injectors in arbitrary finite groups.

In case that G is $\mathcal{N}$-constrained Theorem 1 and Corollary 1 yield the following result, which is originally due to Bender.

Theorem 5 (Bender, cf. [1]) *For $\mathcal{N}$-constrained groups $\mathcal{N}$-injectors and the B-$\mathcal{N}$-injectors coincide. In particular, B-$\mathcal{N}$-injectors are conjugate in $\mathcal{N}$-constrained groups.*

It is to note that Theorem 4 and Theorem 5 require a Theorem of Glauberman, [8], Theorem B: $A \in \mathcal{A}_2(G)$ and A normalizes a nilpotent subgroup B of G, then AB is nilpotent.

Bialostocki conjectures that B-$\mathcal{N}$-injectors are conjugate in in every finite group G. Apart from the case of $\mathcal{N}$-constrained groups mentioned above, this conjecture has been verified for various other classes of groups, namely:

(1) Groups all of whose local subgroups are $\mathcal{N}$-constrained (Flavell, [6]).

(2) Symmetric and Alternating groups (Bialostocki, [2] and [3]).

(3) General Linear groups (Sheu, [11]).

(4) Classical groups (Neumann, A., unpublished).

(5) Sporadic simple groups (al Ali, M. I. M. and Neumann, A., unpublished).

Based on these results the author has developed methods to attack the general problem. It is to be noted that there is no immediate reduction to the case of simple groups since, in general, an intersection of a B-$\mathcal{N}$-injector with a normal subgroup is not a B-$\mathcal{N}$-injector of the normal subgroup.

References

[1] Arad, Z. and Chillag, D.: *Injectors of finite solvable groups*, Comm. Algebra 7(2) (1979), 139-162.

[2] Bialostocki, A.: *Nilpotent injectors in symmetric groups*, Israel J. Math. 41 (1982), 261-273.

[3] Bialostocki, A.: *Nilpotent injectors in alternating groups*, Israel J. Math. 44 (1983), 335-344.

[4] Fischer, B.: *Klassen konjugierter Untergruppen in endlichen auflösbaren Gruppen*, Habilitationsschrift (1966), Universität Frankfurt(M).

[5] Fischer, B., Gaschütz, W. and Hartley, B.: *Injektoren in endlichen auflösbaren Gruppen*, Math. Z. 102 (1967), 337-339.

[6] Flavell, P.: *Nilpotent injectors in finite groups all of whose local subgroups are $\mathcal{N}$-constrained*, J. Algebra 149 (1992), 405-418.

[7] Förster, P.: *Nilpotent injectors in finite groups*, Bull. Austral. Math. Soc. 32 (1985), 293-297.

[8] Glaubermann, G.: *On Burnside's other p^aq^b-Theorem*, Pacific J. Math. 56 (1975), 469-476.

[9] Iranzo, M. J. and Pèrez-Monasor, F.: *Fitting classes F such that all finite groups have F-injectors*, Israel J. Math. 56 (1986), 97-101.

[10] Mann, A.: *Injectors and normal subgroups of finite groups*, Israel J. Math. 9 (1971), 554-558.

[11] Sheu, Tsung-Luen: *Nilpotent injectors in General Linear groups*, J. Algebra 160 (1993), 380-418.

SOME GROUPS WITH RIGHT ENGEL ELEMENTS

WERNER NICKEL

School of Mathematical and Computational Sciences, University of St Andrews, St Andrews, Fife KY16 9SS, Scotland

Abstract

In 1970 I.D. Macdonald exhibited a nilpotent group in which the square and the inverse of a right 3-Engel element need not be 3-Engel and thereby showing that the set of right 3-Engel elements of a group need not form a subgroup. In this note a nilpotent group for each $n \geq 3$ is constructed such that the set of right n-Engel elements in each group is not a subgroup.

1 Introduction

An element a of a group is called a *right n-Engel element*, n a positive integer, if for each element g of the group $[a, {}_n g] = 1$ (cf. [Rob72, p. 40]). Commutators are written left-normed and repeated entries in a commutator are indicated by left subscripts. Clearly, the set of right 1-Engel elements is the centre of the group and therefore a subgroup. W. Kappe [Kap61] proved that the set of right 2-Engel elements of a group is also a subgroup. I.D. Macdonald [Mac70] showed that the set of right 3-Engel elements of a group need not form a subgroup by constructing a group with an element that is right 3-Engel but whose inverse and square are not. In this paper we will construct a group for each $n \geq 3$ with a right n-Engel element whose inverse and square are not right n-Engel. This answers a question raised at the conference by W. Kappe for such an example for $n = 4$. The construction follows the ideas in [NN94] by working in a free nilpotent group with certain commutators made trivial to simplify the calculations involved. The construction of the examples in this note were guided by computer experiments with the author's implementation of a nilpotent quotient algorithm [Nic95]. For details of how this program can be used in this context, see [NN94].

Let F be the free group generated freely by a and b. We fix a basic sequence of commutators as in [Hal59, p. 166]. Let F_n the largest class-$(n+2)$ quotient of F. All basic commutators whose weight with respect to a is at least 3 together with all basic commutators whose last component has weight at least 3 generate a normal subgroup of F_n. Let G_n be the factor group of F_n by this normal subgroup. Denote the terms of the lower central series of G_n by $G_n = \gamma_1(G_n)$ and $\gamma_{i+1}(G_n) = [\gamma_i(G_n), G_n]$. It is clear that $\gamma_3(G_n)$ is abelian and it is not difficult to see that $\gamma_i(G_n)/\gamma_{i+1}(G_n)$ is free abelian of rank 3 for $i \geq 5$. In G_n all commutators with a-weight at least 3 are trivial. To simplify notation, we will not distinguish between elements of F and their images in G_n.

The elements

$$[a, {}_3 b], \quad [a, b, a, b]$$

form a basis for $\gamma_4(G)/\gamma_5(G)$ and, for $i \geq 5$, the elements

$$[a, {}_{i-1}b], \quad [a, b, a, {}_{i-3}b] \quad [a, {}_{i-3}b, [b, a]]$$

form a basis for $\gamma_i(G_n)$ modulo $\gamma_{i+1}(G_n)$. Let U be the normal closure of $[a, {}_n b]$ and $[a, {}_n ab]$ in G_n. Here the inverses (modulo $\gamma_{i+1}(G_n)$) of the corresponding basic commutators are used because they turn up naturally in the proof of the main theorem.

Main Theorem *In G_n/U the element a is a right n-Engel element while a^{-1} and a^2 are not right n-Engel elements.*

2 Preliminaries

This section records some elementary tools for manipulating left-normed commutators for easy reference. Let $g, h, g_1, h_1, \ldots, g_{k+1}, h_{k+1}$ be elements of a group G with $h_i \in \gamma_j(G)$. A left-normed commutator of products can be expanded by repeated applications of the following equation which can be proved by induction on k:

$$[g_1h_1, \ldots, g_{k+1}h_{k+1}] = \prod_{u_i \in \{g_i, h_i\}} [u_1, \ldots, u_{k+1}] \mod \gamma_{k+j+1}(G). \quad (1)$$

By setting $h_1 = 1$ and renumbering the indices the equation

$$[g, g_1h_1, \ldots, g_kh_k] = \prod_{u_i \in \{g_i, h_i\}} [g, u_1, \ldots, u_k] \mod \gamma_{k+j+1}(G) \quad (2)$$

is obtained. Because all but possibly one commutator in the product on the right has weight at least $k + j$, this implies

$$[g, g_1h_1, \ldots, g_kh_k] = [g, g_1, \ldots, g_k] \mod \gamma_{k+j}(G). \quad (3)$$

Another consequence of equation (1) is obtained by setting $h_2 = \ldots = h_k = 1$ and renumbering the indices:

$$[gh, g_1, \ldots, g_k] = [g, g_1, \ldots, g_k][h, g_1, \ldots, g_k] \mod \gamma_{k+j+1}(G). \quad (4)$$

This implies, with $h \in \gamma_j(G)$,

$$[gh, g_1, \ldots, g_k] = [g, g_1, \ldots, g_k] \mod \gamma_{k+j}(G) \quad (5)$$

which will be used frequently, without notice, to drop factors from the first component of a commutator.

If $g \in \gamma_l(G)$, $h \in \gamma_m(G)$ and y an integer, we have

$$[gh, g_1] = [g, g_1][h, g_1] \mod \gamma_{l+m+1}(G) \quad (6)$$

$$[g, h^y] = [g, h]^y [g, h, h]^{\binom{y}{2}} \mod \gamma_{l+3m}(G) \quad (7)$$

Now let $g \in G_n$ be an element with a-weight at least 1. The following two equations are consequences of the Jacobi-Witt identity in G_n and give methods for moving a to the left of b and $[b,a]$ to the right of b in a commutator in G_n.

$$[g,b,a] = [g,a,b][g,[b,a]][g,b,[b,a]]. \tag{8}$$

Note that for $g=[a,b]$ the middle commutator on the right hand side is trivial. If, in addition, $g \in \gamma_3(G)$, then

$$[g,[b,a],b] = [g,b,[b,a]]. \tag{9}$$

3 The normal closure

Each element of G_n can be written as a product $a^x b^y [b,a]^z w$ with $w \in \gamma_3(G_n)$. By equation (3)

$$[a,{}_n a^x b^y [b,a]^z w] = [a,{}_n a^x b^y [b,a]^z]$$

because $\gamma_{n+3}(G_n)$ is trivial. In this section we will show that $[a,{}_n a^x b^y [b,a]^z]$ for $x,y,z \in \mathbf{Z}$ is an element of the normal closure U of $[a,{}_n b]$ and $[a,{}_n ab]$. The elements

$$\begin{array}{ccc} [a,{}_n b], & [a,{}_{n+1} b], & [a,b,a,{}_{n-2} b], \\ [a,{}_{n-2} b,[b,a]], & [a,b,a,{}_{n-1} b], & [a,{}_{n-1} b,[b,a]] \end{array}$$

form a basis for $\gamma_{n+1}(G_n)$. The following theorem, which is proved in Section 4, shows how $[a,{}_n a^x b^y [b,a]^z]$ can be written as an expression in this basis.

Theorem 1 *The following formulae hold in G_n for $x,y,z \in \mathbf{Z}$:*

$$\begin{aligned} &[a,{}_n a^x b^y [b,a]^z] \\ &\quad = [a,{}_n a^x b^y][a,b,a,{}_{n-1} b]^{y^{n-1}z}[a,{}_{n-1} b,[b,a]]^{(n-2)y^{n-1}z} \end{aligned}$$

and

$$\begin{aligned} [a,{}_n a^x b^y] &= [a,{}_n b]^{y^n}[a,{}_{n+1} b]^{ny^{n-1}\binom{y}{2}}[a,b,a,{}_{n-2} b]^{(n-1)xy^{n-1}} \\ &= [a,{}_{n-2} b,[b,a]]^{\binom{n-2}{2}xy^{n-1}}[a,b,a,{}_{n-1} b]^{\lambda} \\ &\quad\ [a,{}_{n-1} b,[b,a]]^{\mu} \end{aligned}$$

where

$$\lambda = (n-1)^2 xy^{n-2}\binom{y}{2} + (n-1)xy^n$$

and

$$\mu = \binom{n-1}{2}xy^{n-1} + \binom{n-2}{2}xy^n + \left(2(n-2) + n\binom{n-2}{2}\right)xy^{n-2}\binom{y}{2}.$$

A generating set for U can be obtained by repeatedly forming commutators of $[a, {}_n b]$ and $[a, {}_n ab]$ with the generators of G_n. Commutators with inverses of generators need not be considered because these give inverses of the commutators with the corresponding generator. Substituting 1 for x and y in the second formula of the theorem gives

$$\begin{aligned}[a, {}_n ab] &= [a, {}_n b]\,[a, b, a, {}_{n-2} b]^{n-1}\,[a, {}_{n-2} b, [b, a]]^{\binom{n-2}{2}} \\ &\quad [a, b, a, {}_{n-1} b]^{n-1}\,[a, {}_{n-1} b, [b, a]]^{\binom{n-1}{2}+\binom{n-2}{2}}.\end{aligned}$$

Using that commutators of weight at least $n+3$ are trivial, the following additional non-trivial generators are obtained:

$$\begin{aligned}[a, {}_n b, a] &= [a, b, a, {}_{n-1} b]\,[a, {}_{n-1} b, [b, a]]^{n-2} \\ [a, {}_n b, b] &= [a, {}_{n+1} b] \\ [a, {}_n ab, a] &= [a, {}_n b, a] \\ [a, {}_n ab, b] &= [a, {}_{n+1} b]\,[a, b, a, {}_{n-1} b]^{n-1}\,[a, {}_{n-1} b, [b, a]]^{\binom{n-2}{2}}.\end{aligned}$$

Their exponents are recorded in the following matrix

$$\begin{bmatrix} 1 & & & & & \\ & 1 & & & & \\ 1 & & n-1 & \binom{n-2}{2} & n-1 & \binom{n-1}{2}+\binom{n-2}{2} \\ & 1 & & & n-1 & \binom{n-2}{2} \\ & & & & 1 & n-2 \end{bmatrix}.$$

The last 3 rows can be slightly simplified to get

$$\begin{bmatrix} & & n-1 & \binom{n-2}{2} & & \binom{n-1}{2} \\ & & & & 1 & n-2 \\ & & & & & \frac{1}{2}(n+1)(n-2) \end{bmatrix}.$$

From this and Theorem 1 it is an easy calculation to show that $[a, {}_n a^x b^y [b, a]^z]$ is an element of U for $x, y, z \in \mathbf{Z}$. Consequently, a is a right n-Engel element in G_n/U.

Expanding $[a^2, {}_n b]$ in terms of the basis for $\gamma_{n+1}(G_n)$ gives:

$$\begin{aligned}[a^2, {}_n b] &= [[a, b][a, b, a][a, b], {}_{n-1} b] \\ &= [[a, b]^2 [a, b, a], {}_{n-1} b] \\ &= [[a, b]^2, {}_{n-1} b][a, b, a, {}_{n-1} b] \\ &= [a, {}_n b]^2 [a, b, a, {}_{n-1} b].\end{aligned}$$

From the above, it is clear that $[a^2, {}_n b]$ is not an element of U. Likewise,

$$[a^{-1}, {}_n b] = [a, {}_n b]^{-1} [a, b, a, {}_{n-1} b]$$

is not an element of U.

4 Expanding the Engel condition

This section sketches a proof of Theorem 1. We will first see that it is sufficient to prove Theorem 1 for positive x, y and z. First note that the formulae in Theorem 1 have the following property, which is easy to check. If a multiple of a positive integer m is added to x or z, then the exponents on the right hand side of the formulae also change by a multiple of m. The same is true if an even multiple of m is added to y.

Assume Theorem 1 for positive x, y and z. As G_n is a polycyclic group, it is residually finite ([Rob82, Theorem 5.4.17]). In any finite image H of G_n, multiples of $|H|$ can be added to the exponents on the left hand-side of the formulae without changing the element in H. By adding suitable multiples of $|H|$ these exponents can be made positive and the formulae apply. The effect is that the exponents on the right hand side change by a multiple of $|H|$ and so the right hand side does not change as an element of H. This shows that Theorem 1 is valid for all $x, y, z \in \mathbf{Z}$.

For the proof of Theorem 1 assume that x, y and z are positive integers. The commutator $[\, a,{}_n a^x b^y [b,a]^z \,]$ needs to be expanded to an expression in the basis for $\gamma_{n+1}(G_n)$.

By the fact that commutators with a-weight larger than 2 are trivial and equation (1), we get

$$[\, a,{}_n a^x b^y [b,a]^z \,]=[\, a,{}_n a^x b^y \,] \prod_{k=0}^{n-1} [\, a,{}_k a^x b^y, [b,a]^z,{}_{n-k-1} a^x b^y \,].$$

For $0 \leq k \leq n-1$ each factor in the product can be expanded further as

$$\begin{aligned} [\, a,{}_k a^x b^y, [b,a]^z,{}_{n-k-1} a^x b^y \,] &= [\, a,{}_k b^y, [b,a]^z,{}_{n-k-1} b^y \,] \\ &= [\, a,{}_k b, [b,a],{}_{n-k-1} b \,]^{y^{n-1}z} \end{aligned}$$

where the first equation uses that commutators with a-weight larger than 2 are trivial. In both equations the linearity in G_n of commutators of weight $n+2$ is used. For $k=0$ the Jacobi-Witt identity and equation (5) imply

$$[\, a, [b,a],{}_{n-1} b \,]=[\, [a,b,a],{}_{n-1} b \,].$$

For $2 \leq k \leq n-1$ applying equation (9) with $g =[\, a,{}_k b \,]$ shows

$$[\, a,{}_k b, [b,a],{}_{n-k-1} b \,]=[\, a,{}_{n-1} b, [b,a] \,].$$

In summary,

$$\begin{aligned} &[\, a,{}_n a^x b^y [b,a]^z \,] \\ &= [\, a,{}_n a^x b^y \,][\, a,b,a,{}_{n-1} b \,]^{y^{n-1}z}[\, a,{}_{n-1} b, [b,a] \,]^{(n-2)y^{n-1}z}. \end{aligned}$$

Expanding $[\, a,{}_n a^x b^y \,]$ will be done in two steps. The first step is to expand it to commutators in a^x and b^y. The second step is to expand the result to commutators in a and b and rewrite these in terms of the basis for $\gamma_{n+1}(G)$.

Lemma 2 *The following equation holds in G_n :*

$$[a,{}_n a^x b^y] = [a,{}_n b^y] \prod_{k=1}^{n-1} [a,{}_k b^y, a^x, {}_{n-k-1} b^y] \prod_{k=1}^{n-1} [a,{}_k b^y, a^x, {}_{n-k} b^y].$$

Proof The proof is by induction on n. For $n = 1$ the statement is clearly true. For $n = 2$ direct calculation shows

$$[a,{}_2 a^x b^y] = [\,[a, b^y], a^x b^y\,] = [a, b^y, b^y][a, b^y, a^x][a, b^y, a^x, b^y].$$

Assume that the statement is true for $n \in \mathbf{N}$. This implies in G_{n+1} that

$$\begin{aligned}
&[a,{}_{n+1} a^x b^y] \\
&= [\,\underbrace{[a,{}_n b^y] \prod_{k=1}^{n-1} [a,{}_k b^y, a^x, {}_{n-k-1} b^y] \prod_{k=1}^{n-1} [a,{}_k b^y, a^x, {}_{n-k} b^y]}_{X}, a^x b^y\,] \\
&= [X, b^y][X, a^x][X, a^x, b^y].
\end{aligned}$$

Because commutators with a-weight larger than 2 are trivial, the last two terms simplify as follows using (4) in G_{n+1}

$$[X, a^x] = [a,{}_n b^y, a^x] \quad \text{and} \quad [X, a^x, b^y] = [a,{}_n b^y, a^x, b^y].$$

The expansion of the first term (using (4) again) in G_{n+1} is

$$\begin{aligned}
[X, b^y] &= [a,{}_n b^y, b^y] \prod_{k=1}^{n-1} [a,{}_k b^y, a^x, {}_{n-k-1} b^y, b^y] \prod_{k=1}^{n-1} [a,{}_k b^y, a^x, {}_{n-k} b^y, b^y] \\
&= [a,{}_{n+1} b^y] \prod_{k=1}^{n-1} [a,{}_k b^y, a^x, {}_{n+1-k-1} b^y] \prod_{k=1}^{n-1} [a,{}_k b^y, a^x, {}_{n+1-k} b^y]
\end{aligned}$$

which together with the previous formulae proves the statement for $n + 1$. □

The second step is done in the following lemma.

Lemma 3 *The following equations hold in G_n :*

1. $[a,{}_n b^y] = [a,{}_n b]^{y^n} [a,{}_{n+1} b]^{n y^{n-1} \binom{y}{2}}$.
2. *For* $k \geq 2$, $[a,{}_k b^y, a^x, {}_{n-k} b^y] = [a, b, a, {}_{n-1} b]^{y^n x} [a, {}_{n-1} b, [b, a]]^{(k-2) y^n x}$.
3. $[a,{}_k b^y, a^x, {}_{n-1-k} b^y] = [a,{}_k b, a, {}_{n-1-k} b^y]^{x y^{n-1}} [a,{}_k b, a, {}_{n-k} b]^{(n-k-1) x y^{n-2} \binom{y}{2}}$ $[a,{}_{k+1} b, a, {}_{n-1-k} b]^{k x y^{n-2} \binom{y}{2}}$.

Proof 1. Induction on n. The case $n = 1$ is equation (7). Assume that the statement is true for $n \in \mathbf{N}$. This gives the following chain of equalities holds in

G_{n+1} repeatedly using (5), (4) and (7)

$$\begin{aligned}
[a, {}_{n+1}b^y] &= [\, [a, {}_n b]^{y^n} [a, {}_{n+1} b]^{ny^{n-1}\binom{y}{2}}, b^y \,] \\
&= [\, [a, {}_n b]^{y^n}, b^y \,][\, [a, {}_{n+1} b]^{ny^{n-1}\binom{y}{2}}, b^y \,] \\
&= [\, [a, {}_n b]^{y^n}, b \,]^y [\, [a, {}_n b]^{y^n}, b, b \,]^{\binom{y}{2}} [\, [a, {}_{n+1} b]^{ny^{n-1}\binom{y}{2}}, b \,]^y \\
&= [a, {}_{n+1} b]^{y^{n+1}} [a, {}_{n+2} b]^{y^n \binom{y}{2}} [a, {}_{n+2} b]^{ny^n \binom{y}{2}} \\
&= [a, {}_{n+1} b]^{y^{n+1}} [a, {}_{n+2} b]^{(n+1)y^n \binom{y}{2}}.
\end{aligned}$$

2. Because $[a, {}_k b^y, a^x, {}_{n-k} b^y]$ has weight $n+2$, we have

$$[a, {}_k b^y, a^x, {}_{n-k} b^y] = [a, {}_k b, a, {}_{n-k} b]^{xy^n}.$$

By equation (8),

$$[a, {}_k b, a, {}_{n-k} b] = [a, b, a, {}_{n-1} b][a, {}_{n-1} b, [b, a]]^{k-2}.$$

3. By part 1 and (5), (4) and (7) the following chain of equations follows

$$\begin{aligned}
&[a, {}_k b^y, a^x, {}_{n-1-k} b^y] \\
&= [\, [a, {}_k b]^{y^k} [a, {}_{k+1} b]^{ky^{k-1}\binom{y}{2}}, a^x, {}_{n-1-k} b^y \,] \\
&= [\, [a, {}_k b]^{y^k}, a^x, {}_{n-1-k} b^y \,][\, [a, {}_{k+1} b]^{ky^{k-1}\binom{y}{2}}, a^x, {}_{n-1-k} b^y \,] \\
&= [\, [a, {}_k b, a^x]^{y^k}, {}_{n-1-k} b^y \,][\, [a, {}_{k+1} b, a^x]^{ky^{k-1}\binom{y}{2}}, {}_{n-1-k} b^y \,] \\
&= [\, [a, {}_k b, a]^{xy^k}, {}_{n-1-k} b^y \,][\, [a, {}_{k+1} b, a]^{kxy^{k-1}\binom{y}{2}}, {}_{n-1-k} b^y \,] \\
&= [\, a, {}_k b, a, {}_{n-1-k} b^y \,]^{xy^k} [\, a, {}_{k+1} b, a, {}_{n-1-k} b^y \,]^{kxy^{k-1}\binom{y}{2}} \\
&= [\, a, {}_k b, a, {}_{n-1-k} b \,]^{xy^{n-1}} [\, a, {}_k b, a, {}_{n-k} b \,]^{(n-k-1)y^{n-k-2}\binom{y}{2}xy^k} \\
&\quad\; [\, a, {}_{k+1} b, a, {}_{n-1-k} b \,]^{kxy^{n-2}\binom{y}{2}}
\end{aligned}$$

proving the last statement of the lemma. □

Repeated application of equations (8) and (9) give the following for $k \geq 2$:

$$[a, {}_k b, a, {}_{n-k-1} b] = [a, b, a, {}_{n-2} b][a, {}_{n-2} b, [b, a]]^{k-2} [a, {}_{n-1} b, [b, a]]^{k-1}.$$

For $k \geq 2$ each of the factors on the right hand-side in part 3 of the previous lemma can be rewritten as follows:

$$\begin{aligned}
&[\, a, {}_k b, a, {}_{n-1-k} b \,] \\
&\quad = [\, a, b, a, {}_{n-2} b \,] [\, a, {}_{n-2} b, [b, a] \,]^{k-2} [\, a, {}_{n-1} b, [b, a] \,]^{k-1} \\
&[\, a, {}_k b, a, {}_{n-k} b \,] \\
&\quad = [\, a, b, a, {}_{n-1} b \,] [\, a, {}_{n-1} b, [b, a] \,]^{k-2} \\
&[\, a, {}_{k+1} b, a, {}_{n-1-k} b \,] \\
&\quad = [\, a, b, a, {}_{n-1} b \,] [\, a, {}_{n-1} b, [b, a] \,]^{k-1}.
\end{aligned}$$

From here it is straightforward to combine these expressions and the two lemmas to obtain the second equation of Theorem 1.

Acknowledgements This paper was motivated by Wolfgang Kappe's talk at the conference of these proceedings and subsequent discussions during the conference. The first computer experiments indicating the existence of these examples were carried out during the conference. I thank the organisers of Groups St Andrews 1997 in Bath for creating a fruitful and creative environment. I also thank M.F. Newman for many helpful comments on drafts of this paper.

References

[Hal59] Marshall Hall, Jr. *The Theory of Groups.* Macmillan Co., New York, 1959.

[Kap61] Wolfgang Kappe. Die A-Norm einer Gruppe. *Illinois J. Math.*, 5 (1961), 187–197.

[Mac70] I.D. Macdonald. Some examples in the theory of groups. In H. Shankar, ed., *Mathematical Essays Dedicated to A.J. Macintyre*, pages 263–269. Ohio University Press, 1970.

[Nic95] Werner Nickel. Computing nilpotent quotients of finitely presented groups. in *Geometric and Computational Perspectives on Infinite Groups*, volume 25 of *Amer. Math. Soc. DIMACS Series*, pages 175–191. (DIMACS, 1994), 1995.

[NN94] M.F. Newman and Werner Nickel. Engel elements in groups. *J. Pure Appl. Algebra*, 96 (1994), 39–45.

[Rob72] Derek J.S. Robinson. *Finiteness Conditions and Generalized Soluble Groups, Part* 2, volume 63 of *Ergeb. Math. Grenzgeb.* Springer-Verlag, 1972.

[Rob82] Derek J.S. Robinson. *A Course in the Theory of Groups*, volume 80 of *Graduate Texts in Math.* Springer-Verlag, New York, Heidelberg, Berlin, 1982.

THE GROWTH OF FINITE SUBGROUPS IN p-GROUPS

A. Yu. OL'SHANSKII[1]

Department of Mechanics and Mathematics, Moscow Lomonosov State University, Moscow 119899, Russia

Abstract

For an infinite p-group with a finite set of generators $\mathcal{A}$, the value of the growth function (of the orders) of finite subgroups $f(r)$ is defined to be the maximal size of a finite subgroup lying in the ball of radius r with respect to $\mathcal{A}$. An equivalence class of the function f does not depend of the choice of a generating system. Growth rates of finite subgroups in p-groups, especially in 2-groups, are under investigation in the paper.

1 Introduction

As Schmidt [1] knew, any infinite 2-group contains finite subgroups of unbounded order. On the other hand, by the Novikov-Adian theorem [2], in the free Burnside group of sufficiently large odd exponent n, every finite subgroup has order at most n. Moreover, it can happen that in an infinite p-group, where p is a sufficiently large prime number, all proper subgroups are cyclic of order p [3].

Thus, the collection of orders of finite subgroups can has exhibit diverse behaviour in various periodic groups, and 2-groups are a special case. For a numerical estimate of the set of finite subgroups, we introduce an asymptotic invariant, namely, the growth of this set. (An analogue could be investigated in other algebraic systems). The case of 2-groups is considered below at greater length.

Let G be a group with a finite set of generators $\mathcal{A} = \{a_1, \dots, a_m\}$. Denote by $|g|_{\mathcal{A}}$ the length of an element $g \in G$ with respect to $\mathcal{A}$, i.e. the number of letters $|W|$ in a shortest word W in $a_1^{\pm 1}, \dots, a_m^{\pm 1}$ representing g. Further denote by

$$\text{Ball}(r) = \text{Ball}_{\mathcal{A}}(r) = \{g \in G \mid |g|_{\mathcal{A}} \leq r\}$$

the ball of radius r in G. In the sequel, all balls will be centered at 1. Let $f(r) = f_{\mathcal{A}}(r)$ be the maximal order of a finite subgroups of G which is contained in Ball(r). Provided it is regarded up to the following equivalence, the function f does not depend of the choice of finite set of generators $\mathcal{A}$, and is invariant under passage to a subgroup of finite index in G.

Given two functions f and g, we write $f \preceq g$ if $f(r) \leq Cg(cr)$ for some constants $c, C \geq 0$. We write $f \sim g$ if both $f \preceq g$ and $g \preceq f$. The equivalence class $\tilde{f}$ is said to be the *growth* (of the orders) of finite subgroups in G. (For $C = 1$ one obtains the stronger equivalence $\sim_1$ where equivalence classes are invariant under changes of

[1]The research is partially supported by Russian Foundation for Fundamental Research (grant N° 96-01-420)

generating set.) In particular, one may speak of bounded growth, of power growth of some exponent $\alpha > 0$, of exponential growth and so on. For example, the growth of finite subgroups in a group G is exponential if G contains a wreath product of a non-trivial finite subgroup and an infinite finitely generated subgroup. Evidently the growth of finite subgroups cannot be superexponential since the growth of the "volumes" of the balls Ball(r) is at most exponential.

Theorem 1 *Let $f : \mathbb{N} \to \mathbb{N} \setminus \{0\}$ be an arbitrary non-decreasing function* ($0 \in \mathbb{N}$) *such that $f(r) \leq a^r$ for some $a > 1$ and every $r \geq 0$. Then for any prime $p \geq 3$, there exists a 2-generated p-group $G(f)$ of finite exponent with growth of finite subgroups equal to $\tilde{f}$.*

As was mentioned above, the growth function of finite subgroups cannot be bounded for an infinite 2-group. A straightforward effective version of Schmidt's argument shows that the growth cannot be arbitrarily slow, but even for 2-groups of bounded exponents it provides rather weak lower bounds such as $\exp\sqrt{\log r}$. To improve such a bound, we introduce the notion of an α-bounded subgroup of a 2-group in §2. In §3, we have the following result based on a bound of the growth rate of α-bounded subgroups.

Theorem 2 (1) *For any infinite finitely generated 2-group G of some exponent 2^k, the growth function of finite subgroups f (respectively to an arbitrary finite generating set) satisfies the inequality*

$$f(r) \geq 2^{\lfloor \frac{1}{k} \log_2((2^{k+1}-2)r+2^k) \rfloor - 1} > \frac{1}{2}\sqrt[k]{r},$$

(Here "$\lfloor\ \rfloor$" is the integer part function.) Moreover, these inequalities can be realized on finite subgroups lying in the balls of radii $r = 0, 1, 2, \ldots$ which form an infinite ascending chain.

(2) *For any infinite 2-group G with a finite generating set $\mathcal{A}$, its finite subgroups growth function $f = f_{\mathcal{A}}$ cannot satisfy the relation $f(r) = o(\Phi(r))$ while $r \to \infty$, where $\Phi(x) = 2^{\sqrt{2\log_2 x}}$.*

Infinite finitely generated 2-groups of finite exponents 2^k were first constructed by Ivanov [4] for $k \geq 48$, and by Lysenok [5] for $k \geq 13$. In §4 we will obtain upper bounds for the growth functions of finite subgroups in 2-groups of bounded exponents from the description of the finite subgroups of the free Burnside groups given in [4]. As for the second claim of the following theorem is obtained by a small modification of the construction of [4].

Theorem 3 (1) *Let $f = f_{m,n}$ be the growth function of finite subgroups for the free Burnside group $B(m,n)$ of exponent $n = 2^k$ with respect to a set of m free generators. Then for any $k \geq 48$ and every $r \geq 0$ the inequality $f(cr) \leq 2nf(r)$ holds for the constant $c = \lfloor 0.441n \rfloor$, whence $f(r) \leq r^{b_k}$ for $b_k = \frac{k+1}{k+\log_2(0.44)}$.*

(2) *There exists no unbounded function g on $\mathbb{N}$ such that one has $g \preceq f$ for the growth function of finite subgroups f of every infinite finitely generated 2-group.*

Thus, there is a gap for 2-groups between the given upper and lower bounds for the possible growth of finite subgroups. The following questions are open. Which functions can be realized (up to the equivalence) as the finite subgroups growth functions in 2-groups? The same question for 2-groups of bounded exponents. Find (up to the equivalence) the growth functions of finite subgroups $f_{m,n}$ for the groups $B(m, 2^k)$. Is this true that the growth of finite subgroups of any infinite finitely generated 2-group of exponent 2^k cannot be slower than $\tilde{f}_{2,2^k}$?

The formulated realizability problem is solved for rapidly increasing functions.

Theorem 4 *Let f be a function of at most exponential growth, and assume that for some $\epsilon > 0$ the ratio $f(r)r^{-1-\epsilon}$ increases while $r \to \infty$. It follows that there is a 2-generated 2-group G of finite exponent with growth function of finite subgroups $\tilde{f}$.*

Theorems 4 and 1 are proved in §6 by constructing a special central extensions of the free Burnside groups. In the case of large odd exponents we use some quotients of the free central extensions in [3]. In contrast to the odd case, the centers of similar extensions of the groups $B(m, 2^k)$ have torsion. The description of such extensions is a separate problem. However, it turns out to be sufficient for our purposes to describe in §5 just the maximal central extensions of $B(m, 2^k)$ by elementary 2-groups.

To formulate Theorem 5, let us recall the construction of defining words [4] for the free Burnside group $B(m, n)$ in the case $n \geq 2^{48}$ and n involves 2^9. A total order is introduced on the set of all words in the alphabet $\mathcal{A}^{\pm 1} = \{a_1^{\pm 1}, \dots, a_m^{\pm 1}\}$, such that $|X| < |Y|$ implies $X < Y$. By definition, $B(0) = F_m$ is the free group with basis $\mathcal{A}$. Suppose the group $B(i-1)$ is already defined for $i > 0$. Then A_i is the first word having infinite order in the group $B(i-1)$ (A_i is called *the period* of rank i). The group $B(i)$ is given by the following defining relations

$$B(i) = \langle \mathcal{A} \mid A_1^n = 1, \dots, A_i^n = 1 \rangle. \tag{1.1}$$

By Theorem B [4] the period A_i does exist for every $i \geq 1$, and $\{A_i^n = 1\}_{i=1}^{\infty}$ is a set of defining relations of the group $B(m, n)$. Let the normal subgroup $R = R(m, n)$ of the group F_m be the kernel of the defined presentation of $B(m, n)$, i.e. R is generated as a normal subgroup by the all defining words A_i^n. Denote by $[F_m, R]$ the mutual commutator subgroup and by R^2 the subgroup generated by the squares of all elements in R. Further denote by $A_2(m, n)$ the quotient $F_m/[F_m, R]R^2$, and by $Z_2(m, n)$ its center $R/[F_m, R]R^2$. (Recall that the center of the group $B(m, n) \cong F_m/R$ is trivial by Theorem A(d) [4].)

Theorem 5 *If $n \geq 2^{48}$ and n involves 2^9 then the subgroup $Z_2(m, n)$ is the direct product of the groups of order 2 generated by the natural images of the words A_i^n in the group $A_2(m, n)$ for $i = 1, 2, \dots$.*

There appear functions bounding the orders of elements of given length in the proofs of Theorems 2, 3. Namely, the value $h(r)$ of the growth function of elements orders h for a finitely generated periodic group G is defined as the maximum of orders of the elements in the ball of radius r (for a fixed finite generating set). Residually finite p-groups with a power growth rate of elements orders are under investigation in R.I. Grigorchuk's paper [6], and A.V. Rozhkov [7] constructs residually finite infinite p-groups, in which the orders of elements grow slower than any multiple logarithm. Note that by E.I. Zelmanov theorem the growth of elements in a residually finite p-group cannot be arbitrarily slow. Proving Theorem 3, we apply

Theorem 6 *Let p be a prime number, and assume $h : \mathbb{N} \to \mathbb{N} \setminus \{0\}$ is a non-decreasing function. Then there exists a 2-generated p-group G_h for which the growth function of elements orders is equivalent to h.*

2 α-bounded subgroups

Let H be a finite subgroup of a group G generated by a set $\mathcal{A}$. By *length $l(H)$* of the subgroup H we mean the maximum of length $|h| = |h|_{\mathcal{A}}$ of the elements $h \in H$.

Assume $\alpha : \mathbf{N} \to \mathbf{N}$ is an increasing function, and $\alpha(0) = 0$. A finite 2-subgroup H is said to be α- *bounded* if either $H = 1$ or the order of the subgroup H equals $2^s \geq 2$, and there is an α-bounded series of subgroups in H, i.e. a series

$$1 = H_0 < H_1 < \ldots < H_s = H, \tag{2.1}$$

where every member of the series has index 2 in the subsequent one, and for every $i = 1, \ldots, s$ there is an element $h_i \in H_i \backslash H_{i-1}$ such that $|h_i| \leq \beta(i-1) = \alpha(i) - \alpha(i-1)$. An obvious induction makes clear that $l(H_i) \leq \alpha(i)$.

Below we suppose in this section that G is an infinite 2-group, that the orders of its elements in the balls Ball(r) are bounded by some function $h(r)$, and the following conditions are fulfilled for the function α and the defined by it above function β

$$\alpha(1) \geq \frac{1}{2}h(1), \tag{2.2}$$

$$\beta(s) \geq h(2\beta(s-1))\beta(s-1), \; s = 1, 2, \ldots \tag{2.3}$$

Lemma 2.1 *Assume K is an α-bounded subgroup of order 2^s in G, and two elements $x_1, x_2 \in G$ satisfy the following conditions for $j = 1, 2$*

$$x_j \notin K, \; x_j K x_j^{-1} = K, \; x_j^2 \in K, \; |x_j| \leq \beta(s). \tag{2.4}$$

Let the quotient L/K has order $2^t \geq 4$ for the subgroup $L = gp\{K, x_1, x_2\}$. Then there is a series

$$K = L_0 < L_1 < \ldots < L_t = L, \tag{2.5}$$

in L, where every subgroup has index 2 in the subsequent one, $L_1 = gp\{K, x_1\}$, and $L_i = gp\{L_{i-1}, (x_1x_2)^{2^{t-i}}\}$ for $i \geq 2$. In addition $|(x_1x_2)^{2^{t-i}}| \leq \beta(s+i-1)$ for $1 < i \leq t$. Therefore series (2.5) extend an α-bounded series of the subgroup K.

Proof The quotient L/K is dihedral, and the first claim follows from elementary properties of dihedral 2-groups. Then

$$|(x_1x_2)^{2^{t-i}}| \le 2^{t-2}(|x_1| + |x_2|) \le 2^{t-1}\beta(s).$$

Note, that the order 2^{t-1} of the coset of the element x_1x_2 in L/K does not exceed its order in G, i.e. $2^{t-1} \le h(2\beta(s))$. Hence

$$|(x_1x_2)^{2t-i}| \le h(2\beta(s))\beta(s) \le \beta(s+1) \le \beta(s+i-1)$$

for $i \ge 2$ by inequality (2.3). □

Lemma 2.2 *Let A and B be α-bounded subgroups in G, such that $B < A$, $B \ne A$, and B has order 2^k. Then there is an element a, such that*

$$a \in A \setminus B,\ aBa^{-1} = B,\ a^2 \in B,\ |a| \le \beta(k). \tag{2.6}$$

Proof Let us induct by the sum of orders of the subgroups A and B. Consider an α-bounded series

$$1 = A_0 < A_1 < \ldots < A_l = A \tag{2.7}$$

in A. If $B = 1$, then the non-identity element of A_1 can be chosen for a, since in this case $|a| \le \alpha(1) - \alpha(0) = \beta(0)$ by the definitions of the subgroup A_1 and the function β.

Further regard B to be non-trivial, and choose the maximal subgroup A_j in the series (2.7) which is contained in B. (Possibly, $j = 0$.) If $A_j = B$ (i.e. $j = k$), one can choose an element $a \in A_{k+1}\setminus A_k$ for a, that has length $\le \beta(k)$ by the definition of series (2.7).

In case $A_j \ne B$ we have $A_{j+1} \cap B = A_j$, and there exists an element $y \in A_{j+1}\setminus B$, such that

$$yA_jy^{-1} = A_j,\ y^2 \in A_j,\ |y| \le \beta(j)$$

by the definition of the α-bounded series (2.7).

Consequently, in this case one gets the non-empty set $\mathcal{X}$ of all α-bounded subgroups $K \le B$ with the following property. If K has an order 2^s, then there is element $y = y(K)$ in $A\setminus B$ such that

$$yKy^{-1} = K,\ y^2 \in K,\ |y| \le \beta(s). \tag{2.8}$$

In fact, $A_j \in \mathcal{X}$.

If $B \in \mathcal{X}$, i.e. $s = k$, the lemma's claim is true for $a = y = y(B)$.

Thus, it is sufficient to obtain a contradiction under the assumption that a maximal subgroup K in $\mathcal{X}$ (of order 2^s) is strictly less than B. In this case the sum of orders of the subgroups K and B is less than that for B and A. Therefore, by inductive hypothesis, there is an element x such that

$$x \in B\setminus K,\ xKx^{-1} = K,\ x^2 \in K,\ |x| \le \beta(s). \tag{2.9}$$

Let us examine the subgroup $L = gp\{K, x, y\}$, in which the subgroup K is normal in view of (2.8) and (2.9), and the quotient L/K has at least four elements because $x \in B \backslash K, y \notin B$. In addition this means that L is not contained in B, and the intersection $M = L \cap B$ strictly contains K.

Lemma 2.1 gives the series (2.5) between K and L for $x_1 = x, x_2 = y$. As far as $M \ni x$, the subgroup M coincides with a subgroup L_i of this series for $i \geq 1$ as is seen from the list of the subgroups of the dihedral group, containing one of the two generating involution. Besides $M = L \cap B = L_i$ is strictly contained in L since B does not contain L. Therefore by Lemma 2.1 the subgroups $L_i = M$ and L_{i+1} are α- bounded, and there is an element z in $L_{i+1} \backslash L_i$ such that $zMz^{-1} = M$, $z^2 \in M$, and $|z| \leq \beta(s+i)$. The subgroup $M = L_i$ has order 2^{s+i}, and $z \in A \backslash B$ because $B \cap L_{i+1} = B \cap L = L_i$. Hence $M \in \mathcal{X}$, contrary to the choice of the subgroup K. The lemma is proved. □

Lemma 2.3 *For an α-bounded subgroup A of order $2^i \geq 1$ there exists an element x such that*

$$x \in G \backslash A,\ xAx^{-1} = A,\ x^2 \in A,\ |x| \leq \beta(i),$$

i.e. $gp\{A, x\}$ is an α-bounded subgroup of order 2^{i+1}.

Proof Denote by $\mathcal{Y}$ the set of α-bounded subgroups $B \leq A$, for which there exists an element $y = y(B)$ such that

$$y \in G \backslash A,\ yBy^{-1} = B,\ y^2 \in B,\ |y| \leq \beta(s), \tag{2.10}$$

if the order of the subgroup B is equal to 2^s. Let us verify that the set $\mathcal{Y}$ is non-empty. Notice for this goal that A does not contain at least one of the generators a_j of the group G because G is infinite. Since the order of the element a_j is at most $h(1)$, some its power y with an exponent $\leq \frac{1}{2}h(1)$ enjoys the property

$$y \notin A,\ y^2 \in A,\ |y| \leq \frac{1}{2}h(1) \leq \alpha(1) = \alpha(1) - \alpha(0) = \beta(0)$$

by condition (2.2), and the cyclic subgroup $B = \langle y^2 \rangle$ of A is α- bounded. Therefore $B \in \mathcal{Y}$ since, if B has order 2^s, $\beta(s) \geq \beta(0)$ by condition (2.3).

The lemma claim is true if $A \in \mathcal{Y}$. Further arguing 'by contradiction' choose a maximal in $\mathcal{Y}$ subgroup B of an order 2^s that less than the order 2^i of the subgroup A.

By Lemma 2.2 there is an element a satisfying conditions (2.6).

For the elements a and $y = y(B)$ set $L = gp\{B, y, a\}$. Then (2.6) and (2.10) implies that the subgroup B is normal in L, and L/B has at least four elements. By Lemma 2.1 we can construct the series $B = L_0 < \ldots < L_t = L$ for $x_1 = a$ and $x_2 = y$.

Let $M = L \cap A$. Then $M \geq L_1$ since $M \ni a$. Therefore (similarly to Lemma 2.2) M coincides with a subgroup L_i for $i \geq 1$. But $i < t$ since $L \backslash M \ni y$. Consequently, by Lemma 2.1 there exists an element z in $L_{i+1} \backslash L_i$ such that

$$zMz^{-1} = M,\ z^2 \in M,\ |z| \leq \beta(s+i).$$

Now let us take into account that the order of the subgroup $M = L_i$ is equal to 2^{s+i}, that $z \notin A$ because $A \cap L_{i+1} = A \cap L = M = L_i$, and that M is α-bounded by Lemma 2.1. Hence $M \in \mathcal{Y}$, contrary to the choice of the subgroup B; and the lemma is proved. □

3 Lower bounds

Proof (of Theorem 2(1)) By the hypothesis of the theorem the orders of all elements in the balls Ball(r) are bounded by the constant: $h(r) \le 2^k$. Introduce $\alpha(s) = 2^{k-1}(2^{ks} - 1)(2^k - 1)^{-1}$ for every $s \ge 0$. Then $\beta(s) = 2^{k(s+1)-1}$, and conditions (2.2), (2.3) obviously hold.

Starting with the identity subgroup, we can, applying Lemma 2.3, inductively construct an infinite increasing series $1 = A_0 < A_1 < \ldots$ of α-bounded subgroups in G, where the subgroups A_i have orders 2^i. By the definition of an α-bounded subgroup and the function α one gets

$$l(A_i) \le \alpha(i) = 2^{k-1}(2^{ki} - 1)(2^k - 1)^{-1}.$$

Thus, a subgroup of order 2^i does exist in the ball of radius r if $r \ge 2^{k-1}(2^{ki} - 1)(2^k - 1)^{-1}$. Consequently a subgroup of order $2^{\lfloor \frac{1}{k} \log_2((2^{k+1}-2)r+2^k)\rfloor-1}$ necessarily exists in every ball of radius $r \ge 0$. □

Proof (of Theorem 2(2)) Assume that $h(r)$-th are upper bounds for the orders of elements in the balls of radii r of the group G. There are two cases.
(a) There exist infinitely many integers r_j, $j = 1, 2, \ldots$, for which

$$h(r_j) \ge \Phi(r_j), \tag{3.1}$$

where Φ is the function given in the formulation of Theorem 2(2).

Then for every r_j the ball of radius $R_j = \frac{1}{2}r_j h(r_j)$ contains a finite cyclic subgroup of order $h(r_j)$, since the length of any element of a cyclic subgroup of order n, generated by an element of length $\le r$, does not exceed $\frac{1}{2}nr$.

It is immediately verifiable that for $x \ge 1$ and any $y \ge \Phi(x)$ the inequality $\Phi(xy) < 2y$ holds. By setting $x = r_j, y = h(r_j)$ we get $f(R_j) \ge h(r_j) \ge \frac{1}{2}\Phi(2R_j)$ from inequality (3.1).

Thus the balls of radii R_j contain subgroups of orders $> \frac{1}{2}\Phi(R_j)$ for infinitely many R_j-th. But this is impossible if $f(r) = o(\Phi(r))$.
(b) $h(r) < \Phi(r)$ whenever $r \ge C$ for some constant C.

Choose an integer $d \ge 1$ so that $2^d \ge h(r)$ for every $r \le C$. Set $\beta(s) = h(1)2^{g(s)}$ for $s \ge 0$, where $g(s) = \frac{1}{2}s(s+1) + ds - 1$, and set

$$\alpha(s) = \beta(0) + \beta(1) + \ldots + \beta(s-1) < 2h(1)2^{g(s-1)}$$

for $s \ge 1$. Then condition (2.2) is fulfilled. Let us check condition (2.3).

Consider $s \ge 1$. If $2\beta(s-1) \le C$ then $h(2\beta(s-1)) \le 2^d$ by the choice of d. Therefore condition (2.3) follows from the definition of the function g, because $g(s) \ge g(s-1) + d$ for any $s \ge 1$.

As for $2\beta(s-1) > C$ one gets

$$h(2\beta(s-1)) < \Phi(2\beta(s-1)) = 2^{\sqrt{2(g(s-1)+\log_2 h(1)+1)}} < 2^{s+d} = 2^{g(s)-g(s-1)}$$

by the choice of the constants C, d and by the definition of the functions Φ, g. Thus inequality (2.3) is true in this case as well.

By induction, Lemma 2.3 allows us to find an infinite increasing series of α-bounded subgroups $1 = A_0 < A_1 < \ldots$, where A_s has order 2^s. The subgroup A_s is situated in the ball of radius $r_s = \alpha(s) < h(1)2^{\frac{1}{2}s^2+(d-\frac{1}{2})s}$. Hence $f(r_s) \geq 2^{\lfloor\sqrt{2\log_2 r_s}\rfloor-d}$ for an infinite set of integers r_s. However this is impossible if $f(r) = o(\Phi(r))$. Theorem 2 is proved. □

4 Upper bounds

To prove Theorem 3 we need, in addition to the definitions from §1, some more notions from the work [4].

By Theorem C [4], for any period A_i there is unique maximal finite subgroup $\mathcal{F}(A_i)$ of the group $B(i-1)$, normalized by the word A_i, and this subgroup isomorphically embeds in $B(m,n)$ by the natural homomorphism $B(i-1) \to B(m,n)$. By the same theorem every non-trivial element g of finite order of the groups $B(i-1)$ and $B(m,n)$ is conjugate in them to a word $A_j^l T$ (and $j < i$ in the case of $B(i-1)$), where l is not divisible by n and $T \in \mathcal{F}(A_j)$. Moreover, a conjugacy of such words $A_{j_1}^{l_1}T_1$ and $A_{j_2}^{l_2}T_2$ in $B(m,n)$ implies $j_1 = j_2$, $k_1 \equiv k_2 (\bmod n)$ and $T_1 = T_2$ in $B(j_1 - 1)$. The uniquely defined index j for the element g (or for a word W representing it) is called the *height* of g (of W). The height of a non-trivial finite subgroup is defined as the maximal height of its elements. The height of the identity subgroup is regarded equal to 0.

Then, by Theorem C [4] all non-trivial elements of $\mathcal{F}(A_j)$ have heights $< j$ in $B(j)$. At last, if a finite subgroup $K \leq B(m,n)$ has height i, it possesses a subgroup H of index $\leq 2n$, which is conjugate to a subgroup of $\mathcal{F}(A_i)$.

The results in lemmas 18.5 and 10.7 [4], can be combined to yield the following result.

Lemma 4.1 (1) *Every element of the subgroup $\mathcal{F}(A_i)$ can be represented by a word of length $< 1.005|A_i|$.*

(2) *For $T \in \mathcal{F}(A_j)$ and $0 < |l| \leq \frac{n}{2}$, the word $A_j^l T$ cannot be conjugate in $B(m,n)$ to a word having length $\leq (0.89l - 1.005)|A_j|$.*

Now we are ready to give another proof.

Proof (Theorem 3(1)) Consider a subgroup K of a height $i > 0$ lying in the ball of radius r of the group $B(m,n)$. Raising an element of height i to an appropriate power, we get an element of K that is conjugate to $A_i^{\frac{n}{2}}T$, $T \in \mathcal{F}(A_i)$. Therefore by Lemma 4.1(2) the subgroup K has an element of length $> 0.444n|A_i|$ since $n \geq 2^{48}$, i.e.

$$r > 0.444n|A_i| \tag{4.1}$$

At the same time, by Lemma 4.1(1) each element of $\mathcal{F}(A_i)$ has length $< 1.005|A_i|$. Therefore K possesses a subgroup of index at most $2n$ conjugated to a subgroup of $B(m,n)$, that is situated in the ball of radius $1,005|A_i|$. From here and from (4.1) we obtain

$$2nf\left(\left\lfloor\frac{r}{0.441n}\right\rfloor\right) \geq f(r) \tag{4.2}$$

for any $r \geq 0$.

Inequality (4.2) an obvious induction show that for any integer $s \geq 1$ a subgroup of order greater than $(2n)^{s-1}$ of the group $B(m,n)$ can occur only in a ball of radius greater than $\lfloor 0.441n \rfloor^s$. Therefore $f(r) < (2n)^{\log_{\lfloor 0.441n \rfloor} r}$ for any $r > 0$, and Theorem 3(1) is proved. □

Proof (Theorem 6) To prove this for $p = 2$, we modify the definition of the group $B(m,n)$ given in §1, according to the preassigned function h. Replacing it by an equivalent function, we can suppose that $h(r) \geq 2^{48}$ for $r > 0$, and that all values of this function are 2-powers. Instead of definition (1.1) of the groups $B(i)$, let now

$$B_h(i) =< A| \; A_1^{n(1)} = 1, \ldots, A_i^{n(i)} = 1 >, \tag{4.3}$$

where $n(i) = h(|A_i|)$. The group given by all relations $A_i^{n(i)} = 1, i = 1, 2, \ldots$ with non-decreasing n_i-th, is denoted by G_h.

This modification either does not change the formulations and proofs of the statements from [4], or changes them in obvious way. For instance, the period A_i will have order $n(i)$ in G_h instead of n; the terms $n|A_j|$ must be substituted by $n(j)|A_j|$ in the estimates; finite subgroups of the group $B_h(i)$ are isomorphically embedded in a direct power of the dihedral group $D(2n(i))$ (instead of $D(2n)$); n must be changed for $n(i)$ in the laws of Lemma 15.10 [4]; and passing to the rank $i+1$ (in §§18, 19 [4]) one has to substitute the exponent n by $n(i+1)$.

After the modification we obtain, in particular, that any finite subgroup K of a height $i \geq 1$ in G_h possesses a subgroup H of index at most $2n(i)$, that is conjugated in G_h to a subgroup of $\mathcal{F}(A_i)$. By Lemma 4.1 (more accurately, by its analogue for G_h) the height j of H is such that $0.44n(j)|A_j| < |A_i|$, if $j > 0$, in particular, $|A_j| < |A_i|$.

As far as by Theorem B [4] (more exactly, by its analogue for G_h) there is a period A_i of arbitrary length ≥ 1, the ball of radius $r \geq 1$ in G_h has an element of order $h(r)$.

By the definition of periods, every word of length at most r has finite order in a rank i, where $|A_i| \leq r$. Then, by Lemmas 10.2 and 10.3 [4] its order does not exceed $n(i) \leq h(r)$.

Thus, $h(r)$ coincides with the maximum of orders of elements in the ball of radius $r \geq 1$, and Theorem 6 is proved for $p = 2$.

In the case of an odd prime p, we may assume that the values of the function h are sufficiently large (for instance, $h(r) > 10^{10}$ for $r \geq 1$) and are p-powers. In the capacity of G_h, we take the 2-generated group $G(\infty)$ from §25 [3], where all periods and relators are of the first type, the sets of periods $\mathcal{X}_i$ are maximal in every rank $i \geq 1$, and the exponent n_A for a period A of length $r \geq 1$ is equal to $h(r)$ by definition. Then G_h is an infinite p-group by Theorems 26.1 26.2 [3], in which the orders of elements in the balls of radii $r \geq 1$ are at most $h(r)$. To finish the proof of Theorem 6, it suffices to find at least one period of length r (i.e. an element of order $h(r)$ in the ball of radius r by Theorem 26.4 [3]) for every sufficiently large r. The existence of such a period is asserted, for instance, by Lemma 10 [8]. □

Proof (Theorem 3(2)) Reasoning 'by contradiction', we define a function h by the function g as follows. Let $h(0) = 1, h(1) = 2^{48}$, and assume the values $h(1), \dots, h(s-1)$ are already defined for $s > 1$. Then unboundness of the function g leads to existence of such integer x_s that

$$g(x_s) > s2^s h(1)h(2)\dots h(s-1). \tag{4.4}$$

Choose for $h(s)$ the minimal 2-power with the property

$$h(s) > \max(h(s-1), sx_s). \tag{4.5}$$

For the function h one can construct the group G_h, as it was done at the proof of Theorem 6 (in the case $p = 2$), so that the growth function of elements orders for G_h should coincide with h, and the statements from [4] (or their natural analogues) should be true. It was noted above, in particular, that a finite subgroup K of height i has a subgroup of a smaller height j and of index at most $2n(i)$; besides $|A_j| < |A_i|$ if $j > 0$. In view of the equality $n(i) = h(|A_i|)$, we obtain by induction the following inequality for the order of K:

$$\mathrm{card}(K) \le 2^{|A_i|} h(1)h(2)\dots h(|A_i|). \tag{4.6}$$

Now consider the ball in G_h of radius sx_s with $s \ge 3$. By Lemma 4.2 and by definition of the element height we conclude that the height i of a finite subgroup lying in the ball, is such that $|A_i| < s$. Indeed, otherwise the ball of radius sx_s were have an element of length less than $0.44n(i)s$ (that conjugate to $A_i^{\frac{n(i)}{2}}T$, $T \in \mathcal{F}(A_i)$), where $n(i) \ge h(s)$ for $|A_i| \ge s$. But this contradicts to inequality (4.5). Then inequalities (4.6) (4.4) make clear that any finite subgroup lying inside the considered ball, has smaller order than $\frac{1}{s}g(x_s)$.

Thus, $sf(sx_s) < g(x_s)$ for the growth function of finite subgroups f of the group G_h, where $s = 3, 4, \dots$. But this contradicts to the assumption that $g \preceq f$. □

5 Central extensions

In addition to the definitions of the words A_i and the subgroups $\mathcal{F}(A_i)$ given in §§1, 4, recall for the proof of Theorem 5, that there are two kinds of periods A_i [4]. For some of them (call them *even* here) there exists a word J which normalizes the subgroup $\mathcal{F}(A_i)$ of the group $B(i-1)$, and $J^2 \in \mathcal{F}(A_i)$ and $J^{-1}A_iJ = A_iT$ for certain $T \in \mathcal{F}(A_i)$ (J is called *A_i-involution*). There are no A_i-involutions for *odd* periods A_i. The following result follows from [4].

Lemma 5.1 (1) *For an even period A_i and any A_i-involution J equality (5.1) is true in the free group $F_m = B(0)$:*

$$J^{-1}A_i^n J A_i^n = \prod_{l=1}^{s} U_l R_l^{\pm 1} U_l^{-1}. \tag{5.1}$$

(2) *For any period A_i and any word $T \in \mathcal{F}(A_i)$ inequality (5.2) is true in F_m:*

$$T^{-1}A_i^nTA_i^{-n} = \prod_{l=1}^{s} U_lR_l^{\pm 1}U_l^{-1}, \tag{5.2}$$

where in both statements, U_l are some words, each of the words R_l has the form A_j^n for some $j = j(l) < i$, and the number of the factors R_l such that R_l coincides with A_j^n is even for every j.

Proof (1) It is easy to see that

$$J^{-1}A_i^nJA_i^n = (J^{-1}A_i^{\frac{n}{2}}JA_i^{\frac{n}{2}})U(J^{-1}A_i^{\frac{n}{2}}JA_i^{\frac{n}{2}})U^{-1},$$

for $U \equiv A_i^{-\frac{n}{2}}$ in the free group F_m. Therefore to prove the statement (1) it suffices to explain that the word $J^{-1}A_i^{\frac{n}{2}}JA_i^{\frac{n}{2}}$ is equal in F_m to a product $\prod V_lR_l^{\pm 1}V_l^{-1}$, where $R_l \equiv A_j^n$ for some $j = j(l) < i$. But this is true indeed since by Lemma 19.2 [4] $J^{-1}A_i^{\frac{n}{2}}J = A_i^{-\frac{n}{2}}$ in the group $B(i-1)$.

(2) We obtain a similar proof of the statement (2) taking into account that by Lemma 18.5(c) [4] $T^{-1}A_i^{\frac{n}{2}}T = A_i^{\frac{n}{2}}$ in $B(i-1)$. □

Further we suppose the reader is familiar with the notion of a diagram over a group presentation. (See [9], [3] or [4].) Below we shall only discuss *disk* diagrams (i.e. simply connected diagrams Δ on the plane with a boundary $\partial\Delta$) or *spherical* diagrams (i.e. closed diagrams on the sphere).

By a diagram of rank i we mean a diagram over presentation (1.1). A pair of 2-cells Π_1 and Π_2 of rank $j \leq i$ (i.e. corresponding to the relation $A_j^n = 1$) is *reducible* (it was called weakly compatible in [4]) in the following two cases.

(1) Starting from some vertices o_1 and o_2 on the boundaries $\partial\Pi_1$ and $\partial\Pi_2$ of these cells and going along them in the clockwise direction one reads the same word A_j^n or A_j^{-n} for the even period A_j, and there is a simple path t connecting o_1 with o_2 in Δ, labelled by a word J, which is an A_j-involution.

(2) Starting with some vertices o_1 and o_2 on the boundaries $\partial\Pi_1$ and $\partial\Pi_2$ and going in the clockwise direction one reads the words A_j^n and A_j^{-n} (or A_j^{-n} and A_j^n) correspondingly, and there is a simple path t connecting o_1 with o_2 in Δ, labelled by a word T representing in $B(j-1)$ an element of $\mathcal{F}(A_j)$.

A diagram Δ is said to be *reduced* (strictly reduced in [4]) if it has no reducible pairs of cells. Every diagram Δ of rank $i \geq 1$ can be transformed into a reduced diagram Δ_0 of rank i with the same boundary label by several reductions. Carrying out a reduction, one cuts out a reducible pair of cells of a rank $j \leq i$ from a diagram, cuts the diagram along the path t (from the definition of a reducible pair), and then pastes up the arising hole by some diagram Γ of rank $j-1$ with the boundary label $J^{-1}A_j^nJA_j^n$ (in the first case) or $T^{-1}A_j^nTA_j^{-n}$ (in the second case). By Lemma 5.1 the diagram Γ can be chosen with an even number of cells labelled by the word $A_s^{\pm 1}$ for any $s \leq j-1$. Thus we obtain:

Lemma 5.2 *For a diagram Δ of rank $i \geq 1$ there exists a reduced diagram Δ_0 of rank i with the same boundary label (in the disk case) and with the same parities of the numbers of cells of rank j for any $j \leq i$.*

Proof (of Theorem 5) Since the subgroup $R = R(m,n)$ is generated as a normal subgroup of F_m by all powers A_i^n for $i \geq 1$, the central subgroup $Z_2(m,n)$ is generated by natural images $\overline{R_i}$ of the words $R_i \equiv A_i^n$ in the quotient $A_2(m,n)$. Thus to prove Theorem 5 it suffices to lead to a contradiction the assumption that for some $t \geq 1$ and some indices $i_1 < \ldots < i_t$ the equality $\overline{R_{i_1}} \ldots \overline{R_{i_t}} = 1$ holds in $A_2(m,n)$. It gives an equality

$$A_{i_1}^n \ldots A_{i_t}^n = V, \tag{5.3}$$

in the free group F_m, where $V \in [F_m, R]R^2$.

It is clear from the definitions of the subgroups $[F_m, R]$ and R^2 that the word V can be written in F_m as a product $\prod_s U_s A_{j_s}^{\pm n} U_s^{-1}$, where for every l the power $A_l^{\pm n}$ occurs as $A_{j_s}^{\pm n}$ an even number of times. Consequently, one can construct a diagram Δ_1 for the equality of V to the identity in the group $B(m,n)$, having an even number of cells of each rank. Meanwhile the equality of the word $A_{i_1}^n \ldots A_{i_t}^n$ to the identity in $B(m,n)$ gives a diagram Δ_2 with the odd ($= 1$) number of cells of each rank $i_1, \ldots, i_t$. Finally, equality (5.3) makes possible getting a spherical diagram Δ by gluing together the diagram Δ_2 and the mirror copy of the diagram Δ_1. Note that Δ has an odd number of cells of the rank i_1.

By Lemma 5.2 there is a reduced spherical diagram Δ_0 with an odd number of cells of the rank i_1. However the claim of Lemma 6.2 [4] means that any reduced spherical diagram over $B(m,n)$ has no cells of positive rank at all. This contradiction proves Theorem 5. □

Lemma 5.3 (1) *For every finite subgroup K of height i in the group $B(m,n)$, there exists such a preimage H under the natural homomorphism $A_2(m,n) \to B(m,n)$ with the kernel $Z_2(m,n)$, that the intersection $H \cap Z_2(m,n)$ is generated by all $\overline{R_j}$-th for $j \leq i$.*

(2) *If the length of this subgroup K in $B(m,n)$ is equal to r, then for some conjugate to it in $B(m,n)$ subgroup K_1 there is a preimage $H_1 \leq A_2(m,n)$ with the same property as in item (1), but in addition, every element of H_1 can be represented in $A_2(m,n)$ by such a word VU that $|V| \leq 1.2\,r$, and $U \in H_1 \cap Z_2(m,n)$.*

Proof (1) By Theorem C(c) [4] every finite subgroup K of height i has a natural preimage in $B(i)$ (1.1) of the same order. Therefore all elements $g_1, \ldots, g_l$ of K are representable by such words $V_1, \ldots, V_l$ that V_1 is empty, and for any V_p, V_q from this list there exist V_r, V_s and some vanishing in $B(i)$ words $U_{p,q}$ and U_p with

$$V_p V_q = V_r U_{p,q} \text{ and } V_p^{-1} = V_s U_p \tag{5.4}$$

in the free group F_m.

Denote by L the (finite) subgroup in $Z_2(m,n)$ generated by the all $\overline{R_j}$ for $j \leq i$. Introduce the set $\mathcal{W}$ of all words VU, where $V \in \{V_1, \ldots, V_l\}$ and U is any word representing an element from the subgroup $L \leq A_2(m,n)$. Denote by H the finite subset of those elements of $A_2(m,n)$ that are representable by the words of $\mathcal{W}$.

Evidently, H is a preimage of K under the homomorphism $A_2(m,n) \to B(m,n)$. Moreover, $H \cap Z_2(m,n) = L$ by the choice of the words U and V_1. Thus to prove Lemma 5.3(1), it suffices to determine that H is a subgroup in $A_2(m,n)$. Since the second factors of the words from $\mathcal{W}$ represent central elements of the group $A_2(m,n)$, we have just to obtain equalities of the form $V_pV_q = V_rU_{p,r}$ and $V_p^{-1} = V_sU_p$ in the group $A_2(m,n)$ for the all p,q and suitable words $U_{p,q}, U_p$ representing elements of the subgroup L. In fact, such equalities arises from equalities (5.4), because the factors $U_{p,q}$ and U_p from (5.4) vanish in $B(i)$ by (1.1), and therefore their images in $A_2(m,n)$ do belong to L.

(2) On the one hand, by Theorem [4] any finite subgroup K of height i in the group $B(m,n)$ is conjugate in it to a subgroup K_1 possessing such a preimage $K_2 \leq B(i-1)$, that every element of K_2 can be represented by a word V_j of the form $A_i^l T$ or $A_i^l J$ with $|l| \leq \frac{n}{2}$, $|T| < 1.005|A_i|$, $|J| < 2.01|A_i|$ (Lemmas 18.5 and 19.3 [4]), i.e. $|V_j| < (\frac{n}{2} + 2.01)|A_j|$. On the other hand, by Lemma 4.1(2) $r > (0.89\frac{n}{2} - 1.005)|A_i|$, because K must contain an element that is conjugate to $A_i^{\frac{n}{2}} T$ for some $T \in \mathcal{F}(A_i)$. Consequently $|V_j| \leq 1.2r$. As for the rest, see item (1). □

6 Groups with prescribed growth of finite subgroups

Theorems 1 and 4 will be proved uniformly. For this purpose we construct some central extensions of the free Burnside groups $B(2,n)$, where $n = p^k$ is a sufficiently large power of the given prime p (for instance, $n \geq 2^{48}$, if $p = 2$ and $n > 10^{10}$ if $p \geq 3$). These extensions $G(f)$ arises as quotients of the groups $A_p(2,n)$ defined by given functions f.

For $p = 2$ the groups $A_p(m,n)$ were defined §1 and investigated in §5. For an odd p the definition is quite similar: $A_p(m,n) = F_m/[F_m,R]R^p$, where R is the kernel of the natural presentation $F_m \to B(m,n)$. By Theorem 19.1 [3] for large even n, the subgroup R is the normal closure of the words A^n for all periods A. (The definition of a period can be found in [3], §18, but we shall not use it explicitly.) By Corollary 31.1 [3] the subgroup $R/[F_m,R]$ of the group $F_m/[F_m,R]$ is free abelian with the basis consisting of the images of all powers A^n of periods in $R/[F_m,R]$. Therefore the natural analogue of Theorem 5 about the quotient $Z_p(m,n) = R/[F_m,R]R^p$, is true for odd p-th as well.

Lemma 6.1 *There exist constants $d > 1$ and l_0 such that for $m > 1$ the number of periods of length l is at least d^l if $l \geq l_0$.*

Proof In the 'odd' case the statement can be found in [8], Lemma 10. Thus, below we consider the case $n = 2^k$.

Let M_l be a maximal set of words of length l in the alphabet $\{a_1^{\pm 1}, \ldots, a_m^{\pm 1}\}$ with the following property. Every word of M_l is cyclically reduced, is not a proper power in the free group and does not contain non-empty subwords of the form D^6;

and if B, C are different words of M_l then B is not a cyclic permutation of the word $C^{\pm 1}$. Then all words of the set M_l can be included in the total list of periods as was explained in Lemma 18.2 [4]. (In [4] the existence of just one period of every length was proved, but used there properties of the word A_{i+1} make possible, with no varying of the proof, to include the whole M_l in the list of periods.) It remains to notice that the set M_l contains at least d^l words for some $d > 1$ and any $l \geq l_0$, but this was proved, for instance, in Lemma 10 [8]. □

Now consider a function $g(l) = p^{s(l)}$, where $s(l)$ is an integral-valued non-decreasing function such that

$$s(l) \geq 0 \textit{ and } p^{s(l)} \leq d^l \textit{ for } l = 0, 1\ldots.. \tag{6.1}$$

Lemma 6.1 and condition (6.1) make possible to get the subsets $S_{l_0}, S_{l_0+1}, \ldots$, where S_l consists of some $g(l)$ periods of length l. Also the definition of the function g allows us to construct an auxiliary series of elementary abelian p-groups

$$T_{l_0} \leq T_{l_0+1} \leq \ldots, \tag{6.2}$$

where $\text{card}(T_i) = g(i)$. Let $T(g) = \cup_{i \geq l_0} T_i$.

Define now an epimorphism $\psi = \psi_g$ of the group $Z_p(2,n)$ onto $T(g)$ with help of Theorem 5 (or its 'odd' analogue) specifying just images of the generating elements $\overline{R_j}$ (which, in their turn, are images of the words A_j^n in $A_p(2,n)$, i.e. are in one-to-one correspondence with the periods) as follows. If a period A_j does not belong to the set $S = \cup_{l \geq l_0} S_l$, then pose $\psi(\overline{R_j}) = 1$. The set of those $\overline{R_j}$ for which the relevant periods belong to S_l for $l \geq l_0$, is one-to-one mapped by ψ onto the subgroup T_l of the group $T(g)$. This is possible since $\text{card}(T_l) = \text{card}(S_l)$.

Denote by N_g the kernel of the homomorphism ψ, and by M_g the quotient $A_p(m,n)/N_g$. (Recall that the subgroup N_g is central in $A_p(m,n)$.) By the construction of ψ, every non-trivial element z of the central subgroup $Z_p(2,n)/N_g \cong T(g)$ can be represented in M_g by a word A_j^n. The minimal length of such periods A_j will be denoted by $\lambda(z)$. The homomorphism ψ induces an identification of the group $Z_2(m,n)/N_g$ with $T(g)$ such that $\lambda(z) = l$ iff $z \in T_l \backslash T_{l-1}$.

Lemma 6.2 *If $\lambda(z) = l \geq l_0$ then the length $|z|$ of the element z in M_g satisfies the inequalities*

$$0.9nl < |z| \leq nl. \tag{6.3}$$

Proof The inequality $|z| \leq nl$ is obvious since the element z is representable in M_g by a word A_j^n for a period A_j of length l.

Assume now, that z can be represented by a word Z with $|Z| \leq 0.9nl$. Since Z vanishes in $B(m,n)$, there is a reduced disk diagram Δ over this group with the boundary label Z. The inequality $|Z| \leq 0.9nl$ implies that the perimeter of every cell of Δ is less than nl (by Corollary 17.1 and Lemma [3] in the 'odd' case, and by Lemmas 6.2 and 9.2 [4] in the 'even' case). Hence these cells correspond to periods of lengths less than l. Therefore the word Z is representable in the free group F_m as a product $\prod U_j A_{i_j}^{\pm n} U_j^{-1}$ with $|A_{i_j}| < l$. Then in the group $A_p(2,n)$, the word Z

is a product of the corresponding $\overline{R_{i_j}}$-th, and in M_g, Z is a product of their images under ψ. Now it follows from the inequalities $|A_{i_j}| < l$ and the construction of the epimorphism ψ that the word Z represents an element of the subgroup T_{l-1}. Hence $\lambda(z) \leq l-1$, contrary to hypothesis. □

Lemma 6.3 *If $nr \geq l_0$ then the maximal order $\phi(nr)$ of subgroups $K \leq T(g)$ with $l(K) \leq nr$ in the group M_g, satisfies the inequalities*

$$g(r) \leq \phi(nr) \leq g(2r).$$

Proof By Lemma 6.2 an element z of $T(g)$ with $\lambda(z) = l$ can have length $\leq nr$ only in case $r > 0.9l \geq 0.5l$. Then z certainly belongs to the subgroup T_{2r}. Therefore $\mathrm{card}(K) \leq \mathrm{card}(T_{2r}) = g(2r)$. Again by Lemma 6.2 the subgroup $T(r)$ is situated in the ball of the radius nr. Thus $\phi(nr) \geq \mathrm{card}(T_r) = g(r)$. □

Proof (Theorem 1) For the function f we construct the function s keeping the rule

$$s(r) = \min(\lfloor \log_p f(r) \rfloor, \lfloor \log_p(d^r) \rfloor). \tag{6.4}$$

Then set $g(r) = p^{s(r)}$ and define $G(f) = M_g$ (where M_g was constructed by given function g above). (The condition (6.1) follows from (6.4).) Its exponent divides $pn = p^{k+1}$ because the exponent of $T(g)$ divides p and $M_g/T(g) \cong B(m,n)$. Since $d > 1$, there exists an integer C such that $d^{Cr} > a^r$ for the fixed in the theorem hypothesis constant a and for all $r \geq 0$. Then by (6.4) $f(r) \leq pg(Cr)$. Therefore for the growth function of finite subgroups γ of the group $G(f)$, by Lemma 6.3 one obtains for any $r \geq \frac{l_0}{n}$ that

$$\gamma(nCr) \geq \phi(nCr) \geq g(Cr) \geq \frac{1}{p} f(r). \tag{6.5}$$

On the other hand, by Theorem 19.6 [3] any finite subgroup of the quotient $M_g/T_g \cong B(m,n)$ has at most $n = p^k$ elements. From here and from Lemma 6.3 $\gamma(nr) \leq ng(2r) \leq nf(2r)$, and together with (6.5) this proves Theorem 1. □

Proof (Theorem 4). By Theorem 3(1), for any $\epsilon > 0$ there is a power $n = 2^k$ such that the growth function of finite subgroups $f_{2,n}$ of the group $B(2,n)$ satisfies the relation $f_{2,n}(r) = o(r^{1+\epsilon})$, because $b_k \to 1$ while $k \to \infty$. Therefore $q(r) = f(r)f_{2,n}(r)^{-1} \geq 1$ for $r \geq r_0$. (Further we may assume that $l_0 \geq r_0$, for the constant l_0 from Lemma 6.1.) Also we may assume that the n is chosen sufficiently large so that

$$\lfloor 0.441n \rfloor^{1+\epsilon} \geq 2n. \tag{6.6}$$

By the hypothesis of the theorem, $f(r)r^{-1-\epsilon} < f(cr)(cr)^{-1-\epsilon}$ for $c = \lfloor 0.441n \rfloor > 1$, and by Theorem 3(1) $f_{2,n}(cr) \leq 2nf_{2,n}(r)$. These two inequalities and inequality (6.6) imply that $q(r) < q(cr)$. Therefore the non-decreasing function $Q(r) = \max_{l \leq r} q(l)$ has the property that for any $r_1 \geq r \geq 0$ the segment $[r_1, cr_1]$ contains a number ρ with

$$Q(r) < q(\rho). \tag{6.7}$$

Now define a non-negative, non-decreasing function s by the rule $s(r) = 0$ for $r < l_0$, and for $r \geq l_0$

$$s(r) = \min(\lfloor \log_2 Q(r) \rfloor, \lfloor \log_2 d^r \rfloor). \tag{6.8}$$

Set $g(r) = 2^{s(r)}$, and for the group G we take the central extension M_g of the group $B(2, 2^k)$ of exponent 2^{k+1} constructed above. (Condition (6.1) follows from (6.8).) The growth function of finite subgroups δ of the group M_g obviously satisfies the inequality $\delta(r) \leq f_{2,n}(r)\phi(r)$, where the function ϕ was defined in Lemma 6.3. Therefore by Lemma 6.3 and by inequality (6.7)

$$\delta(nr) \leq g(2r) f_{2,n}(nr) \leq Q(2r) f_{2,n}(nr) < q(\rho) f_{2,n}(nr)$$

for some $\rho \in [nr, cnr]$. From here

$$\delta(nr) < q(\rho) f_{2,n}(\rho) = f(\rho) \leq f(cnr). \tag{6.9}$$

Then we can choose an integer C such that $d^{Cr} \geq Q(r)$ in view of the similar requirement of Theorem 4 imposed on f. For $Cr \geq l_0$ we get by (6.8)

$$q(r) \leq Q(r) \leq 2g(Cr). \tag{6.10}$$

Consider now a subgroup K of the order $f_{2,n}(Cnr)$ in the ball of radius Cnr of the group $B(2, n)$. Passing to a conjugate subgroup K_1, we can find such a preimage H_1 of it in $A_2(2, n)$ by Lemma 5.3, that any element of H_1 can be represented by a word VU, where $|V| \leq 1.2Cnr$, and U is a product (in $A_2(m, n)$) of the generators $\overline{R_j}$, for which the lengths of corresponding periods A_j are at most $l = |A_i|$, if i is the height of K. Hence under the natural homomorphism of the group $A_2(m, n)$ onto M_g, the image of the word U finds itself in T_l (or in 1 for $l < l_0$). Thus by Lemma 6.2 it represents an element of a length at most nl in M_g.

Lemma 4.1(2) shows that $l = |A_i| < (0.44n)^{-1}Cnr$ in a standard way. This means that any element of the image L of the subgroup H_1 in M_g is representable by a word of a length at most $\leq 1.2Cnr + (0.44)^{-1}Cnr \leq 4Cnr$. We can keep this condition if we enlarge L, if necessarily, regarding T_s as a subgroup of L for $s = Cr$. Then (for $Cr \geq l_0$) the order of the subgroup L is not less than the product of the orders of K and T_s, i.e. it is not less than $f_{2,n}(Cnr)g(Cr)$. From here and from (6.10) for $r \geq \frac{l_0}{C}$ we get

$$\delta(4Cnr) \geq \frac{1}{2} f_{2,n}(Cnr) q(r) \geq \frac{1}{2} f_{2,n}(r) q(r) = \frac{1}{2} f(r).$$

The last inequality together with (6.9) prove Theorem 4. □

References

[1] O. Yu. Schmidt. *Selected Works. Mathematics*, Izdat. Akad. Nauk SSSR, Moscow, 1959 (in Russian).

[2] S. I. Adian. *The Burnside Problem and Identities in Groups*, Nauka, Moscow, 1975; English transl.: Springer-Verlag, New York, 1979.

[3] A. Yu. Ol'shanskii. *Geometry of Defining Relations in Groups*, Nauka, Moscow, 1989; English transl.: Math. Appl. (Soviet Ser.) 70 (1991).
[4] S. V. Ivanov. The free Burnside groups of sufficiently large even exponents, *Internat. J. Algebra Comput.* 4 (1994), 1–308.
[5] I. G. Lysenok. Infinite Burnside groups of even exponent, *Izvestia Ross. Acad Nauk, Ser. Mat.* **60** (1996), 3–224 (in Russian).
[6] R. I. Grigorchuk. The growth rates of finitely generated groups and the invariant mean, *Izvestia Akad. Nauk SSSR, Ser. Mat.* **48** (1984), 572–589 (in Russian).
[7] A. V. Rozhkov. On a question of Grigorchuk and on the Grigorchuk group, in: Abstract of the Int. Algebraic Conf., St.Petersburg, 1997, 266–268 (in Russian).
[8] A. Yu. Olshanskii. Distortion functions for subgroups, *Proc. Conf. Geometric Group Theory, Canberra, 1996*, Berlin, Walter de Gryuter, 1998 (to appear).
[9] R. C. Lyndon and P. E. Schupp. *Combinatorial group theory*, Springer, Berlin, 1977.

SYMPLECTIC AMALGAMS AND EXTREMAL SUBGROUPS

CHRISTOPHER PARKER* and PETER ROWLEY†

*University of Birmingham, Edgbaston, Birmingham B15 2TT, England
†Department of Mathematics, University of Manchester Institute of Science and Technology, P.O. Box 88, Manchester M60 1QD, England

Abstract

We define symplectic amalgams and illustrate the definition with a quick look at some of those amalgams that have already been trapped. To motivate the discussion of the classification of symplectic amalgams we discuss a particular configuration that leads to the amalgam that can be found in $E_6(q)$. This particular example demonstrates how extremal subgroups in Lie type groups come to the fore.

1 Introduction

An amalgam (of rank 2) $\mathcal{A} = (P_1, P_2, B)$ consists of three groups P_1, P_2 and B together with group homomorphisms ϕ_1 , ϕ_2 where

$$\begin{aligned} \phi_1 : B &\to P_1 \text{ and} \\ \phi_2 : B &\to P_2. \end{aligned}$$

We shall only consider amalgams where the ϕ_i are monomorphisms and so will feel free to identify B with its images $\phi_i(B)$. Also the groups P_1 and P_2 will always be assumed to be finite.

Throughout p will denote a prime number.

Definition 1.1 An amalgam $\mathcal{A} = (P_1, P_2, B)$ is a p-constrained amalgam provided

(i) no non-trivial subgroup of B is normal in both P_1 and P_2;

(ii) for $S \in \mathrm{Syl}_p(B)$ we have $B = N_{P_i}(S)$, $i = 1, 2$; and

(iii) $C_{P_i}(O_p(P_i)) \leq O_p(P_i)$, $i = 1, 2$.

Over the past ten or so years p-constrained amalgams have been widely studied. The applications so far are to the revision of the classification of simple groups of characteristic 2-type. In these cases the structure of $P_i/O_p(P_i)$ is usually specified and the aim then is to restrict the structure of P_i. This involves bounding a parameter, b, called the critical distance, which we define shortly. A notable recent success is Stellmacher's [6] revision of part of the N-group paper.

For a p-constrained amalgam, as in Definition 1.1, and for $i = 1, 2$, we define the following subgroups:–

$$L_i = O^{p'}(P_i),\ Z_i := \langle \Omega_1(Z(S))^{P_i} \rangle.$$

Definition 1.2 Let $q = p^a$ be a power of the prime p. An amalgam $\mathcal{A} = (P_1, P_2, B)$ is a symplectic amalgam (over $\mathrm{GF}(q)$) if it is a p-constrained amalgam satisfying the following:–

(i) $L_1/O_p(L_1) \cong \mathrm{SL}_2(q)$;

(ii) There exist $x \in L_1$ such that $O^p(\overline{L_2}) \leq \langle \overline{(O_p(L_2) \cap O_p(L_1))^x}^{\overline{L_2}} \rangle$ (where $\overline{L_2} = L_2/O_p(L_2)$);

(iii) $Z_2 = \Omega_1(Z(S)) \leq \Omega_1(Z(L_2))$; and

(iv) $Z_1 \leq O_p(L_2)$ and there exists $y \in L_2$ such that $[Z_1, Z_1^y] \neq 1$.

Note that (1.2) (i) and (ii) impose restrictions upon $L_i/O_p(L_i)$ though condition (1.2) (ii) is rather weak and allows such things as direct and wreath products.

There is a wide and diverse collection of examples of symplectic amalgams in captivity. Before looking at some of these, we first reinterpret symplectic amalgams into a geometric setting and, along the way, establish some notation.

Let $\mathcal{A} = (P_1, P_2, B)$ be a p-constrained amalgam and let $G = P_1 *_B P_2$ be the free amalgamated product of P_1 and P_2 over B. Since our objective is to uncover the mysteries of P_1 and P_2, there is no loss in working within G. By identifying P_1 and P_2 with their images in G we may regard P_1 and P_2 as subgroups of G. We let Γ be the graph whose vertex set is

$$V(\Gamma) = \{P_i g \mid g \in G, i \in \{1,2\}\}$$

and edge set is

$$E(\Gamma) = \{(P_i g, P_j h) \mid g, h \in P_i g \cap P_j h \neq \emptyset, i \neq j\}.$$

Then G acts upon Γ by right multiplication and, by (1.1) (i), $G \leq \mathrm{Aut}(\Gamma)$. Further, G has two orbits on $V(\Gamma)$, one orbit on $E(\Gamma)$ and for adjacent vertices δ, λ of Γ we have that $\{G_\delta, G_\lambda\}$ is G-conjugate to $\{P_1, P_2\}$ and $G_{\delta\lambda}$ is G-conjugate to B. We now use the action of G on Γ to help us find our way through the undergrowth and trap G_δ and G_λ. Here are some important subgroups of G:–

Notation Suppose that $(\delta, \lambda) \in E(\Gamma)$. Then

$$\begin{aligned}
L_\delta &:= O^{p'}(G_\delta)\\
Q_\delta &:= O_p(L_\delta)\\
S_{\delta\lambda} &:= \text{is the Sylow } p\text{-subgroup of } G_{\delta\lambda}\\
Z_\delta &:= \langle \Omega_1(Z(S_{\delta\lambda}))^{L_\delta} \rangle\\
V_\delta &:= \langle Z_\lambda^{L_\delta} \rangle\\
W_\delta &:= \langle (Q_\delta \cap Q_\lambda)^{L_\delta} \rangle\\
T_\delta &:= \mathrm{core}_{G_\delta}(Q_\delta \cap Q_\lambda).
\end{aligned}$$

The critical distance, b, is defined by

$$\begin{aligned}
b &= \min_{\delta \in V(\Gamma)} b_\delta \text{ where}\\
b_\delta &= \min_{\mu \in V(\Gamma)} \{\mathrm{d}(\delta, \mu) \mid Z_\delta \not\leq Q_\mu\}
\end{aligned}$$

(where d(,) is the distance function on Γ).

In this setting the conditions listed in Definitions 1.1 and 1.2 become:

Lemma 1.3 *For $(\alpha, \beta) \in E(\Gamma)$ with G_α a G-conjugate of P_1, the following hold:*

(i) *no non-trivial subgroup of $G_{\alpha\beta}$ is normal in both G_α and G_β.*

(ii) $N_{G_\alpha}(S_{\alpha\beta}) = G_{\alpha\beta} = N_{G_\beta}(S_{\alpha\beta})$.

(iii) $C_{G_\gamma}(Q_\gamma) \leq Q_\gamma$ *for* $\gamma \in \{\alpha, \beta\}$.

(iv) $L_\alpha/Q_\alpha \cong \mathrm{SL}_2(q)$.

(v) $O^p(L_\beta) \leq Q_\beta \langle W_\alpha^{L_\beta} \rangle$.

(vi) $Z_\beta = \Omega_1(Z(S_{\alpha\beta})) \leq \Omega_1(Z(L_\beta))$.

(vii) $Z_\alpha \leq Q_\beta$ *and there exists* $y \in G_\beta$ *such that* $[Z_\alpha, Z_\alpha^y] \neq 1$.

In (1.3) put $\alpha' = \alpha^y$. Since $Z_{\alpha'} \leq Z(Q_{\alpha'})$, (1.3) (vii) implies that $Z_\alpha \not\leq Q_{\alpha'}$ and so, as G_α is transitive upon the neighbours of α, $b_\alpha = 2$, while (1.3) (vi) gives $b_\beta \geq 2$ (in fact $b_\beta = 3$). Therefore, $b = b_\alpha = 2$.

As mentioned earlier, the overall strategy in amalgam problems is first to bound b (typically $b \leq 3$ is good). Then, using this bound, pin down the structure of G_δ ($\delta \in V(\Gamma)$) generally by *ad hoc* arguments. As a general rule, amalgam examples live in small b land. So here we are concentrating on the latter part of this programme. We aim to give a uniform treatment of the $b = 2$ case.

The approach that we take to classifying symplectic amalgams involves studying the action of L_β on Q_β to first show that typically L_β operates irreducibly on Q_β/Z_β, then to show that typically $O_{p'}(L_\beta/Q_\beta)$ is central in L_β/Q_β and finally to show that typically there is a unique component in L_β/Q_β. There are many exceptions to all these statements and they are classified. After this trek we may assume that L_β/Q_β is almost almost simple. At this stage we invoke our $\mathcal{K}$-group hypothesis and deal with the simple groups class by class. At the present time (July 1997), the classification of the structure of symplectic amalgams is almost complete and here we offer an overview and commentary on certain aspects of this project [4]. This we do by sketching, in Section 4, the proof of the following typical result from [4].

Theorem 1.4 *Assume that (1.3) holds and in addition that*

(i) *L_β/Q_β is isomorphic to a quotient by a central subgroup of $\mathrm{SL}_m(p^r)$ with $m \geq 4$; and*

(ii) *no element of L_β/Q_β acts as a $(1,q)$- or a $(2,q)$-transvection on Q_β/Z_β.*

Then $L_\beta/Q_\beta \cong \mathbb{Z}_{(q-1,2)}\mathrm{L}_6(q)$ and Q_β is an ultraspecial p-group of order q^{1+20}.

We say that $x \in L_\beta$ acts as a $(1,q)$-transvection (respectively $(2,q)$-transvection) upon $\overline{Q_\beta}$, if $[\overline{Q_\beta} : C_{\overline{Q_\beta}}(x)] = q$ (respectively q^2) where $\overline{Q_\beta} = Q_\beta/Z_\beta$.

Next, as promised earlier, we view some of the members of the symplectic amalgam species.

2 The symplectic amalgam zoo

Though there are over fifty 'minimal' symplectic amalgams in captivity at the moment, in this brief tour through the zoo we will only look at some of the more unusual ones. Though we will also peek into the generic enclosure. Among the symplectic amalgams with non-irreducible action of L_β/Q_β on Q_β/Z_β the most remarkable creature is contained in the exceptional group $G_2(4)$. In this case $L_\alpha/Q_\alpha \cong SL_2(4)$, Q_α contains two non-central chief factors and $L_\beta \sim 2^{2+(4\oplus 4)}SL_2(4)$. Her striking feature is that unlike all the other symplectic amalgams the L_β-module Q_β/Z_β has field of definition GF(2) where we would have expected GF(4) as $|Z_\beta| = 4$. This apparent anomaly can be explained when we view the typical $G_2(2^n)$ symplectic amalgam. In general, Q_β/Z_β can be identified with $2 \otimes 2^\sigma$ where 2 represents the natural $GF(2^n)SL_2(2^n)$ module and σ is a primitive element of the Galois group $Gal(GF(2^n) : GF(2))$. Thus when $n = 2$ we see that $|Gal(GF(2^n) : GF(2))| = 2$ and Q_β/Z_β has field of definition GF(2) whereas in all other cases the field of definition is $GF(2^n)$. We also remark that for $q = p^n$ with $p \geq 5$, $G_2(q)$ contains a symplectic amalgam over $GF(q)$ whereas $G_2(3^n)$ does not contain a symplectic amalgam over $GF(3^n)$.

L_α	L_β
$2^{2_1+1_{11}+2_{11}+1_{10}}SL_2(2)$	$2_+^{1+24}Co_1$
$3^{2_1+1_5+2_5+1_2}SL_2(3)$	$3_+^{1+12}2{\cdot}Suz$
$5^{2_1+1_2+2_2}SL_2(5)$	$5_+^{1+6}2{\cdot}J_2$
$7^{2_1+1+2_1}SL_2(7)$	$7_+^{1+4}2{\cdot}Alt(7)$

The Monster cage

In the above description of the L_α-chief series in Q_α we have used $3^{2_1+1_5+2_5+1_2}$ to describe a chief series that has (going from Z_α up to Q_α) first a 2-dimensional chief factor followed by five central chief factors and then five chief factors of dimension 2 and finally two central L_α-chief factors. And similarly for our other examples in the cage.

While investigating the structure of L_β/Q_β we must confront the possibility that there are components of L_β/Q_β which are not normal in L_β/Q_β. It is in just such a situation that the next two critters rear their heads. In fact they are the only 'minimal' symplectic amalgam examples that arise in this situation. They are at glance the same. They have

$$L_\alpha \sim 2^{2_1+1_3+2_3+1_2}SL_2(2)$$

and

$$L_\beta \sim 2_+^{1+8}(Alt(5) \wr 2).$$

A well-known example of such an amalgam lives in the Harada-Norton house at the top of the hill. However, on examining the bowels of the beasts we discover that on restriction to a $2'$-component X of L_β for one of the amalgams Q_β/Z_β is a direct sum of two natural $SL_2(4)$-modules (the Harada-Norton) while the other has $[Q_\beta, X] \cong 2_-^{1+4}$. Not surprisingly the Harada-Norton is a far rarer amalgam. The other occurs in $PSp(8,p)$ for all odd p.

We now tip toe over and have a look in the big game reserve where we expect to see the ten infinite families. To clarify our vision we set $\overline{L_2} = L_2/Q_2$.

$\overline{L_2}/Z(\overline{L_2})$	$O_p(L_2)$	Natural Habitat
$L_2(q)$, $(q,3)=1$, $q \geq 5$	q_+^{1+4}	$G_2(q)$
$L_2(q^3)$,	q_+^{1+8}	${}^3D_4(q)$
$L_2(q) \times P\Omega_{2m}^{\pm}(q)$	q_+^{1+4m}	$\Omega_{2m+4}^{\pm}(q)$
$L_2(q) \times P\Omega_{2m+1}(q)$, q odd	q_+^{1+4m+2}	$\Omega_{2m+5}(q)$
$U_6(q)$	q_+^{1+20}	${}^2E_6(q)$
$L_6(q)$	q_+^{1+20}	$E_6(q)$
$PSp_6(q)$, q odd	q^{1+14}	$F_4(q)$
$P\Omega_{12}^+(q)$	q_+^{1+32}	$E_7(q)$
$E_7(q)$	q_+^{1+56}	$E_8(q)$

The generic enclosure

Below we give some of the subgroup lattice of L_α and L_β for the $E_6(q)$ example.

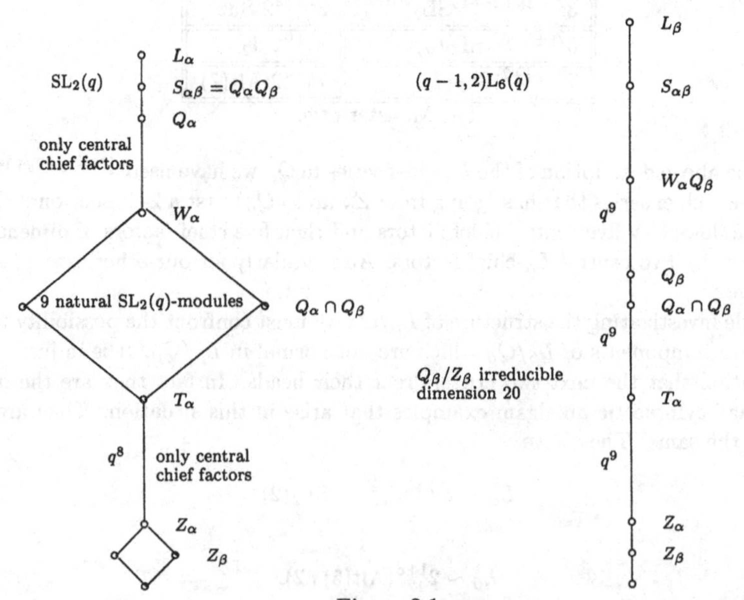

Figure 2.1

Finally, we pass the Lyons house, here is a (bearly visible) further example that we could not resist. We have $L_\beta \sim 2_-^{1+4}\mathrm{Sym}(5)$ and $L_\alpha \sim 2^{2_1+1_1+2_2}SL_2(2)$.

3 Extremal subgroups

In this section we make preparations for the expedition in Section 4 with a few remarks about extremal subgroups in Chevalley groups of type A_n. First we define extremal subgroups. So let $q = p^a$, k denote GF(q) and let U be the unipotent subgroup of the Chevalley group $G(k)$ defined by

$$U = \prod_{\gamma \in \Phi^+} X_\gamma,$$

where $X_\gamma = \langle x_\gamma(t) \mid t \in k \rangle$ is the root subgroup of $G(k)$ associated with γ. Here Φ^+ is an appropriate system of positive roots of the root system associated with $G(k)$. For each fundamental root α we define

$$U_\alpha = \prod_{\gamma \in \Phi^+ \setminus \{\alpha\}} X_\gamma.$$

Definition 3.1 A subgroup A of U is called an extremal subgroup (of U) if
(i) A is an abelian normal subgroup of U; and
(ii) $A \not\leq U_\alpha$ for some fundamental root α.

When faced by a charging Chevalley group it pays to have a labelled Dynkin diagram to hand. For type A_n we use

$$\alpha_1 \quad \alpha_2 \quad \cdots \quad \alpha_n$$

From [1] we extract the following result:

Theorem 3.2 *Suppose that $G(k)$ is of type A_n with n at least 3 and that A is an extremal subgroup of U. Then either*
(i) *$A \not\leq U_\alpha$ for a unique fundamental root α; or*
(ii) *$G(k)$ has type A_3, char $k = 2$, $|A| \leq q^3$ and $A \leq U_{\alpha_2}$, $A \not\leq U_{\alpha_1}$ and $A \not\leq U_{\alpha_3}$.*

Extremal subgroups in the remaining Chevalley groups and in the twisted groups are examined in [1], [2] and [3].

4 Big game hunting

This section aims to give an idea of the methods and strategies employed to capture live symplectic amalgams. From now on we assume that (1.3) holds with $\alpha' = \alpha^y$ and λ is a fixed neighbour of α not equal to β. In our first two results we summarise some of the common properties of symplectic amalgams:

Lemma 4.1 (i) $S_{\alpha\beta} = Z_{\alpha'}Q_\alpha = V_\beta Q_\alpha = Q_\beta Q_\alpha$.
(ii) *Z_α is isomorphic to a natural GF$(q)L_\alpha/Q_\alpha$-module (so $|Z_\alpha| = q^2$) and is the unique minimal normal subgroup of L_α.*

(iii) $Z_\alpha = Z_\beta \times Z_\mu$ *for any neighbour* μ *of* α *with* $\mu \neq \beta$.

(iv) $[Z_\alpha, Z_{\alpha'}] = [Z_\alpha, Q_\beta] = [Z_\alpha, S_{\alpha\beta}] = Z_\beta \cong \mathrm{GF}(q)^+$.

(v) $T_\alpha = Q_\lambda \cap Q_\alpha \cap Q_\beta \trianglelefteq L_\alpha$, T_α *is elementary abelian and all* L_α *chief factors of* T_α/Z_α *are central.*

(vi) $W_\alpha = (Q_\lambda \cap Q_\alpha)(Q_\alpha \cap Q_\beta) \trianglelefteq L_\alpha$ *and all* L_α *chief factors of* Q_α/W_α *are central.*

(vii) $\Omega_1(Z(Q_\alpha)) \cap T_\alpha = Z_\alpha$.

Turning to Q_β we have

Lemma 4.2 *Let* X *be a maximal subgroup of* Z_β *and set* $\widetilde{Q_\beta} = Q_\beta/X$.

(i) $\Phi(Q_\beta) = Q'_\beta = Z_\beta$ *(and so* Q_β/Z_β *may be regarded as a* $\mathrm{GF}(p)L_\beta/Q_\beta$*-module).*

(ii) $\widetilde{Q_\beta}/Z(\widetilde{Q_\beta})$, *as a* $\mathrm{GF}(p)$*-vector space, admits an* L_β*-invariant non-degenerate symplectic form.*

(iii) *Let* m_β *be the p-rank of* L_β/Q_β, $w_{\alpha\beta} = \log_p |W_\alpha Q_\beta/Q_\beta|$, $l_\beta = \log_p |Z(\widetilde{Q_\beta})|$ *and* $a = \log_p(q)$. *Then*

$$\log_p |Q_\beta| \leq 3a + 2w_{\alpha\beta} - l_\beta + 1 \leq 3a + 2m_\beta.$$

It is worth referring back to Figure 2.1 to see what a paradigm the $\mathrm{E}_6(q)$ configuration is of a symplectic amalgam. Lemma 4.2 (iii) underpins much of the analysis of symplectic amalgams. At a very crude level, the aim is, using this result, to bound $|Q_\beta|$ and then to exploit the representation theory of L_β/Q_β to further restrict the structure of L_β/Q_β and Q_β. We also mention that the amalgams satisfying (1.2) were christened symplectic amalgams because of Lemma 4.2 (ii).

Next we prove an elementary, but important result, which points to other factors that influence our overall strategy in dealing with symplectic amalgams.

Lemma 4.3 *Let* $N_{\alpha\beta}$ *be a non-trivial normal subgroup of* $N_{L_\alpha}(S_{\alpha\beta})$ *satisfying* $N_{\alpha\beta} \leq Q_\alpha \cap Q_\beta$ *and* $Z_\alpha \not\leq N_{\alpha\beta}$. *Then* $[Q_\lambda \cap Q_\alpha, N_{\alpha\beta}] \leq Z_\beta$ *and, for* $y \in N_{\alpha\beta}$, $[Q_\lambda : C_{Q_\lambda}(y)] \leq q^2$. *In particular, if* $y \in N_{\alpha\beta} \setminus Q_\lambda$, *then* y *acts as a* $(1,q)$*- or* $(2,q)$*-transvection on* Q_λ/Z_λ.

Proof First we look at $T_\alpha \cap N_{\alpha\beta}$. Put $X = (N_{\alpha\beta} \cap T_\alpha)Z_\alpha$. So, since $Z_\alpha \leq X \leq T_\alpha$, Lemma 4.1 (v) implies that $X \trianglelefteq L_\alpha$. Therefore, using $Z_\alpha \leq Z(Q_\alpha)$,

$$[X, Q_\alpha] = [(N_{\alpha\beta} \cap T_\alpha)Z_\alpha, Q_\alpha] = [N_{\alpha\beta} \cap T_\alpha, Q_\alpha]$$

is also a normal subgroup of L_α. Noting that $[N_{\alpha\beta} \cap T_\alpha, Q_\alpha] \leq N_{\alpha\beta}$ and $Z_\alpha \not\leq N_{\alpha\beta}$, Lemma 4.1 (ii) implies that $[N_{\alpha\beta} \cap T_\alpha, Q_\alpha] = 1$. Hence

$$N_{\alpha\beta} \cap T_\alpha \leq \Omega_1(Z(Q_\alpha)) \cap T_\alpha = Z_\alpha$$

by Lemma 4.1 (vii). Then $N_{\alpha\beta} \cap T_\alpha$ is a proper $N_{L_\alpha}(S_{\alpha\beta})$-invariant subgroup of Z_α, whence, by properties of natural $\mathrm{GF}(q)\mathrm{SL}_2(q)$ modules, $N_{\alpha\beta} \cap T_\alpha \leq Z_\beta$.

Now $[Q_\alpha \cap Q_\lambda, N_{\alpha\beta}] \leq Q_\alpha \cap Q_\lambda \cap N_{\alpha\beta} = Q_\alpha \cap Q_\lambda \cap Q_\beta \cap N_{\alpha\beta} = T_\alpha \cap N_{\alpha\beta} \leq Z_\beta$. Let $y \in N_{\alpha\beta}$. Since $Z_\beta \leq Z(Q_\alpha \cap Q_\lambda)$, the map from $Q_\alpha \cap Q_\lambda$ to Z_β given by $g \mapsto [y, g]$ is a homomorphism and so $[Q_\alpha \cap Q_\lambda : C_{Q_\alpha \cap Q_\lambda}(y)] \leq q$. Hence, as $[Q_\lambda : Q_\lambda \cap Q_\alpha] = q$ by Lemma 4.1 (i), $[Q_\lambda : C_{Q_\lambda}(y)] \leq q^2$. □

Now Lemma 4.2 (i) gives that Q_λ/Z_λ is a $\mathrm{GF}(p)L_\lambda/Q_\lambda$-module, and bounds like $[Q_\lambda : C_{Q_\lambda}(y)] \leq q^2$ given in Lemma 4.3 cannot be very easily exploited from a $\mathrm{GF}(p)$ point of view. Fortunately we have the services of the next result:

Theorem 4.4 *Suppose that V_β/Z_β is an irreducible $\mathrm{GF}(p)L_\beta/Q_\beta$-module. Then as an L_β/Q_β-module V_β/Z_β can be defined over $\mathrm{GF}(q)$ and as a $\mathrm{GF}(q)$-space supports a non-degenerate, L_β/Q_β-invariant symplectic or unitary form.*

In order for Theorem 4.4 to be of any use we must determine under which circumstances Q_β/Z_β is irreducible.

Theorem 4.5 *Either V_β/Z_β is an irreducible $\mathrm{GF}(p)L_\beta/Q_\beta$-module or the amalgam is one of a (short) list of known amalgams.*

It is in the proof of Theorem 4.5 that the $\mathrm{G}_2(4)$ example seen earlier is captured.

There are results in [4] which can only be established when L_β/Q_β does not possess $(1,q)$- or $(2,q)$-transvections and, in one way or another, Lemma 4.3 is at the root of this conundrum. For example, in the absence of $(1,q)$- or $(2,q)$-transvections we can strengthen Lemma 4.2 (iii) (using Theorem 4.4) as follows:

Theorem 4.6 *Assume that Q_β/Z_β is an irreducible $\mathrm{GF}(p)L_\beta/Q_\beta$-module and that L_β/Q_β has no $(1,q)$- or $(2,q)$-transvections on Q_β/Z_β. Then Q_β is an ultraspecial p-group and $\dim_{GF(q)} Q_\beta/Z_\beta = 2 + 2\log_q |W_\alpha Q_\beta/Q_\beta|$.*

Proof of Theorem 1.4 From here on assume the hypothesis of Theorem 1.4 is satisfied. Since there are no amalgams involving $\mathrm{L}_6(p^r)$ on the short list of amalgams in Theorem 4.5, we may assume that Q_β/Z_β is an irreducible $\mathrm{GF}(q)L_\beta/Q_\beta$-module. Set $\mathcal{M} = \{N_{L_\beta}(S_{\alpha\beta}) \leq X \leq L_\beta \mid N_{L_\beta}(S_{\alpha\beta}) \text{ is a maximal subgroup of } X\}$. Notice that if $X \in \mathcal{M}$, then X/Q_β is a minimal parabolic subgroup of L_β and $L_\beta = \langle \mathcal{M} \rangle$. Since, by (1.3) (i), Z_α is not normal in L_β, there is $X_{\alpha\beta} \in \mathcal{M}$ such that $X_{\alpha\beta}$ does not normalize Z_α. Setting $Y = \langle Z_\alpha^{X_{\alpha\beta}} \rangle$ we now see the first appearance of extremal subgroups. We have:

Lemma 4.7 *$W_\alpha Q_\beta/Q_\beta \not\leq O_p(X_{\alpha\beta}/Q_\beta)$ so is an extremal subgroup in $S_{\alpha\beta}/Q_\beta$.*

Proof Suppose that $W_\alpha Q_\beta/Q_\beta \leq O_p(X_{\alpha\beta}/Q_\beta)$ and argue for a contradiction. So $W_\alpha \leq O_p(X_{\alpha\beta}) \cap Q_\alpha$ and hence from Lemma 4.1 (iv)

$$[Y, W_\alpha] \leq \langle [Z_\alpha, O_p(X_{\alpha\beta})]^{X_{\alpha\beta}} \rangle \leq \langle [Z_\alpha, S_{\alpha\beta}]^{X_{\alpha\beta}} \rangle = Z_\beta.$$

If $Y \not\leq T_\alpha$, then the non-trivial elements of YQ_λ/Q_λ operate as $(1,q)$- or $(2,q)$-transvections on Q_λ/Z_λ, which is against our hypothesis.

On the other hand, if $Y \leq T_\alpha$, then, by Lemma 4.1 (v), $Y \trianglelefteq L_\alpha$ and hence so is $[Y, W_\alpha]$. However $[Y, W_\alpha] \leq Z_\beta$ which implies that $[Y, W_\alpha] = 1$. In particular, $Y \leq Z(Q_\alpha \cap Q_\beta)$. Now using the GF($p$)-symplectic form on $\widetilde{Q_\beta}$ (Lemma 4.2 (ii)) we see that $Z(Q_\alpha \cap Q_\beta) = Z_\alpha$. But then $Y = Z_\alpha$, a contradiction to our choice of $X_{\alpha\beta}$. □

Eventually after crossing a few rivers and swamps we prove

Theorem 4.8 (i) $O^{p'}(X_{\alpha\beta}/O_p(X_{\alpha\beta})) \cong \mathrm{SL}_2(q)$; *and*
(ii) Y/Z_β *is a natural* $\mathrm{GF}(q)O^{p'}(X_{\alpha\beta}/O_p(X_{\alpha\beta}))$*-module.*

Together Theorem 3.2 and Lemma 4.7 give us the following two cases
(i) $X_{\alpha\beta}$ is unique; or
(ii) L_β/Q_β is a quotient of $\mathrm{SL}_4(q)$, $p = 2$ and $|W_\alpha| \leq q^3$.

Case (i) From Theorem 4.8 (ii), we have Y/Z_β is a 2-dimensional module and Z_α is normalized by all the other minimal parabolic subgroups of L_β/Q_β. Hence Q_β/Z_β is a minuscule $\mathrm{GF}(q)L_\beta/Q_\beta$-module. Since, by Theorem 4.4, the $\mathrm{GF}(q)L_\beta/Q_\beta$-module Q_β/Z_β is either self-dual or isomorphic to a Galois conjugate of its dual, we conclude that m is even and that $X_{\alpha\beta}$ is the minimal parabolic subgroup which corresponds to the middle node of the Dynkin diagram. Thus we have $|W_\alpha Q_\beta/Q_\beta| \leq q^{m^2/4}$ and $\log_q |Q_\beta/Z_\beta| = \binom{m}{m/2}$. Theorem 4.6 shows us that

$$3 + m^2/2 \geq \binom{m}{m/2}.$$

An easy induction argument shows that the right hand side is strictly greater than the left hand side whenever $m \geq 8$. Thus $m = 6$ or $m = 4$. If $m = 4$, then the module is the 6-dimensional $\mathrm{O}_6^+(q)$-module and as such admits $(2, q)$-transvections. Therefore, $m = 6$ and the result follows.

Case (ii) We have from Theorem 4.6 that $\log_q |Q_\beta/Z_\beta| \leq 2 + 6 = 8$ whereas the Weyl group orbit on $C_{Q_\beta/Z_\beta}(S_{\alpha\beta})$ has length 12. As the Weyl group orbit of $C_{Q_\beta/Z_\beta}(S_{\alpha\beta})$ is a linearly independent set of 1-dimensional subspaces [5] we have a contradiction. □

References

[1] C. Parker, P. Rowley, Extremal Subgroups in Chevalley Groups, *J. London Math. Soc.* **55**(1997) 387-399.

[2] C. Parker, P. Rowley, Extremal Subgroups in Twisted Lie Type Groups, Crelle's Journal (1998), to appear.

[3] C. Parker, P. Rowley, Unique Node Extremal Subgroups in Chevalley Groups, preprint 1996.

[4] C. Parker, P. Rowley, Symplectic Amalgams, manuscript 1997.

[5] R. Steinberg, *Lectures on Chevalley Groups*, Yale University (1967).

[6] B. Stellmacher, An Application of the Amalgam Method: The 2-local Structure of N-Groups of Characteristic 2 type, J. Algebra **190**(1997), 11-67.

PRIMITIVE PRIME DIVISOR ELEMENTS IN FINITE CLASSICAL GROUPS

CHERYL E. PRAEGER

Department of Mathematics, University of Western Australia, Nedlands, W. A. 6907, Australia

Abstract

This is an essay about a certain family of elements in the general linear group $\mathrm{GL}(d, q)$ called primitive prime divisor elements, or ppd-elements. A classification of the subgroups of $\mathrm{GL}(d, q)$ which contain such elements is discussed, and the proportions of ppd-elements in $\mathrm{GL}(d, q)$ and the various classical groups are given. This study of ppd-elements was motivated by their importance for the design and analysis of algorithms for computing with matrix groups over finite fields. An algorithm for recognising classical matrix groups, in which ppd-elements play a central role is described.

1 Introduction

The central theme of this essay is the study of a special kind of element of the general linear group $\mathrm{GL}(d, q)$ of nonsingular $d \times d$ matrices over a finite field $\mathrm{GF}(q)$ of order q. We define these elements, which we call *primitive prime divisor elements* or ppd-*elements*, and give good estimates of the frequencies with which they occur in $\mathrm{GL}(d, q)$ and the various classical matrix groups. Further we describe a classification of the subgroups of $\mathrm{GL}(d, q)$ which contain ppd-elements, and explore their role in the design and analysis of a randomised algorithm for recognising the classical matrix groups computationally.

Perhaps the best way to introduce these ideas, and to explain the reasons for investigating this particular set of research questions, may be to give a preliminary discussion of a generic recognition algorithm for matrix groups. We wish to determine whether a given subgroup G of $\mathrm{GL}(d, q)$ contains a certain subgroup Ω. We design the algorithm to study properties of randomly selected elements from G in such a way that, if G contains Ω then with high probability we will gain sufficient information from these elements to conclude with certainty that G does contain Ω. A skeleton outline of the algorithm could be written as follows.

Algorithm 1.1 To recognise whether a given subgroup of $\mathrm{GL}(d, q)$ contains a certain subgroup Ω.

Input: $G = \langle X \rangle \leq \mathrm{GL}(d, q)$ and possibly some extra information about G.

Output: Either

(a) G contains the subgroup Ω, or

(b) G does not contain Ω.

If Algorithm 1.1 returns option (a) then G definitely contains Ω. However if option (b) is returned there is a possibility that this response is incorrect. In other words Algorithm 1.1 is a *Monte Carlo algorithm*. It proceeds by making a sequence of random selections of elements from the group G, seeking a certain kind of subset E of G, which if found will greatly assist in deciding whether or not G contains Ω. The essential requirements for E are two-fold:

1. If G contains a subset E with the required properties, then either G contains Ω, or G belongs to a short list of other possible subgroups of GL (d, q) (and the algorithm must then distinguish subgroups in this list from subgroups containing Ω).

2. If G contains Ω, then the event of *not* finding a suitable subset E in G after a reasonable number $N(\varepsilon)$ of independent random selections of elements from G has probability less than some small pre-assigned number ε.

In order to make the first requirement explicit, we need a classification of the subgroups of GL (d, q) which contain a suitable subset E. Similarly in order to make the second requirement explicit, we need good estimates for the proportions of "E-type elements" in groups containing Ω. Moreover, if these two requirements are to lead to an efficient algorithm for recognising whether G contains Ω, the proportions of E-type elements in groups containing Ω must be fairly large to guarantee that we have a good chance of finding a suitable subset E after a reasonable number of random selections; and in practice we need good heuristics for producing approximately random elements from a group. Also, among other things, we need efficient procedures to identify E-type elements, and to distinguish between the subgroups on the short list and the subgroups which contain Ω. The aim of this paper is to present and discuss results of these types, and the corresponding recognition algorithms, in the cases where Ω is one of the classical matrix groups. In these cases the subset E consists of certain ppd-elements.

I am grateful to Igor Shparlinski for some very helpful discussions and advice on the analysis in Section 11. Eamonn O'Brien made a careful reading of an early draft of the paper and the current version has been much improved as a result of his detailed comments. Also I thank John Cannon for making available to me the results mentioned in Section 12 of some tests of the MAGMA implementation of the classical recognition algorithm.

2 Classical groups

We consider certain subgroups of GL (d, q) where d is a positive integer and $q = p^a$, a power of a prime p, and we let V denote the underlying vector space of d-dimensional row vectors over GF (q) on which GL (d, q) acts naturally.

The classical groups preserve certain bilinear, sesquilinear or quadratic forms on V. To describe them we adapt some notation from the book of Kleidman and

Liebeck [26]. A subgroup G of $\mathrm{GL}(d,q)$ is said to *preserve a form κ modulo scalars* if there exists a homomorphism $\mu : G \to \mathrm{GF}(q)^{\#}$ such that, in the case of a bilinear or sesquilinear form, $\kappa(ug, vg) = \mu(g) \cdot \kappa(u,v)$, or, in the case of a quadratic form, $\kappa(vg) = \mu(g) \cdot \kappa(v)$, for all $u, v \in V$ and $g \in G$. A matrix g in such a group is said to *preserve κ modulo scalars,* and if $\mu(g) = 1$ then g is said to *preserve κ.* We denote by Δ or $\Delta(V,\kappa)$ the group of all matrices in $\mathrm{GL}(d,q)$ which preserve κ modulo scalars, and by S the subgroup of Δ consisting of those matrices which preserve κ and which have determinant 1.

The subgroup Ω which we shall seek to recognise is equal to S unless κ is a non-degenerate quadratic form, and in this latter case Ω has index 2 in S and is the unique such subgroup of S. There are four families of subgroups which we shall consider, and by a *classical group* in $\mathrm{GL}(d,q)$ we shall mean a subgroup G which satisfies $\Omega \le G \le \Delta$, for Ω, Δ in one of these families. The four families are as follows.

(i) *Linear groups:* $\kappa = 0$, $\Delta = \mathrm{GL}(d,q)$ and $\Omega = \mathrm{SL}(d,q)$;

(ii) *Symplectic groups:* d is even, κ is a non-degenerate alternating bilinear form on V, $\Delta = \mathrm{GSp}(d,q)$ and $\Omega = \mathrm{Sp}(d,q)$;

(iii) *Orthogonal groups:* κ is a non-degenerate quadratic form on V, $\Delta = \mathrm{GO}^{\varepsilon}(d,q)$, and $\Omega = \Omega^{\varepsilon}(d,q)$, where $\varepsilon = \pm$ if d is even, and $\varepsilon = \circ$ if d is odd. If d is odd then also q is odd since κ is non-degenerate;

(iv) *Unitary groups:* q is a square, κ is a non-degenerate unitary form on V, that is a non-degenerate sesquilinear form with respect to the automorphism of $\mathrm{GF}(q)$ of order 2, $\Delta = \mathrm{GU}(d,q)$ and $\Omega = \mathrm{SU}(d,q)$.

The books [26, 40] are good references for information about the finite classical groups.

3 Primitive prime divisors and ppd-elements

Let b, e be positive integers with $b > 1$. A prime r dividing $b^e - 1$ is said to be a *primitive prime divisor* of $b^e - 1$ if r does not divide $b^i - 1$ for any i such that $1 \le i < e$. It was proved by Zsigmondy [44] in 1892 that $b^e - 1$ has a primitive prime divisor unless either the pair (b,e) is $(2,6)$, or $e = 2$ and $b+1$ is a power of 2. Observe that

$$|\mathrm{GL}(d,q)| = q^{\binom{d}{2}} \prod_{1 \le i \le d} (q^i - 1).$$

This means that primitive prime divisors of $q^e - 1$ for various values of $e \le d$ divide $|\mathrm{GL}(d,q)|$, and indeed divide $|\Omega|$ for various of the classical groups Ω in $\mathrm{GL}(d,q)$. We define *primitive prime divisor elements,* sometimes called ppd-*elements,* in $\mathrm{GL}(d,q)$ to be those elements with order a multiple of some such primitive prime divisor. Thus we define an element $g \in \mathrm{GL}(d,q)$ to be a ppd$(d,q;e)$-*element* if its order $o(g)$ is divisible by some primitive prime divisor of $q^e - 1$.

Our interest is mainly in ppd $(d,q;e)$-elements with $e > d/2$ and we shall describe in Section 5 a classification by Guralnick, Penttila, Saxl and the author in [17] of all subgroups of GL (d,q) containing such an element. We shall henceforth reserve the term ppd-elements to refer to elements of GL (d,q) which are ppd $(d,q;e)$-elements for some $e > d/2$. Note that, if $g \in$ GL (d,q) is a ppd $(d,q;e)$-element with $e > d/2$, then there is a unique g-invariant e-dimensional subspace of the underlying vector space V on which g acts irreducibly, and also the characteristic polynomial for g has an irreducible factor over GF (q) of degree e. While neither of these two conditions is sufficient to guarantee that an element is a ppd $(d,q;e)$-element, it turns out that most elements satisfying either of them are in fact ppd $(d,q;e)$-elements. In addition, a large proportion of elements in any of the classical groups are ppd-elements, and this fact has proved to be very important for the development of recognition algorithms for classical groups.

In 1974 Hering [19] investigated subgroups of GL (d,q) containing ppd $(d,q;d)$-elements. Such subgroups act irreducibly on V. Hering was interested in applications of these results to geometry, in particular for constructing finite translation planes. He was also interested in the link between such groups and finite affine 2-transitive permutation groups. If G is a finite affine 2-transitive permutation group acting on a set X, then X may be taken as the set of vectors of a finite vector space, say $V = V(d,q)$ of dimension d over GF (q), and $G = NG_o$ where N is the group of translations of V and G_o is a subgroup of GL (d,q) acting transitively on $V^\#$, that is G_o is a *transitive linear group.* Conversely if G_o is a transitive linear group on V, and N is the group of translations of V, then NG_o is a 2-transitive permutation group of affine type acting on V. Thus the problems of classifying finite affine 2-transitive groups, and classifying finite transitive linear groups are equivalent. Moreover if G_o is transitive on $V^\#$ then $q^d - 1$ divides $|G_o|$ so that G_o contains a ppd $(d,q;d)$-element. Hering's work led to a classification of finite affine 2-transitive permutation groups, see [20] and also [27, Appendix]. In common with most of the classifications we shall mention related to ppd-elements, this classification depends on the classification of the finite simple groups. Merkt [29] extended Hering's work obtaining a better description of certain of the subgroups of GL (d,q) containing a ppd $(d,q;d)$-element.

Dempwolff [12] in 1987 began an investigation of subgroups of GL (d,q) containing a ppd $(d,q;e)$-element for some $e \geq d/2$. His analysis is independent of the work of Aschbacher which we shall describe in the next section, and he made significant progress on describing what we shall call (and shall define in the next section) the "geometric subgroups" containing such ppd-elements. He also did some work on the nearly simple examples. The classification in [17] of all subgroups of GL (d,q) containing a ppd $(d,q;e)$-element for some $e > d/2$ uses the work of Aschbacher to guide both the analysis and the presentation of the examples. Similar results may be obtained if the condition $e > d/2$ is relaxed, but their proofs become more technical.

4 Aschbacher's classification of finite linear groups

Aschbacher's description [2] of subgroups of GL (d, q), where $q = p^a$ with p prime, has been very influential both on the way problems concerning linear groups are analysed and on the way results about such groups are presented. Aschbacher defined eight families of subgroups $\mathcal{C}_1, \ldots, \mathcal{C}_8$ of GL (d, q) as follows. These families are usually defined in terms of some geometrical property associated with the action on the underlying vector space V, and in all cases maximal subgroups of GL (d, q) in the family can be identified. Subgroups of GL (d, q) in these families are therefore called *geometric subgroups*. We indicate in parentheses the rough structure of a typical maximal subgroup in the family. Note that Z denotes the subgroup of scalar matrices in GL (d, q). Also, as in [26], we denote by b a cyclic group of order b, and for a prime r we denote by r^{1+2c} an extraspecial group of that order.

$\mathcal{C}_1$ These subgroups act reducibly on V, and maximal subgroups in the family are the stabilisers of proper subspaces (maximal parabolic subgroups).

$\mathcal{C}_2$ These subgroups act irreducibly but imprimitively on V, and maximal subgroups in the family are the stabilisers of direct sum decompositions $V = \oplus_{i=1}^{t} V_i$ with $\dim V_i = d/t$ (wreath products $\mathrm{GL}(d/t, q) \wr S_t$).

$\mathcal{C}_3$ These subgroups preserve on V the structure of a vector space over an extension field of GF (q), and maximal subgroups in the family are the stabilisers of extension fields of GF (q) of degree b, where b is a prime dividing d (the groups $\mathrm{GL}(d/b, q^b).b$).

$\mathcal{C}_4$ These subgroups preserve on V the structure of a tensor product of subspaces, and maximal subgroups in the family are the stabilisers of decompositions $V = V_1 \otimes V_2$ (central products $\mathrm{GL}(b, q) \circ \mathrm{GL}(c, q)$ where $d = bc$).

$\mathcal{C}_5$ These subgroups preserve on V the structure of a vector space over a proper subfield of GF (q); such a subgroup is said to *be realisable over a proper subfield.* The maximal subgroups in the family are the stabilisers modulo scalars of subfields of GF (q) of prime index b dividing a (central products $\mathrm{GL}(d, q^{1/b}) \circ Z$).

$\mathcal{C}_6$ These subgroups have as a normal subgroup an r-group R of symplectic type (r prime) which acts absolutely irreducibly on V, and maximal subgroups in the family are the normalisers of these subgroups, $(Z_{q-1} \circ R).\mathrm{Sp}(2c, r)$, where $d = r^c$ and R is an extraspecial group r^{1+2c}, or if $r = 2$ then R may alternatively be a central product $4 \circ 2^{1+2c}$.

$\mathcal{C}_7$ These subgroups preserve on V a tensor decomposition $V = \otimes_{i=1}^{t} V_i$ with $\dim V_i = c$, and maximal subgroups in the family are the stabilisers of such decompositions $((\mathrm{GL}(c, q) \circ \ldots \circ \mathrm{GL}(c, q)).S_t$, where $d = c^t$).

$\mathcal{C}_8$ These subgroups preserve modulo scalars a non-degenerate alternating, or sesquilinear, or quadratic form on V, and maximal subgroups in the family are the classical groups.

The main result of Aschbacher's paper [2] (or see [26, Theorem 1.2.1]) states that, for a subgroup G of $\mathrm{GL}(d,q)$ which does not contain $\mathrm{SL}(d,q)$, either G is a geometric subgroup, or the socle S of $G/(G \cap Z)$ is a nonabelian simple group, and the preimage of S in G is absolutely irreducible on V, is not realisable over a proper subfield, and is not a classical subgroup (as defined in Section 2). The family of such subgroups is denoted $\mathcal{S}$, and subgroups in this family will often be referred to as *nearly simple* subgroups. Aschbacher [2] also defined families of subgroups of each of the classical subgroups Δ in $\mathrm{GL}(d,q)$, analogous to $\mathcal{C}_1, \ldots, \mathcal{C}_8, \mathcal{S}$, and proved that each subgroup of a classical group Δ which does not contain Ω belongs to one of these families.

5 Linear groups containing ppd-elements

The analysis in [17] to determine the subgroups of $\mathrm{GL}(d,q)$ which contain a $\mathrm{ppd}(d,q;e)$-element for some $e > d/2$, was patterned on a similar analysis carried out in [30] to classify subgroups of $\mathrm{GL}(d,q)$ which contain both a $\mathrm{ppd}(d,q;d)$-element and a $\mathrm{ppd}(d,q;d-1)$-element. Moreover the results in [17] seek to give information about the smallest subfield over which such a subgroup G is realisable modulo scalars. We say that G is *realisable modulo scalars* over a subfield $\mathrm{GF}(q_0)$ of $\mathrm{GF}(q)$ if G is conjugate to a subgroup of $\mathrm{GL}(d,q_0) \circ Z$.

Suppose that $G \leq \mathrm{GL}(d,q)$ and that G contains a $\mathrm{ppd}(d,q;e)$-element for some $e > d/2$, and let r be a primitive prime divisor of $q^e - 1$ which divides $|G|$. Suppose moreover that $\mathrm{GF}(q_0)$ is the smallest subfield of $\mathrm{GF}(q)$ such that G is realisable modulo scalars over $\mathrm{GF}(q_0)$.

There is a recursive aspect to the description in [17] of such subgroups G which are geometric subgroups. For example, the reducible subgroups G leave invariant some subspace or quotient space U of V of dimension $m \geq e$, and the subgroup G^U of $\mathrm{GL}(m,q)$ induced by G in its action on U contains a $\mathrm{ppd}(m,q;e)$-element. In [17] no further description is given of these examples, though extra information may be obtained about the group G^U by applying the results recursively.

Although the classification of the geometric examples is not difficult, care needs to be taken in order not to miss some of them. For example, while at first sight it might appear that a maximal imprimitive subgroup $\mathrm{GL}(d/t,q) \wr S_t$ (where $t > 1$) cannot contain a $\mathrm{ppd}(d,q;e)$-element since r does not divide $|\mathrm{GL}(d/t,q)|$, it is possible sometimes for r to divide $|S_t| = t!$, so that we do have some examples in the family $\mathcal{C}_2$.

To understand how this can happen, observe that the defining condition for r to be a primitive prime divisor of $q^e - 1$, namely that e is the least positive integer i such that r divides $q^i - 1$, is equivalent to the condition that q has order e modulo the prime r. Thus $r = ke + 1 \geq e + 1$ for some $k \geq 1$. Sometimes we can have $r = e + 1$ (which satisfies $d/2 < r \leq d$) and hence in these cases an imprimitive subgroup $\mathrm{GL}(1,q) \wr S_d$ will contain $\mathrm{ppd}(d,q;e)$-elements.

Both of the above observations come into play in describing the examples in the family $\mathcal{C}_3$. Here either the prime $r = e + 1 = d$ and the group G is conjugate to a subgroup of $\mathrm{GL}(1,q^d).d$, or e is a multiple of a prime b where b is a proper divisor of d and, replacing G by a conjugate if necessary, $G \leq \mathrm{GL}(d/b, q^b).b$ such that $G \cap \mathrm{GL}(d/b, q^b)$ contains a ppd$(d/b, q^b; e/b)$-element.

After determination of the geometric examples there remains the problem of finding the nearly simple examples. So suppose that G is nearly simple and $S \leq G/(Z \cap G) \leq \mathrm{Aut}\, S$ for some nonabelian simple group S. What we need is a list of all possible groups G together with the values of d, e and q_0. Although there is no classification of all the nearly simple subgroups of $\mathrm{GL}(d, q)$ in general, it is possible to classify those which contain a ppd$(d, q; e)$-element. The reason we can do this is that, for each simple group S, the presence of a ppd$(d, q; e)$-element in G leads to both upper and lower bounds for d in terms of the parameters of S strong enough to lead to a complete classification.

On the one hand d is at least the minimum degree of a faithful projective representation of S over a field of characteristic p, and lower bounds are available for this in terms of the parameters of S. On the other hand we have seen that $r = ke + 1 \geq e + 1 \geq (d + 3)/2$, and in all cases we may deduce that r divides $|S|$. Moreover we have an upper bound on the size of prime divisors of S in terms of the parameters of S. For some simple groups S the upper and lower bounds for d obtained in this way conflict and we have a proof that there are no examples involving S. In many cases however this line of argument simply narrows down the range of possible values for d, e and r. Often there are examples involving S, but, in order to complete the classification, we need to have more information about small dimensional representations of S in characteristic p than simply the lower bound for the dimension of such representations.

For example if $S = A_n$ with $n \geq 9$ then $d \geq n - 2$ if p divides n and $d \geq n - 1$ otherwise, by [41, 42, 43]. Moreover $r \leq n$, so $(d + 3)/2 \leq n$, and we obtain $n - 2 \leq d \leq 2n - 3$ and $r = e + 1$. The upper bound for d cannot be improved since we may have $r = n = e + 1$ infinitely often. Thus we need more information about small dimensional representations of A_n in characteristic p. For $n \geq 15$ this is available from a combination of results of James [22] and Wagner [43]. We see that the representations of A_n and S_n of dimension $n-1$ or $n-2$ are those coming from the deleted permutation module in the natural representation. These give an infinite family of examples with $q_0 = p$. All other faithful projective representations of A_n have dimension greater than the upper bound on d. For the remaining cases, where $n < 15$, special arguments are required, making full use of information in [11, 23]. The result of this analysis is an explicit list of examples for alternating groups S.

The list of examples of linear groups containing ppd-elements can be found in [17, Section 2] and is not reproduced here. Note that completing the classification of the nearly simple examples for classical groups S over fields of characteristic different from p involved proving new results about small dimensional representations of such groups over fields of characteristic p.

6 Various applications of the "ppd classification"

The classification of subgroups of GL (d, q) containing ppd-elements has already been used in a variety of applications concerning finite classical groups. In particular the papers [16, 18] make use of it to answer questions concerning the generation of finite classical groups, while in [28] it is used to show that the finite classical groups are characterised by their orbit lengths on vectors in their natural modules. Information about the invariant generation of classical simple groups (see [32, 38]) can be deduced from the classification (in [32], or see Section 7) of subgroups of classical groups containing two different ppd-elements. (Elements $x_1, \ldots, x_s$ of a group G are said to generate G invariably if $\langle x_1^{g_1}, \ldots, x_s^{g_s}\rangle$ is equal to G for all $g_1, \ldots, g_s \in G$.)

Similarly in [4] the ppd classification, or more accurately the more specialised classification based on it (and described in Section 7), can be used to deal with the finite classical groups in an analysis of finite groups with the permutizer property. A group G is said to have the *permutizer property* if, for every proper subgroup H of G, there is an element $g \in G \setminus H$ such that H permutes with $\langle g\rangle$, that is $\langle g\rangle H = H\langle g\rangle$. The main result of [4] is that all finite groups with the permutizer property are soluble. The proof consists of an examination of a minimal counterexample to this assertion, and the ppd classification can be used to show that the minimal counterexample cannot be an almost simple classical group.

7 Two different ppd-elements in linear groups

The principal application up to now of the classification of linear groups containing ppd-elements has been the development by Niemeyer and the author in [32] of a recognition algorithm for finite classical groups in their natural representation. The basic idea of this algorithm is as described in Section 1. Given a subgroup G of a classical group Δ in GL (d, q) (as described in Section 2), we wish to determine if G contains the corresponding classical group Ω. We do this by examining randomly selected elements from G. The elements of G which we seek by random selection are ppd $(d, q; e)$-elements for various values of $e > d/2$, and an appropriate set of such elements will form the subset E mentioned in Section 1.

It turns out that the proportion of ppd $(d, q; e)$-elements in any of the classical groups is very high (as shown in Section 8), so we are very likely to find such an element after a few independent random selections from any subgroup of Δ which contains Ω. Suppose then that we have indeed found a ppd $(d, q; e)$-element in our group G, for some $e > d/2$. The ppd-classification just described then provides a restricted list of possibilities for the group G. The task is to distinguish subgroups containing Ω from the other possibilities, and this task is a nontrivial one.

For the purposes of presenting the basic strategy, we assume that G is irreducible on V and that we have complete information about any G-invariant bilinear, sesquilinear or quadratic forms on V. There are standard tests in practice which may be used to determine whether G is irreducible on V and to find all G-invariant forms (see [21, 35]). Note that in an implementation of the algorithm in [32] a different protocol may be followed for deciding the stage at which to obtain this precise information about G. Nevertheless, we may and shall assume that G does not lie in the Aschbacher classes $\mathcal{C}_1$ or $\mathcal{C}_8$. Then, having found a ppd$(d,q;e)$-element in G for some $e > d/2$, the ppd-classification would still allow the possibility that G lies in one of $\mathcal{C}_2, \mathcal{C}_3, \mathcal{C}_5, \mathcal{C}_6$, or that G is nearly simple, as well as the desired conclusion that G contains Ω. In the nearly simple case, the classification in [17] shows that there are approximately 30 infinite families and 60 individual examples of nearly simple groups in explicitly known representations.

Guided by the original SL-recognition algorithm developed in [30], we decided to seek, in the first instance, *two different* ppd*-elements* in G by which we mean a ppd$(d,q;e)$-element and a ppd$(d,q;e')$-element, where $d/2 < e < e' \leq d$. We also decided to strengthen the ppd-property required of these elements in two different ways, by requiring at least one of the ppd-elements to be large and at least one of them to be basic.

Let $q = p^a$, and let r be a primitive prime divisor of $q^e - 1$. Recall that $r = ke+1$ for some integer k. We say that r is a *basic primitive prime divisor* if r is a primitive prime divisor of $p^{(ae)} - 1$, and that r is a *large primitive prime divisor* if either $r \geq 2e+1$, or $r = e+1$ and $(e+1)^2$ divides $q^e - 1$. Correspondingly we say that a ppd$(d,q;e)$-element g is *basic* if $o(g)$ is divisible by a basic primitive prime divisor of $q^e - 1$, and that g is *large* if $o(g)$ is divisible by a large primitive prime divisor r of $q^e - 1$ and either $r \geq 2e+1$ or $r = e+1$ and $(e+1)^2$ divides $o(g)$. Note that, for $e \geq 2$, if $q^e - 1$ has a primitive prime divisor, then $q^e - 1$ has a basic primitive prime divisor unless $(q,e) = (4,3)$ or $(8,2)$. Similarly an explicit list can be given for pairs (q,e) for which $q^e - 1$ has a primitive prime divisor but does not have a large primitive prime divisor (see [15, 19] or [32, Theorem 2.2]). Thus in most cases $q^e - 1$ has both a large primitive prime divisor and a basic primitive prime divisor; and many ppd-elements will be both large and basic.

We shall see in Section 8 that requiring the additional condition of being large or basic does not alter significantly the very good upper and lower bounds we can give for the proportion of ppd-elements in subgroups of Δ containing Ω.

Suppose that we now have $G \subseteq \Delta$ for some classical group Δ in $\mathrm{GL}(d,q)$, with G irreducible on the underlying vector space V, and suppose also that we have complete information about G-invariant forms so that we can guarantee that G is not contained in the class $\mathcal{C}_8$ of subgroups of Δ. Further we suppose that G contains two different ppd-elements, say a ppd$(d,q;e)$-element g and a ppd$(d,q;e')$-element h, where $d/2 < e < e' \leq d$.

In [32, Theorem 4.7], Niemeyer and the author refined the classification in [17] to find all possibilities for the group G. These possibilities comprise groups containing Ω, members of the Aschbacher families $\mathcal{C}_2, \mathcal{C}_3$ and $\mathcal{C}_5$, and some nearly simple examples. The presence of two different ppd-elements certainly restricts the possibilities within these families, but it is still difficult to distinguish some of them from groups containing Ω.

If we require that at least one of g, h is large and at least one is basic then, as was shown in [32, Theorem 4.8], the possibilities for irreducible subgroups G which do not contain Ω are certain subgroups in $\mathcal{C}_3$ and nearly simple groups in a very short list comprising explicit representations of one infinite family and five individual nearly simple groups.

After our discussion of the proportions of ppd-elements in classical groups in Section 8, we shall return to our discussion of the recognition algorithm. We shall see that the algorithm can be completed by simply seeking a few more ppd-elements of a special kind which, if found, will rule out all but one possibility for G, enabling us to conclude that G contains Ω.

8 Proportion of ppd-elements in classical groups

The questions we wish to answer from our discussion in this section are the following. If $\Omega \leq G \leq \Delta \leq \mathrm{GL}(d,q)$, and G contains two different ppd-elements at least one of which is large and at least one of which is basic, then what is the probability of finding two such elements after a given number N of independent random selections of elements from G? In particular, for a given positive real number ε, is it true that the probability of failing to find such elements after N selections is less than ε provided N is sufficiently large? And if so just how large must N be?

These questions can be answered using simple probability theory provided that we can determine, for a given e (where $d/2 < e \leq d$), the proportion $\mathrm{ppd}(G,e)$ of elements of G which are $\mathrm{ppd}(d,q;e)$-elements. This proportion may depend on the nature of the classical group Δ: that is, on whether Δ is a linear, symplectic, orthogonal or unitary group. In particular $\mathrm{ppd}(G,e) = 0$ if Δ is a symplectic or orthogonal group and e is odd, or if Δ is a unitary group and e is even, or if Δ is of type O^+ and $e = d$. This can be seen easily by examination of the orders of these groups. In all other cases, provided that d and q are not too small, any subgroup of Δ which contains Ω will contain $\mathrm{ppd}(d,q;e)$-elements.

So suppose now that $\Omega \leq G \leq \Delta$, that $d/2 < e \leq d$, and that G contains a $\mathrm{ppd}(d,q;e)$-element g. It is not difficult (see [32, Lemma 5.1]) to show that V has a unique e-dimensional g-invariant subspace W and that g acts irreducibly on W. Moreover, if Δ is a symplectic, orthogonal, or unitary group, then W must be nonsingular with respect to the bilinear, quadratic, or sesquilinear form defining Δ.

Next (see [32, Lemma 5.2]) we observe that the group G acts transitively on the set of all nonsingular e-dimensional subspaces of V (or all e-dimensional subspaces if $\Delta = \mathrm{GL}(d,q)$). Thus the proportion of $\mathrm{ppd}(d,q;e)$-elements in G is the same as the proportion of such elements which fix a particular nonsingular e-dimensional subspace W. Therefore we need to determine the proportion of $\mathrm{ppd}(d,q;e)$-elements in the setwise stabiliser G_W of W in G.

Now consider the natural map $\varphi : g \mapsto g|_W$ which sends $g \in G_W$ to the linear transformation of W induced by g. Then $\Omega(W) \leq \varphi(G) \leq \Delta(W) \leq \mathrm{GL}(W)$, and $\Delta(W)$ has the same type (linear, symplectic, orthogonal, or unitary) as Δ. If $g \in G_W$ and g is a $\mathrm{ppd}(d,q;e)$-element, then every element of the coset $g\mathrm{Ker}\,\varphi$ is also a $\mathrm{ppd}(d,q;e)$-element, since all elements in the coset induce the same linear transformation $g|_W$ of W. Moreover in this case $g|_W$ is a $\mathrm{ppd}(e,q;e)$-element in $\varphi(G)$ and all such elements arise as images under φ of $\mathrm{ppd}(d,q;e)$-elements in G_W.

It follows that $\mathrm{ppd}\,(G, e)$ is equal to the proportion $\mathrm{ppd}\,(\varphi(G), e)$ of $\mathrm{ppd}\,(e, q; e)$-elements in $\varphi(G)$.

Thus it is sufficient for us to determine $\mathrm{ppd}\,(G, d)$ for each of the possibilities for Δ which contain $\mathrm{ppd}\,(d, q; d)$-elements. This was done already by Neumann and the author in [30, Lemmas 2.3 and 2.4] in the case where $\Delta = \mathrm{GL}\,(d, q)$. The techniques used there work also in the other cases although some care is needed. The basic ideas are as follows.

Let g be a $\mathrm{ppd}\,(d, q; d)$-element in G, and let $C := C_G(g)$. Then C is a cyclic group, called a Singer cycle for G, and has order n say, where n divides $q^d - 1$ and n is divisible by some primitive prime divisor of $q^d - 1$. The group C is self-centralising in G. Further each $\mathrm{ppd}\,(d, q; d)$-element in G lies in a unique G-conjugate of C. The number of G-conjugates of C is $|G : N_G(C)|$, and so the number of $\mathrm{ppd}\,(d, q; d)$-elements in G is equal to $|G : N_G(C)|$ times the number of such elements in C. It follows that

$$\mathrm{ppd}\,(G, d) = |G : N_G(C)| \cdot \mathrm{ppd}\,(C, d) \cdot \frac{|C|}{|G|} = \frac{\mathrm{ppd}\,(C, d)}{u},$$

where $\mathrm{ppd}\,(C, d)$ is the proportion of $\mathrm{ppd}\,(d, q; d)$-elements in C, and $u := |N_G(C) : C|$. In the linear, symplectic and unitary cases $u = d$, while in the orthogonal case u is either d or $d/2$ depending on which intermediate subgroup G is $(\Omega \leq G \leq \Delta)$. In the orthogonal case we certainly have $u = d$ if G contains $\mathrm{O}\,(V)$.

Thus we need to estimate $\mathrm{ppd}\,(C, d)$. Let Φ denote the product of all the primitive prime divisors of $q^d - 1$ (including multiplicities), so that $(q^d - 1)/\Phi$ is not divisible by any primitive prime divisor of $q^d - 1$. In all cases Φ divides $n = |C|$. Moreover an element $x \in C$ is a $\mathrm{ppd}\,(d, q; d)$-element if and only if $x^{n/\Phi} \neq 1$, that is if and only if x does not lie in the unique subgroup of C of order n/Φ. Hence

$$\mathrm{ppd}\,(C, d) = \frac{n - n/\Phi}{n} = 1 - \frac{1}{\Phi},$$

and therefore

$$\mathrm{ppd}\,(G, d) = \frac{1}{u}\,(1 - \frac{1}{\Phi}) < \frac{1}{u}.$$

Since each primitive prime divisor of $q^d - 1$ is of the form $kd + 1 \geq d + 1$, the quantity Φ is at least $d + 1$, and hence

$$\mathrm{ppd}\,(G, d) \geq \frac{1}{u}\,(1 - \frac{1}{d+1})$$

so we have

$$\frac{1}{u}\,(\frac{d}{d+1}) \leq \mathrm{ppd}\,(G, d) < \frac{1}{u}.$$

Putting all of this together we see that in almost all cases $\mathrm{ppd}\,(G, d)$ lies between $1/(d+1)$ and $1/d$, with the exception being some orthogonal cases where $\mathrm{ppd}\,(G, d)$ lies between $2/(d+1)$ and $2/d$.

To pull back this result to the general case where $d/2 < e \leq d$, we need to have some particular information about the group $\varphi(G)$ in the orthogonal case in order to know which of the bounds apply. It turns out (see [32, Theorem 5.7]) that for all cases, and all e for which $d/2 < e \leq d$ and Δ contains ppd $(d, q; e)$-elements, we have

$$\frac{1}{e+1} \leq \text{ppd}\,(G, e) < \frac{1}{e}$$

except if Δ is an orthogonal group of minus type, $e = d$ is even, and $G \cap \mathrm{O}^-(d, q)$ is either $\Omega^-(d, q)$ (for any q) or $\mathrm{SO}^-(d, q)$ (for q odd), in which case $2/(d+1) \leq \text{ppd}\,(G, d) < 2/d$.

Further (see [32, Theorem 5.8]), the proportion of large ppd $(d, q; e)$-elements in G and the proportion of basic ppd $(d, q; e)$-elements in G, whenever such elements exist, also lie between the lower and upper bounds we have above for ppd (G, e).

In the classical recognition algorithm in [32] we are not especially interested at first in particular values of e. We simply wish to find ppd-elements for some e between $d/2$ and d. The proportion of such elements in G is

$$\text{ppd}\,(G) := \sum_{d/2<e\leq d} \text{ppd}\,(G, e).$$

In the linear case, where $\Delta = \mathrm{GL}\,(d, q)$, this is approximately equal to $\sum_{d/2<e\leq d} e^{-1}$ which, in turn, is approximately

$$\int_{d/2}^{d} \frac{dx}{x} = \log 2 = 0.693\ldots$$

while in the other cases ppd (G) is approximately equal to the sum of e^{-1} either over all even e, or all odd e between $d/2$ and d; this is approximately equal to $(\log 2)/2$. These computations can be done carefully resulting in very good upper and lower bounds for ppd (G) which differ by a small multiple of d^{-1}, see [32, Theorem 6.1]. Moreover, except for small values of d, these upper and lower bounds for ppd (G) are also upper and lower bounds for the proportions of large ppd-elements and of basic ppd-elements in G.

We can model the process of random selection of N elements from G, seeking ppd-elements, as a sequence of N binomial trials with probability of success on each trial (that is, each selection) being ppd (G). Using this model we can compute the probability of finding (at least) "two different ppd-elements" after N independent random selections. The extent to which this computed probability measures the true probability in a practical implementation depends on whether the assumptions for the binomial model hold for the implementation. In particular the binomial model will give a good fit if the selection procedure is approximately *uniform*, that is the probability of selecting each element of G on each selection is approximately $|G|^{-1}$, and if the selections are approximately independent. For any small positive real number ε, under the binomial model the probability of failing to find "two different ppd-elements" after N independent uniform random selections is less than ε provided that N is greater than a small (specified) multiple of $\log \varepsilon^{-1}$, see [32, Theorem 6.4 and Lemma 6.5].

The same approach (under the same assumptions about uniformity and independence of the random selections) gives good estimates for the number $N = N(\varepsilon)$

of selections needed in order that the probability of failing to find "two different ppd-elements", at least one of which is large and at least one of which is basic, after N random selections is less than ε. Namely $N(\varepsilon)$ is a small (specified) multiple of $\log \varepsilon^{-1}$. For example, if $\Delta = \mathrm{GL}(d,q)$ with $40 \leq d \leq 1000$ and $\varepsilon = 0.1$, then $N(\varepsilon) = 5$.

9 Classical recognition algorithm: an outline

Suppose that $G \subseteq \Delta$ for some classical group Δ in $\mathrm{GL}(d,q)$, with G irreducible on the underlying vector space V, and that we have complete information about G-invariant forms (so that G is not contained in the class $\mathcal{C}_8$ of subgroups of Δ). We wish to determine whether or not G contains the corresponding classical group Ω. Our algorithm is a Monte Carlo algorithm which may occasionally fail to detect that G contains Ω. The probability of this happening is less than a predetermined small positive real number ε.

First we make a number N of independent uniform random selections of elements from G, where $N \geq N(\varepsilon/3)$ as in Section 8. If we fail to find two different ppd-elements in G, with at least one of them large and at least one basic, then we report that G does not contain Ω. There is a possibility that this response is incorrect, but if in this case G does contain Ω then from Section 8, the probability of failing to find suitable elements is less than $\varepsilon/3$. Thus the probability of reporting at this stage that G does not contain Ω, given that G does contain Ω, is less than $\varepsilon/3$.

Suppose now that G contains two different ppd-elements, say a $\mathrm{ppd}(d,q;e)$-element g and a $\mathrm{ppd}(d,q;e')$-element h, where $d/2 < e < e' \leq d$, and that at least one of g,h is large and at least one is basic. As discussed in Section 7, the possibilities for G are that (i) $G \supseteq \Omega$, or that (ii) G is conjugate to a subgroup of $\mathrm{GL}(d/b,q^b).b$ for some prime b dividing d, or that (iii) G is one of a very restricted set of nearly simple groups. In order to distinguish case (i) from cases (ii) and (iii) it turns out that essentially we need to find a few extra ppd-elements.

The "extension field groups" in case (ii) are the most difficult to handle. The basic idea here can be illustrated by considering the linear case where $\Delta = \mathrm{GL}(d,q)$. For a prime b dividing d, the only values of e such that $\mathrm{GL}(d/b,q^b).b$ contains a $\mathrm{ppd}(d,q;e)$-element are multiples of b (apart from the exceptional case where $b = d$ and d is a primitive prime divisor of $q^{d-1}-1$). Thus finding in G a $\mathrm{ppd}(d,q;e)$-element for some e which is not a multiple of b will prove that G is not conjugate to a subgroup of $\mathrm{GL}(d/b,q^b).b$. If $G \supseteq \Omega$, then the proportion of such elements in G is $\mathrm{ppd}(G) - \sum_{d/2<ib\leq d} \mathrm{ppd}(G,ib)$ which is approximately equal to $\mathrm{ppd}(G) - (\sum_{d/(2b)<i\leq d/b}(ib)^{-1})$. This in turn is approximately equal to $\log 2 - b^{-1}\log 2 = (\log 2)(b-1)/b$. By [34, Theorem 8.30], the number $\mu(d)$ of distinct primes dividing d is $O(\log d/\operatorname{loglog} d)$. Arguing as in Section 8, there is an integer $N_b(\varepsilon)$ such that, if $G \supseteq \Omega$, then the probability of failing to find a $\mathrm{ppd}(d,q;e)$-element in G with e coprime to b after $N_b(\varepsilon)$ independent random selections is less than $\varepsilon/3\mu(d)$. If $G \supseteq \Omega$, then we may need to find up to $\mu(d)$ extra ppd-elements to eliminate case (ii) as a possibility, and the probability of failing to eliminate it after N random selections from G, where N is the maximum of the $N_b(\varepsilon)$, is less than $\varepsilon/3$. If we

fail to find the required set of elements after these N further random selections then we report that G does not contain Ω. Thus the probability of reporting at this second stage that G does not contain Ω, given that G does contain Ω, is less than $\varepsilon/3$. The number N of selections we need to make for this second stage is $O(\log \varepsilon^{-1} + \log\log d)$. Eliminating possibility (ii) for the other classical groups is done using these basic ideas, but the details are considerably more complicated for the symplectic and orthogonal groups when $b = 2$.

For each of the nearly simple groups which contain two different ppd-elements g, h as above, there are in fact only two values of e for which the group contains ppd $(d, q; e)$-elements, namely the values corresponding to the elements g and h. To distinguish groups G containing Ω from this nearly simple group we simply need to find in G a ppd $(d, q; e)$-element for a third value of e. For each pair (d, q) there is only a small number of possible nearly simple groups (usually at most 1, and in all cases at most 3). As before there is some $N_{\text{sim}}(\varepsilon)$ such that, if $G \supseteq \Omega$, then the probability of failing to find suitable elements to eliminate these nearly simple groups after $N_{\text{sim}}(\varepsilon)$ random selections from G is less than $\varepsilon/3$. If we fail to find the required elements after $N_{\text{sim}}(\varepsilon)$ further random selections then we report that G does not contain Ω. Thus the probability of reporting at this third and final stage that G does not contain Ω, given that G does contain Ω, is less than $\varepsilon/3$.

Once we have found all the required elements to remove possibilities (ii) and (iii) we may report with certainty that G does contain Ω.

The probability that the algorithm reports that G does not contain Ω, given that G does contain Ω, is less than ε. The requirements to bound the probability of error at the three stages of the algorithm are such that the complete algorithm requires us to make $O(\log \varepsilon^{-1} + \log\log d)$ random selections from G.

10 Computing with polynomials

In this section we describe how we process an element g of a classical group $\Delta \leq \mathrm{GL}(d, q)$ to decide if it is a ppd-element. This is a central part of the algorithm.

The first step is to compute the characteristic polynomial $c_g(t)$ of g, and to determine whether or not $c_g(t)$ has an irreducible factor of degree greater than $d/2$. If no such factor exists then g is not a ppd-element. So suppose that $c_g(t)$ has an irreducible factor $f(t)$ of degree $e > d/2$.

Thus we know that there is a unique g-invariant e-dimensional subspace W of V and that the linear transformation $g|_W$ induced by g on W has order dividing $q^e - 1$; g will be a ppd $(d, q; e)$-element if and only if the order of $g|_W$ is divisible by some primitive prime divisor of $q^e - 1$. By an argument introduced in Section 8, this will be the case if and only if $(g|_W)^{(q^e-1)/\Phi} \neq 1$, where $\Phi = \Phi(e, q)$ and $\Phi(e, q)$ denotes the product of all the primitive prime divisors of $q^e - 1$ (including multiplicities). Determining whether or not this is the case can be achieved by computing within the polynomial ring $\mathrm{GF}(q)[t]$ modulo the ideal $\langle f(t) \rangle$, namely $(g|_W)^{(q^e-1)/\Phi}$ will be a non-identity matrix if and only if $t^{(q^e-1)/\Phi} \neq 1$ in this ring.

We can test whether of not g is a large or basic ppd $(d, q; e)$-element by the same method with $\Phi(e, q)$ replaced by $\Phi_l(e, q)$ or $\Phi_b(e, q)$ respectively. Here $\Phi_l(e, q)$ and

$\Phi_b(e,q)$ are the products of all the large and basic primitive prime divisors of q^e-1 (including multiplicities) respectively.

The idea for checking the ppd-property by determining whether a single power of g is the identity comes from the special linear recognition algorithm in [30], while the idea of deciding this by a computation in the polynomial ring is due to Celler and Leedham-Green [7].

11 Complexity of the classical recognition algorithm

In [31, Section 4] an analysis of the running cost for the classical recognition algorithm was given based on "classical" algorithms for computing in finite fields. For example the cost of multiplying two $d \times d$ matrices was taken to be $O(d^3)$ field operations (that is, additions, multiplications, or computation of inverses). We take this opportunity to re-analyse the algorithm in terms of more modern methods for finite field computations. These methods can lead to improvements in performance over the classical methods. However efficient implementation of the modern methods is a highly nontrival task requiring substantial effort, see for example the paper of Shoup [39] which addresses the problem of efficient factorisation of polynomials over finite fields. I am grateful to Igor Shparlinski for some interesting and helpful discussions concerning such algorithms.

The *exponent of matrix multiplication* is defined as the infimum of all real numbers x for which there exists a matrix multiplication algorithm which requires no more than $O(d^x)$ field operations to multiply together two $d \times d$ matrices over a field of order q. It is denoted by ω or $\omega(d,q)$. Thus, for all positive real numbers ε, there exists such an algorithm which requires $O(d^{\omega+\varepsilon})$ field operations, that is matrix multiplication can be performed with $O(d^{\omega+o(1)})$ field operations. In [6, Sections 15.3, 15.8] an algorithm is given and analysed for which $O(d^x)$ field operations are used with $x < 2.39$ (and hence $\omega < 2.39$), and it was shown there also that ω can depend (if at all) only on the prime p dividing q rather than on the field size q. Moreover the cost of performing a field operation depends on the data structure used to represent the field and is $O((\log q)^{1+o(1)})$ for each field operation, that is, the cost is $O((\log q)^{1+\varepsilon})$ for each $\varepsilon > 0$.

Now let μ be the cost of producing a single random element from the given subgroup $G = \langle X \rangle$ of $\mathrm{GL}(d,q)$. As discussed in [36, p. 190], theoretical methods for producing approximately random elements from a matrix group are not good enough to be translated into practical procedures for use with algorithms such as the classical recognition algorithm. For example, Babai [3, Theorem 1.1 and Proposition 7.2] produces, from a given generating set X for a subgroup $G \leq \mathrm{GL}(d,q)$, a set of $O(d^2 \log q)$ elements of G at a cost of $O(d^{10}(\log q)^5)$ matrix multiplications, from which nearly uniformly distributed random elements of G can be produced at a cost of $O(d^2 \log q)$ matrix multiplications per random element. The practical implementation of the classical recognition algorithm uses an algorithm developed in [9] for producing approximately random elements in classical groups which, when tested on a range of linear and classical groups was found to produce, for each relevant value of e, $\mathrm{ppd}(d,q;e)$-elements in proportions acceptably close to the true proportions in the group. This procedure has an initial phase which costs $O(d^{\omega+o(1)})$ field operations, and then the cost of producing each random element is $O(d^{\omega+o(1)})$ field operations (see also [31, Section 4.1]). Further analysis of the algorithm in [9] may be found in [10, 13, 14].

Testing each random element $g \in G$ involves first finding its characteristic polynomial $c_g(t)$. The cost of doing this deterministically is $O(d^{\omega+o(1)})$ field operations (see [25] or [6, Section 16.6]). Next we test whether $c_g(t)$ has an irreducible factor of degree greater than $d/2$. This can be done deterministically at a cost of $O(d^{\omega+o(1)} + d^{1+o(1)} \log q)$ field operations, see [24]. (Although the full algorithm in [24] for obtaining a complete factorisation of $c_g(t)$ is non-deterministic, we only need the first two parts of the algorithm, the so-called square-free factorisation and distinct-degree factorisation procedures, and these are deterministic.) Suppose now that $c_g(t)$ has an irreducible factor $f(t)$ of degree $e > d/2$. We then need to compute $\Phi(e, q)$, the product of all the primitive prime divisors of $q^e - 1$ (counting multiplicities). A procedure for doing this is given in [30, Section 6]. It begins with setting $\Phi = q^e - 1$ and proceeds by repeatedly dividing Φ by certain integers. The procedure runs over all the distinct prime divisors c of e, and by [34, Theorem 8.30] there are $O(\log e/ \log\log e) = O(\log d/ \log\log d)$ such prime divisors. For each c, the algorithm computes twice the greatest common divisor of two positive integers where the larger of the two integers may be as much as q^e, and then makes up to $d \log q$ greatest common divisor computations for which the larger of the two integers is $O(d)$. By [1, Theorem 8.20 and its Corollary] (or see [6, Note 3.8]), the cost of computing the greatest common divisor of two positive integers less than 2^n, is $O(n(\log n)^{O(1)})$ bit operations. It follows that the cost of computing $\Phi(e, q)$ is $O(d(\log d)^{O(1)}(\log q)^2)$ bit operations. Having found $\Phi(e, q)$, we need to determine whether $t^{(q^e-1)/\Phi(e,q)}$ is equal to 1 in the polynomial ring $\mathrm{GF}(q)[t]$ modulo the ideal $\langle f(t)\rangle$. This involves $O(d \log q)$ multiplications modulo $f(t)$ of two polynomials of degree less than d over $\mathrm{GF}(q)$. Each of these polynomial multiplications costs $O(d \log d \log\log d)$ field multiplications, (see [6, Theorem 2.13 and Example 2.6]). Thus this test costs $O(d^2 \log d \log\log d \log q)$ field operations. Therefore the cost of testing whether a random element g is a ppd-element is $O(d^{\omega+o(1)} + d^2 \log d \log\log d \ \log q)$ field operations plus $O(d(\log d)^{O(1)}(\log q)^2))$ bit operations, and hence is

$$O(d^{\omega+o(1)}(\log q)^{1+o(1)} + d^2 \log d \log\log d\, (\log q)^{2+o(1)})$$

bit operations. This is at most $O(d^{\omega+o(1)}(\log q)^{2+o(1)})$ bit operations. The cost of checking whether g is a large ppd-element is the same as this. To check if g is a basic ppd$(d, q; e)$-element involves computing $\Phi_b(e, q) = \Phi(ae, p)$ (where $q = p^a$) instead of $\Phi(e, q)$. Arguing as above, the cost of computing $\Phi_b(e, q)$ is $O(ad(\log(ad))^{O(1)}(\log p)^2) = O(d(\log d)^{O(1)}(\log q)^2)$ bit operations, and hence the cost of testing whether g is a basic ppd-element is also at most $O(d^{\omega+o(1)}$ $(\log q)^{2+o(1)})$ bit operations.

Since we need to test $O(\log \varepsilon^{-1} + \log\log d)$ elements of G, the total cost of the algorithm is as follows.

Theorem 11.1 *Suppose that $G \subseteq \Delta$ for some classical group Δ in $\mathrm{GL}(d, q)$, with*

G irreducible on the underlying vector space V, and that we have complete information about G-invariant forms (so that G is not contained in the class C_8 of subgroups of Δ). Assume that d is large enough that Ω contains two different ppd*-elements with at least one of them large and at least one basic. Further let ε be a positive real number with $0 < \varepsilon < 1$. Assume that we can make uniform independent random selections of elements from G and that the cost of producing each random element is μ bit operations. Then the classical recognition algorithm in [32] uses $O(\log \varepsilon^{-1} + \log\log d)$ random elements from G to test whether G contains Ω, and in the case where G contains Ω, the probability of failing to report that G contains Ω is less than ε. The cost of this algorithm is*

$$O((\log \varepsilon^{-1} + \log\log d)(\mu + d^{\omega+o(1)}(\log q)^{2+o(1)}))$$

bit operations, where ω is the exponent of matrix multiplication.

12 Classical recognition algorithm: final comments

The classical recognition algorithm in [32] has been implemented and is available as part of the *matrix* share package with the GAP system [37], and is also implemented in MAGMA [5]. In the MAGMA implementation rather large groups have been handled by the algorithm without problems: John Cannon has informed us that, on a SUN Ultra 2 workstation with a 200 MHz processor, recognising SL (5000, 2), for example, took 3214 CPU seconds averaged over five runs, while recognising SL (10000, 2) was possible in 14334 CPU seconds, again averaged over five runs of the algorithm.

The algorithm as described in this paper relies on the presence in the classical group Ω of two different ppd-elements, where at least one is large and at least one is basic. However, for some small values of the dimension d, depending on the type of the classical group and the field order q, Ω may not contain such elements. In these cases a modification of the algorithm has been produced in [33] which makes use of elements which are similar to ppd-elements. The results in [33] demonstrate that, with some effort, it is possible to extend the probability computations in Section 8.

An alternative algorithm to recognise classical groups in their natural representations has been developed by Celler and Leedham-Green in [8]. This algorithm also uses the Aschbacher classification [2] of subgroups of GL (d, q) as its organisational principle. Like the algorithm in [32] it makes use of a search by random selection for certain elements. Although no analysis of the complexity of the algorithm is given in [8], the analysis we give in Section 11 gives a reasonable measure of the complexity of this algorithm also. Finally, as with the algorithm in [32], the algorithm in [8] does not work for certain families of small dimensional classical groups (notably the groups of type $O^+(8, q)$), and the methods of [33] are required to deal with these groups.

References

[1] A. V. Aho, J. E. Hopcroft and J. D. Ullman, *The design and analysis of computer algorithms*, Addison-Wesley, 1975.

[2] M. Aschbacher, On the maximal subgroups of the finite classical groups, *Invent. Math.* **76** (1984), 469-514.

[3] L. Babai, Local expansion of vertex-transitive graphs and random generation in finite groups, *Proc. 23rd ACM STOC* (1991), 164–174.

[4] J. C. Beidleman and D. J. S. Robinson, On finite groups satisfying the permutizer condition, *J. Algebra* **191** (1997), 686–703.

[5] W. Bosma and J. J. Cannon, Handbook of MAGMA functions, *Department of Pure Mathematics, Sydney University*, 1993.

[6] P. Bürgisser, M. Clausen and M. A. Shokrollahi, *Algebraic complexity theory*, Springer, Berlin Heidelberg, 1997.

[7] F. Celler and C.R. Leedham-Green, Calculating the order of an invertible matrix, in *Groups and Computation II*, DIMACS: Series in Discrete Mathematics and Theoretical Computer Science, **28** (ed. L. Finkelstein and W. M. Kantor, American Mathematical Society, 1997), 55–60.

[8] F. Celler and C.R. Leedham-Green, A non-constructive recognition algorithm for the special linear and other classical groups, in *Groups and Computation II*, DIMACS: Series in Discrete Mathematics and Theoretical Computer Science, **28** (ed. L. Finkelstein and W. M. Kantor, American Mathematical Society, 1997), 61–67.

[9] F. Celler, C. R. Leedham-Green, S. H. Murray, A. C. Niemeyer and E. A. O'Brien, Generating random elements of a finite group, *Comm. Algebra* **23** (1995), 4931–4948.

[10] F. R. K. Chung and R. L. Graham, Random walks on generating sets of finite groups, *Electronic J. Combin.* **4(2)** (1997), R7 (14 pp), (http://www.mcs.drexel.edu /EJC/Journal/ejc-wce.html).

[11] J. H. Conway, R. T. Curtis, S. P. Norton, R. A. Parker and R. A. Wilson, *An atlas of finite groups*, Clarendon Press, Oxford, 1985.

[12] U. Dempwolff, Linear groups with large cyclic subgroups and translation planes, *Rend. Sem. Mat. Univ. Padova* **77** (1987), 69-113.

[13] P. Diaconis and L. Saloff-Coste, Walks on generating sets of abelian groups, *Probab. Theory Relat. Fields* **105** (1996), 393–421.

[14] P. Diaconis and L. Saloff-Coste, Walks on generating sets of groups, *Technical Report, Stanford University* **497** (1996).

[15] W. Feit, On large Zsigmondy primes, *Proc. Amer. Math. Soc.* **102** (1988), 29-36.

[16] R. M. Guralnick and W. M. Kantor, Probabilistic generation of finite simple groups, preprint.

[17] R. Guralnick, T. Penttila, C. E. Praeger and J. Saxl, Linear groups with orders having certain large prime divisors, *Proc. London Math. Soc.* (to appear).

[18] R. M. Guralnick and J. Saxl, Generation of classical groups by conjugates, preprint.

[19] C. Hering, Transitive linear groups and linear groups which contain irreducible subgroups of prime order, *Geom. Ded.* **2** (1974), 425-460.

[20] C. Hering, Transitive linear groups and linear groups which contain irreducible subgroups of prime order, II, *J. Algebra* **93** (1985), 151-164.

[21] D. F. Holt and S. Rees, Testing modules for irreducibility, *J. Austral. Math. Soc. (Series A)* **57** (1994), 1–16.

[22] G.D. James, On the minimal dimensions of irreducible representations of symmetric groups, *Math. Proc. Cambridge Philos. Soc.* **94** (1983), 417-424.

[23] C. Jansen, K. Lux, R. Parker and R. Wilson, *An atlas of Brauer characters*, Clarendon Press, Oxford, 1995.

[24] E. Kaltofen and V. Shoup, Subquadratic-time factoring of polynomials over finite

fields, *Math. Comp.* (to appear).

[25] W. Keller-Gehrig, Fast algorithms for the characteristic polynomial, *Theoretical Computer Science* **36** (1985), 309–317.

[26] P. Kleidman and M. Liebeck, *The subgroup structure of the finite classical groups*, London Math. Soc. Lecture Note Series 129, Cambridge University Press, Cambridge, 1990.

[27] M. W. Liebeck, The affine permutation groups of rank three, *Proc. London Math. Soc.* (3) **54** (1987), 477–516.

[28] M. W. Liebeck, Characterization of classical groups by orbit sizes on the natural module, preprint.

[29] B. Merkt, *Zsigmondy-elemente in klassischen Gruppen*, Doctoral dissertation, Universität zu Tübingen, 1995.

[30] P. M. Neumann and C. E. Praeger, A recognition algorithm for special linear groups, *Proc. London Math. Soc.* (3) **65** (1992), 555-603.

[31] A. C. Niemeyer and C. E. Praeger, Implementing a recognition algorithm for classical groups, in *Groups and Computation II*, DIMACS: Series in Discrete Mathematics and Theoretical Computer Science, **28** (ed. L. Finkelstein and W. M. Kantor, American Mathematical Society, 1997), 273–296.

[32] A. C. Niemeyer and C. E. Praeger, A recognition algorithm for classical groups, *Proc. London Math. Soc.* (to appear).

[33] A. C. Niemeyer and C. E. Praeger, A recognition algorithm for non-generic classical groups over finite fields, preprint, 1997.

[34] I. Niven, H. S. Zuckerman and H. L. Montgomery, *An introduction to the theory of numbers*, 5th edn (John Wiley & Sons, New York, 1991).

[35] R. A. Parker, The computer calculation of modular characters (the Meat-Axe), *Computational group theory*, Proceedings of the London Mathematical Society Symposium on Computational Group Theory (ed. Michael Atkinson, Academic Press, London, 1984), 267–274.

[36] C. E. Praeger, Computations with matrix groups over finite fields, *Groups and Computation*, DIMACS: Series in Discrete Mathematics and Theoretical Computer Science, **11** (ed. L. Finkelstein and W. M. Kantor, American Mathematical Society, 1991), 189–195.

[37] M. Schönert *et al.*, GAP–Groups, algorithms and programming, *Lehrstuhl D für Mathematik, Rheinisch Westfälische Technische Hochschule*, Aachen, Germany, fifth edition, 1995, (http://www-gap.dcs.st-and.ac.uk/ gap).

[38] A. Shalev, A theorem on random matrices and some applications, *J. Algebra* (to appear).

[39] V. Shoup, A new polynomial factorization algorithm and its implementation, *J. Symbolic Comp.* **20** (1995), 363-397

[40] D. E. Taylor, *The geometry of the classical groups*, Heldermann Verlag, Berlin, 1992.

[41] A. Wagner, The faithful linear representations of least degree of S_n and A_n over a field of characteristic 2, *Math. Z.* **151** (1976), 127-137.

[42] A. Wagner, The faithful linear representations of least degree of S_n and A_n over a field of odd characteristic, *Math. Z.* **154** (1977), 103-114.

[43] A. Wagner, An observation on the degrees of projective representations of the symmetric and alternating groups over an arbitrary field, *Arch. Math.* **29** (1977) 583-589.

[44] K. Zsigmondy, Zur Theorie der Potenzreste, *Monatsh. für Math. u. Phys.* **3** (1892), 265-284.

ON THE CLASSIFICATION OF GENERALIZED HAMILTONIAN GROUPS

DENISE M. REBOLI

SUNY at Binghamton, Binghamton, NY 13902-6000, U.S.A.

Abstract

A Dedekind group is one in which all subgroups are normal. Hamiltonian groups are nonabelian Dedekind groups. In this paper we consider the characteristic subgroups $CS(G)$ and $S(G)$, which are generated by the nonnormal cyclic subgroups and nonnormal subgroups of G, respectively. It will be shown that $CS(G) = S(G)$. Groups in which $CS(G) \neq G$ are called generalized Dedekind groups. A generalized Dedekind group is either abelian or a torsion group of nilpotency class two. Generalized Hamiltonian groups are groups which are generalized Dedekind, but not Dedekind. For every p, there is a generalized Hamiltonian p-group. The goal of this paper is to classify all two-generator generalized Hamiltonian p-groups.

1 Introduction

A group G is called a Dedekind group if all subgroups of G are normal. For the finite case these groups were classified by Dedekind in 1897 in [5], and for the infinite case by Baer in [2]. Dedekind groups are either abelian or the direct product of a quaternion group of order eight and an abelian torsion group with no elements of order four (see [7], Theorem 5.3.7). The nonabelian Dedekind groups are called Hamiltonian groups. The class of Dedekind groups will be denoted by $\mathcal{DG}$, the class of Hamiltonian groups by $\mathcal{HG}$.

Since 1897, there have been various generalizations of Dedekind groups. Blackburn in [3] considered non-Dedekind finite groups in which the nonnormal subgroups have nontrivial intersection. If a group G is in $\mathcal{DG}$, this intersection is defined to be all of G. In this paper we consider dual notions, namely the subgroup generated by all nonnormal cyclic subgroups and the subgroup generated by all nonnormal subgroups. If H is not normal in G, we will write $H \ntrianglelefteq G$.

Definition 1.1 For any group G, let

$$CS(G) = \langle g \in G \mid \langle g \rangle \ntrianglelefteq G \rangle \text{ and } S(G) = \langle H \leq G \mid H \ntrianglelefteq G \rangle.$$

If there are no nonnormal cyclic subgroups or no nonnormal subgroups, we define $CS(G) = 1$ and $S(G) = 1$, respectively.

As one can easily see, $CS(G)$ and $S(G)$ are characteristic subgroups and $CS(G) \leq S(G)$. We note that the subgroup $S(G)$ was introduced by Cappitt in [4]. Since all cyclic subgroups are normal if and only if all subgroups are normal, one can ask whether $CS(G)$ and $S(G)$ coincide as subgroups. This is answered in the affirmative in Theorem 2.2. Cappitt called groups with $S(G) \neq G$ generalized Dedekind groups. In view of Theorem 2.2, we now make the following definition.

Definition 1.2 If G is a group with $CS(G) \neq G$, then G is called a generalized Dedekind group. The class of generalized Dedekind groups will be denoted by $\mathcal{GDG}$.

Since a Hamiltonian group is a Dedekind group which is not abelian, we consider a parallel notion for generalized Dedekind groups.

Definition 1.3 If G is a generalized Dedekind group, but not a Dedekind group, then G is called a generalized Hamiltonian group. The class of generalized Hamiltonian groups is denoted by $\mathcal{GHG}$.

Cappitt in [4] started the investigation of generalized Dedekind groups and showed that such groups are of nilpotency class at most two and are either periodic or abelian (see Proposition 2.3). It is natural to ask whether generalized Hamiltonian groups can be classified in a way similar to the classification of Hamiltonian groups. In a future publication we will address this question. In view of Proposition 2.3 we can restrict our attention to p-groups. It appears that the structure of generalized Hamiltonian p-groups is fairly limited, but not as limited as that of Hamiltonian groups.

The following theorem, the main result of this paper, is a classification of all 2-generator p-groups which are generalized Hamiltonian groups. This is the first step towards the classification of these groups.

Theorem 1.4 *Let p be a prime and $G = \langle a, b\rangle$ be a finite 2-generator p-group of nilpotency class two. If G is generalized Hamiltonian, then $G \cong \langle a\rangle \rtimes \langle b\rangle$, where $[a,b] = a^{p^{\alpha-\gamma}}$, $|a| = p^\alpha$, $|b| = p^\beta$, $|[a,b]| = p^\gamma$, $\alpha, \beta, \gamma \in \mathbb{N}$, $\alpha \geq 2\gamma$, $\beta \geq \gamma \geq 1$, $\alpha - \beta - \gamma \geq 0$. If $p = 2$, $\alpha + \beta > 3$.*

Hamiltonian p-groups exist only for $p = 2$. There is no restriction on the prime in the case of generalized Hamiltonian p–groups. This is an immediate consequence of Theorem 1.4 which we state here as a corollary.

Corollary 1.5 *There exist generalized Hamiltonian p-groups for every prime p.*

2 Basic results

In this section we give some basic results for generalized Hamiltonian groups. In addition, we state the classification theorem for 2-generator p-groups of class 2 which forms the basis for the classification of 2-generator generalized Hamiltonian p-groups.

In a group, all subgroups are normal if and only if all cyclic subgroups are normal. It is a natural question to ask whether $CS(G) = S(G)$, for any group G. Before establishing that this is true, we need the following lemma, stated here without proof.

Lemma 2.1 *Let G be a group and H a proper subgroup of G. Then there exists a subset $U \subseteq G \backslash H$ such that $G = \langle u \mid u \in U \rangle$.*

We now show that $CS(G)$ is, in fact, the same as $S(G)$. The following method of proof was suggested by S. Gagola.

Theorem 2.2 *Let G be a group. Then $CS(G) = S(G)$.*

Proof Obviously, $CS(G) \leq S(G)$ by definition. To show the reverse containment, we consider $H \leq G$, where H is not a subgroup of $CS(G)$. Let $D = H \cap CS(G)$. Consider $h \in H \setminus D$. Then, since $h \notin CS(G)$, $\langle h \rangle \triangleleft G$. By Lemma 2.1, $H = \langle h \mid h \in H \setminus D \rangle$. Hence, H can be generated by elements, h, where the $\langle h \rangle \triangleleft G$, so $H \triangleleft G$. Thus, by the contrapositive, if K is not normal in G, $K \leq CS(G)$. Therefore, $S(G) \leq CS(G)$. □

D. Cappitt introduced the subgroup $S(G)$ in [4]. We use the following proposition established in his paper.

Proposition 2.3 *If G is a generalized Dedekind group, then G is nilpotent of class at most two, and G is either periodic or abelian.*

The following familiar expansion formulas for groups of nilpotency class two are stated here without proof.

Lemma 2.4 *Let G be a group of nilpotency class two. Then for any $x, y \in G$, and any integer n,*

$$[x, y^n] = [x^n, y] = [x, y]^n; \quad (2.4.1)$$

$$(xy)^n = x^n y^n [y, x]^{\binom{n}{2}}. \quad (2.4.2)$$

Since G is nilpotent of class at most two, we can make use of the following classification theorem for 2-generator p-groups of class 2, found for p odd in [1] and for $p = 2$ in [6].

Theorem 2.5 *Let $G = \langle a, b \rangle$ be a 2-generator p-group of nilpotency class two. Then G is isomorphic to exactly one group of the following four types:*

$$G \cong (\langle c \rangle \times \langle a \rangle) \rtimes \langle b \rangle, \text{where } [a, b] = c,\ [a, c] = [b, c] = 1, \quad (2.5.1)$$
$$|a| = p^{\alpha}, |b| = p^{\beta}, |c| = p^{\gamma},\ \alpha, \beta, \gamma \in \mathbb{N},\ \alpha \geq \beta \geq \gamma \geq 1;$$

$$G \cong \langle a \rangle \rtimes \langle b \rangle, \text{where } [a, b] = a^{p^{\alpha-\gamma}},\ |a| = p^{\alpha},\ |b| = p^{\beta},\ |[a, b]| = p^{\gamma}, \quad (2.5.2)$$
$$\alpha, \beta, \gamma \in \mathbb{N},\ \alpha \geq 2\gamma,\ \beta \geq \gamma \geq 1, \text{ and if } p = 2,\ \alpha + \beta > 3;$$

$$G \cong (\langle c \rangle \times \langle a \rangle) \rtimes \langle b \rangle, \text{where } [a, b] = a^{p^{\alpha-\gamma}} c,\ [c, b] = a^{-p^{2(\alpha-\gamma)}} c^{-p^{\alpha-\gamma}}, \quad (2.5.3)$$
$$|a| = p^{\alpha},\ |b| = p^{\beta},\ |c| = p^{\sigma},\ |[a, b]| = p^{\gamma},\ \alpha, \beta, \gamma, \sigma \in \mathbb{N},$$
$$\gamma > \sigma \geq 1,\ \alpha + \sigma \geq 2\gamma,\ \beta \geq \gamma;$$

$$G \cong (\langle c \rangle \times \langle a \rangle)\langle b \rangle, \text{where } [a, b] = a^2 c,\ [c, b] = a^{-4} c^{-2}, \quad (2.5.4)$$
$$a^{2^{\gamma}} = b^{2^{\gamma}},\ |a| = 2^{\gamma+1} = |b|,$$
$$|[a, b]| = 2^{\gamma},\ |c| = 2^{\gamma-1}, \text{ and } \gamma \in \mathbb{N}.$$

The groups in the above list are pairwise nonisomorphic and have nilpotency class 2 precisely.

3 Preparatory lemmas

In this section we build the framework necessary for proving the main theorem. Since the conclusion states that a generalized Hamiltonian two-generator p-group must be of type (2.5.2), we consider only these groups for now.

Our goal is to determine the subgroup $CS(G)$ for a group G of type (2.5.2) with the added condition $\alpha-\beta-\gamma \geq 0$, so we need to be able to identify the nonnormal cyclic subgroups of such a group. To do this, we consider the following subset of G,

$$A = \{b^{np^j} a^{m'p^{\alpha-\beta-\gamma+2j+1}} \mid 0 \leq j \leq \gamma - 1,\ (n,p) = 1,\ m' \in \mathbb{N}\},$$

and its complement A^C. In Proposition 3.5, it will be shown that A is precisely the set of those elements in G generating nonnormal cyclic subgroups. First, we give a characterization of those elements in a group of type (2.5.2) generating normal cyclic subgroups.

Proposition 3.1 *Let $G = \langle a, b\rangle$ be a group of type (2.5.2) and $g = b^r a^s \in G$. Then the following are equivalent:*

(i) $\langle g\rangle \lhd G$;

(ii) *there exist integers k and l such that $g^a = g^k$ and $g^b = g^l$;*

(iii) *there exist integers k and l such that the following pairs of congruences for k and l, respectively, are solvable:*

$$\left.\begin{aligned} rk &\equiv r \bmod p^\beta, \\ sk + \binom{k}{2} rsp^{\alpha-\gamma} &\equiv s - rp^{\alpha-\gamma} \bmod p^\alpha. \end{aligned}\right\} \tag{3.1.1}$$

$$\left.\begin{aligned} rl &\equiv r \bmod p^\beta, \\ sl + \binom{l}{2} rsp^{\alpha-\gamma} &\equiv s(1+p^{\alpha-\gamma}) \bmod p^\alpha. \end{aligned}\right\} \tag{3.1.2}$$

Proof The equivalence of the first two statements is clear from the definition of normality and the fact that G is generated by a and b. We now show that (ii) and (iii) are equivalent. Using the relations of G we obtain $(b^r a^s)^a = b^r a^{s-rp^{\alpha-\gamma}}$ and $(b^r a^s)^b = b^r a^{s(1+p^{\alpha-\gamma})}$. Since G is nilpotent of class two, it follows by (2.4.2) and the relations of G that $(b^r a^s)^k = b^{rk} a^{sk+\binom{k}{2}rsp^{\alpha-\gamma}}$ and $(b^r a^s)^l = b^{rl} a^{sl+\binom{l}{2}rsp^{\alpha-\gamma}}$. Then, by the above, (ii) is equivalent to

$$\left.\begin{aligned} b^r a^{s-rp^{\alpha-\gamma}} &= b^{rk} a^{sk+\binom{k}{2}rsp^{\alpha-\gamma}}, \\ b^r a^{s(1+p^{\alpha-\gamma})} &= b^{rl} a^{sl+\binom{l}{2}rsp^{\alpha-\gamma}}. \end{aligned}\right\} \tag{3.1.3}$$

We note that in a group of type (2.5.2), representation of elements is unique. Therefore, (3.1.3) is equivalent to (iii). □

In the next four lemmas we will give conditions under which g^a and g^b are in $\langle g\rangle$.

Lemma 3.2 *Let G be a group of type (2.5.2) with p odd and $\alpha-\beta-\gamma\geq 0$. Let $b^r a^s \in G$. Then $(b^r a^s)^b \in \langle b^r a^s\rangle$.*

Proof In light of the previous proposition, we will show the congruences of (3.1.2) are solvable. Set $l = 1+p^{\alpha-\gamma}$. Then l is even and

$$\frac{1}{2} lrsp^{2\alpha-2\gamma} \equiv 0 \bmod p^\alpha$$

since $2\alpha-2\gamma\geq 2\alpha-\alpha=\alpha$. Therefore,

$$\binom{l}{2} rsp^{\alpha-\gamma} \equiv 0 \bmod p^\alpha.$$

Adding sl to both sides yields

$$sl+\binom{l}{2} rsp^{\alpha-\gamma} \equiv s(1+p^{\alpha-\gamma}) \bmod p^\alpha.$$

Thus, l satisfies the second congruence of (3.1.2). Since $\alpha-\gamma\geq\beta$, it follows that $rl\equiv r \bmod p^\beta$ and the first congruence holds as well. By Proposition 3.1, our claim follows. □

In the next three lemmas we consider A^C, the complement of A in G. Let $g = b^r a^s \in G$ where $r = np^j$, $s = mp^i$, $(n,p) = 1 = (m,p)$, $0\leq j\leq\beta$, $0\leq i\leq\alpha$. We will use this presentation throughout the remainder of the paper. Note that if $g\in A^C$, then exactly one of the following conditions holds:

(i) $\gamma\leq j\leq\beta$;

(ii) $0\leq j\leq\gamma-1$ and $i<\alpha-\beta-\gamma+2j+1$.

Lemma 3.3 *Let G be a group of type (2.5.2) with $p=2$, and therefore $\alpha+\beta>3$, with $\alpha-\beta-\gamma\geq 0$. Let $g\in G$. If $g\in A^C$, then $g^b\in\langle g\rangle$.*

Proof Suppose $g\in A^C$ and g satisfies (i) or (ii). If $j=\beta$, then $g=a^s$. By construction, $\langle a^s\rangle \triangleleft G$. Therefore, we assume $\gamma\leq j\leq\beta-1$. We now show that $\alpha-\gamma-1+j>0$. If $\beta>1$, then $\alpha-\gamma-1+j>\alpha-\gamma-\beta\geq 0$ by assumption. If $\beta=1$, then $\gamma=1$. We conclude, as before, $\alpha-\gamma-1+j\geq 1>0$. Thus, $1+n2^{\alpha-\gamma-1+j}$ is odd. Therefore,

$$mt2^{\beta+i-j}(1+n2^{\alpha-\gamma-1+j}) \equiv \begin{cases} 0 \bmod 2^\alpha, & \text{for } i\geq\gamma, \\ m2^{\alpha-\gamma+i} \bmod 2^\alpha, & \text{for } i<\gamma. \end{cases} \tag{3.3.1}$$

is solvable for t. Because $\alpha+2\beta-\gamma-j+i-1\geq\alpha+\beta-j+i-1\geq\alpha+i\geq\alpha$ in case (i) and $\alpha+2\beta-\gamma-j+i-1\geq\alpha+2\beta-\gamma-j-1\geq\alpha+2\beta-2\gamma+1-1\geq\alpha$

in case (ii), we may add multiples of 2^σ, where $\sigma = \alpha + 2\beta - \gamma - j + i - 1$ to the left side of (3.3.1) and arrive at

$$mt2^{\beta+i-j}(1 + n2^{\alpha-\gamma-1+j}) + t^2nm2^\sigma \equiv \begin{cases} 0 \bmod 2^\alpha, & \text{for } i \geq \gamma, \\ m2^{\alpha-\gamma+i} \bmod 2^\alpha & \text{for } i < \gamma. \end{cases} \tag{3.3.2}$$

Setting $s = m2^i$ and $r = n2^j$, we may add multiples of $s2^{\alpha-\gamma}$ to the right side if $i \geq \gamma$ and adding s to both sides in either case. After simplifying, we obtain

$$s(1 + t2^{\beta-j}) + \binom{1 + t2^{\beta-j}}{2} rs2^{\alpha-\gamma} \equiv s(1 + 2^{\alpha-\gamma}) \bmod 2^\alpha.$$

It follows that the congruences of (3.1.2) are solvable with $l = 1+t2^{\beta-j}$. Proposition 3.1 implies that $g^b \in \langle g \rangle$. □

Lemma 3.4 *Let G be a group of type (2.5.2) with $\alpha - \beta - \gamma \geq 0$ and $g \in G$. If $g \in A^C$, then $g^a \in \langle g \rangle$.*

Proof Let $g \in A^C$. As in Lemma 3.3, we may again assume that $j \leq \beta - 1$. Suppose first that (i) holds. Now,

$$st \equiv 0 \bmod p^{\alpha-\beta+j} \tag{3.4.1}$$

is solvable with

$$t \equiv \begin{cases} p^{\alpha-\beta+j-i} \bmod p^{\alpha-\beta+j}, & \text{for } 0 \leq i \leq \alpha - \beta + j, \\ 1 \qquad \bmod p^{\alpha-\beta+j}, & \text{for } i > \alpha - \beta + j. \end{cases} \tag{3.4.2}$$

Multiplying (3.4.1) by $p^{\beta-j}$ and, then adding s to both sides yields

$$s(1 + tp^{\beta-j}) \equiv s \bmod p^\alpha.$$

Because $\alpha - \gamma + j \geq \alpha - j + j = \alpha$, we may add multiples of $p^{\alpha-\gamma+j}$ to both sides of the above and obtain

$$s(1 + tp^{\beta-j}) + \binom{1 + tp^{\beta-j}}{2} snp^{\alpha-\gamma+j} \equiv s - np^{\alpha-\gamma+j} \bmod p^\alpha.$$

We see that the congruences of (3.1.1) are satisfied with $k = 1+tp^{\beta-j}$. Proposition 3.1 then implies that $g^a \in \langle g \rangle$, proving our claim in case (i).

Now assume $g \in A^C$ and (ii) holds. Suppose, first, that p is odd. As in Lemma 3.3, it can be shown that $2+np^{\alpha-\gamma-1+j} \not\equiv 0 \bmod p$. Noting that $i \leq \alpha-\beta-\gamma+2j$, the congruence

$$mt(2 + np^{\alpha-\gamma+j}) \equiv -2np^{\alpha-\beta-\gamma+2j-i} \bmod p^{\alpha-\beta-i+j}$$

is solvable for t. Since $\beta + i - j > 0$, we can multiply both sides of the above by $p^{\beta+i-j}$. Expanding the left side yields

$$2mtp^{\beta+i-j} + nmtp^{\alpha+\beta-\gamma+i} \equiv -2np^{\alpha+j-\gamma} \bmod p^\alpha. \tag{3.4.3}$$

Without loss of generality, t is even. (If t is odd, then $t' = t + p^\alpha$ is even.) After factoring out 2 from (3.4.3), we have

$$2(mtp^{\beta+i-j} + nm\frac{t}{2}p^{\alpha-\gamma+\beta+i}) \equiv -2np^{\alpha-\gamma+j} \bmod p^\alpha.$$

Since $(2,p) = 1$ and the above is solvable, the following congruence is also solvable for t

$$mtp^{\beta+i-j} + nm\frac{t}{2}p^{\alpha-\gamma+\beta+i} \equiv -np^{\alpha-\gamma+j} \bmod p^\alpha. \tag{3.4.4}$$

Because $j \leq \gamma - 1$ and $\beta \geq \gamma$, we have $\tau = \alpha + 2\beta - \gamma - j + i \geq \alpha + 2\beta - 2\gamma \geq \alpha$. Thus, setting $r = np^j$ and $s = mp^i$, we may add multiples of p^τ to the left side of (3.4.4) and s to both sides. After simplifying, we obtain

$$s(1 + tp^{\beta-j}) + \binom{1 + tp^{\beta-j}}{2} rsp^{\alpha-\gamma} \equiv s - rp^{\alpha-\gamma} \bmod p^\alpha.$$

It follows that the congruences of (3.1.1) are solvable with $k = 1 + tp^{\beta-j}$. Proposition 3.1 implies that $g^a \in \langle g \rangle$.

Now, suppose $p = 2$. As in Lemma 3.3, $1 + n2^{\alpha-\gamma-1+j}$ is odd. Noting that $i \leq \alpha - \beta - \gamma + 2j$, the congruence

$$mt(1 + n2^{\alpha-\gamma-1+j}) \equiv -n2^{\alpha-\beta-\gamma+2j-i} \bmod 2^{\alpha-\beta-i+j}$$

is solvable for t. The remainder of the proof follows in a similar way to that of p odd and is left to the reader. Thus, the congruences of (3.1.1) are solvable with $k = 1 + t2^{\beta-j}$. Proposition 3.1 implies that $g^a \in G$. □

We are now in a position to actually determine the non-normal subgroups in a group of type (2.5.2) with $\alpha - \beta - \gamma \geq 0$.

Proposition 3.5 *Let G be a group of type (2.5.2) with $\alpha - \beta - \gamma \geq 0$. Then*

$$\{g \mid \langle g \rangle \ntriangleleft G\} = A = \{b^{np^j} a^{m'p^{\alpha-\beta-\gamma+2j+1}} \mid 0 \leq j \leq \gamma - 1,\ (n,p) = 1, m' \in \mathbb{N}\}.$$

Proof Suppose $g \in A$. Then $g = b^{np^j} a^{m'p^{\alpha-\beta-\gamma+2j+1}}$, where $0 \leq j \leq \gamma - 1$, and $n, m' \in \mathbb{N}$ with $(n,p) = 1$. We will show $g^a \notin \langle g \rangle$, or, equivalently, the congruences of (3.1.1) are not solvable for k. As before, if the first congruence of (3.1.1) is solvable, we have $k = 1 + tp^{\beta-j}$. Recall that if p is odd, we may assume that t is even. We first substitute k, $r = np^j$, and $s = m'p^{\alpha-\beta-\gamma+2j+1}$ into the second congruence of (3.1.1). Let $\lambda = 2\alpha - 2\gamma + 2j$ and $\mu = 2\alpha + \beta + j - 2\gamma$. Then, expanding the binomial coefficient, and subtracting s on both sides yields

$$\left.\begin{array}{rcl} m't2^{\alpha-\gamma+j+1} + nm't2^\lambda + t^2nm'2^\mu & \equiv & -n2^{\alpha-\gamma+j} \bmod 2^\alpha, \\ m'tp^{\alpha-\gamma+j+1} + \frac{1}{2}tnm'p^{\lambda+1} + \frac{1}{2}t^2nm'p^{\mu+1} & \equiv & -np^{\alpha-\gamma+j} \bmod p^\alpha. \end{array}\right\} \tag{3.5.1}$$

Since $\lambda > \alpha$, and $\mu > \alpha$, the second and third terms of (3.5.1) are congruent to zero mod p^α. Thus, we are left in both cases with

$$p^{\alpha-\gamma+j+1}tm' \equiv -np^{\alpha-\gamma+j} \bmod p^\alpha.$$

However, this is not solvable for t since $p^{\alpha-\gamma+j+1}$ divides the left side, but does not divide the right side. Thus, the congruences of (3.1.1) cannot be simultaneously satisfied. So, by Lemma 3.1, $\langle g\rangle$ is not normal in G.

Conversely, let $g \in A^C$. We will show that $\langle g\rangle \triangleleft G$. By Proposition 3.1, we must show g^a and g^b are elements of $\langle g\rangle$. If p is odd, by Lemma 3.2, $g^b \in \langle g\rangle$, and by Lemma 3.4, $g^a \in \langle g\rangle$. Thus, if p is odd, $\langle g\rangle \triangleleft G$ by Proposition 3.1. If $p = 2$, $g^b \in \langle g\rangle$ by Lemma 3.3 and $g^a \in \langle g\rangle$ by Lemma 3.4. Hence, by Proposition 3.1, $\langle g\rangle \triangleleft G$. We conclude that nonnormal cyclic subgroups are precisely those generated by elements in A. □

4 Proof of the main result and corollaries

With the results of the previous sections, we are now ready to prove our main theorem.

Proof of Theorem 1.4 By Theorem 2.5, we know a finite 2-generator p-group of class two is one of four types. If G is a group of type (2.5.1), then by construction $\langle b\rangle \ntriangleleft G$. Likewise, $\langle a\rangle \ntriangleleft G$, since $a^b = a[a,b] = ac \notin \langle a\rangle$. So, $G = \langle a,b\rangle \leq CS(G)$. Thus, $CS(G) = G$. Next, suppose G is of type (2.5.3). Again, by construction, $\langle b\rangle \ntriangleleft G$. Similarly, $\langle a\rangle \ntriangleleft G$, since $a^b = a[a,b] = a^{1+p^{\alpha-\gamma}}c \notin \langle a\rangle$. So $G = \langle a,b\rangle \leq CS(G)$. We conclude that $CS(G) = G$. Now, let G be of type (2.5.4). If $\gamma = 1$, then $G \cong Q_8$. So G is a Dedekind group. Thus $CS(G) = 1$. If $\gamma \geq 2$ then observe that $\langle b\rangle \ntriangleleft G$ by construction. We have $a^b = a[a,b] = a^3c$. Since $\langle a\rangle \cap \langle c\rangle = 1$, $a^3c \notin \langle a\rangle$. Hence, $\langle a\rangle \ntriangleleft G$. So, $G = \langle a,b\rangle \leq CS(G)$, which implies $CS(G) = G$.

Finally, suppose G is of type (2.5.2). We first consider the case that $\alpha - \beta - \gamma < 0$. Again, by construction, $\langle b\rangle \ntriangleleft G$. Consider the cyclic subgroup $\langle ba\rangle$. Then $(ba)^a = ba^{1-p^{\alpha-\gamma}}$. Suppose $(ba)^a \in \langle ba\rangle$. By Proposition 3.1, the congruences in (3.1.1) with $r = s = 1$ must be solvable for k. If the first congruence of (3.1.1) holds, then $k = 1 + tp^\beta$ for some $t \in \mathbb{Z}$, and, without loss of generality, t is even for p odd.

First, let $p = 2$. Substituting this into the second congruence of (3.1.1) and simplifying lead to

$$2^\beta(t + t2^{\alpha-\gamma-1} + t^2 2^{\beta-1+\alpha-\gamma}) \equiv -2^{\alpha-\gamma} \bmod 2^\alpha. \tag{4.0.1}$$

Likewise, for p odd we obtain

$$p^\beta(t + \frac{1}{2}tp^{\alpha-\gamma}(1 + tp^\beta)) \equiv -p^{\alpha-\gamma} \bmod p^\alpha. \tag{4.0.2}$$

So the left hand sides of (4.0.1) and (4.0.2) are divisible by 2^β and p^β, respectively, but the right hand sides are not, since $\beta > \alpha - \gamma$. Thus, neither (4.0.1) nor (4.0.2) is solvable for t. So, by Proposition 3.1, $\langle ba\rangle \ntriangleleft G$. Therefore, $G = \langle b, ba\rangle \leq CS(G)$. So, $G = CS(G)$.

Finally, we consider a group of type (2.5.2), where $\alpha - \beta - \gamma \geq 0$. By Lemma 3.5,

$$A = \{g | \langle g\rangle \ntriangleleft G\} = \{b^{np^j} a^{m'p^{\alpha-\beta-\gamma+2j+1}} | (n,p) = 1,\ m' \in \mathbb{Z},\ 0 \leq j \leq \gamma - 1\}.$$

Thus, $CS(G) = \langle A \rangle$. Let $j = 0$ and $m' = p^{\beta+\gamma-1}$. This gives us that $b^n \in CS(G)$ for any $(n,p) = 1$, so $b \in CS(G)$. Then $a^{m'p^{\alpha-\beta-\gamma+2j+1}} \in CS(G)$ for any $m' \in \mathbb{Z}$ and $0 \leq j \leq \gamma - 1$. So $a^{p^{\alpha-\beta-\gamma+1}} \in CS(G)$. Therefore $\langle b, a^{p^{\alpha-\beta-\gamma+1}} \rangle \leq CS(G)$. On the other hand, $g \in \langle b, a^{p^{\alpha-\beta-\gamma+1}} \rangle$ for any $g \in A$. It follows that $CS(G) = \langle b, a^{p^{\alpha-\beta-\gamma+1}} \rangle \neq G$, and $G \in \mathcal{GHG}$. □

In light of the above result, we observe the following structural properties of a two-generator generalized Hamiltonian p-group.

Corollary 4.1 *Let $G = \langle h, g \rangle$ be a p-group. If $G \in \mathcal{GHG}$, then at most one of $\langle g \rangle, \langle h \rangle$ is not normal in G.*

Proof Suppose both $\langle g \rangle$ and $\langle h \rangle$ are not normal in G. Then $g, h \in CS(G)$ and $G = \langle g, h \rangle \leq CS(G)$, a contradiction. □

Corollary 4.2 *If $G \in \mathcal{GHG}$ is a two-generator p-group, then $CS(G)$ is abelian if and only if $\alpha - \beta - 2\gamma + 1 \geq 0$.*

Proof We observe that $CS(G)$ is abelian if and only if the generators of $CS(G)$ commute. By Lemma 2.4 , $[a^{p^{\alpha-\beta-\gamma+1}}, b] = [a,b]^{p^{\alpha-\beta-\gamma+1}}$. Now, $[a,b]$ has order p^γ. Therefore, $a^{p^{\alpha-\beta-\gamma+1}}$ and b commute if and only if $\alpha-\beta-\gamma+1 \geq \gamma$, or equivalently, $\alpha - \beta - 2\gamma + 1 \geq 0$. □

References

[1] M. Bacon, L.C. Kappe, *The nonabelian tensor square of a 2-generator p-group of class 2*, Arch.Math. **61** (1993), 508-516.

[2] R. Baer, *Situation der Untergruppen und Struktur der Gruppen*, Sitz. Heidelberg Akademie Wiss. Math.-Natur.Kl.no.2, (1933), 12-17.

[3] N. Blackburn, *Finite Groups in which the Nonnormal Subgroups Have Nontrivial Intersection*, J. Algebra **3** (1966), 30-37.

[4] D. Cappitt, *Generalized Dedekind Groups*, J. Algebra **17** (1971), 310-316.

[5] R. Dedekind, *Über Gruppen, deren sämtliche Teiler Normalteiler sind*, Math. Ann. **48** (1897), 548-561.

[6] L.C. Kappe, N. Sarmin, M. Visscher, *Two-generator 2-groups of class 2 and their nonabelian tensor squares*, Glasgow Math. J., to appear.

[7] D. Robinson, *A Course in the Theory of Groups*, Spring-Verlag, New York-Berlin-Heidelberg, 1982.

PERMUTABILITY PROPERTIES OF SUBGROUPS

DEREK J. S. ROBINSON

Department of Mathematics, University of Illinois, 1409 West Green Street, Urbana, Illinois 61801, U.S.A.

1 Introduction

Subgroups H and K of a group G *permute* if $HK = KH$. If H permutes with every subgroup of G, then H is said to be *permutable* (or *quasi-normal*) in G. Permutability can be an awkward property to study, perhaps because subgroups can permute for a variety of reasons.

Here we are interested in the structure of groups which have lots of subgroup permutability built in. The extreme case is where every subgroup is permutable in G; then G is called a *quasi-Dedekind group.* We briefly review what is known about finite quasi-Dedekind groups – for more details see [10].

First there is the well-known result of Ore that permutable subgroups of finite groups are subnormal. Hence finite quasi- Dedekind groups are nilpotent.

Next a group G is called *modular* if the modular law is valid, i.e. if

$$\langle H, K\rangle \cap L = \langle H, K \cap L\rangle$$

where $H, K, L \leq G$ and $H \subseteq L$. By the usual proof *every quasi-Dedekind group is modular.* For finite nilpotent groups the converse holds; therefore we have:

Theorem 1 *A finite group is quasi-Dedekind if and only if it is a nilpotent modular group.*

The structure of finite modular p-groups was determined by Iwasawa [4].

Theorem 2 *The finite modular p-groups which are not Dedekind groups are the groups of the form* $G = \langle t, N\rangle$ *where* $N \lhd G$, N *is abelian and* $a^t = a^{1+p^s}$, $(a \in N)$, *with* $s > 1$ *if* $p = 2$.

From Theorems 1 and 2 the structure of finite quasi-Dedekind groups can be read off. For infinite quasi-Dedekind groups see [10].

We will now discuss permutability properties that lead to wider classes of groups than quasi-Dedekind groups.

2 The permutizer condition

If H is a subgroup of a group G, the *permutizer* of H in G is defined to be

$$P_G(H) = \langle g \in G \mid H\langle g\rangle = \langle g\rangle H\rangle.$$

Thus $H \leq N_G(H) \leq P_G(H)$.

A group G satisfies the *permutizer condition* **P** if $H < P_G(H)$ whenever $H < G$. Compare this with the *normalizer condition* **N**: here G satisfies **N** if $H < G$ implies that $H < N_G(H)$. It is well-known that for finite groups **N** is equivalent to nilpotency, and by a result of Plotkin **N**-groups are locally nilpotent.

Some examples of groups satisfying the permutizer condition are: (i) groups satisfying **N**; (ii) supersoluble groups; (iii) S_4, (but not A_4).

The permutizer condition was introduced by Deskins and Venzke [3]; it has also been studied by Zhang [11] and, more recently, by Beidleman and Robinson [1], [2].

Results

While **P**-groups need not be supersoluble, evidence that they are quite close to supersoluble groups soon surfaced. For example, Deskins and Venzke [3] proved that a **P**-group of odd order is supersoluble.

The main structure theorem for finite **P**-groups shows just how close these groups are to being supersoluble.

Theorem 3 ([1]) *Let G be a finite group with the permutizer condition* **P**. *Then every chief factor of G has order 4 or a prime. Also, if F is a chief factor of order 4, then $G/C_G(F) \simeq S_3$.*

At present it is not known if the converse holds. Notice that the last assertion of the theorem distinguishes between S_4 and A_4.

Definition Let G be a group and X a non-trivial group. If $H < G$, then H is said to be *constrained by X in G* if there exist subgroups L, M such that $L < H < M$, $L \lhd M$ and $M/L \simeq X$.

This notion enables us to formulate a criterion for **P**.

Theorem 4 ([1], [2]) *Let G be a finite group. Then G has* **P** *if and only if each proper self-normalizing subgroup is constrained by S_4 or by a non-abelian subgroup of the holomorph* $\mathrm{Hol}(\mathbb{Z}_p)$, *where p is an odd prime.*

Here the sufficiency of the condition follows at once from the fact that S_4 and the subgroups of $\mathrm{Hol}(\mathbb{Z}_p)$ have **P**; the necessity can be deduced from Theorem 3.

Method of Proof The main step in the proof of Theorem 3 is to show that *a finite* **P**-*group is soluble.* Assume this is false and let G be a counterexample of minimal order. Then one argues successively as follows:

(i) G is semisimple, i.e. G has no non-trivial abelian normal subgroups;

(ii) G is *almost simple*, i.e. $R \leq G \leq \mathrm{Aut}(R)$ where R is a (non-abelian) simple group;

(iii) let M be maximal in R and put $L = N_G(M)$. Then there is a subgroup G_M satisfying the following:

(a) $R \leq G_M \leq LR$;

(b) there is a *maximal factorization* $G_M = A_M B_M$ with $G_M = A_M R$, (so A_M and B_M are maximal in G_M);

(c) $A_M \cap R = M$ and B_M contains an element of order $|R : M|$.

In applying (iii) it is usually advantageous to choose for M a small maximal subgroup, so that B_M will contain an element of large order.

Now the maximal factorizations of the almost simple groups have been determined by Liebeck, Praeger and Saxl [7]. The idea of the proof is to show that M can be chosen to avoid all the listed factorizations. Exceptional cases are then dealt with by using order arguments.

3 Infinite groups with the permutizer condition

Of course the techniques described above apply only to finite **P**-groups, and little is known about the infinite case. For example, it would be of interest to determine the structure of locally finite **P**-groups.

One approach is to look at groups with many finite quotients; for these are of known structure. A good starting point is where G is a polycyclic **P**-group. The essential fact to prove is this.

Every rationally irreducible factor of G is infinite cyclic.

Let $A \lhd G$ with A rationally irreducible. Then the result is true for G/A by induction on the Hirsch length. Let p be an odd prime. By a theorem of Hirsch, G/A^p has a torsion-free normal subgroup of finite index, say N/A^p. Then $A/A^p \cap N/A^p = 1$ and $A/A^p \overset{G}{\simeq} AN/N \lhd G/N$. Since p-chief factors of G/N have order p, there is a series of G-submodules in A/A^p with factors of order p. One can deduce from this that A must be cyclic, using some non-trivial algebraic number theory (specifically, what one needs is this: let $f \in \mathbb{Z}[t]$ be a product of linear factors mod p for almost all primes p; then f is a product of linear factors).

Definition Let $\sigma(G)$ denote the unique maximum supersolubly embedded normal subgroup of a polycyclic group G; so there is a G-invariant series in $\sigma(G)$ with cyclic factors, but $G/\sigma(G)$ has no non-trivial cyclic normal subgroups.

It follows from the previous discussion that if G is a polycyclic group with **P**, then $G/\sigma(G)$ is finite. Hence we obtain:

Theorem 5 **([2])** *A polycyclic group G has* **P** *if and only if $G/\sigma(G)$ is a finite* **P***-group.*

One consequence of this theorem is an algorithm which can detect whether a given polycyclic group satisfies **P**. We mention as well that there is a generalization of Theorem 5 to soluble minimax groups ([2]).

Finally, a somewhat surprising result.

Theorem 6 **([2])** *A finitely generated soluble group with* **P** *is polycyclic.*

The proof proceeds by showing that a soluble just-non-polycyclic group does not satisfy **P**. It uses the well-known theorem of Lennox and Roseblade [6] that a soluble product of two polycyclic groups is polycyclic.

4 More general permutability conditions

If **C** is a class of groups, a group G satisfies the permutability condition $\mathbf{P_C}$ if $H < G$ implies that $HK = KH$ for some **C**-subgroup K not contained in H. Thus $\mathbf{P_C} = \mathbf{P}$ if **C** is the class of cyclic groups. We are interested in the successively weaker properties

$$\mathbf{P}_{\mathrm{ab}},\ \mathbf{P}_{\mathrm{nil}},\ \mathbf{P}_{\mathrm{sol}},$$

obtained by taking for **C** the classes of abelian, nilpotent, soluble groups.

Examples (i) All finite soluble groups satisfy $\mathbf{P}_{\mathrm{nil}}$, but not $\mathbf{P}_{\mathrm{ab}}$. (The extension of Q_8 by its automorphism group of order 3 does not have $\mathbf{P}_{\mathrm{ab}}$.)
(ii) $PSL_2(7)$ has $\mathbf{P}_{\mathrm{nil}}$, but $PGL_2(7)$ does not.
(iii) $PSL_2(7)$ does not have $\mathbf{P}_{\mathrm{ab}}$.
(iv) S_5 has $\mathbf{P}_{\mathrm{sol}}$, but not $\mathbf{P}_{\mathrm{nil}}$; also A_5 does not have $\mathbf{P}_{\mathrm{sol}}$.

The natural question to raise is: how close are these properties to solubility for finite groups?

At present the best result known is:

Theorem 7 ([9]) *A finite group satisfies* $\mathbf{P}_{\mathrm{nil}}$ *if and only if it is soluble or soluble-by-*$PSL_2(7)$.

As a consequence one obtains the following characterization of $PSL_2(7)$.

Corollary 8 $PSL_2(7)$ *is the only non-trivial finite semisimple group which satisfies* $\mathbf{P}_{\mathrm{nil}}$.

Since $PSL_2(7)$ does not have $\mathbf{P}_{\mathrm{ab}}$, we also obtain:

Corollary 9 *A finite group satisfying* $\mathbf{P}_{\mathrm{ab}}$ *is soluble.*

This provides some evidence for the truth of an old conjecture of Kegel [5]. *If G is a finite group in which every maximal subgroup has an abelian supplement, is G soluble?* Two other open questions here are:

(i) Which finite soluble groups satisfy $\mathbf{P}_{\mathrm{ab}}$?

(ii) Which finite insoluble groups satisfy $\mathbf{P}_{\mathrm{sol}}$?

Some comments on the proof of Theorem 7

The main burden of proof is to show that a finite group with $\mathbf{P}_{\text{nil}}$ is soluble or soluble-by-$PSL_2(7)$. Let G be a counterexample of minimal order; then G is semisimple. Let R be the maximum normal completely reducible subgroup of G.

Suppose first that R is simple, so that G is almost simple. Many of the techniques in the proof of Theorem 3 can be applied. In addition, in the case where R is a classical simple group, *primitive prime divisors* provide a powerful tool, as we briefly explain.

A ppd for $a^n - 1, (a, n > 1)$, is a prime dividing $a^n - 1$ but not $a^i - 1$ for $1 \leq i < n$. It is known that ppd's exist if $(a, n) \neq (2, 6)$ or $(2^s - 1, 2)$, by a theorem of Zsigmondy. Let $q = p^m$ where p is prime. A *basic ppd* for $q^e - 1$ is a ppd of $p^{em} - 1$. Assuming that ppd's exist for $q^e - 1$, it is easy to see that basic ppd's exist if $(q, e) \neq (4, 3)$ or $(8, 2)$. (For an account of ppd's and their importance for linear groups see [8]).

Now let $R \leq G \leq \text{Aut}(R)$ where R is a classical simple group of degree n other than $U_3(q)$. Assume that G has $\mathbf{P}_{\text{nil}}$. Let M be maximal in R and put $L = N_G(M)$. It can be shown that L is maximal in $G_M = LR$ and $G_M = LK$ where K is nilpotent.

The key observation is:

Lemma 10 *If q_e, q_f are basic ppd's of $q^e - 1$, $q^f - 1$ where $\frac{1}{2}n < e < f$ and $q_e q_f$ divides $|R|$, then q_e or q_f divides $|M|$.*

Our strategy is to try to choose M, e, f to contradict Lemma 10. In fact this leaves just 15 low degree groups R. These are dealt with by using the tables of maximal factorizations and order arguments.

Now let $R = S_1 \times S_2 \times \ldots \times S_r$ where the S_i are simple groups and $r > 1$. Put $N_i = N_G(S_i), C_i = C_G(S_i)$. The critical fact to establish is:

Lemma 11 *N_i/C_i has $\mathbf{P}_{\text{nil}}$ and is almost simple with completely reducible radical isomorphic with S_i.*

This shows that $S_i \simeq PSL_2(7)$ by the almost simple case. When $r = 2$, a contradiction is obtained by directly constructing a maximal subgroup without a nilpotent supplement in G.

Finally, let $r > 2$ and write $N = N_G(S_{r-1} \times S_r)$, $C = C_G(S_{r-1} \times S_r)$. One can prove that N/C has $\mathbf{P}_{\text{nil}}$ and its completely reducible radical is isomorphic with $PSL_2(7) \times PSL_2(7)$, a contradiction by the case $r = 2$.

Postscript Recently B. Baumeister has proved Kegel's conjecture and obtained results relevant to the condition $\mathbf{P}_{\text{sol}}$ (preprint).

References

[1] J. C. Beidleman and D. J. S. Robinson, *On finite groups satisfying the permutizer condition*, J. Algebra **191** (1997), 686–703.

[2] J. C. Beidleman and D. J. S. Robinson, *The permutizer condition for infinite soluble groups*, preprint.

[3] W. E. Deskins and P. Venzke, *Supersolvable groups*, in "Between Nilpotent and Solvable", by M. Weinstein , Polygonal, Passaic, N.J., 1982.

[4] K. Iwasawa, *Über die endlichen Gruppen und die Verbände ihrer Untergruppen*, J. Fac. Sci. Imp. Univ. Tokyo Sect I. **4** (1941), 171–199.

[5] O. H. Kegel, *On Huppert's characterization of finite supersoluble groups*, in Proc. Intermat. Conf. Theory of Groups, Canberra 1965, pp. 209–215, Gorden and Breach, New York, 1967.

[6] J. C. Lennox and J. E. Roseblade, *Soluble products of polycyclic groups*, Math. Z. **170** (1980), 153–154.

[7] M. W. Liebeck, C. E. Praeger and J. Saxl, *The maximal factorizations of the finite simple groups and their automorphism groups*, Mem. Amer. Math. Soc. **86** (1990).

[8] A. C. Niemeyer and C. E. Praeger, *A recognition algorithm for classical groups over finite fields*, preprint.

[9] D. J. S. Robinson, *Permutability properties of finite groups*, preprint.

[10] R. Schmidt, *Subgroup Lattices of Groups*, de Gruyter, Berlin, 1994.

[11] J. Zhang, *A note on finite groups satisfying the permutizer condition*, Science Bulletin **31** (1986), 363–365.

WHEN SCHREIER TRANSVERSALS GROW WILD

AMNON ROSENMANN

School of Mathematical Sciences, Tel-Aviv University, Tel-Aviv 69978, Israel

1 Introduction

The Schreier formula for the rank of a subgroup of finite index of a finitely generated free group F is generalised to an arbitrary (even infinitely generated) subgroup H through the Schreier transversal of H in F. The rank formula may also be expressed in terms of the cogrowth $\Gamma_{F/H}$ of H:

$$rank(H) = 1 + \lim_{i\to\infty}\left((n-1)\Gamma_{F/H}(i) - \frac{1}{2}\gamma_{F/H}(i+1)\right). \tag{1}$$

The cogrowth of H is the growth function of the cosets of H, each coset represented by a minimal element with respect to the generators of F. The rank formula is given through the limit of the values of expressions e_i. Each e_i has a part $1+(n-1)\Gamma_{F/H}(i)$ which refers to the number of cosets of length less or equal to i. This part is similar to the classical Schreier formula. The other part of e_i, $e_i' = \frac{1}{2}\gamma_{F/H}(i+1)$ involves the number of cosets of length $i+1$. When H is of finite index in F then $e_i' = 0$ for i large enough, and we get Schreier formula as a special case.

We introduce the rank-growth function $rk_H(i)$ of a subgroup H of a finitely generated free group F. $rk_H(i)$ is defined to be the rank of the subgroup of H generated by elements of length less than or equal to i (with respect to the generators of F), and it equals the rank of the fundamental group of the subgraph of the cosets graph of H, which consists of the paths starting at 1 that are of length $\leq i$. When H is supnormal, i.e. contains a non-trivial normal subgroup of F, we show that its rank-growth is equivalent to the cogrowth of H:

$$rk_H(i) \sim \Gamma_{F/H}(i). \tag{2}$$

A special case of this is the known result of Greenberg from 1960 (see [2], [6]) that a supnormal subgroup of F is of finite index if and only if it is finitely generated. The particular case of H being normal tells us that the growth of any finitely generated non-trivial group G is equivalent to the rank-growth of the kernel H of the presentation of G as the image of a finitely generated free group F: $G = F/H$.

Let us remark that similar results to the ones appearing here hold in free group algebras. This is by the similarity of Schreier bases and Schreier transversals of subgroups of free groups to these structures of one-sided ideals in free group algebras (see [7], [8]), [9]).

2 Generalised Schreier formula

Let H be a non-trivial subgroup of a free group F. By the Nielsen-Schreier Theorem H is free too (see, for example, [6]), and explicit free generators for it can be given.

Suppose that F is freely generated on a set X (not necessarily finite). The Cayley graph of F (with respect to X) has the form of a tree and is the universal covering of a space $\mathcal{Q}$ which is a bouquet of $|X|$ loops ($|.|$ denotes cardinality throughout the paper). The covering space of $\mathcal{Q}$ with regard to H is the *cosets graph* $\mathcal{G}$ of H, and it is obtained as the quotient of the Cayley graph of F under the left action of H. Thus H is the fundamental group of $\mathcal{G}$. The set of vertices of $\mathcal{G}$ is the set of right cosets of H in F. A (double) edge which is labelled with $x \in X$ goes in the direction from the coset Ht_1 to the coset Ht_2 if and only if $Ht_2 = Ht_1x$, and it is labelled with x^{-1} in the direction from Ht_2 to Ht_1. This gives a connected graph with $|X \cup X^{-1}|$ edges at each vertex. It is more convenient to label the vertices of $\mathcal{G}$ with specific coset representatives in the following way. Let $\mathcal{T}$ be a spanning tree of $\mathcal{G}$. The identity element 1 is chosen to represent the coset H and defined to be the root of $\mathcal{T}$, and each other vertex is labelled with the group element one gets by reading off the edge labels in a path in $\mathcal{T}$ that starts at the root and ends at the given vertex. We also denote by $\mathcal{T}$ the set of (the labels of) the vertices $V(\mathcal{T})$ of the tree $\mathcal{T}$, that is the coset representatives of H. This set is a *Schreier transversal* for H in F, which is characterised by the property that every initial segment of an element of $\mathcal{T}$ is also in $\mathcal{T}$. For each $1 \neq w \in H$ there exist $u, v \in \mathcal{T}$ of maximal lengths such that u is a prefix of w and v is a prefix of w^{-1}. Since $t_1t_2^{-1} \notin H$ for every $t_1 \neq t_2$ in $\mathcal{T}$, then $l(u) + l(v) < l(w)$, where l denotes the length of the (reduced) element in F. The Schreier generators for H with respect to $\mathcal{T}$ are those $w \in H$ for which

$$l(u) + l(v) = l(w) - 1. \tag{3}$$

Moreover, if ϕ is the coset map associated with $\mathcal{T}$ then H is freely generated by the non-trivial elements

$$tx(\phi(tx))^{-1}, \tag{4}$$

where t ranges over $\mathcal{T}$ and x ranges over X (see [6]). This set is called a Schreier basis for H. Since $tx = \phi(tx)$ only when $tx \in \mathcal{T}$ then by (4) the rank of H equals the *cyclomatic number* of $\mathcal{G}$, the cardinality of the "missing" edges in the directions of X in $\mathcal{T}$, that is

$$\operatorname{rank}(H) = |\{e \in E(\mathcal{G}) - E(\mathcal{T})\}|, \tag{5}$$

where $E(\mathcal{G})$, $E(\mathcal{T})$ denote the set of edges of $\mathcal{G}$, $\mathcal{T}$ respectively. This is because each edge is labelled with some $x \in X$ in exactly one direction, and thus counted exactly once.

Suppose now that F is finitely generated with $\operatorname{rank}(F) = n$, and H is of finite index m in F. Then $|E(\mathcal{G})| = nm$ and $|E(\mathcal{T})| = m - 1$ (since $\mathcal{T}$ is a tree). By (5) we get that

$$\operatorname{rank}(H) = 1 + (n-1)m. \tag{6}$$

This is Schreier Formula (see [6]). When H is not necessarily of finite index in F and also not necessarily finitely generated, we give in the proposition below a formula that generalises the above one. The rank is computed on a Schreier transversal, and the simpler form of the formula is given in Corollary 2.2, which expresses the rank in terms of the cogrowth (see below) of the subgroup. The common way of computing the rank of the subgroup as a limit of the ranks of the fundamental groups (the cyclomatic numbers) of finite subgraphs deals with counting *edges*. Whereas, what we are doing here is counting only *vertices*.

We use the following terminology and notation on graphs. A *path* in a graph $\mathcal{G}$ is a sequence $v_0, e_1, v_1, e_2, \ldots$, $v_i \in V(\mathcal{G})$, $e_i \in E(\mathcal{G})$, such that e_i starts at the vertex v_{i-1} and terminates at v_i. The length of a path $v_0, e_1, v_1, e_2, \ldots, v_n$ is n. A *simple path* is a path in which the vertices along it are distinct, except possibly for the first and last one, in which case it is a *simple closed path* or a *simple circuit*. We assume that each path is *reduced*, i.e. it is not homotopic to a shorter one when the initial and terminal vertices are kept fixed.

If $\mathcal{H} \subseteq \mathcal{G}$, i.e. $\mathcal{H}$ is a collection of vertices and edges of the graph $\mathcal{G}$, then we denote by $< \mathcal{H} >$ the subgraph *generated* by $\mathcal{H}$. It is the smallest subgraph of $\mathcal{G}$ which contains $\mathcal{H}$. That is, we add to $\mathcal{H}$ the endpoint vertices of all the edges in $\mathcal{H}$. On the other hand, the subgraph of $\mathcal{G}$ *induced* by $\mathcal{H}$ is the one whose vertices are those of $\mathcal{H}$ and whose edges are all the edges which join these vertices in $\mathcal{G}$. An induced subgraph is a subgraph which is induced by some $\mathcal{H} \subseteq \mathcal{G}$. If $\mathcal{H}_1, \mathcal{H}_2 \subseteq \mathcal{G}$ then $\mathcal{H}_1 - \mathcal{H}_2$ is the collection of vertices $V(\mathcal{H}_1) - V(\mathcal{H}_2)$ and edges $E(\mathcal{H}_1) - E(\mathcal{H}_2)$, and it does not necessarily form a subgraph of $\mathcal{G}$, even when $\mathcal{H}_1$ and $\mathcal{H}_2$ are subgraphs of $\mathcal{G}$. The *boundary* of the subgraph $\mathcal{H}$ of $\mathcal{G}$ is $\partial\mathcal{H} = \mathcal{H} \cap < \mathcal{G} - \mathcal{H} >$, and its *interior* is $\dot{\mathcal{H}} = \mathcal{H} - \partial\mathcal{H}$. The *outer boundary* of $\mathcal{H}$ (in $\mathcal{G}$) is the set of vertices of $\mathcal{G} - \mathcal{H}$ which are adjacent to $\mathcal{H}$ in $\mathcal{G}$. Assume now that each edge of $\mathcal{G}$ is labelled with some $x \in X$ in one direction and with $x^{-1} \in X^{-1}$ in the other direction. Then we define $E^X_{out}(\mathcal{H})$ to be the set of edges of $\mathcal{G} - \mathcal{H}$ whose initial vertices with respect to the directions X are in $\mathcal{H}$. If $\mathcal{H}_i \subseteq \mathcal{G}$, $i = 1, 2, \ldots$, then $\mathcal{H} = \liminf \mathcal{H}_i$ if the vertices of $\mathcal{H}$ are $V(\mathcal{H}) = \bigcup_{i \geq 1} \bigcap_{j \geq i} V(\mathcal{H}_j)$, and its edges are $E(\mathcal{H}) = \bigcup_{i \geq 1} \bigcap_{j \geq i} E(\mathcal{H}_j)$.

Finally, let $\alpha(\mathcal{H}) = |\pi_0(\mathcal{H})|$ be the cardinality of the (connected) components of $\mathcal{H}$.

Proposition 2.1 *Let F be a free group of rank n and let $H < F$. Let $\mathcal{T}$ be a Schreier transversal for H in F and let $\mathcal{T}_i$ be finite subgraphs of $\mathcal{T}$ such that $\mathcal{T} = \liminf \mathcal{T}_i$. Then*

$$\operatorname{rank}(H) = \lim_{i \to \infty} \left(\alpha(\mathcal{T}_i) + (n-1)|V(\mathcal{T}_i)| - \frac{1}{2} \sum_{j=1}^{\alpha(\mathcal{T}_i)} |V(\partial_{out} \mathcal{T}_{i,j})| \right), \tag{7}$$

where, for a fixed i, $\partial_{out} \mathcal{T}_{i,j}$ is the outer boundary (in $\mathcal{T}$) of the component $\mathcal{T}_{i,j}$ of $\mathcal{T}_i$, for $j = 1, \ldots, \alpha(\mathcal{T}_i)$.

Proof If H is of finite index m in F then there exists i_0 such that $\mathcal{T}_i = \mathcal{T}$ for every $i \geq i_0$, and then $\alpha(\mathcal{T}_i) = 1$, $|V(\mathcal{T}_i)| = m$ and $|V(\partial_{out} \mathcal{T}_i)| = 0$. Thus (7) reduces to the Schreier Formula.

Assume that H is finitely generated but of infinite index. Denote as before by $\mathcal{G}$ the cosets graph of H, which contains the Schreier transversal tree $\mathcal{T}$. Let $C(\mathcal{G})$ be the *core* of $\mathcal{G}$ (see [11]), that is the minimal deformation retract of $\mathcal{G}$.

It is the minimal connected subgraph of $\mathcal{G}$ which contains all its simple circuits. Since H is finitely generated $C(\mathcal{G})$ is finite, and there exists i_0 such that, after possibly renaming the components of each $\mathcal{T}_i$, $V(C(\mathcal{G}))$ is contained in $V(\mathcal{T}_{i,1})$ for each $i \geq i_0$. Let us denote by $\mathcal{G}_{i,j}$ the subgraph of $\mathcal{G}$ induced by $\mathcal{T}_{i,j}$. Then $E(\mathcal{G}) - E(\mathcal{T}) = E(\mathcal{G}_{i,1}) - E(\mathcal{T}_{i,1})$, for each $i \geq i_0$, and by (5) the cardinality of this set equals the rank of H. Hence it suffices to show that for each $i \geq i_0$ and for each j

$$\begin{aligned} |E(\mathcal{G}_{i,j}) - E(\mathcal{T}_{i,j})| &= |E(\mathcal{G}_{i,j})| - |E(\mathcal{T}_{i,j})| & (8) \\ &= 1 + (n-1)|V(\mathcal{T}_{i,j})| - \frac{1}{2}|V(\partial_{out}\mathcal{T}_{i,j})|. & (9) \end{aligned}$$

So assume $i \geq i_0$. Then $|E^X_{out}(\mathcal{G}_{i,j})| = |E^X_{out}(\mathcal{T}_{i,j})| = |V(\partial_{out}\mathcal{T}_{i,j})|/2$ for each j, since all simple circuits of $\mathcal{G}$ are in $\mathcal{G}_{i,1}$. Each vertex in $\mathcal{G}$ is the initial vertex of exactly n edges in the directions X. Therefore

$$|E(\mathcal{G}_{i,j})| = n|V(\mathcal{G}_{i,j})| - |E^X_{out}(\mathcal{G}_{i,j})|. \qquad (10)$$

As for $\mathcal{T}_{i,j}$, since it is a tree then

$$|E(\mathcal{T}_{i,j})| = |V(\mathcal{T}_{i,j})| - 1. \qquad (11)$$

Substituting in (8) gives (9).

Assume now that H is not finitely generated. Then, because in general

$$|E^X_{out}(\mathcal{G}_{i,j})| \geq |V(\partial_{out}\mathcal{T}_{i,j})|/2, \qquad (12)$$

we get that for each i, j

$$|E(\mathcal{G}_{i,j}) - E(\mathcal{T}_{i,j})| \leq 1 + (n-1)|V(\mathcal{T}_{i,j})| - \frac{1}{2}|V(\partial_{out}\mathcal{T}_{i,j})|. \qquad (13)$$

Since $\operatorname{rank}(H) = \lim_{i\to\infty}\left(\sum_{j=1}^{\alpha(\mathcal{T}_i)} |E(\mathcal{G}_{i,j}) - E(\mathcal{T}_{i,j})|\right) = \infty$, equation (7) follows. □

We remark that instead of taking finite subgraphs $\mathcal{T}_i$ such that $\mathcal{T} = \liminf \mathcal{T}_i$, the rank formula can be clearly given as the supremum, over all finite subgraphs of $\mathcal{T}$, of the expression appearing in (7).

A special case of Proposition 2.1 is when each component $\mathcal{T}_{i,j}$ is a *ball.* That is, there exist a vertex $v_{i,j}$ and an integer $k_{i,j} \geq 0$ such that $\mathcal{T}_{i,j}$ contains all vertices of $\mathcal{T}$ of distance $\leq k_{i,j}$ from $v_{i,j}$. If $\mathcal{H}$ is a subgraph of $\mathcal{G}$ and $|V(\mathcal{G})| > 1$ then we define $\delta(\mathcal{H})$ to be the number of components of $\mathcal{H}$ which consist of a single vertex, i.e. balls of radius 0. When $|V(\mathcal{G})| = 1$ then $\delta(\mathcal{G})$ is defined to be 0.

When the $\mathcal{T}_i$ are concentric balls centred at the identity 1 then the values $|V(\mathcal{T}_i)|$, $i = 0, 1, 2, \ldots$ relate to the *growth* function $\Gamma_{\mathcal{T}}$ of $\mathcal{T}$, as is defined below. By $l(g)$ we denote the *length* of $g \in F$, and we always assume that the group elements are written in reduced form with respect to the generating set X of F. Then define

$$\begin{aligned} \gamma_{\mathcal{T}}(i) &= |\{v \in \mathcal{T} \mid l(v) = i\}|, & (14) \\ \Gamma_{\mathcal{T}}(i) &= |\{v \in \mathcal{T} \mid l(v) \leq i\}|. & (15) \end{aligned}$$

When $\mathcal{T}$ is a *minimal* Schreier transversal tree, that is when it has also the property that every coset of H is represented by an element of minimal length, then $\Gamma_{\mathcal{T}}(i)$ is the *cogrowth* function of H, relative to the generating set of F, and is denoted by $\Gamma_{F/H}(i)$ (see [10]). We may look at $\Gamma_{F/H}(i)$ as representing the "volume" of the ball of radius i with centre 1 in the cosets graph of H (with the metric induced by the word metric on F). If, in addition, H is a normal subgroup of F then the cogrowth function of H equals the growth function of the group F/H, relative to the the generating set which is the canonical image of the generating set of F. (In this case the Schreier transversal for H which is minimal with regard to a fixed ShortLex order on F is also suffix-closed.)

Corollary 2.2 *Let F be a free group of rank n, let H be a subgroup of F and let $\mathcal{T}$ be a Schreier transversal for H in F. Let $\mathcal{T}_i$ be induced finite subgraphs of $\mathcal{T}$, whose components $\mathcal{T}_{i,j}$ are balls, such that $\mathcal{T} = \liminf \mathcal{T}_i$. Then*

$$\begin{aligned} \operatorname{rank}(H) &= \lim_{i\to\infty} \left(\alpha(\mathcal{T}_i) - \frac{1}{2}\delta(\mathcal{T}_i) + (n-1)|V(\mathcal{T}_i)| - \frac{2n-1}{2}|V(\partial \mathcal{T}_i)| \right) \quad (16) \\ &= \lim_{i\to\infty} \left(\alpha(\mathcal{T}_i) - \frac{1}{2}\delta(\mathcal{T}_i) + (n-1)|V(\dot{\mathcal{T}}_i)| - \frac{1}{2}|V(\partial \mathcal{T}_i)| \right). \quad (17) \end{aligned}$$

In particular,

$$\begin{aligned} \operatorname{rank}(H) &= 1 + \lim_{i\to\infty} \left((n-1)\Gamma_{\mathcal{T}}(i) - \frac{1}{2}\gamma_{\mathcal{T}}(i+1) \right) \quad (18) \\ &= 1 + \lim_{i\to\infty} \left((n-1)\Gamma_{F/H}(i) - \frac{1}{2}\gamma_{F/H}(i+1) \right). \quad (19) \end{aligned}$$

Proof The corollary follows from the fact that when the core $C(\mathcal{G})$ is finite then for each i large enough every vertex of $\partial \mathcal{T}_{i,j}$ is adjacent to $2n-1$ vertices of $\mathcal{T} - \mathcal{T}_{i,j}$, unless $\mathcal{T}_{i,j}$ is a single vertex and then it is adjacent to $2n$ vertices of $\mathcal{T} - \mathcal{T}_{i,j}$. When H is not finitely generated then we first notice that the expression we calculate for each ball is non-negative. Secondly, since $\mathcal{T} = \liminf \mathcal{T}_i$, then for every r there exists i_r such that, for every $i \geq i_r$, $\mathcal{T}_i$ has a component (ball) which contains the ball of radius r around the identity. But the expression calculated on these balls tends to infinity whenever H is of infinite rank, as shown below. This can also be concluded directly from Proposition 2.1. □

3 Rank-growth

Given a Schreier transversal $\mathcal{T}$, let us define

$$\begin{aligned} r_{\mathcal{T}}(i) &= 1 + (n-1)\Gamma_{\mathcal{T}}(i) - \frac{1}{2}\gamma_{\mathcal{T}}(i+1) \quad (20) \\ &= 1 + \frac{2n-1}{2}\Gamma_{\mathcal{T}}(i) - \frac{1}{2}\Gamma_{\mathcal{T}}(i+1). \quad (21) \end{aligned}$$

$r_{\mathcal{T}}(i)$ is an upper bound to the cyclomatic number of the subgraph of $\mathcal{G}$ which is induced by the vertices of $\mathcal{T}$ of distance at most i from the root. In case $\mathcal{T}$ is a minimal Schreier transversal then $r_{\mathcal{T}}(i)$ is also denoted by $r_H(i)$:

$$r_H(i) = 1 + (n-1)\Gamma_{F/H}(i) - \frac{1}{2}\gamma_{F/H}(i+1). \tag{22}$$

The sequence $r_{\mathcal{T}}(i)$, $i = 1, 2, \ldots$ is non-decreasing. This is because

$$r_{\mathcal{T}}(i) - r_{\mathcal{T}}(i-1) \quad = \quad \frac{2n-1}{2}\gamma_{\mathcal{T}}(i) - \frac{1}{2}\gamma_{\mathcal{T}}(i+1), \tag{23}$$

and each vertex of $\mathcal{T}$ of level i is adjacent to at most $2n-1$ vertices of level $i+1$. Thus $r_{\mathcal{T}}(i)$ becomes eventually constant if and only if either $\mathcal{T}$ is finite, or for some i_0 each vertex of $\mathcal{T}$ of level $i \geq i_0$ has degree exactly $2n$, and this happens if and only if there are only finitely many edges in $E(\mathcal{G}) - E(\mathcal{T})$, or equivalently when H is finitely generated.

It is interesting to know also the rate in which the function $r_{\mathcal{T}}(i)$ grows. A pre-order is defined on growth functions by

$$f_1(i) \preceq f_2(i) \iff \exists c > 0 \ \forall i \ [f_1(i) \leq c f_2(ci) \ ]. \tag{24}$$

Then an equivalence relation is given by

$$f_1(i) \sim f_2(i) \iff f_1(i) \preceq f_2(i) \text{ and } f_2(i) \preceq f_1(i). \tag{25}$$

(we refer to [4] for a survey on growth functions of groups and to Gromov's [5] rich and beautiful geometric theory.) In Theorem 3.1 below we show that when the subgroup H of the free group F is *supnormal*, i.e. contains a non-trivial subgroup which is normal in F, then for every Schreier transversal $\mathcal{T}$ of H, its growth function $\Gamma_{\mathcal{T}}(i)$ is equivalent to the function $r_{\mathcal{T}}(i)$. This implies that the cogrowth of H is also equivalent to what we call the rank-growth of H. We look at H as the direct limit of the subgroups

$$H_i = < \{h \in H \mid l(h) \leq i\} >, \tag{26}$$

where $l(h)$ is measured with respect to the generating set of F. Then the *rank-growth* of H (with respect to the generators of F) is

$$rk_H(i) = \text{rank}(H_i). \tag{27}$$

Clearly, if we choose another generating set for F, we get an equivalent rank-growth function. Notice that H_i is the fundamental group of the subgraph of the cosets graph $\mathcal{G}$ of H which contains all paths starting at 1 of length $\leq i$. Thus $rk_H(i)$ is a non-decreasing function. If we define

$$\rho_H(i) = \text{rank}(\pi_1(\mathcal{B}_i)), \tag{28}$$

where $\mathcal{B}_i$ is (the induced subgraph which is) the ball of radius i centred at the vertex 1 of $\mathcal{G}$, then

$$\rho_H(i) = rk_H(2i+1). \tag{29}$$

Therefore $rk_H(i)$ and $\rho_H(i)$ are equivalent. Also $\rho_H(i) \sim r_H(i)$. In fact,

$$\rho_H(i) \leq r_H(i) \leq \rho_H(i+1). \tag{30}$$

More precisely,

$$r_H(i) = \rho_H(i) + \frac{1}{2}(|E^X_{out}(\mathcal{B}_i)| - \gamma_{F/H}(i+1)) \leq \rho_H(i+1). \tag{31}$$

Theorem 3.1 *Let H be a supnormal subgroup of a finitely generated free group F, and let $\mathcal{T}$ be a Schreier transversal for H in F. Then*

$$r_{\mathcal{T}}(i) \sim \Gamma_{\mathcal{T}}(i). \tag{32}$$

Thus, with respect to a minimal Schreier transversal we get

$$rk_H(i) \sim \Gamma_{F/H}(i). \tag{33}$$

In fact, a sufficient condition for H is to contain a non-trivial subgroup A with $|F : N_F(A)| < \infty$.

Proof For every Schreier transversal of a subgroup of F we have $r_{\mathcal{T}}(i) \preceq \Gamma_{\mathcal{T}}(i)$. This follows immediately from the definition of $r_{\mathcal{T}}(i)$ - see (21).

Suppose now that H is supnormal. Let h be a non-trivial element of a subgroup of H which is normal in F, and let $m = l(h)$ (as usual, the length is with respect to the generators of F). Then at every vertex v of the cosets graph $\mathcal{G}$ of H, if we follow the path defined by h we form a circuit. Therefore at every vertex of $\mathcal{T}$ of level at most i, by following the path defined by h we reach a vertex of $\mathcal{T}$ of level at most $i+m$ where we must stop because the next edge is missing. The number of these missing edges is less then or equal to $r_{\mathcal{T}}(i+m)$. Since at most m vertices are the starting point of a tour defined by h which reaches the same missing edge then

$$\Gamma_{\mathcal{T}}(i) \leq m r_{\mathcal{T}}(i+m). \tag{34}$$

By the two inequalities we have

$$r_{\mathcal{T}}(i) \sim \Gamma_{\mathcal{T}}(i). \tag{35}$$

Applying this result to a minimal Schreier transversal yields

$$rk_H(i) \sim r_H(i) \sim \Gamma_{F/H}(i). \tag{36}$$

The condition of H being supnormal can be weakened. It suffices to demand that H contains a non-trivial subgroup A such that $|F : N_F(A)| < \infty$, because then the cogrowth of H is equivalent to the growth (with respect to the generators of F) of the minimal coset representatives of $H \cap N_F(A)$ in $N_F(A)$. □

Since $\Gamma_{\mathcal{T}}(i) \preceq \Gamma_{F/H}(i)$ for every Schreier transversal $\mathcal{T}$ of a subgroup H of F, then by Theorem 3.1 when H is supnormal in F then $r_{\mathcal{T}}(i) \preceq rk_H(i)$. We also notice that a special case of Theorem 3.1 is the known result stating that a supnormal subgroup of a finitely generated group is of finite index if and only if it is finitely generated. And when H is normal in F, then the growth $\Gamma_G(i)$ of the group $G = F/H$ is equivalent to the rank-growth of H and to the growth of

$$r_H(i) = 1 + (n-1)\Gamma_G(i) - \frac{1}{2}\gamma_G(i+1). \tag{37}$$

The growth of the subgroup H is always exponential when it is of rank greater than 1, since it is free. But Grigorchuk showed ([3]) that when H is normal then its "growth exponent" $\limsup_{i\to\infty} \Gamma_H^{(F)}(i)^{1/i} = 2n-1$, if and only if $G = F/H$ is amenable, (in fact, Grigorchuk [3] obtained more: a formula which connects the growth exponent of G with the spectral radius of a random walk on G), where $n = \mathrm{rank}(F)$ and $\Gamma_H^{(F)}(i)$ represents the growth of H with respect to the generators of F. (Recall that a group G is amenable if there exists an invariant mean on $B(G)$, the space of all bounded complex-valued functions on G with the sup norm $\| f \|_\infty$). When G is non-amenable then the growth exponent of H is less than $2n-1$. But then the group G has exponential growth, and we have shown that in this case the rank-growth of H is also exponential, i.e. the maximal possible (up to equivalence).

Although the rank of the subgroup of a free group can be expressed, as we have seen in Corollary 2.2, in terms of the growth function of any Schreier transversal of it, the growth function itself of one Schreier transversal of an infinitely generated subgroup may in general differ completely from that of another Schreier transversal. This is shown in the next proposition.

Proposition 3.2 *There exists a subgroup of the free group of rank 2 with exponential cogrowth which has a Schreier transversal $\mathcal{T}$ whose growth is $\Gamma_{\mathcal{T}}(i) = i+1$.*

Proof We will construct the cosets graph of such a subgroup inductively. Let c be a positive integer which is large enough. First we make a simple circuit λ_1 of length c that starts at the root 1. Then at the n-th step we construct a path λ_n of length $2nc$, whose vertices, apart from the initial and terminal ones, are new. The initial vertex of λ_n is the one before the last vertex in the path λ_{n-1}. The terminal vertex of λ_n is chosen to be of minimal distance from the root among the vertices whose degree is less than 4.

If we delete the last edge of each path λ_n, then we get a linear Schreier transversal $\mathcal{T}$, i.e. $\Gamma_{\mathcal{T}}(n) = n+1$. On the other hand, if we delete the middle edge of each λ_n, then we get a tree $\mathcal{T}'$ with exponential growth, because each vertex of it has degree 4, except for a sequence of vertices v_n of distances $\geq cn$ respectively from the root. Since the cogrowth function is greater than or equal to the growth function of any Schreier transversal of the subgroup, the result follows. □

It is shown in [10] that when $H = H_1 \cap H_2$ the cogrowth of H satisfies

$$\Gamma_{F/H}(i) \leq \Gamma_{F/H_1}(i)\Gamma_{F/H_2}(i) \quad \text{for every } i. \tag{38}$$

The rank-growth of the intersection of two subgroups behaves similarly.

Proposition 3.3 *Let H_1, H_2 be non-trivial subgroups of a finitely generated free group F and let $H = H_1 \cap H_2$. Then*

$$rk_H(i) \leq 1 + 2(rk_{H_1}(i) - 1)(rk_{H_2}(i) - 1) - \min(rk_{H_1}(i), rk_{H_2}(i)) \tag{39}$$

for every i. Hence

$$rk_H(i) \preceq rk_{H_1}(i) rk_{H_2}(i). \tag{40}$$

Proof This follows immediately from the best general bound for the intersection of finitely generated subgroups in free groups, which is due to Burns ([1]). □

References

[1] Burns, R.G. (1971). On the intersection of finitely generated subgroups of a free group. *Math. Z.* **119**, 121-130.

[2] Greenberg, L. (1960). Discrete groups of motion. *Canad. J. Math.* **12**, 414-425.

[3] Grigorchuk, R.I. (1980). Symmetrical random walks on discrete groups. In: (Dobrushin, R.L., Sinai, Ya.G. eds.) *Multicomponent Random Systems*. Advances in Probability and Related Topics, Vol. 6, 285-325. Marcel Dekker.

[4] Grigorchuk, R.I. (1990). On growth in group theory. *Proc. of the International Congress of Mathematicians*, Kyoto, 1990, 325-338.

[5] Gromov, M. (1993). Asymptotic invariants of infinite groups. In: (Niblo, G.A., Roller, M.A. eds) *Geometric Group Theory*, Vol. 2. London Math. Soc. Lecture Note Ser. 182. Cambridge University Press, 1–295.

[6] Lyndon, R.C., Schupp, P.E. (1977). *Combinatorial Group Theory*. Springer-Verlag.

[7] Lewin, J. (1969). Free modules over free algebras and free group algebras: the Schreier technique. *Trans. Amer. Math. Soc.* **145**, 455-465.

[8] Rosenmann, A. (1993). An algorithm for constructing Gröbner and free Schreier bases in free group algebras. *J. Symbolic Comput.* **16**, 523-549.

[9] Rosenmann A., Rosset, S. (1994). Ideals of finite codimension in free algebras and the fc-localization. *Pacific J. Math.* **162**, 351-371.

[10] Rosenmann, A. (1993). Cogrowth and essentiality in groups and algebras. In: (Duncan, A.J., Gilbert, N.D., Howie, J. eds.) *Proc. ICMS Workshop on Geometric and Combinatorial Methods in Group Theory.*, Edinburgh, 1993, 284–293.

[11] Stallings, J.R. (1983). Topology of finite graphs. *Invent. Math.* **71**, 551-565.

PROBABILISTIC GROUP THEORY

ANER SHALEV

Institute of Mathematics, The Hebrew University, Jerusalem 91904, Israel

Dedicated to the memory and heritage of Paul Erdős, 1913-1996

Abstract

In recent years there have been several developments in the study of probabilistic aspects of certain finite and profinite groups, and various conjectures in this field were settled. Moreover, the probabilistic approach led to the solution of interesting problems whose formulation had nothing to do with probability; these include problems regarding the modular group, free groups, as well as conjectures on finite permutation groups. In this lecture series I will try to survey these developments and discuss directions for further research.

Contents:

1 Finite simple groups: random generation

Group theory and measure theory seem to intersect highly non-trivially, and so there are many branches in mathematics which could be referred to as probabilistic group theory. In this lecture series I would like to focus on a relatively young area, which concerns probabilistic aspects of finite groups and their inverse limits. I shall also demonstrate how probabilistic ideas can be used to solve classical problems in finite and infinite groups.

A classical scheme, applied successfully in combinatorics, number theory, and other areas, is to prove existence theorems using a probabilistic approach. The idea is to show that most objects have a certain property, and then to deduce that an object with that property exists. This may sound strange at first, since it seems we are proving a statement much stronger than the one which we are required to establish to begin with. Still, experience shows that in some circumstances stronger statements are easier to prove; more specifically, explicit constructions of objects of certain types can be harder than certain counting arguments showing that such objects are abundant.

This approach is mostly associated with Paul Erdős, who developed and applied it in combinatorics and number theory with remarkable achievements. In particular, Erdős, in a joint work with Rényi, introduced the notion of the random graph, and used it (and variants of it) to solve a whole spectrum of problems – typical examples being the existence of graphs with large girth and large chromatic number, and estimates for Ramsey numbers. See Alon and Spencer [AS] and the references therein. Of course, the use of sieve methods in analytic number theory to prove the existence of primes in some intervals or arithmetic progressions can also be regarded as part of the same philosophy. See Halberstam and Richert [HR]. Yet, it is somewhat unexpected that the probabilistic approach may also be applied successfully in highly non-linear contexts, such as the theory of finite (nonabelian) groups, and finite simple groups in particular.

It is quite striking that the roots of the subject we refer to here as probabilistic group theory also lie in Erdős' work. In the years 1965-1972 Erdős and Turán published a series of seven papers, all entitled "On some problems of a statistical group theory", which focused on probabilistic aspects of the symmetric group S_n. Let me start this survey by briefly describing some of their results.

1.1 Statistical theory of S_n

What does a random permutation $\sigma \in S_n$ look like? To begin with, it is not even clear that this question is meaningful. We can, for instance, meaningfully speak of the average order of a permutation in S_n, but this does not imply that "most" permutations have order "close" to this average. The study of the orders of permutations goes back to E. Landau, who showed that the maximal order of a permutation in S_n is about $e^{\sqrt{n \log n}}$. In the first paper in the series [ET1] Erdős and Turán show that almost all permutations have a much lower order, namely $n^{(1/2+o(1)) \log n}$. This means that, for any $\epsilon, \delta > 0$ there exists $n_0 = n_0(\epsilon, \delta)$ such that, if $n > n_0$ then at least $(1 - \delta)n!$ permutations in S_n have order between $n^{(1/2-\epsilon) \log n}$ and $n^{(1/2+\epsilon) \log n}$. This "concentration" phenomenon is rather surprising, and was analyzed more deeply in subsequent papers in the series. For example, it was shown in [ET3] that for large n, if $\sigma \in S_n$ is randomly chosen, then $\log |\sigma|$ is well approximated by a normal distribution, with mean $\frac{1}{2} \log^2 n$ and standard deviation $3^{-1/2} \log^{3/2} n$.

Of course, the order of a permutation is determined by its cycle structure, and a detailed analysis of the cycle structure of random permutations was required in proving the above stated results. For example, most permutations in S_n have about $\log n$ cycles; and the probability of not having cycles of lengths $a_1, \ldots, a_k$ is bounded above by $(\sum_{i=1}^{k} a_i^{-1})^{-1}$. These, and other results on cycle structure have interesting applications, some of which will be mentioned below.

1.2 Random generation of S_n

One of the recurring themes in this lecture series is the study of generating sets for the finite simple groups, an area where probabilistic methods have proved rather useful. The story seems to begin in the 19th century with the following conjecture from [Ne].

Conjecture 1 (Netto, 1882) Two randomly chosen permutations in S_n generate

S_n or A_n with probability $\to 1$ as $n \to \infty$.

In 1969 J.D. Dixon settled Netto's conjecture in the affirmative [Di1]. His proof, which is briefly sketched below, made use of the Erdős-Turán statistical theory of S_n. To describe it, suppose $x, y \in S_n$ are chosen at random, and let $H = \langle x, y \rangle$ be the subgroup they generate. The first (and elementary) step of the argument is to show that the probability that H is primitive tends to 1 as $n \to \infty$. In fact the argument shows that the probability that H is primitive is of the form $1 - n^{-1} + O(n^{-2})$.

The next question is, then, what forces a primitive subgroup $H \leq S_n$ to contain A_n? This question has been studied extensively in the literature since the days of Jordan, and several useful criteria have been given. We list some below.

(a) If H contains a p-cycle ($p \leq n-3$ a prime) then $H \supseteq A_n$.

(b) If H contains an m-cycle ($1 < m < (n-m)!$) then $H \supseteq A_n$.

(c) If H contains a permutation with support $< n^{1/2-\epsilon}$ then $H \supseteq A_n$.

(d) If $|H| > n^{\sqrt{n}}$ then $H \supseteq A_n$.

Criteria (a) and (c) date back to Jordan in the eighteen seventies; criterion (b) was obtained by Williamson a century later, and (d) follows from the Classification of finite simple groups. Dixon's proof used criterion (a) above. Applying the Erdős-Turán theory he showed that the probability that $\langle x \rangle$ contains a p-cycle as in (a) tends to 1 as $n \to \infty$. In particular it follows that, with probability tending to 1, H contains a p-cycle as in (a), in which case H must be A_n or S_n.

In order to discuss generation probabilities, set

$$p_n = \mathrm{Prob}(\langle x, y \rangle = A_n, S_n).$$

Dixon's argument showed that $p_n \geq 1 - 2/(\log\log n)^2$. Using criterion (b) instead of (a), Bovey and Williamson showed in 1978 that $p_n \geq 1 - e^{-\sqrt{\log n}}$ [BW]. This was improved in 1980 by Bovey [Bo]. Using criterion (c) above he showed that, for any $\epsilon > 0$ and sufficiently large n we have $p_n \geq 1 - n^{-1+\epsilon}$. The definitive result was finally given by Babai in 1989 using the Classification of finite simple groups and criterion (d) above. He showed that $p_n = 1 - n^{-1} + O(n^{-2})$, an estimate which was conjectured by Dixon in his original 1969 paper. In section 1.4 below we will encounter some analogues of this result for simple groups of Lie type.

More subtle results concerning random generation of symmetric groups were proved in recent years. For example, L. Pyber and T. Müller showed that, if $A, B \neq 1$ are finite groups (not both C_2), then random copies of A and B in S_n generate A_n or S_n with probability $\to 1$. This confirms a conjecture of A. Lubotzky [Lu2], posed during the 4th St Andrews conference (Galway, 1993). Other results on the so called invariable generation of S_n can be found in Dixon [Di2] and in Łuczak and Pyber [ŁP].

1.3 Dixon's conjecture

For a group G, let $P(G)$ denote the probability that two randomly chosen elements x, y of G generate G:

$$P(G) = \mathrm{Prob}(\langle x, y \rangle = G).$$

It follows from the proof of Netto's conjecture that $P(A_n) \to 1$ as $n \to \infty$. Dixon conjectured in [Di1] that the same phenomenon holds for all finite simple groups (throughout this lecture series simple groups are assumed to be nonabelian).

Conjecture 2 (Dixon, 1969) Let G be a finite simple group. Then $P(G) \to 1$ as $|G| \to \infty$.

At the time this was a rather daring conjecture, as the Classification was not yet completed. Assuming the Classification Theorem, it remains to verify that the conjecture holds for the simple groups of Lie type. A breakthrough was achieved in 1990, when Kantor and Lubotzky proved that Dixon's conjecture holds for classical groups, as well as for some small rank exceptional groups of Lie type [KL]. The proof of the conjecture was completed in my joint work [LiSh1] with Martin Liebeck, so we have:

Theorem 1 *Dixon's conjecture holds.*

The starting point of the proof is the following basic, though surprisingly useful, observation, made in [KL]. Suppose $x, y \in G$ do not generate G. Then there exists a maximal subgroup $M \max G$ with $x, y \in M$. Now,

$$\mathrm{Prob}(x, y \in M) = (|M|/|G|)^2 = |G : M|^{-2}$$

Hence

$$1 - P(G) \leq \sum_{M \max G} |G : M|^{-2}.$$

In some situations it will be useful to reformulate this simple inequality. Let $\mathcal{M}$ be a set of representatives for the conjugacy classes of the maximal subgroups of G. Since $N_G(M) = M$ for all maximal subgroups M of G we have

$$\sum_{M \max G} |G : M|^{-2} = \sum_{M \in \mathcal{M}} |G : M|^{-1}.$$

It will be useful to define, for a real number $s > 0$,

$$\zeta_G(s) = \sum_{M \max G} |G : M|^{-s} = \sum_{M \in \mathcal{M}} |G : M|^{1-s}.$$

Then we have $1 - P(G) \leq \zeta_G(2)$, so to prove Theorem 1 it suffices to show that $\zeta_G(2) \to 0$ as $|G| \to \infty$.

Before embarking on the general argument, let us examine a simple example. Suppose $G = \mathrm{PSL}_2(p)$. The subgroup structure of G is well known (see Dickson [D, Chapter VII]). In particular it is known that G has at most 7 conjugacy classes of maximal subgroups. Furthermore, according to a result which dates back to Galois, all maximal subgroups of G have index at least p. We conclude that

$$\zeta_G(2) \leq |\mathcal{M}|p^{-1} \leq 7p^{-1} \to 0 \quad \text{as } p \to \infty.$$

This confirms Dixon's conjecture for the family $\mathrm{PSL}_2(p)$.

As for classical groups in general, the main tool is Aschbacher's work [A1] on the subgroup structure of these groups. A rough version of his main result is given below.

Aschbacher Theorem *Let M be a maximal subgroup of a finite simple classical group. Then either*

(i) *M is a known geometric subgroup (stabilizer of a subspace, a direct sum decomposition, a tensor product decomposition, a subfield, an extension field, a form, etc), or*

(ii) *M is almost simple, absolutely irreducible.*

The geometric subgroups in (i) split naturally into 8 classes, the so called Aschbacher classes, which are denoted by $\mathcal{C}_1, \dots, \mathcal{C}_8$. The set of the remaining maximal subgroups is often denoted by $\mathcal{S}$. See Kleidman and Liebeck [KLi] for full details. To sketch the proof of Theorem 1 for classical groups it is convenient to deviate from Kantor-Lubotzky's original paper, and to apply more recent results by Guralnick, Kantor and Saxl [GKS], counting maximal subgroups in classical groups. Let G be a classical group of dimension n over a field with q elements. Let $\mathcal{M}_1$, $\mathcal{M}_2$ denote sets of representatives for the conjugacy classes of the maximal subgroups of G satisfying conditions (i), (ii) in Aschbacher's Theorem respectively. Then it is shown in [GKS] that

$$|\mathcal{M}_1| \le an^2 \log q,$$

where a is some absolute constant. We also have $|G : M| \ge q^{bn}$ for some $b > 0$ and for all maximal subgroups M of G, and this shows that the contribution of the subgroups in $\mathcal{M}_1$ to $\zeta_G(2)$ is at most $an^2 \log q \cdot q^{-bn}$, which tends to zero as n or q tends to infinity.

It remains to deal with maximal subgroups in $\mathcal{M}_2$. We will sketch the argument in the case where n tends to infinity. If $M \in \mathcal{M}_2$ then (excluding the case where M is symmetric or alternating in a minimal representation, which could be dealt with as above) we have $|M| < q^{3n}$ by a result of Liebeck [Li]. This yields $|G : M| \ge q^{cn^2}$ for some constant $c > 0$. Now we count the possibilities for M as an abstract group. Since there are at most two simple groups of any given order, there are at most $2q^{3n}$ possibilities for the simple socle $\mathrm{Soc}(M)$. Using known facts about outer automorphism groups of simple groups, one easily concludes that there are at most q^{dn} almost simple groups of order $\le q^{3n}$, where d is a suitable constant. Now, given M as an abstract group, there are at most $|\widetilde{M}|$ absolutely irreducible projective representations of M in characteristic p (where $\widetilde{M}$ is the covering group of M), hence at most q^{en} embeddings of M in G up to conjugacy, for some constant e. We see that the contribution of the subgroups $M \in \mathcal{M}_2$ to $\zeta_G(2)$ is bounded above by $q^{dn} \cdot q^{en} \cdot q^{-cn^2}$, which tends to 0 as $n \to \infty$.

A few remarks on the proof for exceptional groups. There are theorems on the subgroup structure of these groups, see Liebeck-Seitz [LiSe1], [LiSe2], Liebeck-Saxl-Testerman [LST], etc. The known geometric subgroups as in (i) can be handled easily. The difficulty is that there are no known effective ways to enumerate the conjugacy classes of the (unknown) almost simple maximal subgroups M. It turns out that this difficulty, which occurs in many situations, can sometimes be overcome (or bypassed) by using special ad-hoc enumeration methods. The method used in [LiSh1] was to apply the following.

Theorem (Malle-Saxl-Weigel, 1994) *All finite simple groups can be generated by an involution and another element.*

See [MSW] and [LiSh1, 2.2]. To explain the relevance of this theorem to our context, let $i_2(G)$ denote the number of involutions in the group G. Now, suppose G is some exceptional group of Lie type whose subgroup structure is not completely known, e.g. $G = E_8(q)$. If $M < G$ is some unknown maximal subgroup, then $S = \mathrm{Soc}(M)$ is simple and $M = N_G(S)$. By the above theorem we have $S = \langle u, v \rangle$ where $u^2 = 1$, so the number of choices for the subgroup S is at most $i_2(G)|G|$. In fact, since the $|S|$ pairs (u^s, v^s) where $s \in S$ generate the same subgroup, we see (counting multiplicities) that there are at most $i_2(G)|G|/|S|$ possibilities for S, and hence for M.

This counting method, combined with some new results on maximal subgroups of exceptional groups, completes the proof of Dixon's conjecture.

It is interesting that the existence of a nice generating pair for simple groups (such as an involution and an additional element) eventually implies that most pairs are generating pairs. This is part of a two-sided interplay between "probabilistic" and "deterministic" results which we will witness throughout this lecture series.

1.4 Estimating generation probabilities

Let $m(G)$ denote the minimal index of a proper subgroup of G. The values of $m(G)$ are known (see [KLi, 5.2.2] and [LiSal]). For example, we have (with few exceptions)

$$m(\mathrm{PSL}_n(q)) = \frac{q^n - 1}{q - 1}, \quad m(E_8(q)) \sim q^{57}.$$

It is shown in [LiSh3] that the generation probabilities $P(G)$ for finite simple groups G are almost determined by the parameter $m(G)$, as follows.

Theorem 2 *There exist constants* $0 < c_1 < c_2$ *such that for every finite simple group* G *we have*

$$1 - c_1 m(G)^{-1} \leq P(G) \leq 1 - c_2 m(G)^{-1}.$$

In fact

$$P(G) = 1 - c(G)m(G)^{-1} + O(m(G)^{-16/15}),$$

where $c(G)$ *is known and lies in the closed interval* $[1, 3]$.

For classical groups, results of this type were proved by Kantor [K, 3.3]. The proof of Theorem 2 is based not only on the crude bound $1-P(G) \leq \zeta_G(2)$ which we have encountered before, but also on more delicate bounds derived using inclusion and exclusion principles.

Note that Theorem 2 extends Babai's $P(A_n) = 1 - n^{-1} + O(n^{-2})$ estimate.

Results from [K] and [LiSh3] also provide a likely geometric explanation for non-generation. More precisely, it is shown that, if two randomly chosen elements x, y of the simple group G do not generate G, then, with probability tending to 1, x, y have a common fixed point in some prescribed geometric action of G.

2 Applications: free groups, the modular group, symmetric groups

Is probability useful in real life (namely, group theory)? In Lecture 1 I discussed solutions to some probabilistic questions concerning finite groups. But can the probabilistic approach help to solve non-probabilistic problems in group theory? The purpose of this lecture is to demonstrate that this may be done in several different contexts.

2.1 Free groups

Let F_d be the free group on d generators ($d \geq 2$). It is well known that F_d is residually finite, and even residually p for any prime p. The following problem concerning residual properties of free groups was raised by Magnus [Ma], and then by Gorchakov and Levchuk [GL].

Magnus problem Is F_d residually X for any infinite collection X of finite simple groups?

In other words, suppose X is an infinite collection of finite simple groups; does it follow that

(a) $\cap\{N \lhd F_d : F_d/N \in X\} = 1$?

Since F_d is residually F_2, the question is reduced to the case $d = 2$. Several partial answers were given in the past three decades. For example, work by Katz and Magnus [KM] and by Wiegold [Wi] provided an affirmative answer in the case where X consists of alternating groups. Lubotzky used strong approximation to obtain some results for simple groups of Lie type [Lu1]. An affirmative answer for groups of type PSL was provided by J.S. Wilson [W1]. Recently, in a series of three papers [We1], [We2], [We3], T. Weigel completed the positive solution of the Magnus problem. His proof is rather long and involved. It turns out that, using a probabilistic approach, one can obtain a short and elegant solution. Let me summarize this approach, which is a joint work with J.D. Dixon, L. Pyber and Á. Seress [DPSSh].

First note that it suffices to show the following.

(b) Let $1 \neq w = w(u, v) \in F_2$; then for almost all finite simple groups G there is a generating pair x, y such that $w(x, y) \neq 1$.

Indeed, suppose condition (b) holds, let X be an infinite collection of finite simple groups, and let $w \in F_2$ be a non-identity element. By (b) we know that there is a simple group $G \in X$ and a generating pair x, y for G such that $w(x, y) \neq 1$ in G. We can therefore construct a homomorphism $\phi : F_2 \to G$ by sending u to x and v to y. Then ϕ is an epimorphism and we have $\phi(w) \neq 1$. This shows that condition (b) implies condition (a) above.

In order to prove (b), we establish a stronger result of probabilistic nature.

Theorem 3 *Fix* $1 \neq w = w(u, v) \in F_2$. *Let* G *be a finite simple group, and let* $x, y \in G$ *be randomly chosen elements. Then, as* $|G| \to \infty$ *we have*

$$\text{Prob}(\langle x, y\rangle = G \wedge w(x, y) \neq 1) \to 1.$$

The proof of Theorem 3 starts with the following reduction. Applying Theorem 1, we know that $\text{Prob}(\langle x,y\rangle = G) \to 1$. Hence it is enough to prove

(c) $\text{Prob}(w(x,y) \neq 1) \to 1$ as $|G| \to \infty$,

since then the probability of the intersection of the two events would also tend to 1. Statement (c) has the advantage that it no longer deals with generating pairs. We just have to show that (as $|G| \to \infty$) most pairs in G do not satisfy a given relation. This can be proved using some combinatorial tricks for alternating groups and classical groups of unbounded rank. The case of groups of Lie type of bounded Lie rank can be dealt with using some algebraic geometry. The idea is to regard G as the q-rational points in some irreducible algebraic variety, and to note that the equation $w(x,y) = 1$ defines a proper subvariety (see [J1]). Results on the number of points on subvarieties can then be applied. While this approach resolves most of the cases, some modifications are needed for Suzuki groups and Ree groups.

It is noteworthy that Weigel's original solution of the Magnus problem also has some probabilistic ingredients. Theorem 3 and extensions of it were recently used by Pyber to obtain results on free dense subgroups of profinite groups.

2.2 The modular group

What are the finite simple quotients of the modular group $\text{PSL}_2(\mathbb{Z})$? The study of this problem goes back to the beginning of this century, and possibly earlier. Since $\text{PSL}_2(\mathbb{Z})$ is isomorphic to the free product of C_2 with C_3, a group G is a quotient of $\text{PSL}_2(\mathbb{Z})$ if and only if $G = \langle x, y\rangle$ with $x^2 = y^3 = 1$. Such groups are termed $(2,3)$-*generated.*

The study of $(2,3)$-generated finite groups was usually carried out with geometric and number-theoretic motivations. Let S_g denote a compact connected Riemann surface of genus $g \geq 2$. Suppose G acts faithfully on S_g by conformal automorphisms. Then, by the Hurwitz-Schwartz inequality, we have $|G| \leq 84(g-1)$. If equality holds G is termed a *Hurwitz group* (see [Hu]). Hurwitz groups are known to be $(2,3)$-generated (in fact they are characterized by having generators of orders 2 and 3 whose product has order 7). Thus, if we have to determine whether a given group is a Hurwitz group, we could first test it for $(2,3)$-generation.

Another motivation for the study of $(2,3)$-generated finite groups is the study of the degree of failure of the Congruence Subgroup Property for $\text{PSL}_2(\mathbb{Z})$. In this context, see [J2].

For background, let us now discuss some major examples of known $(2,3)$- generated finite simple groups. The alternating groups A_n ($n \geq 9$) were shown to be $(2,3)$-generated by G.A. Miller back in 1901 [Mi]. Partial results concerning the groups of type $\text{PSL}_2(q)$ were obtained by Brahana and Sinkov in the 20s and 30s [Br1], [Br2], [Si1], [Si2], [Si3]. In 1967 Macbeath showed that all simple groups of type $\text{PSL}_2(q)$ ($q \neq 9$) are $(2,3)$-generated [Mac]. Some other results for groups of small Lie rank were later established, and a breakthrough was achieved in 1987 by Tamburini, who showed that $\text{PSL}_n(q)$ is $(2,3)$-generated for all $n \geq 25$ [T]. In 1995 DiMartino and Vavilov showed that for odd q all the groups $\text{PSL}_n(q)$ are $(2,3)$-generated [DiMV1], [DiMV2]. Some other classical groups of large rank were dealt with by Tamburini, Wilson and Gavioli [TWG], [TW]. For more details, see the recent surveys [DiMT] and [W2].

The proofs of the above results, which become increasingly difficult, are based on explicit constructions of generators of orders 2 and 3. This approach seems to fail for various families of classical groups, e.g. those with "intermediate" Lie rank. While some simple groups are not $(2,3)$-generated, the following conjecture was recently posed (see [W2]).

Conjecture 3 (DiMartino-Vavilov, Wilson) All finite simple groups of Lie type except some of low rank in characteristic 2 or 3 are quotients of $\mathrm{PSL}_2(\mathbb{Z})$.

In [LiSh2] we address this problem for classical groups, proving the following.

Theorem 4 *All finite simple classical groups except* $\mathrm{PSp}_4(2^k)$, $\mathrm{PSp}_4(3^k)$ *and finitely many others are quotients of* $\mathrm{PSL}_2(\mathbb{Z})$.

Some remarks are in order. First, the groups $\mathrm{PSp}_4(2^k)$ and $\mathrm{PSp}_4(3^k)$ are genuine exceptions: they are not $(2,3)$-generated. The existence of infinite families of non $(2,3)$-generated classical groups was somewhat unexpected. Of course, there exists an infinite family of simple groups which are not $(2,3)$-generated, namely the Suzuki groups $\mathrm{Sz}(q)$; indeed these groups do not have elements of order 3. However, the 4-dimensional symplectic groups above do contain 3-elements, and so a less obvious argument should be used (see section 6 in [LiSh2]).

Let us now discuss the way Theorem 4 is proved. We need some notation. For $k \geq 1$ let x_k denote a randomly chosen element of order k in G. Set

$$P_{2,3}(G) = \mathrm{Prob}(\langle x_2, x_3 \rangle = G).$$

It is now possible to formulate the probabilistic result behind Theorem 4.

Theorem 5 *Let* $G \neq \mathrm{PSp}_4(q)$ *be a finite simple classical group. Then* $P_{2,3}(G) \to 1$ *as* $|G| \to \infty$. *If* $G = \mathrm{PSp}_4(p^k)$ $(p \geq 5)$ *then* $P_{2,3}(G) \to 1/2$ *as* $|G| \to \infty$.

Clearly, Theorem 5 implies Theorem 4. In fact this is the classical way in which probabilistic methods are applied prove existence theorems using probability estimates instead of explicit constructions.

The proof of Theorem 5 boils down to the subtle inequality

$$\frac{2}{5} + \frac{8}{13} > 1.$$

This will be explained below in several steps. Let $i_k(G)$ denote the number of elements of order k in G.

STEP 1. Show that

$$1 - P_{2,3}(G) \leq \sum_{M \max G} \frac{i_2(M) i_3(M)}{i_2(G) i_3(G)}.$$

Indeed, if $\langle x_2, x_3\rangle \neq G$, then $x_2, x_3 \in M$ for some $M \max G$, and $\mathrm{Prob}(x_k \in M) = i_k(M)/i_k(G)$.

STEP 2. Count involutions and elements of order 3 in classical groups G and in their maximal subgroups M (the latter is rather painstaking). Deduce that, for some absolute constant c we have

$$\frac{i_2(M)}{i_2(G)} \leq c|G:M|^{-2/5},$$

and (with few exceptions such as $\mathrm{PSp}_4(q)$)

$$\frac{i_3(M)}{i_3(G)} \leq c|G:M|^{-8/13}.$$

STEP 3. Conclude that

$$1 - P_{2,3}(G) \leq c^2 \sum_{M \max G} |G:M|^{-2/5}|G:M|^{-8/13}$$
$$= c^2\zeta_G(2/5 + 8/13) = c^2\zeta_G(66/65).$$

STEP 4. Prove that for any fixed $s > 1$, $\zeta_G(s) \to 0$ as $|G| \to \infty$.
While we prove it for classical groups, we conjecture that this holds for G exceptional as well.

Steps 3 and 4 readily imply $P_{2,3}(G) \to 1$, completing the proof of (the main part of) Theorem 5.

It is easy to see that $s \leq 1$ implies $\zeta_G(s) \to \infty$. Therefore the theorem indeed depends on the inequality $2/5 + 8/13 > 1$.

Recall that the proof of Theorem 1 (Dixon's conjecture) was based on estimating $\zeta_G(2)$, while the proof of Theorems 4 and 5 is based on estimating $\zeta_G(66/65)$. It is curious that other generation results can be obtained by studying $\zeta_G(s)$ for other values of s. For instance, the probability that a randomly chosen involution and a randomly chosen additional element generate a finite simple classical group G is at least $1 - c\zeta_G(7/5)$, which tends to 1 by Step 4 above; and the probability that G is generated by three randomly chosen involutions is at least $1 - c\zeta_G(6/5)$, which also tends to 1. See [LiSh2, Section 2] for more details.

Note that Theorem 4, in spite of its generality, does not tell us anything about any particular group. This is a typical drawback of the probabilistic approach. For example, it is still unknown whether the groups $\mathrm{PSL}_5(2^k)$ are $(2,3)$-generated; Theorem 4 reveals that the answer is positive for almost all values of k. However, by carefully examining the proof in [LiSh2] it should be possible to provide an explicit number c such that all finite simple classical groups of order at least c except the 4-dimensional symplectic groups in characteristic $2, 3$ are $(2,3)$-generated. We leave this unpleasant exercise to the ambitious reader.

So far we have discussed $(2,3)$-generation of classical groups. But what about exceptional groups of Lie type? G. Malle has obtained various partial results on the problem (see for instance [Mal]), and recently, in a joint work with Lübeck, the problem was solved completely [LM].

Theorem (Lübeck-Malle, 1997) *Except the Suzuki groups and finitely many other groups, all exceptional groups of Lie type are quotients of* $\mathrm{PSL}_2(\mathbb{Z})$.

The proof for the large rank exceptional groups $F_4(q), E_6^\epsilon(q), E_7(q)$ and $E_8(q)$ applies character theory and requires the use of a computer to determine the structure of constants in the respective groups.

To sum up, the problem of $(2,3)$-generation of the finite simple groups, studied since 1901, is now virtually solved: *the non* $(2,3)$*-generated simple groups are* $\mathrm{PSp}_4(2^k)$, $\mathrm{PSp}_4(3^k)$, $\mathrm{Sz}(2^k)$, *and finitely many others.*

Finally, let me briefly describe some applications. The above $(2,3)$-generation results can be used in several contexts. A typical use is to bound the number of unknown maximal subgroups M in groups G of Lie type: if $M = N_G(S)$ then the simple group S is usually $(2,3)$-generated, and so the number of choices for it is bounded above by $i_2(G)i_3(G)$, and even by $(1+\epsilon)i_2(G)i_3(G)/i_2(S)i_3(S)$ provided $P_{2,3}(S) \to 1$.

In fact the estimates on $P(G)$ presented in Lecture 1 (see Theorem 2) rely on this enumeration method. Another by-product is the proof in [LiSh3] of the following conjecture from [KL].

Conjecture 5 (Kantor-Lubotzky, 1990) A finite simple group is almost surely generated by a randomly chosen involution and a randomly chosen additional element.

In the next section we describe a third application, which is of some combinatorial flavour.

2.3 Symmetric groups

How many maximal subgroups does the symmetric group S_n have up to conjugacy? Without using the Classification Theorem there seems to be no hope of getting reasonable upper bounds. The following bound is obtained in [Ba2] using the Classification.

Theorem (Babai, 1989) S_n *has at most* $n^{(1+o(1))\log^3 n}$ *conjugacy classes of maximal subgroups.*

In [LiSh4] this is improved as follows.

Theorem 6 S_n *has* $(\frac{1}{2}+o(1))n$ *conjugacy classes of maximal subgroups.*

Note that the intransitive subgroups, which have the form $S_k \times S_{n-k}$, already yield $n/2$ classes of maximal subgroups. Therefore Theorem 6 asserts that, in some sense, almost all maximal subgroups of S_n are the obvious intransitive ones.

The maximal subgroups which are transitive but imprimitive have the form $S_k \wr S_{n/k}$ (where k is a proper divisor of n), and they split into $d(n) - 2 = n^{o(1)}$ classes (here $d(n)$ denotes the number of divisors of n).

So the main task is to count the primitive maximal subgroups of S_n. A standard tool here is

Theorem (O'Nan-Scott) *The primitive maximal subgroups of S_n split into four types:*

(a) *affine: $AGL(V)$, $|V| = p^k$;*

(b) *product type: $S_k \wr S_l$, $n = k^l$;*

(c) *diagonal type: description omitted;*

(d) *almost simple.*

It is not difficult to verify that types (a)-(c) yield at most $\log n$ classes, so the hard core of the proof is to enumerate subgroups of type (d), namely, the primitive almost simple ones.

Suppose $G \leq S_n$ is primitive and almost simple, and let $M \leq G$ be a point-stabilizer. Theorem 5 and the enumeration method described subsequently help to count the possibilities for M and deduce: *S_n has at most $n^{6/11+o(1)}$ primitive maximal subgroups up to conjugacy.*

Altogether we see that the number of conjugacy classes of maximal subgroups of S_n is bounded above by

$$n/2 + d(n) - 2 + n^{6/11+o(1)} = n/2 + n^{6/11+o(1)}.$$

In particular we obtain Theorem 6. We conjecture that the number of conjugacy classes of primitive maximal subgroups of S_n is in fact of the form $n^{o(1)}$.

This discussion demonstrates that the connection between maximal subgroups and random generation is two-sided: results on maximal subgroups help to prove results on random generation, and vice versa.

We close this section with a corollary concerning the number of maximal subgroups of S_n (counted not just up to conjugacy). There are 2^{n-1} intransitive ones, and the transitive ones split into at most $n^{6/11+o(1)}$ classes. Each such class has size at most $(n-1)!$ (since the order of a transitive subgroup is at least n). Summing up we obtain at most $2^{n-1} + (n-1)!n^{6/11+o(1)}$ subgroups.

Corollary 7 *S_n has at most $n!n^{-5/11+o(1)}$ maximal subgroups.*

This gives a partial solution to the following.

Conjecture 6 (G.E. Wall, 1961) The number of maximal subgroups of a finite group G is less than $|G|$.

This conjecture was confirmed by Wall [Wa] for soluble groups. We now see that it also holds for symmetric groups of sufficiently large degree. The next challenge seems to be the groups $\mathrm{PSL}_n(q)$. For large n (say $n > 27$) this seems to be very hard.

3 Profinite groups I: Hausdorff dimension

In this talk, as well as the next one, we are going to focus on infinite groups. The main theme of the present talk is the adaptation of a fundamental concept in fractal geometry – that of Hausdorff dimension – for profinite groups. As we shall see below, the study of Hausdorff dimension in certain profinite groups is related to some Lie-theoretic problems concerning affine Kac-Moody algebras. For a general background on fractal dimensions and their various properties, see Falconer [F].

3.1 Hausdorff dimension

Though Hausdorff dimension is related to certain measures (namely Hausdorff measures), it is defined in metric spaces. Let (X, d) be a metric space, let $Y \subseteq X$ and let $\delta > 0$.

Definition

1) a *δ-cover* of Y is a cover of Y by balls U_i of diameter $\leq \delta$.

2) For $s \geq 0$ let

$$\mu_\delta^s(Y) = \inf\{\sum_{i=1}^{\infty} (\mathrm{diam} U_i)^s : \quad (U_i) \text{ a } \delta\text{-cover of } Y\},$$

and set

$$\mu^s(Y) = \lim_{\delta \to 0} \mu_\delta^s(Y).$$

The function μ^s is referred to as the *s-dimensional Hausdorff measure on X.*

3) The *Hausdorff dimension* of Y is defined by

$$\mathrm{Dim}(Y) = \inf\{s : \mu^s(Y) = 0\} = \sup\{s : \mu^s(Y) = \infty\}.$$

Thus, if $s_0 = \mathrm{Dim}(Y)$, then $\mu^s(Y) = 0$ for $s > s_0$, $\mu^s(Y) = \infty$ for $s < s_0$. The value of $\mu^{s_0}(Y)$ can be finite, infinite, or zero.

For nice subsets Y we often have $\mathrm{Dim}(Y) = \dim(Y)$, where dim denotes the topological dimension. However, $\mathrm{Dim}(Y)$ can well exceed $\dim(Y)$, as demonstrated by various fractals (Cantor sets, snow flakes, Julia sets, etc). For example, the Hausdorff dimension of the famous Koch snow flake is $\frac{\log 4}{\log 3}$. See [F] for more details and examples.

3.2 Profinite groups

Profinite groups are compact Hausdorff topological groups, where the open subgroups form a base for the neighbourhoods of the identity. Equivalently, profinite groups are obtained as inverse limits of finite groups.

The basic examples are Galois groups of infinite field extensions, the additive groups of the p-adic integers $\mathbb{Z}_p$, and of the ring of formal power series $F_p[[t]]$, as well as various matrix groups over these rings, such as $\mathrm{SL}_d(\mathbb{Z}_p)$ and $\mathrm{SL}_d(F_p[[t]])$.

Let G be a profinite group. Then G admits an invariant measure, namely the Haar measure, which (when normalized) turns G into a probability space. This probability space will be discussed in Lecture 4. At the moment we are more interested in metric structures on G. Suppose G is infinite and countably based. Then there is a filtration

$$G = G_0 > G_1 > \ldots > G_n > \ldots,$$

where G_n are open normal subgroups which form a base for the neighbourhoods of 1. Given the filtration $\{G_n\}$ we can construct an invariant metric d on G as follows. For $x, y \in G$ set

$$d(x,y) = \inf\{|G : G_n|^{-1} : xy^{-1} \in G_n\}.$$

Note that the coset xG_n is a closed ball of diameter $|G : G_n|^{-1}$ around x (with respect to this metric), and that G has diameter 1. Once G is equipped with a metric, the Hausdorff dimension function on subsets of G is well defined. We shall particularly be interested in the Hausdorff dimension of closed subgroups H of G. The following result, which is essentially due to Abercrombie, provides us with an algebraic formula for the computation of this dimension. For the proof see [Ab] and [BSh].

Theorem 8 *Let $H \leq G$ be a closed subgroup. Then*

$$\mathrm{Dim}(H) = \liminf \frac{\log|HG_n/G_n|}{\log|G/G_n|}.$$

Thus, if $|HG_n/G_n|$ is approximately $|G/G_n|^\alpha$, then $\mathrm{Dim}(H) = \alpha$.

It follows from Theorem 8 that G, as well as open subgroups of G, have Hausdorff dimension 1. On the other hand finite subgroups of G have Hausdorff dimension 0. Commensurable subgroups have the same Hausdorff dimension.

While the Haar measure of G is useful for studying open subgroups of G, the Hausdorff dimension function seems to be useful for studying intermediate subgroups, namely subgroups which are neither finite nor of finite index.

Definition The *spectrum* of G is the set of Hausdorff dimensions of closed subgroups of G, namely

$$\mathrm{Spec}(G) = \{\mathrm{Dim}(H) : H \leq_c G\}.$$

Then we have

$$\{0,1\} \subseteq \mathrm{Spec}(G) \subseteq [0,1].$$

The spectrum of G can be regarded as a picture reflecting the subgroup structure of G; it may give a useful insight, especially in cases where the subgroup structure cannot be determined completely.

3.3 Pro-p groups

A pro-p group is an inverse limit of finite p-groups. The theory of pro-p groups and its applications has expanded greatly during recent years; for background see, for instance, [DDMS], [Sh1], [Sh2]. Let G be a (topologically) finitely generated pro-p group. Then the subgroups

$$G_n = \langle x^{p^n} : x \in G\rangle$$

are open and form a filtration of G. We shall usually use this filtration (or refinements of it) to define the metric on G. Our first result on Hausdorff dimension in pro-p groups deals with a well-behaved class of pro-p groups: the p-adic analytic ones (see [DDMS]); it is included in the joint work [BSh] with Yiftach Barnea.

Theorem 9 *Let G be a p-adic analytic pro-p group, and let $H \leq G$ be a closed subgroup. Then*

$$\mathrm{Dim}(H) = \dim H / \dim G,$$

where dim *denotes dimension as p-adic Lie groups.*

It follows from the theorem that the notion of Hausdorff dimension extends the notion of dimension in Lie groups (of course it is more widely applicable).

Corollary 10 *If G is a p-adic analytic pro-p group then* $\mathrm{Spec}(G)$ *is finite and consists of rational numbers.*

It would be interesting to find out whether the converse also holds. We conjecture that finitely generated pro-p groups with finite spectrum are p-adic analytic. This is unknown even if we assume $\mathrm{Spec}(G) = \{0, 1\}$! However, we do have some sort of a characterization of p-adic analytic groups in terms of their Hausdorff dimension function (see [BSh, 1.3]).

Theorem 11 *Let G be a finitely generated pro-p group. Then G is p-adic analytic if and only if only finite subgroups of G have Hausdorff dimension zero.*

The 'only if' part follows from Theorem 9 using the fact that zero-dimensional p-adic analytic pro-p groups are finite. The proof of the 'if' part combines Lazard's theory of p-adic analytic pro-p groups with Zelmanov's theorem on the local finiteness of torsion pro-p groups [Z].

We now turn to a pro-p group whose subgroup structure is far from clear, namely

$$G = \mathrm{SL}_d(F_p[[t]]).$$

Note that G is only virtually pro-p; in the discussion below we can replace it by an open pro-p subgroup, namely its first congruence subgroup. These groups are not p-adic analytic, but they can be regarded as analytic over $F_p[[t]]$. For a preliminary investigation of these generalized analytic groups, see [LuSh].

What can be said of the spectrum of the group G above? We use the congruence filtration to define the metric on G. We conjecture that $\mathrm{Spec}(G)$ is a finite union of closed intervals (including points). This holds for example for $d = 2, p > 2$, in which case $\mathrm{Spec}(G) = [0, 2/3] \cup \{1\}$.

In general we can show the following.

(a) $\mathrm{Spec}(G) \supset [0, 1/2]$.

(b) 1 is an isolated point in $\mathrm{Spec}(G)$.

While part (a) is relatively easy (the construction uses subgroups of the Borel subgroup of G), the proof of (b) is rather involved. It relies on the following more precise result from [BSh].

Theorem 12 *Let $G = \mathrm{SL}_d(F_p[[t]])$ $(p > 2)$ and let $H \leq G$ be a closed non-open subgroup. Then*

$$\mathrm{Dim}(H) \leq 1 - \frac{1}{d+1}.$$

It can be verified that this upper bound is best-possible: taking H to be a parabolic subgroup which is the stabilizer of a line or a hyperplane, we have $\mathrm{Dim}(H) = 1 - 1/(d+1)$. Therefore the theorem asserts that among all non-open subgroups of G, these parabolic subgroups are of largest size (namely Hausdorff dimension). Similar results for finite groups of type $\mathrm{SL}_d(q)$ have been known since the seventies [Pa], but the proof in the infinite case is rather different, using some unanticipated tools which are described below.

3.4 Kac-Moody algebras

We can associate with our group $G = \mathrm{SL}_d(F_p[[t]])$ the Lie algebras $sl_d(F_p[t])$ and

$$L = sl_d(F_p[t, t^{-1}]) \cong sl_d(F_p) \otimes F_p[t, t^{-1}].$$

With close subgroups of G one can associate graded subalgebras of these Lie algebras. The subalgebra structure of Lie algebras of this type is therefore relevant in the context of Theorem 12.

Let me provide some background on the subalgebra structure of certain Lie algebras. In 1957 Dynkin described the maximal subalgebras of the finite-dimensional simple complex Lie algebras [D1], [D2]. The subalgebra structure of the infinite-dimensional Lie algebras we are considering here can be studied in the more general framework of affine Kac-Moody algebras (see [Kac]). Recall that affine Kac-Moody algebras are natural infinite-dimensional generalizations of the classical simple Lie algebras. They have similar generators e_i, f_i, h_i $(1 \leq i \leq n)$, with relations given by an extended Cartan matrix of rank $n-1$ (instead of n). Modulo the center we get simple $\mathbb{Z}$-graded algebras of the form

$$L(\mathcal{G}) = \mathcal{G} \otimes F[t, t^{-1}],$$

where $\mathcal{G}$ is a finite-dimensional simple Lie algebra over the ground field F.

The representation theory of affine Kac-Moody algebras is quite developed. However, not much seems to be known about their subalgebra structure. For the group-theoretic purpose we have in mind it suffices to study the graded maximal subalgebras of $L(\mathcal{G})$. Some examples of nice graded subalgebras are given below.

(a) $L^+(\mathcal{G}) = \mathcal{G} \otimes F[t]$, $L^-(\mathcal{G}) = \mathcal{G} \otimes F[t^{-1}]$.

(b) $L(\mathcal{H}) = \mathcal{H} \otimes F[t, t^{-1}]$ for a subalgebra $\mathcal{H} \leq \mathcal{G}$.

(c) Let $\mathcal{G} = \mathcal{G}_0 \oplus \ldots \oplus \mathcal{G}_{m-1}$ be a $\mathbb{Z}/m\mathbb{Z}$-grading of $\mathcal{G}$ (denoted by α). The subalgebra

$$L(\mathcal{G}, m, \alpha) = \oplus_{n \in \mathbb{Z}} \mathcal{G}_{n \bmod m} \otimes t^n$$

is called a *loop algebra of period m associated with $\mathcal{G}$.*

In the joint work [BShZ] with Barnea and Zelmanov it is shown that the above examples essentially exhaust all the graded maximal subalgebras of $L(\mathcal{G})$.

Theorem 13 *Let $\mathcal{G}$ be a central simple Lie algebra over a field F, and M a maximal graded subalgebra of $L(\mathcal{G})$. Then one of the following holds:*
(a) *M is commensurable to $L^+(\mathcal{G})$ or to $L^-(\mathcal{G})$.*
(b) *$M = L(\mathcal{H})$ for some maximal subalgebra $\mathcal{H}$ of $\mathcal{G}$.*
(c) *M is a loop algebra of prime period associated with $\mathcal{G}$.*

Recall that a simple Lie algebra $\mathcal{G}$ over a field F is called *central simple* if the centroid of $\mathcal{G}$ coincides with F. For example, $\mathcal{G} = sl_d(F_p)$ is central simple over F_p, and so our theorem applies for $L = sl_d(F_p) \otimes F_p[t, t^{-1}]$.

In the important case $F = \mathbb{C}$ (the complex numbers) we have the following.

1. $\mathcal{G}$ is simple if and only if it is central simple.
2. The subalgebras M in (b) are parameterized by the maximal subalgebras $\mathcal{H}$ of $\mathcal{G}$; these are known ([D1], [D2]).
3. The subalgebras M in (c) are parameterized by gradings of $\mathcal{G}$ of prime period, hence by elements of prime order in $\mathrm{Aut}(\mathcal{G})$; these are known (see [Kac, Chapter 8]).

There is a version of Theorem 13 for simple (not necessarily central simple) Lie algebras $\mathcal{G}$ (see [BShZ, 1.2]).

It is interesting that the proof of Theorem 13 relies on tools from the theory of associative algebras, namely central polynomials for matrix algebras. Returning to Hausdorff dimension, Theorem 13 for $F = F_p$ and $\mathcal{G} = sl_d(F_p)$ is instrumental in proving Theorem 12: *the maximal Hausdorff dimension of a non-open subgroup of* $\mathrm{SL}_d(F_p[[t]])$ *is* $1 - 1/(d+1)$.

We shall see in the next lecture that this result has some interesting applications concerning random generation in profinite groups.

4 Profinite groups II: random generation

4.1 Positively finitely generated groups

In this lecture we finally deal with genuine probability spaces, namely those arising from profinite groups and their Haar measures. We shall also apply a useful tool from probability theory, namely the Borel-Cantelli Lemma (see [R]).

So let G be a (topologically) finitely generated profinite group, and let μ be the normalized Haar measure on G. Then (G, μ) and the Cartesian powers (G^k, μ^k) ($k \geq 1$) are probability spaces.

Definition $P(G, k)$ denotes the μ^k-measure of the set of k-tuples in G^k which generate G. In other words, $P(G, k) = \mathrm{Prob}(\langle x_1, \ldots, x_k\rangle = G)$, where $x_1, \ldots, x_k \in G$ are randomly chosen.

In particular, for G finite, $P(G, 2)$ is the function $P(G)$ discussed in Lecture 1. Note that (assuming G is non-trivial) $P(G, k) < 1$ for all k; indeed, if H is a proper open subgroup of G, then with probability $|G : H|^{-k} > 0$ the elements $x_1, \ldots, x_k$ all lie in H.

The following key definition is due to Mann [M].

Definition G is termed *positively finitely generated* (PFG) if $P(G,k) > 0$ for some k.

Let me provide some examples. It was observed by Fried and Jarden in 1986 (in the context of field arithmetic) that procyclic groups are positively finitely generated [FJ]; this was extended for abelian profinite groups by Kantor and Lubotzky [KL], who also showed that (nonabelian) free profinite groups are not PFG. In his 1996 paper [M] Mann proved that prosoluble groups are PFG, and so are the profinite completions of the groups $\mathrm{SL}_d(\mathbb{Z})$ ($d \geq 3$). In the recent joint work [BPSh] with Borovik and Pyber we show that profinite groups not involving all finite groups as quotients of open subgroups are PFG. This extends all examples of PFG groups discussed above. Additional constructions of PFG groups are given by Bhattacharjee [Bh].

Can one find conditions which are necessary and sufficient for a profinite group G to be positively finitely generated?

Definition $m_n(G)$ denotes the number of index n maximal subgroups of G. We say that G has PMSG (polynomial maximal subgroup growth) if $m_n(G) \leq n^\alpha$ for some α and for all n.

As observed by Mann, if G has PMSG, then G is PFG. Indeed, we have

$$1 - P(G,k) \leq \sum_{M \max G} |G:M|^{-k} = \sum_{n>1} m_n(G) n^{-k}.$$

Now, if $m_n(G) \leq n^\alpha$, then by choosing say $k \geq \alpha + 2$ we see that

$$1 - P(G,k) \leq \sum_{n>1} n^{\alpha-k} \leq \sum_{n>1} n^{-2} < 1,$$

so $P(G,k) > 0$.

The following result from [MSh] shows that the reverse implication also holds.

Theorem 13 *A profinite group G is positively finitely generated if and only if it has PMSG.*

We have to prove the 'only if' part. Let me sketch the main stages of the proof. Suppose G is PFG and fix k with $P(G,k) > 0$. We aim to show that $m_n(G)$ grows polynomially with n.

STEP 1. Use the Classification Theorem and Lecture 1 to show: there exists c such that $m_n(T) \leq n^c$ for all finite simple groups T and for all n.

Note that

$$\sum_{n>1} m_n(T) n^{-2} = \zeta_T(2) \to 0 \text{ as } |T| \to \infty$$

by results from Lecture 1. This implies $m_n(T) \leq Cn^2$. Sharper bounds of this type are obtained in [MSh] and in [LiSh4].

STEP 2. Use Step 1 and the O'Nan-Scott Theorem to deduce: the number of index n maximal subgroups of G with a given core grows polynomially with n.

It therefore remains to show that the number $c_n(G)$ of cores of index n maximal subgroups of G grows polynomially with n.

STEP 3. Apply:

Borel-Cantelli Lemma *Let X_i be a series of events in a probability space X with probabilities p_i $(i \geq 1)$.*

(i) *if $\sum p_i = \infty$ and X_i are pairwise independent then with probability 1 infinitely many of the events X_i happen.*

(ii) *if $\sum p_i < \infty$ then with probability 1 only finitely many of the events X_i happen.*

To explain how to apply the lemma, let N_i $(i \geq 1)$ be a list of all different cores of maximal subgroups of G. For each i let M_i be a maximal subgroup with core N_i. Let X be G^k considered as a probability space, and set $X_i = M_i^k$ $(i \geq 1)$. It is easy to see that, if the maximal subgroups M, L of G have different cores, then M^k, L^k are independent events in G^k. Therefore the events X_i are pairwise independent. Obviously, the probability that X_i holds is $p_i = |G : M_i|^{-k}$.

We claim that $\sum p_i < \infty$. Suppose otherwise. Then, by part (i) of the Borel-Cantelli Lemma, with probability 1 infinitely many of the events X_i happen. Hence, with probability 1, $x_1, \ldots, x_k$ lie in infinitely many subgroups M_i. This certainly implies $P(G, k) = 0$, a contradiction. Therefore $\sum p_i < \infty$.

Now, note that

$$\sum p_i = \sum |G : M_i|^{-k} = \sum c_n(G) n^{-k}.$$

It follows that $\sum c_n(G) n^{-k} < \infty$, so $c_n(G) = o(n^k)$. This completes the proof.

Theorem 14 provides a characterization of PFG groups in terms of maximal subgroup growth. But can we find a structural characterization of profinite groups with PMSG? This seems to be a very difficult problem; at the moment even a conjecture would be welcome.

For comparison, let me make some remarks on subgroup growth and PSG groups. Recall that G is termed a PSG group if the number $a_n(G)$ of index n subgroups of G grows polynomially with n. Subgroup growth was the subject of Lubotzky's lecture series in the 4th St Andrews meeting in Galway in 1993 (see [Lu2]). In 1993 Lubotzky, Mann and Segal characterized residually finite finitely generated PSG groups as virtually soluble groups of finite rank [LMS]. A characterization of profinite PSG groups (and indirectly of PSG groups in general) has just been given by Segal and myself. Roughly speaking we show that these groups are an extension of a prosoluble group of finite rank by a virtually 'quasisemisimple group' of certain type. For more details, see the paper [SSh] in the Hartley memorial JLMS issue.

4.2 Generation up to commensurability

Let G be a (topologically) finitely generated profinite group. From some points of view it is natural to identify commensurable subgroups of G, and to study generation properties via this identification. Thus, instead of studying the probability of generating G, we can study the probability of generating a finite index subgroup. While the probability that k elements generate G is always less than 1, it often happens that k elements generate an open subgroup with probability 1.

Definition $Q(G,k) = \text{Prob}(\langle x_1,\ldots,x_k\rangle \text{ is open})$, where $x_1,\ldots,x_k \in G$ are randomly chosen.

We start with the following result from [M].

Proposition 15 (Mann) *If G is a PSG group then $Q(G,k) = 1$ for some k.*

Proof This time we apply part (ii) of the Borel-Cantelli Lemma. Suppose $a_n(G) \leq n^\alpha$ for all n, and let H_i ($i \geq 1$) be all open subgroups of G. Take $k > \alpha + 1$ and let $X = G^k$, $X_i = H_i^k$, $p_i = \text{Prob}(X_i) = |G : H_i|^{-k}$. Then

$$\sum p_i = \sum |G : H_i|^{-k} = \sum a_n(G)n^{-k} \leq \sum n^{\alpha-k},$$

which is finite (since $\alpha - k < -1$).
By the Borel-Cantelli Lemma, with probability 1 only finitely many of the events X_i happen. Hence, with probability 1, the subgroup $\langle x_1,\ldots,x_k\rangle$ lies in only finitely many open subgroups. This shows that $\langle x_1,\ldots,x_k\rangle$ is open with probability 1. □

For some time it was unclear whether the converse holds. However, examples of non-PSG groups which satisfy $Q(G,k) = 1$ for some k were constructed in [MSh]. The pro-p version of the problem is formulated in [M]: let G be a pro-p group satisfying $Q(G,k) = 1$ for some k; does it follow that G is p-adic analytic? It was conjectured by the present author that the answer is negative, and that the groups $SL_d(F_p[[t]])$ should be a counter-example (see the survey paper [Sh5]). This has just been verified.

Theorem 16 (Barnea-Larsen, 1997) *Let $G = SL_d(F_p[[t]])$ where $d, p > 2$. Then two randomly chosen elements of G generate an open subgroup with probability 1.*

The proof uses two main tools. One is a result from Lecture 3 on Hausdorff dimensions, and the other is a brand new theorem by Richard Pink on certain subgroups of algebraic groups over local fields. Roughly speaking, Pink's theorem asserts that *if F is a non-archimedian local field, G a semisimple algebraic group over F, and Γ a compact Zariski-dense subgroup of $G(F)$, then the commutator subgroup Γ' is open in some algebraic subgroup H over some subfield of F.*
Let me now sketch the proof of Theorem 16. I should stress that there is no preprint available yet, and this sketch may be subject to some future modifications.
Fundamental to the proof is the following concept (introduced in [Sh5]).

Definition A closed subgroup $H \leq G$ is called *weakly maximal* if it is non-open and is maximal with respect to this property.

It is easy to verify that every (closed) non-open subgroup of a finitely generated pro-p (or virtually pro-p) group is contained in a weakly maximal one. Incidentally, this is not true in profinite groups in general.

Let W denote the set of weakly maximal subgroups of $G = \mathrm{SL}_d(F_p[[t]])$. Suppose $x, y \in G$. Then $\langle x, y\rangle$ is non-open if and only if $x, y \in H$ for some $H \in W$.

Let $\bar{G} = G \rtimes \mathrm{Aut}(F_p[[t]])$, the split extension of G by the group of field automorphisms $t \mapsto a_1 t + a_2 t^2 + \ldots + a_n t^n + \ldots$.

STEP 1. Use Pink's Theorem to show: G has only countably many $\bar{G}$-orbits of weakly maximal subgroups.

STEP 2. Conclude that it suffices to show that for each $H \in W$,

$$\mathrm{Prob}(x, y \in \text{ some } \bar{G}\text{-conjugate of } H) = 0.$$

STEP 3. Fix $H \in W$ and let $G_n \leq G$ be the congruence subgroup modulo t^n. Then it is easy to see that HG_n has at most p^n $\mathrm{Aut}(F_p[[t]])$-conjugates. Clearly, HG_n has at most $|G : HG_n|$ G-conjugates. It follows that HG_n has at most $p^n|G : HG_n|$ $\bar{G}$-conjugates. Now, the probability that x, y lie in some $\bar{G}$-conjugate of H is at most the probability that x, y lie in some $\bar{G}$-conjugate of HG_n. This probability is in turn bounded above by $p^n|G : HG_n| \cdot |G : HG_n|^{-2} = p^n|G : HG_n|^{-1}$.

STEP 4. Apply results on Hausdorff dimension from Lecture 3.

Since $H \leq G$ is non-open, it follows from Theorem 12 that $\mathrm{Dim}(H) \leq 1 - 1/(d+1)$. In view of Theorem 8 this shows that for any $\epsilon > 0$ and for all large n we have

$$|G : HG_n| \geq p^{(d-1)(1-\epsilon)n} \geq p^{(2-2\epsilon)n}.$$

It follows that

$$p^n|G : HG_n|^{-1} \to 0,$$

completing the argument proving Theorem 16.

Let me end this lecture with a wild speculation. Theorem 16 looks somewhat analogous to results from Lecture 1 on random generation of finite simple groups by two elements. We can therefore try to formulate a Dixon-type conjecture for simple algebraic groups instead of finite simple groups. More specifically, *let R be a valuation ring of a local field k, and G a simply connected semisimple algebraic group defined over k. Do two random elements of the group $G(R)$ generate an open subgroup with probability 1?*

5 Finite permutation groups: minimal degree, genus, base size

I started this lecture series with finite permutation groups, and then turned to finite groups of Lie type. I then discussed some infinite groups of Lie type over profinite rings. In the final lecture I would like to return to finite permutation groups.

5.1 Fixed points and minimal degrees

Let $G \leq S_n$ be a primitive permutation group, and let $x \in G$. Denote by $\mathrm{fix}(x)$ the number of fixed points of x, and let $\mathrm{rfix}(x) = \mathrm{fix}(x)/n$ be the corresponding fixed point ratio. The *fixity* of the group G is defined by $\text{fix}(G) = \max\{\text{fix}(x) : 1 \neq x \in G\}$, and the *fixity ratio* of G is given by $\text{rfix}(G) = \max\{\text{rfix}(x) : 1 \neq x \in G\} = \text{fix}(G)/n$. The *minimal degree* of G is defined by $m(G) = n - \text{fix}(G)$.

The fixity and the minimal degree of primitive permutation groups have been the subject of intensive research since the days of Jordan, with numerous applications to other subjects. A central problem was to find upper bounds on $\text{fix}(G)$, assuming $G \neq A_n, S_n$. Jordan proved that, if $\epsilon > 0$ and $n > n(\epsilon)$, then $\text{fix}(G) \leq n - n^{1/2-\epsilon}$. In 1981 Babai showed that $\text{fix}(G) \leq n - (\sqrt{n}-1)/2$ [Ba1]. This bound is best possible, but if one allows further exceptions (in addition to A_n and S_n) then better bounds have since been obtained. Indeed, using the Classification Theorem Liebeck and Saxl proved in 1991 that $\text{fix}(G) \leq 2n/3$ with known exceptions [LiSa2]. The core of their argument is the case where the primitive group G is almost simple of Lie type. So suppose

$$G = X_l(q),$$

a finite group of Lie type of rank l over F_q. Then it is shown in [LiSa2] that $\text{rfix}(G) \leq 4/3q$ with known exceptions. This yields good bounds for large q. However, the case where q is small and l large seems to require further analysis. In a yet unpublished work [LiSh5] with Liebeck we show that, by allowing a few more exceptions, significantly better bounds on $\text{fix}(G)$ can be obtained. For $x \in G$ denote by $|x^G|$ the size of the conjugacy class of x in G.

Theorem 17 *There exists $\epsilon > 0$ such that whenever $G = X_l(q)$ acts primitively on a set Ω and $x \in G$ has prime order, then*
(i) $\text{rfix}(x) \leq |x^G|^{-\epsilon}$, *and*
(ii) $\text{rfix}(G) \leq q^{-\epsilon l}$,
with known exceptions.

The exceptions in the theorem (which happen to be genuine) are subspace actions of classical groups of large rank. For recent results on the fixity of such actions, see Guralnick and Kantor [GK]. Note that part (ii) of the theorem (with a possibly smaller ϵ) follows from part (i), since $|x^G| \geq q^{l/2}$ for all $x \neq 1$. The proof of part (i) is too long and technical to be sketched here. The rest of this lecture is devoted to various applications of Theorem 17.

5.2 The genus conjecture

We start with a few definitions. Suppose $\sigma \in S_n$ is a permutation with k cycles. We define the *index* $\text{ind}(\sigma)$ of σ by $n - k$. A transitive subgroup $G \leq S_n$ is said to have *genus* $\leq g$ if there exist $x_1, \ldots, x_r \in G$ generating G with $x_1 \cdot \ldots \cdot x_r = 1$, and $\sum_{i=1}^r \text{ind}(x_i) = 2n - 2g + 2$. In this case G is the monodromy group of a cover $X \to Y$, where X is compact connected Riemann surface of genus g, and Y is the Riemann sphere.

Conjecture 7 (Guralnick-Thompson, 1990) Only finitely many groups of Lie type occur as composition factors of groups of genus $\leq g$.

This conjecture from [GT] has attracted considerable attention in the past few years. Partial positive results have been obtained by many authors. See for instance [GT], [A2], [Gu], [Shi], [LiP].

Guralnick reduced the conjecture to the case of primitive almost simple groups of Lie type [Gu]. Liebeck and Saxl bounded the field size of such a group in terms of the genus [LiSa2, Cor. 2], and thus reduced the problem to the case of almost simple classical groups of large rank.

Let G be such a group, acting on a set Ω of size n. It is known that only finitely many pairs (G, Ω) with $\mathrm{rfix}(G) \le 1/86$ have genus $\le g$. So it suffices to consider pairs (G, Ω) with $\mathrm{rfix}(G) > 1/86$.

Now, by Theorem 17, $\mathrm{rfix}(G, \Omega) \le q^{-cl}$ in non subspace actions, and this bound is less than $1/86$ except for finitely many groups G. We can therefore deduce the following.

Corollary 18 *In order to complete the proof of the genus conjecture it suffices to consider subspace actions of classical groups.*

Recall that, if G is a classical group in some primitive permutation representation, and $M < G$ is a point-stabilizer, then by Aschbacher's Theorem

$$M \in \mathcal{C}_1 \cup \mathcal{C}_2 \cup \mathcal{C}_3 \cup \mathcal{C}_4 \cup \mathcal{C}_5 \cup \mathcal{C}_6 \cup \mathcal{C}_7 \cup \mathcal{C}_8 \cup \mathcal{S},$$

where $\mathcal{C}_i$ are the Aschbacher classes (see Lecture 1). Theorem 17 settles all cases except $M \in \mathcal{C}_1$. Some progress, also in the case $M \in \mathcal{C}_1$, has recently been made by Frohardt and Magaard, and it is hoped that the full conjecture will relent in the forseeable future.

5.3 Base conjectures

So far probabilistic methods were not mentioned in this lecture. I shall now show how Theorem 17, combined with probabilistic arguments, can be used to settle some conjectures regarding bases for permutation groups.

Definition Let G be a permutation group on a set Ω of size n. A *base* for G is a subset $Y \subseteq \Omega$ whose pointwise stabilizer $G_{(Y)}$ is trivial. The *base size* $b(G)$ of G is the minimal size of a base for G.

There are several motivations for studying bases. One stems from computational group theory. In permutation group algorithms devised by Sims (see [S]) and others, bases play a key role, and small bases can speed up the performance of various algorithms. A more classical motivation is the on-going study of orders of primitive permutation groups. Since a permutation $g \in G$ is determined by its action on any base for G, we clearly have

$$|G| \le n(n-1)\cdots(n-b(G)+1) \le n^{b(G)}.$$

Thus permutation groups with small (e.g. bounded) base have small (e.g. polynomial) order. Classical bounds on the orders of primitive permutation groups (from Bochert [Boc] to Babai [Bal]) were obtained using this formula, namely by constructing small bases for the respective groups.

In this section we will focus on the reverse implication. Is it true that primitive permutation groups of small order have small bases? Several conjectures suggest that this is the case. We start with a conjecture from [Cal] (see also [CK] and [Ca2]).

Conjecture 8 (Cameron, 1990) Let G be an almost simple primitive permutation group. Then $b(G) \leq c$ with known exceptions.

Here c denotes an absolute constant (not depending on G). The exceptions are essentially that of A_m, S_m acting on subsets or partitions, and of subspace actions of classical groups. Conjecture 8 has just been settled in [LiSh5].

Theorem 19 *Cameron's conjecture holds.*

Our approach is probabilistic: we show that for some absolute constant c, almost all c-tuples from the permutation domain Ω form a base for G. This probabilistic version of the conjecture is posed in the paper [CK] by Cameron and Kantor, where the cases $G = A_m, S_m$ are settled.

It remains to deal with groups G of Lie type. The main steps of the argument are sketched below.

STEP 1. Let $B(G, k)$ denote the probability that k randomly chosen letters from Ω form a base (we note that the letters are chosen independently, so there may be repetitions).

Fix $x \in G$ and choose $\omega \in \Omega$ at random. Then the probability that x fixes ω is $\mathrm{fix}(x)/n = \mathrm{rfix}(x)$. Therefore, if $\omega_1, \ldots, \omega_c \in \Omega$ are chosen at random, we have

$$\mathrm{Prob}(x \text{ fixes } \omega_1, \ldots, \omega_c) = \mathrm{rfix}(x)^c.$$

Let P denote the set of elements of prime order in G. If $\omega_1, \ldots, \omega_c$ are not a base for G, then there exists $x \in P$ fixing $\omega_1, \ldots, \omega_c$. This yields

$$1 - B(G, c) \leq \sum_{x \in P} \mathrm{rfix}(x)^c.$$

STEP 2. Since subspace actions of classical groups are among the exceptions in Cameron's conjecture, we can apply Theorem 17, and deduce that $\mathrm{rfix}(x) \leq |x^G|^{-\epsilon}$. Hence

$$1 - B(G, c) \leq \sum_{x \in P} |x^G|^{-\epsilon c}.$$

Let $c \geq 11\epsilon^{-1}$ be an integer, and let $x_1, \ldots, x_r$ be a set of representatives of the conjugacy classes in G of the elements of P. Then

$$1 - B(G, c) \leq \sum_{x \in P} |x^G|^{-11} = \sum_{i=1}^{r} |x_i^G|^{-10}.$$

STEP 3. Use estimates on the number and the sizes of conjugacy classes in G to conclude:

$$\sum_{i=1}^{r} |x_i^G|^{-10} \to 0 \quad \text{as } |G| \to \infty.$$

Therefore $B(G,c) \to 1$, completing the proof.

Can we have a constructive proof of Cameron's conjecture? Recently Carmit Benbenisty obtained some partial results, using explicit base constructions [Be]. It is still unclear whether the constructive approach could work in general.

Next, we turn to groups which are not necessarily almost simple. Let Γ_d denote the set of finite groups not involving the alternating group A_d as a section. The following result from [BCP] is well known, and has many applications.

Theorem (Babai-Cameron-Pálfy, 1982) *Let $G \leq S_n$ be a primitive subgroup and suppose $G \in \Gamma_d$. Then $|G| \leq n^{f(d)}$, for some function f of d alone.*

The original proof yields $f(d) = O(d \log d)$. A new (yet unpublished) proof by Pyber yields $f(d) = O(d)$.

Can we find a structural explanation for the Babai-Cameron-Pálfy Theorem? Babai suggested one, namely the following.

Conjecture 9 (Babai, 1982) If $G \leq S_n$ is a primitive group in Γ_d, then $b(G) \leq f(d)$.

See Pyber's excellent survey [Py]. First positive evidence was provided in 1996 by Seress, who showed that $b(G) \leq 4$ for G soluble [Se]. Then, in the joint work [GSSh] with Gluck and Seress, we show the following.

Theorem 20 *Babai's conjecture holds.*

The original proof in [GSSh] yields $f(d) = O(d^2)$. A modified proof from [LiSh5] which is based on Theorem 17 yields $f(d) = O(d)$. This provides a structural explanation for the Babai-Cameron-Pálfy Theorem with a best possible bound.

The proof of Babai's conjecture combines the O'Nan-Scott Theorem and bounds on fixed point ratios with some representation theory, and its core is still probabilistic.

Let me conclude this section with an even more ambitious conjecture.

Conjecture 10 (Pyber) There exists an absolute constant c such that, if $G \leq S_n$ is primitive, then $b(G) \leq c \log |G| / \log n$.

Note that we always have $b(G) \geq \log |G| / \log n$ (since $|G| \leq n^{b(G)}$). Hence the conjecture bounds the base size $b(G)$ in a very narrow cage. Pyber's conjecture is still very much open, and is the focus of some current activity.

5.4 Epilogue?

It is hopefully too early to write an epilogue while describing such a young research area as probabilistic group theory. Let me, however, say a few words on some new results and possible future directions. I should start with an apology. The material presented here is by no means exhaustive, and many interesting aspects were omitted due to lack of time and space. Fortunately, lecture series by Babai and Praeger during this conference discuss some of these aspects in detail.

We started this lecture series by briefly describing the Erdős-Turán statistical theory of S_n. Can one develop a similar theory for finite groups of Lie type? Various results in this direction were established in the past few years by Schmutz [Sc], Stong, and others. These results deal with the order of a typical matrix in $\mathrm{GL}_n(q)$ where $n \to \infty$, and related properties of the characteristic polynomial, etc. A natural next step would be to extend this to other classical groups. In this context, see J. Fulman's preprint [Fu]. Another natural goal is to try to apply such a statistical theory in the solution of some remaining open problems. For example, Schmutz' work [Sc] is the main tool in [Sh3], where a problem posed in [ŁP] is solved. It is shown in [Sh3] that, as $n \to \infty$, a random matrix in $\mathrm{GL}_n(q)$ does not lie in any proper irreducible subgroup (not containing $\mathrm{SL}_n(q)$). This can be regarded as a matrix analogue of a similar phenomenon for permutation groups (where irreducible is replaced by transitive) which was conjectured by Cameron and settled in [ŁP]. This result on random matrices has, in turn, some other applications. For example, it can be used in the study of invariable generation of some matrix groups, and random generation by conjugate elements.

Let me say a few words about the latter problem. G. Robinson asked whether random generation of simple groups by two elements still holds if we insist that the two generators are conjugate. In the recent preprint [GLSSh] with Guralnick, Liebeck and Saxl we show that, if G is a finite simple group and $x, y \in G$ are randomly chosen, then $\mathrm{Prob}(\langle x, x^y \rangle = G) \to 1$ as $|G| \to \infty$. Earlier partial results were obtained in [Sh4] and [LiSh5], and also by Pyber (in a slightly different interpretation of the problem). The proof of this theorem uses many of the results and methods described throughout this lecture series, including Theorem 4, Theorem 17, and new results from [LiSe2]. Incidentally, this result on random generation by two conjugate elements, and the result on random generation by two elements which opened this lecture series, were both obtained at Oberwolfach meetings. We are grateful to the Mathematisches Forschungsinstitut Oberwolfach for its inspiration.

5.5 More thanks

It is a great pleasure to thank Colin Campbell, Ed Robertson, Geoff Smith and Olga Tabachnikova for their devoted work which contributed so much to the success of this conference. I am just as pleased to thank Edna Wigderson for her crucial help in producing my latex coloured slides for these talks.

References

[Ab] J.L. Abercrombie, Subgroups and subrings of profinite rings, *Math. Proc. Cambridge Philos. Soc.* **116** (1994), 209-222.

[AS] N. Alon and J.H. Spencer, *The Probabilistic Method*, Wiley-Interscience, New York, 1992.

[A1] M. Aschbacher, On the maximal subgroups of the finite classical groups, *Invent. Math.* **76** (1984), 469-514.

[A2] M. Aschbacher, On conjectures of Guralnick and Thompson, *J. Algebra* **135** (1990), 277-343.

[Ba1] L. Babai, On the order of uniprimitive permutation groups, *Ann. Math. (2)* **113** (1981), 553-568.

[Ba2] L. Babai, The probability of generating the symmetric group, *J. Combin. Theory Ser. A* **52** (1989), 148-153.

[BCP] L. Babai, P.J. Cameron and P.P. Pálfy, On the orders of primitive groups with restricted nonabelian composition factors, *J. Algebra* **79** (1982), 161-168.

[BSh] Y. Barnea and A. Shalev, Hausdorff dimension, pro-*p* groups, and Kac-Moody algebras, *Trans. Amer. Math. Soc.* **349** (1997), 5073-5091.

[BShZ] Y. Barnea, A. Shalev and E.I. Zelmanov, Graded subalgebras of affine Kac-Moody algebras, *Israel J. Math.* **104** (1998), 321-334.

[Be] C. Benbenishty, *Small Bases for some Primitive Actions of $GL_n(q)$*, M.Sc. Thesis, The Hebrew University, Jerusalem, 1998.

[Bh] M. Bhattacharjee, The probability of generating certain profinite groups by two elements, *Israel J. Math.* **86** (1994), 311-320.

[Boc] A. Bochert, Ueber die Transitivitätsgrenze der Substitutionengruppen, welche die Alternierende ihres Grades nicht einhalten, *Math. Ann.* **33** (1889), 572-583.

[BPSh] A. Borovik, L. Pyber and A. Shalev, Maximal subgroups in finite and profinite groups, *Trans. Amer. Math. Soc.* **348** (1996), 3745-3761.

[Bo] J.D. Bovey, The probability that some power of a permutation has small degree, *Bull. London Math. Soc.* **12** (1980), 47-57.

[BW] J.D. Bovey and A. Williamson, The probability of generating the symmetric group, *Bull. London Math. Soc.* **10** (1978), 91-96.

[Br1] H.R. Brahana, Certain perfect groups generated by two operators of orders two and three, *Amer. J. Math.* **50** (1928), 345-356.

[Br2] H.R. Brahana, Pairs of generators of the known simple groups whose orders are less than one million, *Ann. of Math.* (2) **31** (1930), 529-549.

[Ca1] P.J. Cameron, Some open problems on permutation groups, in *Groups, Combinatorics and Geometry* (eds: M.W. Liebeck and J. Saxl), London Math. Soc. Lecture Note Series **165**, Cambridge University Press, Cambridge, 1992, 340-350.

[Ca2] P.J. Cameron, Permutation groups, in *Handbook of Combinatorics* (eds: R.L. Graham et al.), Elsevier Science B.V., Amsterdam, 1995, 611-645.

[CK] P.J. Cameron and W.M. Kantor, Random permutations: some group-theoretic aspects, *Combinatorics, Probability and Computing* **2** (1993), 257-262.

[DiMT] L. Di Martino and M.C. Tamburini, 2-generation of finite simple groups and related topics, in *Generators and Relations in Groups and Geometry* (eds: A. Barlotti et al.), Kluwer, 1991, 235-278.

[DiMV1] L. Di Martino and N. Vavilov, (2,3)-generation of $SL(n,q)$. I, cases $n = 5,6,7$, *Comm. Algebra* **22** (1994), 1321-1347.

[DiMV2] L. Di Martino and N. Vavilov, (2,3)-generation of $SL(n,q)$. II, cases $n \geq 8$, *Comm. Algebra* **24** (1996), 487-515.

[D] L.E. Dickson, *Linear Groups with an Exposition of The Galois Field Theory*, Teubner, 1901, and Dover, 1958.

[Di1] J.D. Dixon, The probability of generating the symmetric group, *Math. Z.* **110** (1969), 199-205.

[Di2] J.D. Dixon, Random sets which invariably generate the symmetric group, *Discrete Math.* **105** (1992), 25-39.

[DDMS] J. Dixon, M.P.F. Du Sautoy, A. Mann and D. Segal, *Analytic Pro-p Groups*, London Math. Soc. Lecture Note Series **157**, Cambridge University Press, Cambridge, 1991.

[DPSSh] J.D. Dixon, L. Pyber, Á. Seress and A. Shalev, Residual properties of free groups: a probabilistic approach, in preparation.

[D1] E.B. Dynkin, Semisimple subalgebras of semisimple Lie algebras, *Amer. Math. Soc. Transl.* (2) **6** (1957), 111-244.

[D2] E.B. Dynkin, Maximal subgroups of the classical groups, *Amer. Math. Soc. Transl.* (2) **6** (1957), 245-378.

[ET1] P. Erdős and P. Turán, On some problems of a statistical group theory. I, *Z. Wahrscheinlichkeitstheorie Verw. Gabiete* **4** (1965), 175-186.

[ET2] P. Erdős and P. Turán, On some problems of a statistical group theory. II, *Acta Math. Acad. Sci. Hungar.* **18** (1967), 151-163.

[ET3] P. Erdős and P. Turán, On some problems of a statistical group theory. III, *Acta Math. Acad. Sci. Hungar.* **18** (1967), 309-320.

[ET4] P. Erdős and P. Turán, On some problems of a statistical group theory. IV, *Acta Math. Acad. Sci. Hungar.* **19** (1968), 413-435.

[ET5] P. Erdős and P. Turán, On some problems of a statistical group theory. V, *Period. Math. Hungar.* **1** (1971), 5-13.

[ET6] P. Erdős and P. Turán, On some problems of a statistical group theory. VI, *J. Indian Math. Soc.* **34** (1970), 175-192.

[ET7] P. Erdős and P. Turán, On some problems of a statistical group theory. VII, *Period. Math. Hungar.* **2** (1972), 149-163.

[F] K.J. Falconer, *Fractal Geometry: mathematical foundations and applications*, John Wiley & Sons, New York, 1990.

[FJ] M.D. Fried and M. Jarden, *Field Arithmetic*, Springer, Berlin, 1986.

[Fu] J. Fulman, Cycle indices for the finite classical groups, Preprint, 1997.

[GSSh] D. Gluck, Á. Seress and A. Shalev, Bases for primitive permutation groups and a conjecture of Babai, *J. Algebra* **199** (1998), 367-378.

[GL] Yu.M. Gorchakov and V.M. Levchuk, On approximation of free groups, *Algebra i Logika* **9** (1970), 415-421.

[Gu] R.M. Guralnick, The genus of a permutation group, in *Groups, Combinatorics and Geometry* (eds: M.W. Liebeck and J. Saxl), London Math. Soc. Lecture Note Series **165** (1992), 351-363.

[GK] R.M. Guralnick and W.M. Kantor, Probabilistic generation of finite simple groups, to appear.

[GKS] R.M. Guralnick, W.M. Kantor and J. Saxl, The probability of generating a classical group, *Comm. Algebra* **22** (1994), 1395-1402.

[GLSSh] R.M. Guralnick, M.W. Liebeck, J. Saxl and A. Shalev, Random generation of finite simple groups, Preprint, 1998.

[GT] R.M. Guralnick and J.G. Thompson, Finite groups of genus zero, *J. Algebra* **131** (1990), 303-341.

[HR] H. Halberstam and H.-E. Richert, *Sieve Methods*, Academic Press, London, 1974.

[Hu] A. Hurwitz, Über algebraische Gebilde mit eindeutigen Transformationen in sich, *Math. Ann.* **41** (1893), 408-442.

[J1] G.A. Jones, Varieties and simple groups, *J. Austral. Math. Soc.* **17** (1974), 163-173.

[J2] G.A. Jones, Congruence and non-congruence subgroups of the modular group: a survey, *Groups '85 – St Andrews* (eds: E.F. Robertson and C.M. Campbell), London Math. Soc. Lecture Note Series **121**, Cambridge University Press, Cambridge, 1986, 223-234.

[Kac] V.G. Kac, *Infinite Dimensional Lie Algebras*, Cambridge University Press, Cambridge, 1990.

[K] W.M. Kantor, Some topics in asymptotic group theory, in *Groups, Combinatorics and Geometry* (eds: M.W. Liebeck and J. Saxl), London Math. Soc. Lecture Note Series **165**, Cambridge University Press, Cambridge, 1992, 403-421.

[KL] W.M. Kantor and A. Lubotzky, The probability of generating a finite classical group, *Geom. Ded.* **36** (1990), 67-87.

[KM] R. Katz and W. Magnus, Residual properties of free groups, *Comm. Pure Appl. Math.* **22** (1969), 1-13.

[KLi] P.B. Kleidman and M.W. Liebeck, *The Subgroup Structure of the Finite Classical Groups*, London Math. Soc. Lecture Note Series **129**, Cambridge University Press, 1990.

[Li] M.W. Liebeck, On the orders of maximal subgroups of the finite classical groups, *Proc. London Math. Soc.* **50** (1985), 426-446.

[LiP] M.W. Liebeck and C.W. Purvis, On the genus of a finite classical group, *Bull. London Math. Soc.* **29** (1997), 159-164.

[LiSa1] M.W. Liebeck and J. Saxl, On the orders of maximal subgroups of the finite exceptional groups of Lie type, *Proc. London Math. Soc.* **55** (1987), 299-330.

[LiSa2] M.W. Liebeck and J. Saxl, Minimal degrees of primitive permutation groups, with an application to monodromy groups of covers of Riemann surfaces, *Proc. London Math. Soc.* (3) **63** (1991), 266-314.

[LST] M.W. Liebeck, J. Saxl and D. Testerman, Simple subgroups of large rank in groups of Lie type, *Proc. London Math. Soc.* **72** (1996), 425-457.

[LiSe1] M.W. Liebeck and G.M. Seitz, Maximal subgroups of exceptional groups of Lie type, finite and algebraic, *Geom. Ded.* **35** (1990), 353-387.

[LiSe2] M.W. Liebeck and G.M. Seitz, On the subgroup structure of exceptional groups of Lie type, to appear in *Trans. Amer. Math. Soc.*

[LiSh1] M.W. Liebeck and A. Shalev, The probability of generating a finite simple group, *Geom. Ded.* **56** (1995), 103-113.

[LiSh2] M.W. Liebeck and A. Shalev, Classical groups, probabilistic methods, and the (2,3)-generation problem, *Ann. Math.* **144** (1996), 77-125.

[LiSh3] M.W. Liebeck and A. Shalev, Simple groups, probabilistic methods, and a conjecture of Kantor and Lubotzky, *J. Algebra* **184** (1996), 31-57.

[LiSh4] M.W. Liebeck and A. Shalev, Maximal subgroups of symmetric groups, *J. Combin. Theory Ser. A* **75** (1996), 341-352.

[LiSh5] M.W. Liebeck and A. Shalev, Permutation groups, simple groups, and probability, submitted.

[LM] F. Lübeck and G. Malle, $(2,3)$-generation of exceptional groups, to appear in *J. London Math. Soc.*

[Lu1] A. Lubotzky, On a problem of Magnus, *Proc. Amer. Math. Soc.* **98** (1986), 583-585.

[Lu2] A. Lubotzky, Counting finite index subgroups, *Groups '93 – Galway/St Andrews* (eds: C.M. Campbell et al.), London Math. Soc. Lecture Note Series **212**, Cambridge University Press, Cambridge, 1995, 368-404.

[LMS] A. Lubotzky, A. Mann and D. Segal, Finitely generated groups of polynomial subgroup growth, *Israel. J. Math.* **82** (1993), 363-371.

[LuSh] A. Lubotzky and A. Shalev, On some Λ-analytic pro-p groups, *Israel J. Math.* **85** (1994), 307-337.

[ŁP] T. Łuczak and L. Pyber, On random generation of the symmetric group, *Combina-*

torics, Probability and Computing **2** (1993), 505-512.

[Mac] A.M. Macbeath, Generators of the linear fractional groups, *Proc. Sympos. Pure Math.* **12** (1967), 14-32.

[Ma] W. Magnus, *Non-Euclidean Groups and their Tesselations*, Academic Press, New York - London, 1974.

[Mal] G. Malle, Hurwitz groups and G_2, *Canad. Math. Bull.* **33** (1990), 349-357.

[MSW] G. Malle, J. Saxl and T. Weigel, Generation of classical groups, *Geom. Ded.* **49** (1994), 85-116.

[M] A. Mann, Positively finitely generated groups, *Forum Math.* **8** (1996), 429-459.

[MSh] A. Mann and A. Shalev, Simple groups, maximal subgroups, and probabilistic aspects of profinite groups, *Israel. J. Math.* **96** (1997), 449-468 (Amitsur memorial issue).

[Mi] G.A. Miller, On the groups generated by two operators, *Bull. Amer. Math. Soc.* **7** (1901), 424-426.

[Ne] E. Netto, *Substitutionentheorie und ihre Anwendungen auf die Algebra*, Teubner, Leipzig, 1882.

[Pa] W.H. Patton, *The Minimum Index for Subgroups in some Classical Groups: a generalization of a theorem of Galois*, Ph.D. Thesis, Univ. of Illinois at Chicago Circle, 1972.

[Py] L. Pyber, Asymptotic results for permutation groups, in *Groups and Computation* (eds: L. Finkelstein and W.M. Kantor), DIMACS Series on Discrete Math. and Theor. Computer Science **11**, AMS 1993, 197-219.

[R] A. Rényi, *Probability Theory*, North Holland, Amsterdam, 1970.

[SaSh] J. Saxl and A. Shalev, The fixity of permutation groups, *J. Algebra* **174** (1995), 1122-1140.

[Sc] E. Schmutz, The order of a typical matrix with entries in a finite field, *Israel J. Math.* **91** (1995), 349-371.

[SSh] D. Segal and A. Shalev, Profinite groups with polynomial subgroup growth, *J. London Math. Soc.* **55** (1997), 320-334 (Hartley memorial issue).

[Se] Á. Seress, The minimal base size of primitive solvable permutation groups, *J. London Math. Soc.* **53** (1996), 243-255.

[Sh1] A. Shalev, Some problems and results in the theory of pro-p groups, *Groups '93 - Galway/St Andrews* (eds: C.M. Campbell et al.), London Math. Soc. Lecture Note Series **212**, Cambridge University Press, Cambridge, 1995, 528-542.

[Sh2] A. Shalev, Finite p-groups, in *Finite and Locally Finite Groups* (eds: B. Hartley et al.), NATO ASI Series, Kluwer, 1995, 401-450.

[Sh3] A. Shalev, A theorem on random matrices and some applications, *J. Algebra* **199** (1998), 124-141.

[Sh4] A. Shalev, Random generation of simple groups by two conjugate elements, *Bull. London Math. Soc.* **29** (1997), 571-576.

[Sh5] A. Shalev, Subgroup structure, Fractal dimension, and Kac-Moody algebras, to appear in *Groups and Geometries*, Proc. of the 1996 Siena Conference, Birkhäuser, 1998.

[Shi] T. Shih, *Bounds of Fixed Point Ratios of Permutation Representations of $GL_n(q)$ and Groups of Genus Zero*, Ph.D. Thesis, California Institute of Technology, Pasadena, 1990.

[S] C.C. Sims, Computation with permutation groups, in *Proc. Second Symp. on Symbolic and Algebraic Manipulation* (ed: S.R. Petrick), ACM, 1971.

[Si1] A. Sinkov, Necessary and sufficient conditions for generating certain simple groups by two operators of periods two and three, *Amer. J. Math.* **59** (1937), 67-76.

[Si2] A. Sinkov, On the group-defining relations $(2,3,7;p)$, *Ann. of. Math.* **38** (1937),

577-584.

[Si3] A. Sinkov, On generating the simple group $LF(2, 2^n)$ by two operators of periods two and three, *Bull. Amer. Math. Soc.* **44** (1938), 449-455.

[T] M.C. Tamburini, Generation of certain simple groups by elements of small order, *Rend. Istit. Lombardo Sci. (A)* **121** (1987), 21-27.

[TW] M.C. Tamburini, J.S. Wilson, On the (2,3)-generation of some classical groups. II, *J. Algebra* **176** (1995), 667-680.

[TWG] M.C. Tamburini, J.S. Wilson and N. Gavioli, On the (2,3)-generation of some classical groups. I, *J. Algebra* **168** (1994), 353-370.

[Wa] G.E. Wall, Some applications of the Eulerian function of a finite group, *J. Austral. Math. Soc.* **2** (1961), 35-59.

[We1] T. Weigel, Residual properties of free groups, *J. Algebra* **160** (1993), 16-41.

[We2] T. Weigel, Residual properties of free groups, II, *Comm. Algebra* **20** (1992), 1395-1425.

[We3] T. Weigel, Residual properties of free groups, III, *Israel J. Math.* **77** (1992), 65-81.

[Wi] J. Wiegold, Free groups are residually alternating of even degree, *Arch. Math. (Basel)* **28** (1977), 337-339.

[W1] J.S. Wilson, A residual property of free groups, *J. Algebra* **138** (1991), 36-47.

[W2] J.S. Wilson, Economical generating sets for finite simple groups, in *Groups of Lie type and Their Geometries* (eds: W.M. Kantor and L. Di Martino), London Math. Soc. Lecture Note Series **207**, Cambridge University Press, Cambridge, 1995, 289-302.

[Z] E.I. Zelmanov, On periodic compact groups, *Israel J. Math.* **77** (1992), 83-95.

COMBINATORIAL METHODS: FROM GROUPS TO POLYNOMIAL ALGEBRAS

VLADIMIR SHPILRAIN

Department of Mathematics, The City College of New York, New York, NY 10031, U.S.A.

Abstract

Combinatorial methods (or methods of elementary transformations) came to group theory from low-dimensional topology in the beginning of the century. Soon after that, combinatorial group theory became an independent area with its own powerful techniques. On the other hand, combinatorial commutative algebra emerged in the sixties, after Buchberger introduced what is now known as Gröbner bases. The purpose of this survey is to show how ideas from one of those areas contribute to the other.

1 Introduction

Let $F = F_n$ be the free group of a finite rank $n \geq 2$ with a set $X = \{x_1, ..., x_n\}$ of free generators. Let $Y = \{y_1, ..., y_m\}$ and $\widetilde{Y} = \{\widetilde{y}_1, ..., \widetilde{y}_m\}$ be arbitrary finite sets of elements of the group F. Consider the following elementary transformations that can be applied to Y:

(N1) y_i is replaced by one of the following elements: $y_i y_j^{\pm 1}$, $y_j^{\pm 1} y_i$, $y_j y_i y_j^{-1}$, or $y_j^{-1} y_i y_j$ for some $j \neq i$;

(N2) y_i is replaced by y_i^{-1}.

(N3) y_i and y_j are interchanged, $j \neq i$.

In (N1), (N2), it is understood that y_j does not change if $j \neq i$.

One might notice that some of these transformations are redundant, i.e., are compositions of other ones. There is a reason behind that which we are going to explain a little later.

We say that two sets Y and $\widetilde{Y}$ are Nielsen equivalent if one of them can be obtained from another by applying a sequence of transformations (N1)–(N3). It was proved by Nielsen that two sets Y and $\widetilde{Y}$ generate the same subgroup of the group F if and only if they are Nielsen equivalent. This result is now one of the central points in combinatorial group theory.

Note however that this result alone does not give an *algorithm* for deciding whether or not Y and $\widetilde{Y}$ generate the same subgroup of F. To obtain an algorithm, we need to somehow define the *complexity* of a given set of elements, and then to show that a sequence of Nielsen transformations (N1)–(N3) can be arranged so that this complexity decreases (or, at least, does not increase) *at every step* (this is where we may need "redundant" elementary transformations !).

This was also done by Nielsen; the complexity of a given set $Y = \{y_1, ..., y_m\}$ is just the sum of the lengths of the words $y_1, ..., y_m$. We refer to [17, Theorem 3.1] for details.

In particular, Nielsen's method therefore yields an algorithm for deciding whether or not a given endomorphism of a free group of finite rank is in fact an automorphism.

A somewhat more difficult problem is, given a pair of elements of a free group F, to find out if one of them can be taken to the other by an automorphism of F. We call this problem *the automorphic conjugacy problem.* It was addressed by Whitehead who came up with another kind of elementary transformations in a free group:

(W1) For some j, let $g = x_j$ or x_j^{-1}, and replace every x_i, $i \neq j$, by one of the elements $x_i g$, $g^{-1}x_i$, $g^{-1}x_i g$, or x_i.

(W2) x_i is replaced by x_i^{-1}.

(W3) x_i and x_j are interchanged, $j \neq i$.

One might notice a similarity of Nielsen and Whitehead transformations. However, they differ in one essential detail: Nielsen transformations are applied to arbitrary sets of elements, whereas Whitehead transformations are applied to a *basis* of the group F.

Using (informally) matrix language, we can say that Nielsen transformations correspond to elementary transformations of rows of a matrix (this correspondence can be actually made quite formal – see [21]), whereas Whitehead transformations correspond to conjugations (via changing the basis). This latter type of matrix transformation is known to be more complex, and the corresponding structural results are deeper.

There is very much the same relation between the Nielsen and Whitehead transformations in a free group.

Note also that the Whitehead transformation (W1) is somewhat more complex than its analog (N1). This is – again – in order to be able to arrange a sequence of elementary transformations so that the complexity of a given element (in this case, just the lexicografic length of a cyclically reduced word) would decrease (or, at least, not increase) *at every step* – see [16, Proposition I.4.17].

This arrangement still leaves us with a difficult problem - to find out if one of two elements *of the same complexity* (= of the same length) can be taken to the other by an automorphism of F. This is actually the most difficult part of Whitehead's algorithm.

In one special case however this problem does not arise, namely, when one of the elements is *primitive*, i.e., is an automorphic image of x_1. If we have managed to reduce an element of a free group (by Whitehead transformations) to an element of length 1, we immediately conclude that it is primitive; no further analysis is needed.

Thus, the problem of distinguishing primitive elements of a free group is a relatively easy case of the automorphic conjugacy problem. As we shall see in Section 2, this is also the situation in a polynomial algebra.

In Section 2, we review the results of various attempts to create something similar to Nielsen's and Whitehead's methods for a polynomial algebra in two variables. For a polynomial algebra in more than two variables, these problems are unapproachable so far, since we don't even know what the generators of the automorphism group of an algebra like that look like.

In Section 3, we talk about *retracts* of a polynomial algebra in two variables. Basic properties of retracts of a free group are given in [17]. Since then, retracts have not been getting much attention until very recently, when Turner [26] and, independently, Bergman [4] brought them back to life by employing them in various interesting research projects in combinatorial group theory. Here we show the relevance of polynomial retracts to several well-known problems about polynomial mappings, in particular, to the notorious Jacobian conjecture.

In the concluding Section 4, we have gathered some open combinatorial problems about polynomial mappings that are motivated by similar issues in combinatorial group theory.

2 Elementary transformations in polynomial algebras

Let $P_n = K[x_1, ..., x_n]$ be the polynomial algebra in n variables over a field K of characteristic 0. We are going to concentrate here mainly on the algebra P_2.

The first description of the group $Aut(P_2)$ was given by Jung [14] back in 1942, but it was limited to the case $K = \mathbb{C}$ since he was using methods of algebraic geometry. Later on, van der Kulk extended Jung's result to arbitrary ground fields. In the form we give it here, the result appears as Theorem 6.8.5 in P.M. Cohn's book [6]; this form is consistent with the idea of elementary transformations as described in the Introduction.

Theorem 2.1 ([6]) *Every automorphism of $K[x_1, x_2]$ is a product of linear automorphisms and automorphisms of the form $x_1 \to x_1 + f(x_2)$; $x_2 \to x_2$. More precisely, if $(x_1 \to g_1,\ x_2 \to g_2)$ is an automorphism of $K[x_1, x_2]$ such that $deg(g_1) \geq deg(g_2)$, say, then either $(x_1 \to g_1,\ x_2 \to g_2)$ is a linear automorphism, or there exists a unique $\mu \in K^*$ and a positive integer d such that $deg(g_1 - \mu g_2^d) < deg(g_1)$.*

The proof given in [6] is attributed to Makar-Limanov (unpublished), with simplifications by Dicks [8].

Note that the "More precisely, ..." statement serves algorithmic purposes: upon defining the complexity of a given pair of polynomials (g_1, g_2) as the sum $deg(g_1) + deg(g_2)$, we see that Theorem 2.1 allows one to arrange a sequence of elementary automorphisms (these are linear automorphisms and automorphisms of the form $x_1 \to x_1 + f(x_2)$; $x_2 \to x_2$) so that this complexity decreases at every step, until we either get a pair of polynomials that represents a linear automorphism, or conclude that $(x_1 \to g_1,\ x_2 \to g_2)$ was not an automorphism of $K[x_1, x_2]$. The parallel with Nielsen's method described in the Introduction is obvious.

We also mention here another proof of this result (in case $char\ K = 0$) due to Abhyankar and Moh [1]. In fact, their method is even more similar to Nielsen's method in a free group. Many of their results are based on the following fundamental theorem which we give here only in the characteristic 0 case (it will also play an essential role in our Section 3):

Theorem 2.2 **([1])** *Let* $u(t), v(t) \in K[t]$ *be two one-variable polynomials of degree* $n \geq 1$ *and* $m \geq 1$. *Suppose* $K[t] = K[u, v]$. *Then either* n *divides* m, *or* m *divides* n.

Thus, for pairs of one-variable polynomials that generate the whole algebra $K[t]$, there is a reduction process (similar to Nielsen's process) that brings any of those pairs to the canonical form $(t, 0)$.

Now let's see how one can adopt a more sophisticated Whitehead's method in a polynomial algebra situation. It appears that elementary basis transformations (see Theorem 2.1), when applied to a polynomial $p(x_1, x_2)$, are mimicked by Gröbner transformations of a basis of the ideal of P_2 generated by partial derivatives of this polynomial. To be more specific, we have to give some background material first.

In the course of constructing a Gröbner basis of a given ideal of P_n, one uses "reductions", i.e., transformations of the following type (see [2], p.39-43): given a pair (p, q) of polynomials, set $S(p,q) = \frac{L}{l.t.(p)} \cdot p - \frac{L}{l.t.(q)} \cdot q$, where $l.t.(p)$ is the *leading term* of p, i.e., the *leading monomial* together with its coefficient; $L = l.c.m.(l.m.(p), l.m.(q))$ (here, as usual, $l.c.m.$ means the least common multiple, and $l.m.(p)$ denotes the leading monomial of p). In this survey, we'll always consider what is called "deglex ordering" in [2] – where monomials are ordered first by total degree, then lexicographically with $x_1 > x_2 > ... > x_n$.

Now a crucial observation is as follows. These Gröbner reductions appear to be of two essentially different types:

(i) *regular*, or *elementary*, transformations. These are of the form $S(p, q) = \alpha \cdot p - r \cdot q$ or $S(p, q) = \alpha \cdot q - r \cdot p$ for some monomial r and scalar $\alpha \in K^*$. This happens when the leading monomial of p is divisible by the leading monomial of q (i.e., $l.m.(p) = r \cdot l.m.(q)$) or vice versa. After performing this kind of transformation, we replace p with $S(p, q)$, i.e., this transformation *does not produce an extra polynomial*, and, in fact, it is *invertible*. That is why we call it *regular*.

The reason why we call these transformations *elementary* is that they can be written in the form $(p,\ q) \to (\alpha_1 p,\ \alpha_2 q) \cdot M$, where M is an *elementary matrix*, i.e., a matrix which (possibly) differs from the identity matrix by a single element outside the diagonal. In case where we have more than 2 polynomials $(p_1, ..., p_k)$, we also can write $(p_1, ..., p_k) \to (\alpha_1 p_1, ..., \alpha_k p_k) \cdot M$, where M is a $k \times k$ elementary matrix; elementary reduction here is actually applied to a pair of polynomials (as usual) while the other ones are kept fixed. Sometimes, it is more convenient for us to get rid of the coefficients α_i and write $(p_1, ..., p_k) \to (p_1, ..., p_k) \cdot M$, where M belongs to the group $GE_k(P_n)$ generated by all elementary **and** diagonal matrices from $GL_k(P_n)$. It is known [25] that $GE_k(P_n) = GL_k(P_n)$ if $k \geq 3$, and $GE_2(P_n) \neq GL_2(P_n)$ if $n \geq 2$ – see [5].

(ii) *singular* transformations – these are non-regular ones. More specifically, these transformations produce an *extra* polynomial (a "baby" polynomial) as follows: $(p,\ q) \to S(p, q) = \alpha \cdot r_1 \cdot q - r_2 \cdot p$, where the monomials r_1, r_2 and the coefficient $\alpha \in K^*$ are chosen so that the leading terms of $\alpha \cdot r_1 \cdot q$ and $r_2 \cdot p$ cancel out.

Denote by $I_{d(p)}$ the ideal of P_2 generated by the partial derivatives of p. As usual, by the *rank* of an ideal we mean the minimal number of generators. We say that a polynomial $p \in P_n$ has a *unimodular gradient* if $I_{d(p)} = P_n$ (in particular, the ideal $I_{d(p)}$ has rank 1 in this case). Note that if the ground field K is algebraically closed, then this is equivalent, by Hilbert's Nullstellensatz, to the gradient being nowhere-vanishing.

Furthermore, define the *outer rank* of a polynomial $p \in P_n$ to be the minimal number of generators x_i on which an automorphic image of p can depend. The outer rank was introduced in [20] for elements of a free group. Later, Umirbaev [27] established a remarkable duality by proving that the outer rank of a free group element is equal to the rank of the right ideal of the free group ring generated by partial (non-commutative) Fox derivatives of this element. For details on Fox derivatives we refer to [12].

In [18], similar result was proved for an element of a free Lie algebra.

These results might tempt one to assume that also in the case of a polynomial algebra P_n, the outer rank of a polynomial p equals the rank of the ideal of P_n generated by the partial derivatives of p. This however is not the case; the situation in a polynomial algebra appears to be more subtle:

Theorem 2.3 ([22]) *Let a polynomial $p \in P_2$ have a unimodular gradient. Then the outer rank of p equals* 1 *if and only if one can get from $(d_1(p), d_2(p))$ to $(1, 0)$ by using only elementary transformations. Or, in the matrix form: if and only if $(d_1(p),\ d_2(p)) \cdot M = (1,\ 0)$ for some matrix $M \in GE_2(P_2)$.*

By $d_1(p)$ and $d_2(p)$ we denote the partial derivatives of p with respect to x_1 and x_2.

The proof [22] of Theorem 2.3 is based on a generalization of Wright's Weak Jacobian Theorem [28].

Remark 2.4 Elementary transformations that reduce $(d_1(p),\ d_2(p))$ to $(1,\ 0)$, can be actually chosen to be Gröbner reductions, i.e., to decrease the maximum degree of monomials *at every step* – the proof [22] is based on a recent result of Park [19].

Now we show how one can apply this result to the study of so-called coordinate polynomials.

We call a polynomial $p \in P_n$ *coordinate* if it can be included in a generating set of cardinality n of the algebra P_n. Thus, coordinate polynomials correspond to *primitive elements* in a free group context.

It is clear that the outer rank of a coordinate polynomial equals 1 (the converse is not true!). It is easy to show that a coordinate polynomial has a unimodular gradient, and again – the converse is not true! On the other hand, we have:

Proposition 2.5 (**[22]**) *A polynomial $p \in P_n$ is coordinate if and only if it has outer rank 1* **and** *a unimodular gradient.*

Combining this proposition with Theorem 2.3 yields the following

Theorem 2.6 (**[22]**) *A polynomial $p \in P_2$ is coordinate if and only if one can get from $(d_1(p),\ d_2(p))$ to $(1,0)$ by using only elementary Gröbner reductions.*

This immediately yields an algorithm for detecting coordinate polynomials in P_2 (see [22]) which is similar to Whitehead's algorithm for detecting primitive elements in a free group. This algorithm is very simple and fast: it has quadratic growth with respect to the degree of a polynomial. In case p is revealed to be a coordinate polynomial, the algorithm also gives a polynomial which completes p to a basis of P_2.

In the case where $K = \mathbb{C}$, the field of complex numbers, an alternative, somewhat more complicated algorithm, has been recently reported in [9]. It is not known whether or not there is an algorithm for detecting coordinate polynomials in P_n if $n \geq 3$.

Theorems 2.3 and 2.6 also suggest the following conjecture which is relevant to an important problem known as "effective Hilbert's Nullstellensatz" (see [24]):

Conjecture "G" Let a polynomial $p \in P_2$ have a unimodular gradient. Then one can get from $(d_1(p),\ d_2(p))$ to $(1,\ 0)$ by using *at most one* singular Gröbner reduction.

Remark 2.7 For $n \geq 3$, Theorem 2.3 is no longer valid since in this case, by a result of Suslin [25], the group $GL_n(P_n) = GE_n(P_n)$ acts transitively on the set of all unimodular polynomial vectors of dimension n, yet there are polynomials with unimodular gradient, but of the outer rank 2, for example, $p = x_1 + x_1^2 x_2$. The "only if" part however is valid for an arbitrary $n \geq 2$ - see [22]. It is also easy to show that one always has $orank\ p \geq rank(I_{d(p)})$.

Finally, we mention that our method also yields an algorithm which, given a coordinate polynomial $p \in P_2$, finds a sequence of elementary automorphisms that reduces p to x_1.

3 Polynomial retracts

Let $K[x,y]$ be the polynomial algebra in two variables over a field K of characteristic 0. A subalgebra R of $K[x,y]$ is called a *retract* if it satisfies either of the following equivalent conditions:

(R1) There is an idempotent homomorphism (a *retraction*, or *projection*) $\varphi : K[x,y] \to K[x,y]$ such that $\varphi(K[x,y]) = R$.

(R2) There is a homomorphism $\varphi : K[x,y] \to R$ that fixes every element of R.

(R3) $K[x,y] = R \oplus I$ for some ideal I of the algebra $K[x,y]$.

(R4) $K[x, y]$ is a projective extension of R in the category of K-algebras. In other words, there is a splitting exact sequence $1 \to I \to K[x, y] \to R \to 1$, where I is the same ideal as in (R3) above.

Examples K; $K[x, y]$; any subalgebra of the form $K[p]$, where $p \in K[x, y]$ is a *coordinate* polynomial (i.e., $K[p, q] = K[x, y]$ for some polynomial $q \in K[x, y]$). There are other, less obvious, examples of retracts: if $p = x + x^2 y$, then $K[p]$ is a retract of $K[x, y]$ (take y to 0, and x to p), but p is not coordinate since it is reducible.

The very presence of several equivalent definitions of retracts shows how natural these objects are.

In [7], Costa has proved that every proper retract of $K[x, y]$ (i.e., a one different from K and $K[x, y]$) has the form $K[p]$ for some polynomial $p \in K[x, y]$, i.e., is isomorphic to a polynomial K-algebra in one variable. A natural problem now is to characterize somehow those polynomials $p \in K[x, y]$ that generate a retract of $K[x, y]$. Since the image of a retract under any automorphism of $K[x, y]$ is again a retract, it would be reasonable to characterize retracts up to an automorphism of $K[x, y]$, i.e., up to a "change of coordinates". We give an answer to this problem in the following

Theorem 3.1 ([23]) *Let $K[p]$ be a retract of $K[x, y]$. There is an automorphism ψ of $K[x, y]$ that takes the polynomial p to $x+y \cdot q$ for some polynomial $q = q(x, y)$. A retraction for $K[\psi(p)]$ is given then by $x \to x + y \cdot q$; $y \to 0$.*

Geometrically, Theorem 3.1 says that (in case $K = \mathbb{R}$ or $\mathbb{C}$) every polynomial retraction of a plane is a "parallel" projection (sliding) on a fiber of a coordinate polynomial (which is isomorphic to a line) along the fibers of another polynomial (which generates a retract of $K[x, y]$).

The proof [23] of this result is based on the Abhyankar-Moh theorem (see Theorem 2.2).

Theorem 3.1 yields another characterization of retracts of $K[x, y]$:

Corollary 3.2 ([23]) *A polynomial $p \in K[x, y]$ generates a retract of $K[x, y]$ if and only if there is a polynomial mapping of $K[x, y]$ that takes p to x. The "if" part is actually valid for a polynomial algebra in arbitrarily many variables.*

Theorem 3.1 has several interesting applications, in particular, to the notorious

Jacobian conjecture If for a pair of polynomials $p, q \in K[x, y]$, the corresponding Jacobian matrix is invertible, then $K[p, q] = K[x, y]$.

This problem was introduced in [15], and is still unsettled. For a survey and background, the reader is referred to [3].

Now we establish a link between retracts of $K[x, y]$ and the Jacobian conjecture by means of the following

Conjecture "R" If for a pair of polynomials $p, q \in K[x,y]$, the corresponding Jacobian matrix is invertible, then $K[p]$ is a retract of $K[x,y]$.

This statement is formally much weaker than the Jacobian conjecture since, instead of asking for p to be a coordinate polynomial, we only ask for p to generate a retract, and this property is much less restrictive as can be seen from Theorem 3.1. However, the point is that these conjectures are actually equivalent:

Theorem 3.3 ([23]) *Conjecture "R" implies the Jacobian conjecture.*

Another application of retracts to the Jacobian conjecture (somewhat indirect though) is based on the "φ^∞-trick" familiar in combinatorial group theory (see [26]). For a polynomial mapping $\varphi : K[x,y] \to K[x,y]$, denote by $\varphi^\infty(K[x,y]) = \bigcap_{k=1}^{\infty} \varphi^k(K[x,y])$ the *stable image* of φ. Then we have:

Theorem 3.4 ([23]) *Let φ be a polynomial mapping of $K[x,y]$. If the Jacobian matrix of φ is invertible, then either φ is an automorphism, or $\varphi^\infty(K[x,y]) = K$.*

The proof [23] of Theorem 3.4 is based on a recent result of Formanek [11].

Obviously, if φ fixes a polynomial $p \in K[x,y]$, then $p \in \varphi^\infty(K[x,y])$. Therefore, we have:

Corollary 3.5 ([23]) *Suppose φ is a polynomial mapping of $K[x,y]$ with invertible Jacobian matrix. If $\varphi(p) = p$ for some non-constant polynomial $p \in K[x,y]$, then φ is an automorphism.*

This yields the following promising re-formulation of the Jacobian conjecture: if φ is a polynomial mapping of $K[x,y]$ with invertible Jacobian matrix, then for some automorphism α, the mapping $\alpha \cdot \varphi$ fixes a non-constant polynomial.

4 Some open problems

In this section, we have gathered a few combinatorial problems about polynomial mappings that are motivated by similar issues in combinatorial group theory. Two most important problems however – Conjectures "G" and "R" – appear earlier in the text (in Sections 2 and 3, respectively).

Throughout, $P_n = K[x_1, ..., x_n]$ is the polynomial algebra in n variables, $n \geq 2$, over a field K of characteristic 0.

(1) [10] Is it true that every endomorphism of P_n taking any coordinate polynomial to a coordinate one, is actually an automorphism? (It is true for $n = 2$ – see [10]). [*Added in proof:* Jelonek [13] has recently settled this problem in the affirmative for $K = \mathbb{C}$.]

(2) Is there a polynomial $p \in P_n$ with the following property: whenever $\varphi(p) = \psi(p)$ for some non-degenerate endomorphisms φ, ψ of P_n, it follows that $\varphi = \psi$? In other words, is it true that every non-degenerate endomorphism of P_n is completely determined by its value on just 1 polynomial? (We call an endomorphism φ non-degenerate if there is a non-constant polynomial in the image of φ.)

(3) Suppose $\varphi(p) = x_1$ for some *monomorphism* (i.e., injective endomorphism) φ of the algebra P_n. Is it true that p is a coordinate polynomial?

(4) Let $p \in P_n$ be a polynomial such that $K[p]$ is a retract of P_n. Is it true that $\varphi(p) = x_1$ for some endomorphism φ of the algebra P_n ? (It is true for $n = 2$ – see [23]).

(5) Is it true that for any endomorphism φ of the algebra P_n, its stable image $\varphi^\infty(P_n)$ is a retract of P_n ? (The answer to this question might depend on properties of the ground field K).

References

[1] S. S. Abhyankar, T.-T. Moh, *Embeddings of the line in the plane*, J. Reine Angew. Math. **276** (1975), 148–166.

[2] W. Adams and P. Loustaunau, *An introduction to Gröbner bases*, Graduate Studies in Mathematics **3**, American Mathematical Society, 1994.

[3] H. Bass, E. Connell and D. Wright, *The Jacobian conjecture: reduction of degree and formal expansion of the inverse*, Bull. Amer. Math. Soc. **7** (1982), 287–330.

[4] G. Bergman, *Supports of derivations, free factorizations, and ranks of fixed subgroups in free groups*, Trans. Amer. Math. Soc., to appear.

[5] P. M. Cohn *On the structure of the GL_2 of a ring*, Inst. Hautes Études Sci. Publ. Math. **30** (1966), 365–413.

[6] P. M. Cohn, *Free rings and their relations*, Second edition, Academic Press, London, 1985.

[7] D. Costa, *Retracts of polynomial rings*, J. Algebra **44** (1977), 492–502.

[8] W. Dicks, *Automorphisms of the polynomial ring in two variables*, Publ. Sec. Mat. Univ. Autonoma Barcelona **27** (1983), 155–162.

[9] A. van den Essen, *Locally nilpotent derivations and their applications, III*, J. Pure Appl. Algebra **98** (1995), 15–23.

[10] A. van den Essen and V. Shpilrain, *Some combinatorial questions about polynomial mappings*, J. Pure Appl. Algebra **119** (1997), 47-52.

[11] E. Formanek, *Observations about the Jacobian conjecture*, Houston J. Math. **20** (1994), 369–380.

[12] N. Gupta, *Free group rings*, Contemp. Math. **66** (1987).

[13] Z. Jelonek, *A solution of the problem of van den Essen and Shpilrain*, Preprint IMUJ 1996/19, Krakow.

[14] H. W. E. Jung, *Über ganze birationale Transformationen der Ebene*, J. Reine Angew. Math. **184** (1942), 161–174.

[15] O. Keller, *Ganze Cremona-Transformationen*, Monatsh. Math. Phys. **47** (1939), 299–306.

[16] R. Lyndon, P. Shupp, *Combinatorial Group Theory*, Series of Modern Studies in Math. **89**. Springer-Verlag, 1977.

[17] W. Magnus, A. Karrass, and D. Solitar, *Combinatorial Group Theory*, Wiley, New York, 1966.

[18] A. A. Mikhalev and A. A. Zolotykh, *The rank of an element of the free color Lie (p-)superalgebra*, Russian Acad. Sci. Dokl. Math. **49** (1994), 189–193.

[19] H. Park, *A Computational Theory of Laurent Polynomial Rings and Multidimensional FIR Systems*, PhD thesis, University of California at Berkeley, 1995.

[20] V. Shpilrain, *Recognizing automorphisms of the free groups*, Arch. Math. **62** (1994), 385–392.

[21] V. Shpilrain, *On monomorphisms of free groups*, Arch. Math. **64** (1995), 465–470.

[22] V. Shpilrain and J.-T. Yu, *Polynomial automorphisms and Gröbner reductions*, J. Algebra **197** (1997), 546–558.

[23] V. Shpilrain and J.-T. Yu, *Polynomial retracts and the Jacobian conjecture*, Trans. Amer. Math. Soc., to appear.

[24] M. Shub, S. Smale, *On the intractability of Hilbert's Nullstellensatz and an algebraic version of "$NP \neq P$?"*, Duke Math. J. **81** (1996), 47–54.

[25] A. A. Suslin, *On the structure of the special linear group over polynomial rings*, Math. USSR Izv. **11** (1977), 221–238.

[26] E. C. Turner, *Test words for automorphisms of free groups*, Bull. London Math. Soc. **28** (1996), 255–263.

[27] U. U. Umirbaev, *On ranks of elements of free groups*, Fundamentalnaya i Prikladnaya Matematika **2** (1996), 313–315. (Russian)

[28] D. Wright, *The amalgamated free product structure of $GL_2(k[X_1, ..., X_n])$ and the weak Jacobian theorem for two variables*, J. Pure Appl. Algebra **12** (1978), 235–251.

FORMAL LANGUAGES AND THE WORD PROBLEM FOR GROUPS

IAIN A. STEWART and RICHARD M. THOMAS

Department of Mathematics and Computer Science, University of Leicester, Leicester LE1 7RH, England

1 Introduction

The aim of this article is to survey some connections between formal language theory and group theory with particular emphasis on the word problem for groups and the consequence on the algebraic structure of a group of its word problem belonging to a certain class of formal languages. We define our terms in Section 2 and then consider the structure of groups whose word problem is regular or context-free in Section 3. In Section 4 we look at groups whose word-problem is a one-counter language, and we move up the Chomsky hierarchy to briefly consider what happens above context-free in Section 5. In Section 6, we see what happens if we consider languages lying in certain complexity classes. For general background material on group theory we refer the reader to [25, 26], and for formal language theory to [7, 16, 20].

2 Word problems and decidability

In this section we set up the basic notation we shall be using and introduce the notions of "word problems" and "decidability".

In order to consider the word problem of a group as a formal language, we need to introduce some terminology. Throughout, if Σ is a finite set (normally referred to in this context as an *alphabet*) then we let Σ^* denote the set of all finite words of symbols from Σ, including the *empty word* ϵ, and Σ^+ denote the set of all non-empty finite words of symbols from Σ. To put this another way, Σ^* is the free monoid generated by Σ, and Σ^+ is the free semigroup generated by Σ. A subset of Σ^* is known as a *language*.

One of the central notions in combinatorial group theory is that of the *word problem* of a group G. Given a (finite) generating set X for G, every word over the alphabet $\Sigma = X \cup X^{-1}$ represents a unique element of G, and we have a natural homomorphism from the free monoid Σ^* onto G. The word problem $W_X(G)$ of G (with respect to the generating set X) is defined to be the set of all words of Σ^* that represent the identity element of G. This is a slightly different approach to the more usual one which is to think of the word problem as being the question as to whether or not a word α represents the identity: by defining the word problem to be a set of words, we transfer the question to deciding membership of this set, and this approach fits more naturally in the context of formal language theory.

A classic question was that of the *decidability* of the word problem: given a finitely presented group G, is there an algorithm that decides whether or not a given word $\alpha \in \Sigma^*$ belongs to $W_X(G)$? In order to make sense of this, we should briefly explain what we mean by "decidable".

There are many (equivalent) models of computation, but we shall take the Turing machine here since it is in the spirit of the other models we will be considering. Further, there are many versions of a Turing machine, and we choose one that will suit our needs. Our Turing machines have a read-only input tape which, for an input of length n, consists of $n+2$ tape cells, the first of which contains the "left-end-of-tape" marker $\triangleright$, the last of which contains the "right-end-of-tape" marker $\triangleleft$ and such that the cells between hold the input word. They also have a fixed number of work tapes on which we perform the computation. The work tapes initially contain $\triangleright$ in their leftmost cells, are blank everywhere else and are unbounded in length to the right. We may move freely over these work tapes and read from and write to the cells as we wish. Initially, the work heads are positioned over the leftmost cells of the work tapes (i.e., those initially holding the symbol $\triangleright$) and the input head is positioned over the cell adjacent to the leftmost cell of the input tape (i.e., over the cell holding the first symbol of the input string, if the input string is non-empty, or over the cell holding the symbol $\triangleright$ otherwise). More formally, a *deterministic Turing machine* (DTM) M is a tuple $(Q, \Sigma, \Gamma, \delta, s, h)$ where Q is a finite set of "states", Σ is a finite set of input symbols, and Γ is a finite set of work tape symbols, including $\triangleright$, $\triangleleft$ and Δ (where Δ is the blank symbol). In addition, s and h are designated states (the "start state" and the "halt state") and δ is a partial function

$$Q \times \Sigma \times \Gamma^m \to Q \times \{L, R, N\} \times (\Gamma \times \{L, R, N\})^m$$

where m is the number of work tapes.

If $\delta(q, x, g_1, \ldots, g_m) = (r, d, g'_1, d_1, \ldots, g'_m, d_m)$, we imagine that, when M is in state q reading a symbol x on the input tape and $g_1, \ldots, g_m$ on the respective work tapes, then: M erases $g_1, \ldots, g_m$ and writes $g'_1, \ldots, g'_m$ in their place; changes state to r; moves the input head in the direction indicated by d (left if $d = L$, right if $d = R$ and not at all if $d = N$); and moves the heads on the work tapes in the directions indicated by $d_1, \ldots, d_m$. If ever M attempts to move left on any tape when reading the leftmost cell then M crashes, as it does if ever it attempts to move the input head right when reading the rightmost cell of the input tape; there is no transition defined from the halt state. If M is set up in the start state s with input word α then α is said to be *accepted* if M reaches the halt state h and *rejected* otherwise (i.e., if the machine either hangs, crashes or else runs indefinitely without entering the halt state). The *language* $L(M)$ of M is the set of all words accepted by M: we say that a language L is *semi-decidable* or *recursively enumerable* if $L = L(M)$ for some DTM M.

One way of looking at this is to imagine that the input word α represents a question to which we want an answer, either "yes" or "no", and that the DTM M accepts α if, and only if, the answer is "yes". However, with this notion, we may never know if the answer is "no" as a Turing machine may fail to halt on some input. We can modify our definition of a Turing machine so as to have two halt states h_y and h_n, and insist that such a machine halts if it enters either of these states and that, for any input α, it necessarily reaches one of these two states. We define the *yes-language* $Y(M)$ of such a Turing machine M to be the set of all input words such that M reaches h_y (and the *no-language* $N(M) = \Sigma^* - Y(M)$ of M to be the set of all input words such that M reaches h_n) and call such a machine a *decision-making DTM.* We say that a language L is *decidable* or *recursive* if $L = L(M)$ for some decision-making DTM M. Two standard results in computability theory state that a recursive language is necessarily recursively enumerable, but not conversely, and that given a (encoding of a) description of a Turing machine as input, there does not exist a decision-making Turing machine which answers "yes" if the input machine is a decision-making Turing machine and "no" otherwise.

One should comment here, in passing, that the computational power of our Turing machines is not increased if one introduces "non-determinism". We define a non-deterministic Turing machine (NTM) $M = (Q, \Sigma, \Gamma, \delta, s, h)$ in the same way as a DTM, except that, as opposed to a partial function $\delta : Q \times \Sigma \times \Gamma^m \to Q \times \Gamma \times \{L, R, N\} \times (\Gamma \times \{L, R, N\})^m$, we allow δ to be a subset of $Q \times \Sigma \times \Gamma^m \times Q \times \Gamma \times \{L, R, N\} \times (\Gamma \times \{L, R, N\})^m$. The idea here is that there may be choice at some instant as to which transition the machine performs, and we say that a word α is *accepted* if at least one possible computation path reaches the halt state h. This does not lead to an increase in power (in the sense of what can be computed), in that any language accepted by a NTM is also accepted by a DTM (and, similarly, any language which is the yes-language of a decision making NTM, where a word is accepted if some computation path leads to the halt state h_y, is the yes-language of a decision-making DTM).

It is not too difficult to see that in any finitely presented group, the word problem is semi-decidable: one systematically enumerates all words representing the identity and compares each in turn with the input word, stopping if and when we get a match. However, a more interesting question is that of whether or not finitely presented groups necessarily have a decidable word problem, or, to put this question another way, is the set $W_X(G)$ necessarily a decidable language for a finitely presented group G (with generator set X)?

The answer to this question was shown to be "no" by Novikov [30] and Boone [8] (independently) in the 1950's, and much subsequent research tended to focus on whether or not the word problem is decidable if we impose restrictions on our class of groups (e.g., to the class of automatic groups, where the word problem is decidable [13], or to the class of solvable groups of derived length three, where the word problem is undecidable [22]), and also the degrees of undecidability of the word problem for various groups or classes of groups, along with an analogous analysis of related problems like the conjugacy problem, the generalized word problem and the isomorphism problem (see [27] for an elegant survey of such results).

We will be looking at degrees of decidability, focusing on what happens when the word problem can be decided by a computing device less powerful than a Turing machine. We will be particularly interested in the consequences on the algebraic structure of such a group. As far as groups with a decidable word problem are concerned, we have the following fascinating result of Boone and Higman [9]:

Theorem 2.1 *Let G be a finitely generated group. Then the word problem of G is decidable if, and only if, there exists a simple group H and a finitely presented group K with $G \leq H \leq K$.*

This was generalized in [37]:

Theorem 2.2 *Let G be a finitely generated group. Then the word problem of G is decidable if, and only if, there exists a finitely generated simple group H and a finitely presented group K with $G \leq H \leq K$.*

We will survey some similar results when we start restricting the range of languages for the word problem.

3 Regular and context-free word problems

In the previous section we discussed the idea of decidability with respect to word problems. However, an alternative approach to classifying word problems was suggested by work of Anisimov [1] who classified groups with a regular word problem. Essentially a *regular language* is one accepted by a *finite state automaton*, i.e., a Turing machine that has no work tapes and where the head on the input tape always moves right on every move. It is conventional here to replace the idea of a halt state with a set F of *accept states*: we continue reading the input α, accepting α if we end up in a state in F after reading the last symbol in α and rejecting α otherwise. As with Turing machines, allowing non-determinism does not increase the range of languages accepted. We let $\mathcal{R}$ denote the class of all regular languages. It turns out that no matter which finite generating set X we choose for a group G, either $W_X(G) \in \mathcal{R}$ for all such generating sets or for none. We can now state the result from [1]:

Theorem 3.1 *Let G be a finitely generated group. Then $W(G)$ is regular if, and only if, G is finite.*

In this way, the focus was shifted from degrees of undecidability to degrees of decidability. Anisimov's result implied a more general research plan of linking the structure of a group with the nature of its word problem as a formal language. To this end, another significant related result was obtained by Muller and Schupp [28] who classified (if we blend in a deep result of Dunwoody [12]) groups with a context-free word problem. In order to explain the term "context-free", we need some more definitions.

We can extend our finite state automaton by adding a *stack*. We now assume that our finite state automaton has a single work tape where access to the symbols of this work tape is restricted. The idea is that we may only read the rightmost non-blank cell of the work tape at any point in time. We may delete the contents of this cell and move the work head left, if we are not scanning the leftmost cell of the work tape, i.e., "pop an element off the stack", or we may move the work head one cell to the right and write a new symbol into this cell, i.e., "push an element onto the stack". In this way the contents of the stack are always of the form $\triangleright\gamma$

for some (possibly empty) string γ with the head positioned over the rightmost symbol in $\triangleright\gamma$. An extension of a non-deterministic finite state automaton in this way is known as a *pushdown automaton.* Unlike Turing machines and finite state automata, insisting on determinism *does* restrict the range of languages accepted (see [20] for example), and we have the class $\mathcal{CF}$ of *context-free languages* accepted by pushdown automata and the class $\mathcal{DCF}$ of *deterministic context-free languages* accepted by deterministic pushdown automata. It is clear that $\mathcal{R} \subseteq \mathcal{DCF}$, and, in fact, $\mathcal{R} \subset \mathcal{DCF}$.

The fact that $\mathcal{R} \subset \mathcal{CF}$ is a standard result in formal language theory, and is reflected here by the fact that recognizing word problems by pushdown automata allows a wider class of groups than the finite groups. We say that a finitely-generated group G is *context-free* if $W(G) \in \mathcal{CF}$ and *deterministic context-free* if $W(G) \in \mathcal{DCF}$ (as for regular languages, it turns out that as to whether $W_X(G) \in \mathcal{CF}$ or $W_X(G) \in \mathcal{DCF}$ is independent of the choice of generating set X). For instance, it is reasonably clear that a finitely generated free group is deterministic context-free: as one reads the input word, the generators are pushed on the stack, unless one reads a generator x with x^{-1} on the top of the stack, in which case we merely pop x^{-1} off the stack. At the end of the input we check whether the stack is empty. One can extend this to show that any finitely generated virtually free group (i.e., a finitely generated group with a free subgroup of finite index) is deterministic context-free.

Let G be a finitely generated group and let Γ be the Cayley graph of G with respect to some finite generating set. Let $\Gamma^{(n)}$ denote the set of vertices in Γ at distance at most n from the identity element. Then the number of *ends* of G is defined to be

$$\lim_{n\to\infty} (\text{the number of infinite connected components of } \Gamma - \Gamma^{(n)})$$

(which may be infinite). This number is, in fact, independent of the generating set chosen. An *accessible series* for G is a series $G = G_0 \geq G_1 \geq G_2 \geq \ldots \geq G_n$ of subgroups such that each G_i has a decomposition as a non-trivial free product with amalgamation, or as an HNN extension, where one of the factors or the base is G_{i+1}, and the amalgamated or associated subgroups are finite. A group G is *accessible* if there is a finite upper bound on the length n of such a series. In [28], it was shown that an infinite context-free group has more than one end, and then, using Stallings' classification of finitely generated groups with more than one end [36], that G is virtually free if, and only if, G is context-free and accessible. It was proved in [12] that all finitely presented groups are accessible. Since every context-free group is finitely presented [2] and deterministic context-free [29], we have the following:

Theorem 3.2 *If G is a finitely-generated group, then:*

G is context-free $\Leftrightarrow$ G is deterministic context-free $\Leftrightarrow$ G is virtually free.

See also [23, 35]. It is interesting that, despite the fact that there are languages that are context-free but not deterministic context-free, there are no such languages that are the word problems of groups!

For the classes of languages encountered so far, we have stated that whether the word problem of some group is in the class is independent of the generating set chosen for the group. We now give a general condition for this to be so.

Let $\mathcal{F}$ be a class of languages; we say that $\mathcal{F}$ is *closed under homomorphism* if

$$L \in \mathcal{F}, L \subseteq \Sigma^*, \phi : \Sigma^* \to \Omega^* \text{ a monoid homomorphism } \Rightarrow L\phi \in \mathcal{F},$$

and that $\mathcal{F}$ is *closed under inverse homomorphism* if

$$L \in \mathcal{F}, L \subseteq \Omega^*, \phi : \Sigma^* \to \Omega^* \text{ a monoid homomorphism } \Rightarrow L\phi^{-1} \in \mathcal{F}.$$

We also say that $\mathcal{F}$ is *closed under intersection with regular languages* if

$$L_1 \in \mathcal{F}, L_1 \subseteq \Sigma^*, L_2 \in \mathcal{R}, L_2 \subseteq \Sigma^* \Rightarrow L_1 \cap L_2 \in \mathcal{F}.$$

We have the following well-known result (see [19], for example):

Proposition 3.3 *If $\mathcal{F}$ is a family of languages closed under inverse homomorphism, S and T are finite subsets of the group G, and $G = \langle S \rangle = \langle T \rangle$, then $W_S(G) \in \mathcal{F}$ if, and only if, $W_T(G) \in \mathcal{F}$.*

A *cone* is a family $\mathcal{F}$ of languages closed under homomorphism, inverse homomorphism and intersection with regular languages. It is well known that $\mathcal{R}$ and $\mathcal{CF}$ are both cones (see [20], for example). In particular, they are closed under inverse homomorphism, and so we can talk about the word problem of a group being "regular" or "context-free" without ambiguity. Whereas $\mathcal{DCF}$ is not a cone, it is closed under inverse homomorphisms.

4 One-counter groups

Returning to machines, a *one-counter automaton* is a pushdown automaton where there are only two stack symbols, $\triangleright$ and g say. At any stage, the stack contains $\triangleright g^n$ for some $n \geq 0$, and so is effectively described by a single natural number n; hence the title "one-counter". If, in addition, the pushdown automaton is deterministic then we say that we have a *deterministic one-counter automaton*. We say that L is a *one-counter language* if $L = L(M)$ for some one-counter automaton M, and a *deterministic one-counter language* if $L = L(N)$ for some deterministic one-counter automaton N.

We let $\mathcal{OC}$ be the class of one-counter languages and $\mathcal{DOC}$ be the class of deterministic one-counter languages. Whereas $\mathcal{OC}$ is a cone (see [7] for example), $\mathcal{DOC}$ is not but it is closed under inverse homomorphisms (see [17], for example). We say that a finitely generated group is *one-counter* if its word problem lies in $\mathcal{OC}$ and *deterministic one-counter* if its word problem lies in $\mathcal{DOC}$. The family of one-counter languages is particularly significant when studying the word problem in groups, as the next result from [17] shows:

Theorem 4.1 *Let $\mathcal{F}$ be a cone with $\mathcal{F} \subseteq \mathcal{CF}$, and let Υ be the class of all finitely-generated groups G such that $W(G) \in \mathcal{F}$. Then Υ is the class of finite groups, one-counter groups or context-free groups.*

As far as one-counter groups are concerned, we have a result from [17] analogous to Theorem 3.2:

Theorem 4.2 *If G is a finitely-generated group, then:*

$$G \text{ is one-counter} \Leftrightarrow G \text{ is deterministic one-counter} \Leftrightarrow G \text{ is virtually cyclic.}$$

This result completes the classification of classes of groups whose word problem lies in a cone $\mathcal{F}$ with $\mathcal{F} \subseteq \mathcal{CF}$. Theorem 4.1 shows that the class of one-counter groups is a very natural one when one considers characterizations of groups in terms of their word problems. In addition, the infinite one-counter groups are precisely the groups with two ends. Notice that, in Theorem 4.2, as in Theorem 3.2, we have that the class of languages accepted by the deterministic model of the machine is a proper subclass of the class of languages accepted by the non-deterministic model, but this is not preserved when we restrict ourselves to considering the word problems of groups. One is tempted to ask to what extent this is a general phenomenon.

We saw in Proposition 3.3 that we really only need to consider classes of languages $\mathcal{F}$ closed under inverse homomorphisms to remove the problem that the word problem for a group lying in $\mathcal{F}$ might depend on the particular generating set chosen. This leads to the natural question as to whether there are any interesting classes $\mathcal{F}$ of languages closed under inverse homomorphism and contained in $\mathcal{CF}$ such that the class of groups whose word problem lies in $\mathcal{F}$ is not the class of recognizable, one-counter or context-free groups.

5 Above context-free

The situation higher up the Chomsky Hierarchy, above the context-free languages, seems to be much less clear. In particular, while many examples of groups with a *context-sensitive* word problem, i.e., a word problem that can be recognised by a non-deterministic Turing machine which uses (at most) n cells on its single work tape for an input of length n), or even a *deterministic context-sensitive word problem* (where the Turing machine is necessarily deterministic), are known (see [18], for example), as yet we have no structural classification as to precisely which groups have a context-sensitive word problem. As pointed out in [38], this may be quite a challenging problem.

Our discussion of interactions between formal language theory and group theory brings us naturally to the study of automatic groups. These groups were defined by Thurston as a class of finitely presented groups: he noticed that a result of Cannon [10] on the combinatorial structure of discrete hyperbolic groups could be re-expressed in terms of finite automata. We will want to consider automata accepting pairs (α, β) of words with $\alpha, \beta \in A^*$. If $\alpha = a_1 a_2 \dots a_n$ and $\beta = b_1 b_2 \dots b_m$, this is accomplished by having an automaton with input alphabet $A \times A$ and reading pairs (a_1, b_1), (a_2, b_2), and so on. To deal with the case where $n \neq m$, we introduce a *padding symbol* \$. More formally, we define a mapping $\delta_A : A^* \times A^* \to A(2, \$)^*$, where $\$ \notin A$ and $A(2,\$) = ((A \cup \{\$\}) \times (A \cup \{\$\})) - \{(\$,\$)\}$, by:

$$(\alpha,\beta)\delta_A = \begin{cases} (a_1,b_1)\dots(a_n,b_n) & \text{if } n = m \\ (a_1,b_1)\dots(a_n,b_n)(\$,b_{n+1})\dots(\$,b_m) & \text{if } n < m \\ (a_1,b_1)\dots(a_m,b_m)(a_{m+1},\$)\dots(a_n,\$) & \text{if } n > m. \end{cases}$$

Given this, if G is a group, A is a finite set, L is a regular subset of A^*, and $\phi : A^* \to G$ is a homomorphism with $L\phi = G$ then we say that (A,L) is an *automatic structure* for G if $\{(\alpha,\beta) : \alpha,\beta \in L, \alpha = \beta\}\delta_A$ is regular, and $\{(\alpha,\beta) : \alpha,\beta \in L, \alpha a = \beta\}\delta_A$ is regular for each $a \in A$. If a group G has an automatic structure (A,L), for some A and L, then we say that G is *automatic* (an equivalent definition can be expressed in terms of the Cayley graph of the group).

The class of automatic groups includes finite groups, free groups, free abelian groups, various small-cancellation groups, word-hyperbolic groups and co-compact discrete groups of isometries of negatively curved space. The study of automatic groups is a thriving research area and the book [13] provides a standard reference (see also [6]). A significant result in this area is the following theorem from [32]:

Theorem 5.1 *A finitely generated subgroup of an automatic group has deterministic context-sensitive word problem.*

Since a direct product of finitely generated free groups is automatic, this means that any finitely generated subgroup of $F_2 \times \ldots \times F_2$ has a deterministic context-sensitive word problem, so that there are examples of groups with a deterministic context-sensitive word problem that are not finitely presented. It would be interesting to know if context-sensitive is the optimal result here, i.e., is there a proper subfamily $\mathcal{F}$ of the deterministic context-sensitive languages which is closed under inverse homomorphism and such that the word problem for finitely-generated subgroups of automatic groups lies in $\mathcal{F}$? It is even an open question as to whether or not the deterministic context-sensitive languages form a proper subclass of the context-sensitive languages.

6 Complexity

The advent of electronic computers led to a realization that "decidable" need not mean "feasibly decidable", and a theory, *computational complexity theory*, has been developed whose aim is the precise classification of the degrees of decidability of problems, with particular emphasis on the resources needed. One motivation for studying groups whose word problem lies in a particular class of formal languages is that computation in the group might actually be practically, rather than just theoretically, possible. As can be seen above, there is a close relationship between classes of formal languages and the languages accepted by resource-bounded models of computation (for a general account of complexity theory, see [20, 31]).

Roughly, complexity theory has to it two aspects: the study of upper bounds and the study of lower bounds. The former are essentially concrete algorithms that have specific bounds on the resources used, and the latter are proofs that some specific problem can not be solved by some model of computation within some resource bounds. Lower bound results are notoriously difficult to obtain and often we make do with showing some problem to be "complete" for some "complexity class", as we now explain.

As in the rest of this paper, let us restrict ourselves here to the question of membership of a language L. We say that this problem can be solved in *deterministic time* $f(n)$ if there is a decision-making DTM M such that, given any input of length n, M terminates in at most $f(n)$ steps: we similarly say that the problem can be solved in *deterministic space* $g(n)$ if there is a decision-making DTM M such that, given any input of length n, M terminates using at most $g(n)$ cells on the work tape. We have similar definitions for *non-deterministic time* and *non-deterministic space*. Bounding resources in this way yields "complexity classes" such as: DTIME(*poly*), the class of languages L for which the membership problem is solvable in deterministic time $O(f(n))$, for some polynomial $f(n)$; NSPACE(*log*), the class of languages L for which the membership problem is solvable in non-deterministic space $O(\log n)$; and so on (recall, the complexity classes NSPACE(n) and DSPACE(n) are the classes of context-sensitive languages and deterministic context-sensitive languages, respectively). Complexity classes such as DTIME(*poly*) and NSPACE(*log*) are very robust in the sense that the same complexity class arises by restricting different (reasonable) models of computation in the same way.

At the fulcrum of complexity theory is the open question as to whether DTIME(*poly*) and NTIME(*poly*) are identical (also known as the P = NP question), although there are a great number of similar open questions concerning other complexity classes. For example, the basic hierarchy of complexity theory is:

$$\begin{aligned} &\text{DSPACE}(log) \subseteq \text{NSPACE}(log) \subseteq \text{DTIME}(poly) \\ &\quad \subseteq \text{NTIME}(poly) \subseteq \text{DSPACE}(poly) = \text{NSPACE}(poly), \end{aligned}$$

and the only additional information known about these containments is that NSPACE(*log*) $\neq$ DSPACE(*poly*). The questions as to whether the containments are proper are generally regarded as being extraordinarily difficult to answer, with most researchers of the opinion that the ones above are indeed proper!

Each of these complexity classes has *complete languages* with the property that if some complete language for NTIME(*poly*), say, is in DTIME(*poly*) then NTIME(*poly*) = DTIME(*poly*). That is, the difficulty of a complexity-theoretic containment question is entirely encapsulated in the question of whether one particular language is in some complexity class or not. Complexity classes tend to have numerous complete languages and the proof that a language is complete for some complexity class, such as NTIME(*poly*), is usually interpreted as providing strong evidence that the particular complete language is not in (in this case) DTIME(*poly*).

The precise complexity of word problems of groups and classes of groups (in terms of in which complexity classes they lie and whether they are complete for these complexity classes or not) has not really been investigated in a systematic fashion. However, some results do exist, amongst them the following: Avenhaus and Madlener [3, 4, 5] showed that a group may have a decidable word problem of varying degrees of space complexity; Lipton and Zalcstein [24] and Simon [34] showed that the word problem for linear groups over fields of zero and prime characteristic, respectively, can be solved in the complexity class DSPACE(*log*); Weispfenning [39] showed that the word problem for *abelian lattice-ordered groups* (an abelian group which is also a lattice such that the group operations are compatible with the lattice operations) is complete for the complexity class co-NTIME(*poly*) (problems whose complement is solvable in non-deterministic polynomial-time); Garzon and Zalcstein [15] showed that if certain space complexity classes (containing DSPACE(*log*)) can be separated at all then there are separating word problems of certain Grigorchuk groups witnessing this separation; and Immerman and Landau [21] exhibited a number of groups for which the word problem is complete for complexity classes "contained in" DTIME(*poly*). Surprisingly, there seems to be no group or class of groups for which the word problem is known to be complete for DTIME(*poly*) or NTIME(*poly*) (see [14, 38], for example).

A strong link between automatic groups and computational complexity is that the word problem for automatic groups can be solved in $O(n^2)$ time; moreover, the word problem for word-hyperbolic groups (i.e., groups for which Dehn's algorithm solves the word problem) can be solved in $O(n)$ time (when the Turing machine is allowed to have at least two work tapes: see [11]). This leads to the question (as in [33]) as to whether there is a natural characterization of the class of groups whose word problem can be solved in $O(n)$ time (respectively $O(n^2)$) time. Also, might the word problem for automatic groups be complete for DTIME(*poly*)?

Apart from those hinted at above, other directions for research in this general area present themselves. For example, there has been an explosion of interest in the model theory of finite structures in recent years, and it is an interesting question as to the extent to which one can classify (word) problems in group theory according to whether they can be defined in certain logics or not.

Acknowledgements We are grateful to Sarah Rees for several helpful and constructive comments on an early version of this paper and to the referee for his/her constructive suggestions. The second author would also like to thank Hilary Craig for all her help and encouragement.

References

[1] A. V. Anisimov, Group languages, *Kibernetika* **4** (1971) 18–24.

[2] A. V. Anisimov, Some algorithmic problems for groups and context-free languages, *Kibernetika* **8** (1972) 4–11.

[3] J. Avenhaus and K. Madlener, Subrekursive Komplexität bei Gruppen I, *Acta Informat.* **9** (1977) 87–104.

[4] J. Avenhaus and K. Madlener, Subrekursive Komplexität bei Gruppen II, *Acta Informat.* **9** (1978) 183–193.

[5] J. Avenhaus and K. Madlener, Algorithmische Probleme bei Einrelatorgruppen und ihre Komplexität, *Arch. Math. Logic* **19** (1978) 3–12.

[6] G. Baumslag, S. M. Gersten, M. Shapiro and H. Short, Automatic groups and amalgams, *J. Pure Appl. Algebra* **76** (1991) 229–316.
[7] J. Berstel, *Transductions and Context-Free Languages*, Teubner (1979).
[8] W. W. Boone, The word problem, *Ann. Math.* **70** (1959) 207–265.
[9] W. W. Boone and G. Higman, An algebraic characterization of groups with a solvable word problem, *J. Australian Math. Soc.* **18** (1974) 41–53.
[10] J. W. Cannon, The combinatorial structure of cocompact discrete hyperbolic groups, *Geom. Dedicata* **16** (1984) 123–148.
[11] B. Domanski and M. Anshel, The complexity of Dehn's algorithm for word problems in groups, *J. Algorithms* **6** (1985) 543–549.
[12] M. J. Dunwoody, The accessibility of finitely presented groups, *Invent. Math.* **81** (1985) 449–457.
[13] D. B. A. Epstein, J. W. Cannon, D. F. Holt, S. Levy, M. S. Patterson and W. Thurston, *Word Processing in Groups*, Jones and Bartlett, London/Boston (1992).
[14] M. Garzon and Y. Zalcstein, On isomorphism testing of a class of 2-nilpotent groups, *J. Comput. System Sci.* **42** (1991) 237–248.
[15] M. Garzon and Y. Zalcstein, The complexity of Grigorchuk groups with application to cryptography, *Theoret. Comput. Sci.* **88** (1991) 83–98.
[16] M. A. Harrison, *Introduction to Formal Language Theory*, Addison-Wesley (1978).
[17] T. Herbst, On a subclass of context-free groups, *Theor. Informatics and Applications* **25** (1991) 255 -272.
[18] T. Herbst, Some remarks on a theorem of Sakarovitch, *J. Comput. System Sci.* **44** (1992) 160–165.
[19] T. Herbst and R. M. Thomas, Group presentations, formal languages and characterizations of one-counter groups, *Theoret. Comput. Sci.* **112** (1993) 187–213.
[20] J. E. Hopcroft and J. D. Ullman, *Introduction to Automata Theory, Languages, and Computation*, Addison-Wesley (1979).
[21] N. Immerman and S. Landau, The complexity of iterated multiplication, *Inform. Computat.* **116** (1995) 103–116.
[22] O. Kharlampovich, A finitely presented soluble group with unsoluble word problem, *Izvestia Akad. Nauk. Ser. Math* **45** (1981) 852–873.
[23] A. A. Letičevskiĭ and L. B. Smikun, On a class of groups with solvable problem of automata equivalence, *Dokl. Acad. Nauk SSSR* **227** (1976), 36–38; translated in *Soviet Math. Dokl.* **17** (1976) 341–344.
[24] R. J. Lipton and Y. Zalcstein, Word problems solvable in logspace, *J. Assoc. Comput. Mach.* **24** (1977) 522–526.
[25] R. C. Lyndon and P. E. Schupp, *Combinatorial Group Theory*, Ergebnisse der Mathematik und ihrer Grenzgebiete **89**, Springer-Verlag (1977).
[26] W. Magnus, A. Karrass and D. Solitar, *Combinatorial Group Theory*, Dover Publications (1976).
[27] C. F. Miller, Decision problems for groups - a survey, in: G. Baumslag and C. F. Miller (eds.), *Algorithms and Classification in Combinatorial Group Theory*, MSRI Publications **23**, Springer Verlag (1992) 1–59.
[28] D. E. Muller and P. E. Schupp, Groups, the theory of ends and context-free languages, *J. Comput. System Sci.* **26** (1983) 295–310.
[29] D. E. Muller and P. E. Schupp, The theory of ends, pushdown automata, and second-order logic, *Theor. Comp. Sci.* **37** (1985) 51–75.
[30] P. S. Novikov, On the algorithmic unsolvability of the word problem in group theory, *Trudy. Mat. Inst. Steklov* **44** (1955) 1–143.
[31] C. H. Papadimitriou, *Computational Complexity Theory*, Addison-Wesley (1995).
[32] M. Shapiro, A note on context-sensitive languages and word problems, *Internat. J.*

Algebra Comput. **4** (1994) 493–497.
[33] H. Short, An introduction to automatic groups, in: J. Fountain (ed.), *Semigroups, Formal Languages and Groups*, NATO ASI Series **C466**, Kluwer (1995) 233–253.
[34] H.-U. Simon, Word problems for groups and context-free recognition, in: L. Budach (ed.), *Fundamentals of Computation Theory*, Math. Research **2**, Akademie-Verlag, Berlin (1979) 417–422.
[35] L. B. Smihun, Relations between context-free groups and groups with a decidable problem of automata equivalence, *Kibernetica* **5** (1976) 33–37.
[36] J. Stallings, *Group Theory and Three-dimensional Manifolds*, Yale University Press (1971).
[37] R. J. Thompson, Embeddings into finitely generated groups which preserve the word problem, in: S. I. Adian, W. W. Boone and G. Higman (eds.), *Word Problems II*, North Holland, Amsterdam (1980) 401–441.
[38] C. Tretkoff, Complexity, combinatorial group theory and the language of palutators, *Theoret. Comput. Sci.* **56** (1988) 253–275.
[39] V. Weispfenning, The complexity of the word problem for abelian l-groups, *Theoret. Comput. Sci.* **48** (1986) 127–132.

PERIODIC COHOMOLOGY AND FREE AND PROPER ACTIONS ON $\mathbb{R}^n \times S^m$

OLYMPIA TALELLI

Department of Mathematics, University of Athens, Athens 15784, Greece

To Karl W. Gruenberg for his 70th birthday

Introduction

It was realized by Killing, Klein and others that the sphere and real projective space offered two different global models for geometry which were locally the same. In 1891, Killing formulated the problem of determining all such models. In 1926 Hopf revived the problem and also raised the more general question of studying manifolds covered by spheres, hence groups which act freely and properly on spheres.

If a group acts freely and properly on a sphere then the group is finite, since the sphere is a compact space, and it has periodic (Tate) cohomology [C-E, Ch. XVI §9]. The structure of finite groups with periodic cohomology is well known and is essentially based on the fact that if a group G has periodic cohomology then so does every subgroup of G, and if C_p is the cyclic group of order a prime p, then $C_p \times C_p$ does not have periodic cohomology. So if a finite group G has periodic cohomology then G does not contain subgroups of the form $C_p \times C_p$ for any prime p. It turns out that this is a sufficient condition for a finite group G to have periodic cohomology [Artin-Tate (unpublished), [C-E, Ch. XII] or [B, Ch. VI]].

Not all finite groups with periodic cohomology can act freely on a sphere. Note that if a finite group acts freely on a sphere then the action is proper. In 1957 Milnor [Mil] showed that a group which acts freely on a sphere has at most one element of order 2. However, R. G. Swan in 1960 [Sw_1] showed that any finite group with periodic cohomology acts freely and simplicially on a finite simplicial homology sphere.

Madsen, Thomas and Wall [M-T-W] proved in 1976 that if a finite group G has periodic cohomology and satisfies Milnor's condition then G acts freely on a sphere.

In this paper we give a brief survey of some results concerning the phenomenon of periodicity in the cohomology of infinite groups and we obtain:

Theorem A *If a countable group G has periodic cohomology after* 1*-step then G acts freely and properly on $\mathbb{R}^n \times S^m$, for some n and m.*

Note that if an infinite group G has periodic cohomology then the periodicity appears after some steps and every finite subgroup of G has periodic Tate cohomology . Moreover, if a group G acts freely and properly on $\mathbb{R}^n \times S^m$, for some n and m, then G must be countable since $\mathbb{R}^n \times S^m$ is a separable metric space. Theorem A follows essentially from Theorem 2 in [T_5] and Theorem 10 of Prassidis in [P].

In [T_2, Corollary 2.3] we show that if $(\mathcal{G}, X)$ is a graph of groups (in the sense of Bass-Serre theory) which are finite with periodic cohomology and the first Betti number of the graph X is finite, then the fundamental group G of $(\mathcal{G}, X)$ has periodic cohomology after 1-step.

So we can construct examples of groups which act freely and properly on $\mathbb{R}^n \times S^m$, for some n and m.

In 1979, C.T.C. Wall [W$_1$] conjectured that if a countable group G has finite virtual cohomological dimension and periodic Farrell cohomology then G acts freely and properly on $\mathbb{R}^n \times S^m$, for some n and m. A group has finite virtual cohomological dimension if it contains a subgroup of finite index which has finite cohomological dimension and the Farrell cohomology is a generalization of the Tate cohomology to groups of finite virtual cohomological dimension. F.X. Connolly and S. Prassidis proved the conjecture in 1989 [C-P], while a special case of the conjecture was shown by F.E.A. Johnson in 1985 [J]. In 1992 S. Prassidis in [P] showed that there are groups of infinite virtual cohomological dimension which act freely and properly on $\mathbb{R}^n \times S^m$, for some n and m. He used a notion of generalized Farrell cohomology, defined by Ikenaga in 1984 [I] for a class of groups larger than the class of groups of finite virtual cohomological dimension. In 1993 Farrell and Stark gave examples of groups of infinite virtual cohomological dimension which act freely and properly on $\mathbb{R}^n \times S^m$ with compact quotient [F-S].

Clearly the problem of classifying all free and proper actions of discrete groups on the product of a sphere and a Euclidean space amounts to classifying, up to homeomorphism, all manifolds which admit the product of a sphere and a Euclidean space as universal covering. We believe :

Conjecture A The following statements are equivalent for a countable group G.

(i) G acts freely and properly on $\mathbb{R}^n \times S^m$ for some n and m.

(ii) G has periodic cohomology after k-steps, for some k.

(iii) There is an exact sequence

$$0 \to \mathbb{Z} \to T \to P_{q-2} \to \cdots \to P_0 \to \mathbb{Z} \to 0$$

for some q, where $\mathbb{Z}$ is regarded as a trivial $\mathbb{Z}G$-module, P_i, $0 \leq i \leq q-2$ are projective $\mathbb{Z}G$-modules and $p \cdot d_{\mathbb{Z}G}T < \infty$.

(iv) G has periodic complete cohomology (Definition3.2).

Conjecture A is true for groups of finite virtual cohomological dimension and locally finite groups.

The contents of the paper are :

§ 1 Periodicity in Tate cohomology of finite groups.

§ 2 Periodicity in cohomology after some steps.

§ 3 Generalized Tate cohomologies.

§ 4 Free and proper actions on $\mathbb{R}^n \times S^m$.

We conclude with a few conjectures.

The author is indebted to helpful discussions with S. Prassidis.

1 Periodicity in (Tate) cohomology of finite groups

The phenomenon of periodicity in the cohomology of finite groups was first studied by E. Artin and J. Tate (unpublished) and an account of their work can be found in [C-E, Ch. XII] or [B, Ch. VI].

Let G be a finite group. All modules considered are left $\mathbb{Z}G$-modules and $\mathbb{Z}$ is always regarded as a trivial $\mathbb{Z}G$-module.

A complete resolution for G is an acyclic complex of projective $\mathbb{Z}G$-modules

$$\mathcal{P}_c : \cdots \to P_2 \xrightarrow{\partial_2} P_1 \xrightarrow{\partial_1} P_0 \xrightarrow{\partial_0} P_{-1} \to \cdots$$

together with a map $\varepsilon : P_0 \to \mathbb{Z}$ such that

$$\mathcal{P}_c^+ : \cdots \to P_1 \xrightarrow{\partial_1} P_0 \xrightarrow{\varepsilon} \mathbb{Z} \to 0$$

is a resolution for G. Clearly $\partial_0 : P_0 \to P_{-1}$ factors uniquely as $\partial_0 = i\varepsilon$ where $i : \mathbb{Z} \to P_{-1}$ is a monomorphism.

Complete resolutions exist for any group G. For example, let

$$\cdots \to P_n \to P_{n-1} \to \cdots P_1 \to P_0 \xrightarrow{\varepsilon} \mathbb{Z} \to 0$$

be a projective resolution of G and consider the map $\iota : \mathbb{Z} \to \mathbb{Z}G$ with $\iota(1) = \sum_{g \in G} g$. Then ι is a $\mathbb{Z}$-split $\mathbb{Z}G$-map and we have the following short exact sequence of $\mathbb{Z}G$-modules which is $\mathbb{Z}$-split.

$$(*) \quad 0 \to \mathbb{Z} \xrightarrow{\iota} R_{-1} \to 0 \text{ where } R_{-1} = \text{coker } \iota.$$

If A is any $\mathbb{Z}$-free $\mathbb{Z}G$-module then by tensoring $(*)$ with A we obtain a short exact sequence of $\mathbb{Z}G$-modules (diagonal action) which is $\mathbb{Z}$-split

$$0 \to A \to \mathbb{Z}G \otimes A \to R_{-1} \otimes A \to 0.$$

Since $\mathbb{Z}G \otimes A$, with the diagonal action, is isomorphic to the induced module $\mathbb{Z}G \otimes A$ and A is free it follows that $\mathbb{Z}G \otimes A$ is $\mathbb{Z}G$-is projective. Hence we obtain the following complete resolution for G

$$\cdots \longrightarrow P_1 \xrightarrow{\partial_1} P_0 \longrightarrow \mathbb{Z}G \longrightarrow \mathbb{Z}G \otimes R_{-1} \longrightarrow \mathbb{Z}G \otimes R_{-2} \longrightarrow \cdots$$

(with $P_0 \xrightarrow{\varepsilon} \mathbb{Z} \to \mathbb{Z}G$, $\mathbb{Z}G \to R_{-1} \to \mathbb{Z}G \otimes R_{-1}$, $\mathbb{Z}G \otimes R_{-1} \to R_{-2} \to \mathbb{Z}G \otimes R_{-2}$)

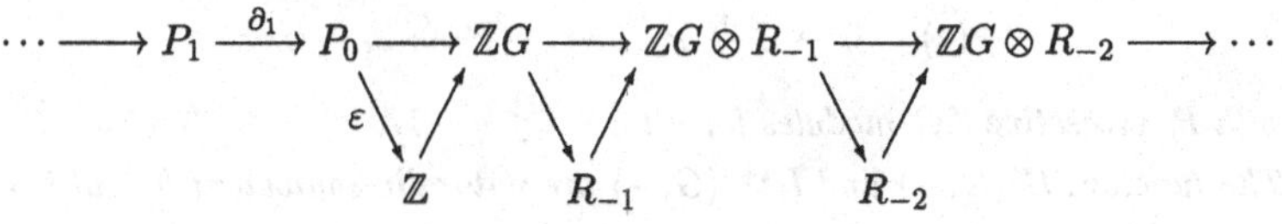

where for $n \geq 2$ we put $R_{-n} = R_{-1} \otimes \cdots \otimes R_{-1}$ n-times.

Any two complete resolutions for G are canonically homotopy equivalent and the Tate cohomology of a finite group G with coefficients in a $\mathbb{Z}G$-module A is now defined as

$$\widehat{H}^i(G, A) = H^i(\text{Hom}_{\mathbb{Z}G}(\mathcal{P}_c, A)) \; i \in \mathbb{Z}$$

where $\mathcal{P}_c$ is a complete resolution for G.

The Tate cohomology has the following properties :

$$\widehat{H}^i(G,A) = \begin{cases} H^i(G,A) & i > 0 \\ H_{-i-1}(G,A) & i < 0 \end{cases}$$

and there is an exact sequence

$$0 \to \widehat{H}^{-1}(G,A) \to H_0(G,A) \xrightarrow{N} H^0(G,A) \to \widehat{H}^0(G,A) \to 0$$

where N is the norm map.

Since the Tate cohomology is defined via complete resolutions it follows easily that $\{\widehat{H}^i(G,-)\}_{i\in\mathbb{Z}}$ is a cohomological functor which has all the nice cohomological properties i.e. Shapiro's lemma, restriction and transfer maps, and cup products. In particular, $\widehat{H}^*(G,\mathbb{Z})$ is an anti-commutative graded ring with identity element $1 \in \mathbb{Z}/|G|\mathbb{Z} = \widehat{H}^0(G,\mathbb{Z})$.

Moreover,

a) $\widehat{H}^*(G, \text{induced}) = 0$; hence the functors are both effaceable and coeffaceable (which allows dimension shifting in both directions).

b) $\widehat{H}^*(G,A) = 0$ if G is the trivial group

c) $\widehat{H}^*(G,A)$ are torsion groups.

Definition A group G is said to have periodic (Tate) cohomology if for some positive integer q there is an element of $\widehat{H}^q(G,\mathbb{Z})$ which is invertible in the ring $\widehat{H}^*(G,\mathbb{Z})$. Cup product with such an element then defines periodicity isomorphisms

$$\widehat{H}^i(G,A) \cong \widehat{H}^{i+q}(G,A)$$

for any $\mathbb{Z}G$-module A and any integer i.

The integer q is called a period of G and the anticommutativity of the cup product shows that a period is even.

Theorem 1.1 *The following statements are equivalent for a finite group G and a positive integer q.*

(a) *G has periodic cohomology with period q.*

(b) *There is an exact sequence*

$$(*) \quad 0 \to \mathbb{Z} \xrightarrow{i} P_{q-1} \to \cdots P_0 \xrightarrow{\varepsilon} \mathbb{Z} \to 0$$

with P_i projective $\mathbb{Z}G$-modules for all $0 \le i \le q-1$.

(c) *The functors $H^i(G,-)$ and $H^{i+q}(G,-)$ are naturally equivalent for all $i > 0$.*

Proof Clearly by splicing together copies of $(*)$

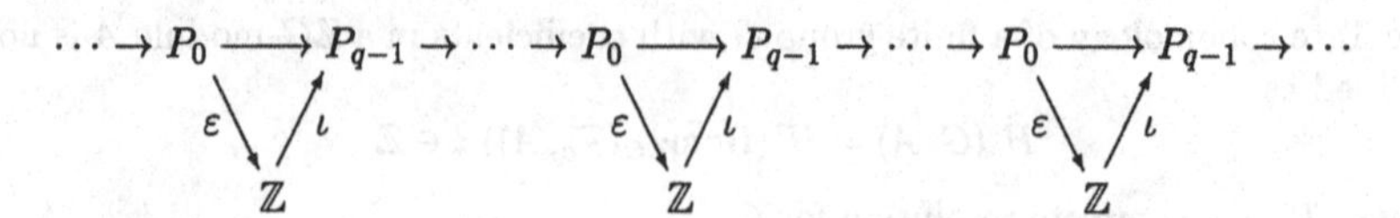

one obtains a periodic complete resolution.

The proof of (a) $\Leftrightarrow$ (b) i.e. that periodicity in cohomology is equivalent to the existence of a periodic projective resolution was given by Swan in [Sw1]. Another proof was given by Hilton and Rees in [H-R] essentially in the form (b)$\Leftrightarrow$ (c) and it was obtained as a consequence of their result that every natural transformation $\tau : \text{Ext}(A, -) \to \text{Ext}(B, -)$ is induced by a homomorphism $\theta_\tau : B \to A$. □

Corollary 1.2 *If G has periodic cohomology with period q then so does every subgroup of G.*

Proof By Theorem 1.1 (a) $\Leftrightarrow$ (b) and the result follows since if $K \leq G$, then every projective $\mathbb{Z}G$-module is $\mathbb{Z}K$-projective. □

Examples

1) If G acts freely on a sphere S^n, then G has periodic cohomology with period $n+1$ [C-E, Ch. XVI]. By the fixed point theorem of Brower, an orientation preserving homeomorphism of S^n has a fixed point if n is even. Thus if G acts freely, the orientation preserving subgroup (of index 1 or 2) is trivial. As C_2 can act freely on any sphere, one can assume that n is always odd.

2) Cyclic groups have periodic cohomology with period 2. If $C_m = \langle x \rangle$, then we have the following exact sequence

$$0 \to \mathbb{Z} \xrightarrow{\iota} \mathbb{Z}\langle x \rangle \xrightarrow{\Phi} \mathbb{Z}\langle x \rangle \xrightarrow{\varepsilon} \mathbb{Z} \to 0$$

where $\iota(1) = \sum x^i$ and $\Phi(1) = x - 1$.

3) Generalized quaternion groups have periodic cohomology with period 4. A periodic resolution is given in [C-E, page 253].

Of course 2) and 3) are special cases of 1) since cyclic groups act freely on S^1 and generalized quaternions as they can be considered as subgroups of the group S^3 of unit quaternions, they act freely on S^3 by left multiplication.

4) If p is a prime then $C_p \times C_p$ does not have periodic cohomology.

If $\mathbb{Z}_p$ is the field of p elements then it follows from Künneth's formula that $d_{\mathbb{Z}_p} H^n(C_p \times C_p, \mathbb{Z}_p) = n + 1$ for $r \geq 0$.

Corollary 1.2 and Example 4 imply that if G has periodic cohomology, then G does not contain any subgroup of the form $C_p \times C_p$ for any prime p. It turns out that this group-theoretic condition is equivalent to the periodicity of Tate cohomology:

Theorem 1.3 ([C-E, Ch XII] or [B, Ch VI]) *The following statements are equivalent for a group G:*

1) *G has periodic Tate cohomology.*
2) *G does not contain subgroups of the form for any prime p.*
3) *Every abelian subgroup of G is cyclic.*
4) *The Sylow subgroups of G are cyclic or generalized quaternion groups.*

A classification of the groups will periodic cohomology has been obtained by Suzuki [Su] extending earlier work of Zassenhaus. It can also be found in [J.W, pages 179, 195-8].

The following theorem gives cohomological conditions equivalent to the periodicity in the Tate cohomology.

Theorem 1.4 *The following statements are equivalent for a group G and a positive integer q.*

1) *G has periodic cohomology with period q.*
2) *The functors $\widehat{H}^i(G,-)$ and $\widehat{H}^{i+q}(G,-)$ are naturally equivalent for some $i \in \mathbb{Z}$.*
3) *$H^q(G,\mathbb{Z})$ is cyclic of order $|G|$.*
4) *$H^q(G,\mathbb{Z})$ contains an element of order $|G|$.*
5) *$H^i(G,\mathbb{Z}) = 0$ for almost all odd i.*
6) *There is a complete resolution for G which has an automorphism of degree q.*

Proof 1) $\Leftrightarrow$ 2) $\Leftrightarrow$ 6) follow from Theorem 1.1; 1)$\Leftrightarrow$ 3) $\Leftrightarrow$ 4) is in [C-E, Ch. XII, Proposition 11.1]; 1) $\Leftrightarrow$ 5) was proved by Swan in [Sw$_2$]. □

Remark It follows from the proof of Theorem 1.4 (1) $\Leftrightarrow$ (3), that if $H^q(G,\mathbb{Z})$ is cyclic of order $|G|$ and g is a generator of $H^q(G,\mathbb{Z})$, then g is invertible in $\widehat{H}^*(G,\mathbb{Z})$ and hence for every $\mathbb{Z}G$-module A, cup product with g

$$\cup g : \widehat{H}^i(G,A) \to \widehat{H}^{i+q}(G,A)$$

is an isomorphism for all $i \in \mathbb{Z}$.

The element g is called a maximal generator.

Note that in Yoneda's [Yo] interpretation of $H^q(G,\mathbb{Z}) = \mathrm{Ext}^q_{\mathbb{Z}G}(\mathbb{Z},\mathbb{Z})$ the element g is represented by a q-extension of the form

$$0 \to \mathbb{Z} \to P_{q-1} \to \cdots \to P_0 \to \mathbb{Z} \to 0$$

with P_i $\mathbb{Z}G$-projective for all $0 \le i \le q-1$ [W$_1$].

2 Periodicity in cohomology after some steps

Definition (**[T$_1$]**) A group G is said to have periodic cohomology after k-steps, where k is a nonnegative integer, if there is a positive integer q such that the functors $H^i(G,-)$ and $H^{i+q}(G,-)$ are naturally equivalent for all $i > k$. We then say that G has period q after k-steps.

This definition coincides with the classical one for finite groups, since for finite groups we have dimension shifting in both directions, so if a finite group G has period q after k-steps then it has period q after 0-steps. If, however, G is an infinite group and it has period q after k-steps then $k \ge 1$. [T$_1$]

The following result asserts that periodicity in cohomology after some steps is equivalent to the existence of a projective resolution which is periodic after some steps.

Theorem 2.1 ([T_1]) *The following statements are equivalent for a group G*

1) *G has period q after k-steps*
2) *There is an exact sequence*

$$0 \to R_{k+q} \to P_{k+q-1} \to \cdots \to P_k \longrightarrow P_{k-1} \to \cdots \to P_0 \to \mathbb{Z} \to 0$$

(with $P_k \to P_{k-1}$ factoring through R_k: $P_k \to R_k \to P_{k-1}$)

where $R_{k+q} \cong R_k$ and P_i, $0 \le i \le k+q-1$ are projective $\mathbb{Z}G$-modules.

We take $R_0 = \mathbb{Z}$. The periodicity isomorphisms are given by Yoneda product with the q-extension

$$(*) \qquad 0 \to R_k \to P_{k+q-1} \to \cdots \to P_k \to R_k \to 0.$$

It is now clear that:

Corollary 2.2 *If a group G has period q after k-steps then so does every subgroup of G.*

Examples

1) finite groups with periodic cohomology have periodic cohomology after 0-steps.
2) If G has cohomological dimension m then G has periodic cohomology after m-steps.
3) The infinite locally cyclic p-group, C_{p^∞}, has period 2 after 1-step.
4) It follows from Lyndon's Identity Theorem that 1-relator groups have period 2 after 2-steps.

Remark The notion of periodicity after some steps generalizes both the notion of finite cohomological dimension and the notion of periodicity in cohomology of finite groups.

The following result is a structure theorem for groups with periodic cohomology after 1-step.

Theorem 2.3 ([T_5]) *If G has periodic cohomology after 1-step then G acts on a tree with finite vertex stabilizers (or in the language of Bass-Serre theory, G is the fundamental group of a graph of finite groups).*

The proof uses the Almost Stability Theorem of Dicks and Dunwoody [D-D] which generalizes the Stallings-Swan theorem on the structure of groups of cohomological dimension one.

Corollary 2.4 *Let G be a group with periodic cohomology after 1-step.*

a) *If G is torsion free, then G is free*

b) *If G is torsion, then G is a countable locally finite group.*

If G acts on a tree with finite vertex stabilizers then $H^i(G, \text{projective}) = 0$ for $i > 1$ (e.g.[T_5]), hence if G has period q after 1-step then the q-extension (*) of Theorem 2.1 represents an element $g \in H^q(G, \mathbb{Z})$ since $\text{Ext}^q_{\mathbb{Z}G}(R_1, R_1) = H^q(G, \mathbb{Z})$ and we obtain the following:

Corollary 2.5 *The following statements are equivalent for a group G and a positive integer q.*

1) *G has period q after 1-step.*
2) *There is an element $g \in H^q(G, \mathbb{Z})$ such that for any $\mathbb{Z}G$-module A, cup product with g*
$$\cup g : H^i(G, A) \to H^{i+q}(G, A)$$
is an isomorphism for $i > 1$ and an epimorphism for $i = 1$.
3) *There is an exact sequence*
$$0 \to \mathbb{Z} \to T \to P_{q-2} \to \cdots \to P_0 \to \mathbb{Z} \to 0$$
with P_i, $0 \le i \le q-2$ projective $\mathbb{Z}G$-modules and $p \cdot d_{\mathbb{Z}G}T \le 1$.

That (2) $\Leftrightarrow$ (3) is a special case of:

Theorem 2.6 *Let R be any ring and A, B be R-modules. The following statements are equivalent for an element $g \in Ext^q_R(A, B)$.*

1) *For any R-module M, cup product (Yoneda product) with g*
$$\cup g : Ext^i_R(B, M) \to Ext^{i+q}_R(A, M)$$
is an isomorphism for $i > k$ and an epimorphism for $i = k$.
2) *The element g is represented by a q-extension*
$$0 \to B \to T \to P_{q-2} \to \cdots \to P_0 \to A \to 0$$
with P_i, $0 \le i \le q-2$ projective $\mathbb{Z}G$-modules and $p \cdot d_R T \le k$.

The proof is along the lines of the proof of Theorem 1.2 in [W_1] or Theorem 1.2 in [T_1].

Remarks a) If G is finite in Corollary 2.5 then T is a projective $\mathbb{Z}G$-module, since it is $\mathbb{Z}$-free and of finite projective dimension. If G is infinite, however, then $p \cdot d_{\mathbb{Z}G}T = 1$.

b) In [T_3] we introduced property $\mathcal{P}_1$ i.e. a group G is said to have property $\mathcal{P}_1$ if there is a $\mathbb{Z}$-free $\mathbb{Z}G$-module A such that $H^0(g, A) \ne 0$ and $p \cdot d_{\mathbb{Z}G}A \le 1$. In [K-T] and [$T_4$] it is shown that a group G has property $\mathcal{P}_1$ if and only if G acts on a tree with finite vertex stabilizers.

So by Theorem 2.3, Corollary 2.2 and Corollary 2.5, if a group G has period q after 1-step then G is the fundamental group of a graph of finite groups with period q and the periodicity isomorphisms are given by cup product with an element of $H^q(G, \mathbb{Z})$.

Now let $(\mathcal{G}, X)$ be a graph of finite groups and G its fundamental group. So we are given a connected non-empty graph X, a group G_ν for each $\nu \in \text{vert}X$, and a group G_e for each $e \in \text{edge}X$, together with a monomorphism $G_e \to G_\tau(e)$; it is also required that $G_e = G_{\bar{e}}$. Hence if $\mathcal{A}$ is an orientation of the graph X, then for every $e \in \mathcal{A}$ there are given two monomorphisms

$$i_{e,o} : G_e \to G_{o(e)} \text{ and } i_{e,\tau} : G_e \to G_{\tau(e)}.$$

Assume that every vertex group G_ν has period q. If $g_\nu \in H^q(G_\nu, \mathbb{Z})$ is a maximal generator for G_ν, then the family $\{g_\nu; \nu \in \text{vert}X\}$ of maximal generators is called compatible if

$$i^*_{e,o}(g_{o(e)}) = i^*_{e,\tau}(g_{\tau(e)}) \text{ for every } e \in \mathcal{A},$$

where $i^*_{e,o}$, $i^*_{e,\tau}$ are the maps of cohomology induced by $i_{e,o}$, $i_{e,\tau}$ respectively.

Theorem 2.7 **([T$_2$])** *Let $(\mathcal{G}, X)$ be a graph of finite groups and G its fundamental group. Then G has period q after 1-step if and only if every vertex group G_ν has period q and there is a compatible family $\{g_\nu; \nu \in \text{vert}X\}$ of maximal generators.*

Corollary 2.8 *There is a graph of finite cyclic groups whose fundamental group G does not have periodic cohomology after any finite number of steps. So although every finite subgroup of G is cyclic, hence it has period 2, there does not exist a compatible family of maximal generators.*

However, if X has only finitely many loops, then starting with a family of maximal generators we can construct a compatible family of maximal generators i.e.

Theorem 2.9 **([T$_2$])** *Let $(\mathcal{G}, X)$ be a graph of finite groups with period q and G its fundamental group. If the first Betti number of X is finite, then G has periodic cohomology after 1-step. If X is a tree then G has period q after 1-step.*

Since a countable locally finite group is the fundamental group of a graph $(\mathcal{G}, X)$ of finite groups with X a tree we obtain :

Corollary 2.10 **([T$_1$])** *A countable locally finite group G has period q after 1-step if and only if every finite subgroup of G has period q.*

The structure of locally finite groups with periodic cohomology after some steps is studied in [G-T]. It turns out that those are countable and have periodic cohomology after 1-step.

One of the results in [G-T] asserts that if a countable locally finite group G has periodic cohomology after 1-step then any two Sylow p-subgroups of G (i.e. maximal p-subgroups of G) are conjugate.

Moreover a Sylow p-subgroup of G is either cyclic or the infinite locally cyclic group C_{p^∞}, or generalized quaternion or the infinite locally generalized quaternion group Q_{2^∞}.

3 Generalized Tate cohomologies

α Farrell cohomology

In [F] Farrell constructed a cohomology theory for groups of finite virtual cohomological dimension which generalizes the Tate cohomology theory of finite groups.

Let G be a group which is virtually torsion-free, i.e. which has a torsion-free subgroup of finite index. By a theorem of Serre [Se] all such subgroups have the same cohomological dimension, and this common dimension is called the virtual cohomological dimension of G denoted vcd G (vcd $G = 0$ if and only if G is finite).

Definition A complete resolution for any group G is an acyclic complex $\mathcal{F} = (F_i, \partial_i)_{i \in \mathbb{Z}}$ of projective $\mathbb{Z}G$-modules, together with a projective resolution $\mathcal{P}$ of G such that $\mathcal{F}$ and $\mathcal{P}$ coincide in sufficiently high dimensions : 999

$$\begin{array}{ccl} & & F_{\lambda-1} \to \cdots \to F_0 \to F_{-1} \to \cdots \\ \cdots F_{\lambda+1} \to F_\lambda & \nearrow \atop \searrow & \\ & & P_{\lambda-1} \to P_{\lambda-2} \to \cdots \to P_0 \to \mathbb{Z} \to 0. \end{array}$$

Clearly this definition generalizes the notion of a complete resolution for a finite group.

Now let G be a group such that vcd $G = n < \infty$, then by [B, ch. X] we have the following.

1) G has a complete resolution.
2) Any two complete resolutions for G are canonically homotopy equivalent, so if M is a $\mathbb{Z}G$-module and $(\mathcal{F}, \mathcal{P})$ a complete resolution for G, then the groups $H^i(\mathrm{Hom}_{\mathbb{Z}G}(\mathcal{F}, \mathcal{M}))$, $i \in \mathbb{Z}$ are independent of the choice of complete resolution and these are the Farrell cohomology groups of G with coefficients in M
$$\widehat{H}^i(G, M) = H^i(\mathrm{Hom}_{\mathbb{Z}G}(\mathcal{F}, \mathcal{M})),\ i \in \mathbb{Z}.$$
3) The group G has a complete resolution $(\mathcal{F}, \mathcal{P})$ such that $\mathcal{F}$ coincides with $\mathcal{P}$ in dimensions $\geq n$.
If G is a finite group then the Farrell cohomology coincides with the Tate cohomology.
4) Clearly there is a canonical map
$$\iota : H^i(G, M) \to \widehat{H}^i(G, M)$$
which is an isomorphism for $i > n$ and an epimorphism for $i = n$.
The Farrell theory has properties analogous to the properties of Tate theory listed in §1.
5) $\{\widehat{H}^i(G, -)\}_{i \in \mathbb{Z}}$ is a cohomological functor which has all the nice cohomological properties i.e.
6) Shapiro's lemma, restriction and transfer maps

7) cup product, which is compatible with that in ordinary cohomology.

In particular, $\widehat{H}^*(G,\mathbb{Z})$ is an anticommutative graded ring with identity element the image in $\widehat{H}^0(G,\mathbb{Z})$ of $1 \in H^0(G,\mathbb{Z}) = \mathbb{Z}$.

Moreover,

a) $\widehat{H}^*(G, \text{projective}) = 0$ hence the functors $\widehat{H}^i(G,-)$ are both effaceable and coeffaceable (which allows dimension shifting in both directions).

b) $\widehat{H}^*(G,-) = 0$ if G is torsion free.

c) $\widehat{H}^*(G,M)$ are torsion groups.

Definition[B, Ch. X] A group G of finite virtual cohomological dimension is said to have periodic (Farrell) cohomology if for some positive integer q there is an element of $\widehat{H}^q(G,\mathbb{Z})$ which is invertible in the ring $\widehat{H}^*(G,\mathbb{Z})$.

Cup product with such an element defines periodicity isomorphisms

$$\widehat{H}^i(G,M) \cong H^{i+q}(G,M),\ i \in \mathbb{Z}$$

for any $\mathbb{Z}G$-module M.

Theorem 3.1 *The following statements are equivalent for a group G with vcd $G = n < \infty$ and G not torsion free.*

(i) *G has periodic Farrell cohomology.*

(ii) *G has periodic cohomology after n-steps.*

(iii) *There is an exact sequence of $\mathbb{Z}G$-modules*

$$0 \to \mathbb{Z} \to A \to P_{q-2} \to \cdots \to P_0 \to \mathbb{Z} \to 0$$

for some q with P_i, $0 \le i \le q-2$ projective $\mathbb{Z}G$-modules and $p\cdot d_{\mathbb{Z}G}A \le n$.

(iv) *Every finite subgroup of G has periodic cohomology.*

Proof By property (4) of Farrell cohomology $\widehat{H}^j(G,-) = H^j(G,-)$ for $j >$ vcd G, hence (i) $\Rightarrow$ (ii). Since $H^j(G,\text{projective}) = 0$ for $j >$ vcd G it is easy to see that if G has periodic cohomology after n-steps then there is a period q such that the q-extension $(*)$ of Theorem 2.1 represents an element g of $H^q(G,\mathbb{Z})$ [e.g. Proposition 1.1 in [T2]]. By Theorem 2.6 the element g is represented by a q-extension as in (iii), so (ii) $\Rightarrow$ (iii). From Theorem 2.6 follows that (iii) represents a unit in $\widehat{H}^*(G,\mathbb{Z})$ if G is not torsion free so (iii) $\Rightarrow$ (i). It was shown by Brown [B, Ch. X, Theorem 6.7] that (i) $\Leftrightarrow$ (iv).

Note that (ii) $\Leftrightarrow$ (iii) for any group G with vcd $G = n < \infty$. □

β Generalized Farrell cohomology

In [W1] Wall suggested generalizing Farrell theory to rings where projectives have finite injective dimension and injectives have finite projective dimension.

Ikenaga in [I] extended Farrell cohomology to the class of groups which admit complete resolutions and for which

$$\sup\{i : \mathrm{Ext}^i_{\mathbb{Z}G}(M,F) \neq 0,\ M\ \mathbb{Z}\text{-free},\ F\ \mathbb{Z}G\text{-free}\}\text{ is finite.}$$

The quantity $\sup\{i : \mathrm{Ext}^i_{\mathbb{Z}G}(M,F) \neq 0,\ M\ \mathbb{Z}\text{-free},\ F\ \mathbb{Z}G\text{-free}\}$ is called the generalized cohomological dimension of G and is denoted by $\underline{\mathrm{cd}}\, G$.

It is easy to show that $\underline{\mathrm{cd}}\, G < \infty$ if and only if projectives have bounded injective dimension.

If G is a group such that $\mathrm{vcd}\, G = n < \infty$, then $\mathrm{vcd}\, G = \underline{\mathrm{cd}}\, G$. There are groups for which $\mathrm{vcd}\, G$ is infinite and $\underline{\mathrm{cd}}\, G$ finite e.g. if G is a countable locally finite group then $\mathrm{vcd}\, G$ is infinite but $\underline{\mathrm{cd}}\, G \leq 1$.

The condition $\underline{\mathrm{cd}}\, G < \infty$ ensures that if G admits a complete resolution, then any two such are canonically homotopy equivalent, so if M is a $\mathbb{Z}G$-module and $(\mathcal{F}, \mathcal{P})$ a complete resolution for G, then the groups $H^i(\mathrm{Hom}_{\mathbb{Z}G}(F, M))$, $i \in \mathbb{Z}$ are independent of the choice of complete resolution and these are the generalized Farrell cohomology groups of G with coefficients in M

$$\widehat{H}^i(G,M) = H^i\left(\mathrm{Hom}_{\mathbb{Z}G}(\mathcal{F}, M)\right), \quad i \in \mathbb{Z}.$$

So for groups G which admit a complete resolution and for which $\underline{\mathrm{cd}}\, G < \infty$, the generalized Farrell cohomology is defined and it is shown that it has properties analogous to the properties of the Farrell theory listed above, where the role of vcd G is played by $\underline{\mathrm{cd}}\, G$:

(3)′ The group G has a complete resolution $(\mathcal{F}, \mathcal{P})$ such that $\mathcal{F}$ coincides with $\mathcal{P}$ in dimensions $\geq \underline{\mathrm{cd}}\, G$.

(4)′ There is a canonical map

$$i : H^i(G,M) \to \widehat{H}^i(C,M)$$

which is an isomorphism for $i > \underline{\mathrm{cd}}\, G$ and an epimorphism for $\underline{\mathrm{cd}}\, G$.

(5)′ $\{\widehat{H}^i(G,-)\}_{i\in\mathbb{Z}}$ is a cohomological functor for which we have

(6)′ Shapiro's lemma, restriction and transfer maps and

(7)′ cup product which is compatible with that in ordinary cohomology.

Moreover,

(a)′ $\widehat{H}^*(G, \text{projective}) = 0$, hence the functors $\widehat{H}^i(G,-)$ are both effaceable and coeffaceable.

Ikenaga then constructs a class of groups, C_∞ , via actions on finite dimensional acyclic simplicial complexes and he shows that if G is in C_∞, then G has a complete resolution $\underline{\mathrm{cd}}\, G < \infty$, hence the generalized Farrell cohomology is defined.

The class C_∞ is defined as follows:

Let C_0 = finite groups, C_n = groups G such that there is an acyclic simplicial complex X on which G acts simplicially so that

(i) $G_\sigma \in C_{n-1}$ for all simplices σ of X

(ii) $\sup_{\sigma\in\Sigma}\{\dim\sigma + \underline{\mathrm{cd}}\, G_\sigma\} < \infty$ where Σ is a set of representatives for the simplices of X mod G and $C_\infty = \cup_n C_n$.

By a theorem of Serre [Se] if $\mathrm{vcd}\, G < \infty$, then $G \in C_1$. Clearly if G acts on a tree with finite vertex stabilizers then $G \in C_1$. In particular a countable locally finite group is in C_1.

If G is in C_∞ and G is torsion free then it has finite cohomological dimension. Hence

(b)′ $\widehat{H}^*(G,-)=0$ if G is in C_∞ and is torsion-free.
However, $\widehat{H}^*(G,M)$ are not necessarily torsion any more.

Remark Some of Ikenaga's results overlap with results of Gedrich and Gruenberg in [G-G].

Definition ([P]) A group G which admits a complete resolution and for which $\underline{\text{cd}} < \infty$ is said to have periodic generalized Farrell cohomology if for some positive integer q there is an element of $\widehat{H}^q(G,\mathbb{Z})$ which is invertible in the ring $\widehat{H}^*(G,\mathbb{Z})$.

Clearly we have

Theorem 3.1′ *The following statements are equivalent for a group G in C_∞ which is not torsion-free*
(i) *G has periodic generalized Farrell cohomology.*
(ii) *G has periodic cohomology after k-steps, for some k.*
(iii) *There is an exact sequence of $\mathbb{Z}G$-modules*

$$0 \to \mathbb{Z} \to A \to P_{q-2} \to \cdots \to P_0 \to \mathbb{Z} \to 0$$

for some q, with P_i projective $\mathbb{Z}G$-modules, $0 \le i \le q-2$ and $p \cdot d_{\mathbb{Z}G} A < \infty$.

Note that (ii) $\Leftrightarrow$ (iii) for any group G which admits a complete resolution and for which $\underline{\text{cd}}\, G < \infty$.

Remark Condition (i) of Theorem 3.1′ is not equivalent any more to every finite subgroup of G having period q, e.g. 2.7. So, although one feels that the periodicity in cohomology of infinite groups derives from the periodicity of the finite subgroups, clearly this is not sufficient.

γ Complete cohomology

For any ring R and A, B left R-modules Mislin in [Mis] defined complete Ext groups associated to A and B as

$$\widehat{\text{Ext}}^n_R(A,B) = \varinjlim_{j \ge 0} S^{-j}\text{Ext}^{n+j}_R(A,B), \quad n \in \mathbb{Z}$$

where $S^{-j}\text{Ext}^{n+j}_R(A,-)$ denotes the j-th left satellite of the functor $\text{Ext}^{n+j}_R(A,-)$.
Alternative equivalent definitions were given by Benson and Carlson [B, B-C] and Vogel [G].
It follows that $\widehat{\text{Ext}}^*_R(A,-) = \{\widehat{\text{Ext}}^n_R(A,-)\}_{n\in\mathbb{Z}}$ is a family of covariant additive functors from R-modules to abelian groups such that
(i) $\widehat{\text{Ext}}^*_R(A,-)$ is a cohomological functor.
(ii) $\widehat{\text{Ext}}^*_R(A,\text{projective}) = 0$.

(iii) There is a map of cohomological functors

$$\iota : \mathrm{Ext}^*_R(A, -) \to \widehat{\mathrm{Ext}}^*_R(A, -)$$

such that if $T^* = \{T_n\}_{n\in\mathbb{Z}} : {}_R\mathcal{M}od \to \mathcal{A}b$ is any other cohomological functor which vanishes on projective modules, then any map of cohomological functors $\widehat{\mathrm{Ext}}^*_R(A, -) \to T^*$ factors uniquely through the given map ι.

There is a nice property of complete groups which is due to Kropholler [K]:

$$p \cdot d_R A < \infty \text{ if and only if } \widehat{\mathrm{Ext}}^0_R(A, A) = 0.$$

In [C-K] Cornick and Kropholler showed that the complete Ext groups can be calculated, under certain conditions, using complete resolutions but not always.

In particular they proved that if there is a complete resolution $(\mathcal{F}, \mathcal{P})$ for an R-module A, i.e. an acyclic complex $\mathcal{F} = (F_i, \partial_i)_{i\in\mathbb{Z}}$ of projective R-modules which coincides with a projective resolutions $\mathcal{P}$ of A in sufficiently high dimensions, with the property that $\mathrm{Hom}_R(\mathcal{F}, Q)$ is acyclic for every projective R-module Q then

$$\widehat{\mathrm{Ext}}^n_R(A, B) = H^n(\mathrm{Hom}_R(\mathcal{F}, \mathcal{B})), \ n \in \mathbb{Z}$$

for any R-module B.

Clearly it follows that if a group G admits a complete resolution and $\underline{\mathrm{cd}}\, G < \infty$, then the generalized Farrell cohomology of G coincides with the complete

$$\widehat{H}^*(G, -) = \widehat{\mathrm{Ext}}^*_{\mathbb{Z}G}(\mathbb{Z}, -)$$

cohomology.

Definition 3.2 A group G is said to have periodic complete cohomology if there is a positive integer q such that the functors $\widehat{H}^i(G, -)$ and $H^{i+q}(G, -)$ are naturally equivalent for all $i \in \mathbb{Z}$.

Remark 3.3 It is easy to see that we have an analogue of Theorem 3.1′ with respect to complete cohomology, for any group G which admits a complete resolution $(\mathcal{F}, \mathcal{P})$ such that $\mathrm{Hom}_{\mathbb{Z}G}(\mathcal{F}, \mathcal{P})$ is acyclic.

4 Free and proper actions on $\mathbb{R}^n \times S^m$

In [W$_1$] Wall conjectured that if a countable group G of finite virtual cohomological dimension has periodic Farrell cohomology then G acts freely and properly on a product $\mathbb{R}^n \times S^m$, for some m and n.

A group G is said to act properly (equivalently properly discontinuously) on $X = \mathbb{R}^n \times S^m$ if the isotropy group G_x of any $x \in X$ is finite and for every $x \in X$ there is a neighborhood U_x such that $gU_x \cap U_x = \emptyset$ for $g \in G \setminus G_x$. If G acts freely and properly on $\mathbb{R}^n \times S^m$, then the orbit space is a manifold which admits $\mathbb{R}^n \times S^m$ as universal covering. Connoly and Prassidis proved the conjecture in 1989 in [C-P] while in 1985 Johnson [J] proved a special case of the conjecture, namely when G

is an extension $1 \to K \to G \to H \to 1$ with K a finitely presented group of type FP and H a finite group of periodic cohomology. Clearly Milnor's condition is not necessary for free and proper actions on $\mathbb{R}^n \times S^m$, for some m and n.

In [P] Prassidis showed that there are groups G of infinite virtual cohomological dimension for which the generalized Farrell cohomology is defined and the periodicity in the generalized Farrell cohomology of G implies the existence of a free and proper action of G on $\mathbb{R}^n \times S^m$, for some m and n.

In particular he shows [P, Proposition 7] that if a group G acts on a tree with finite vertex stabilizers and G has periodic generalized Farrell cohomology then there is a free G-complex E which is homotopy equivalent to a sphere.

He then shows [P, Lemma 9] that if G is countable and the tree on which G acts is countable, then the free G-complex E can be chosen so that the orbit space E/G has the homotopy type of a countable finite dimensional simplicial complex. Then standard methods [e.g. [C-P, Lemma 2.8 and the Proof of Theorem A] produce a free and proper action of G on $\mathbb{R}^n \times S^m$ for some m and n. So he obtains :

Theorem (Prassidis) *If a countable group G acts on a countable tree with finite vertex stabilizers and G has periodic generalized Farrell cohomology then G acts freely and properly on $\mathbb{R}^n \times S^m$, for some m and n.*

Let G be a group which has periodic cohomology after 1-step. Then by Theorem 2.3 G acts on a tree T with finite vertex stabilizers. Moreover, it follows from the proof of Theorem 2.3 that if the group G is countable, then the tree T can be chosen to be a countable tree. Hence by Prassidis's Theorem and Theorem 3.1′, if G is not torsion free then G acts freely and properly on $\mathbb{R}^n \times S^m$, for some m and n. If, however, G is torsion free then G is free and hence it acts freely and properly on $\mathbb{R}^2$. So we obtain:

Theorem A *If a countable group G has periodic cohomology after 1-step then G acts freely and properly on a product $\mathbb{R}^n \times S^m$, for some m and n.*

It follows from Theorem 2.9 that we can construct countable groups G of infinite virtual cohomological dimension which have periodic cohomology after 1-step and hence they act freely and properly on $\mathbb{R}^n \times S^m$, for some m and n.

Conjectures

Conjecture A The following statements are equivalent for a countable group G:

(i) G acts freely and properly on $\mathbb{R}^n \times S^m$, for some n and m.

(ii) G has periodic cohomology after k-steps, for some k.

(iii) There is an exact sequence

$$0 \to \mathbb{Z} \to A \to P_{q-2} \to \cdots \to P_0 \to \mathbb{Z} \to 0$$

for some q with P_i, $0 \leq i \leq q-2$ projective $\mathbb{Z}G$-modules and $p \cdot d_{\mathbb{Z}G} A < \infty$.

(iv) G has periodic complete cohomology (Definition 3.2).

That (i) ⇒ (ii) was essentially shown by Wall in [W1].

Let G be a group of finite virtual cohomological dimension. Then by Remark 3.3 (ii) ⇔ (iii) ⇔ (iv). If G is not torsion free then by the Connoly-Prassidis [C-P] theorem (ii)⇒ (i). If G is torsion free then it has finite cohomological dimension and Johnson showed that then G acts freely and properly on $\mathbb{R}^n$ for some n [e.g., [W2], Theorem 1], hence (ii) ⇒ (i) and thus Conjecture A holds for groups which have finite virtual cohomological dimension. Moreover, it holds for locally finite groups since by Remark 3.3 (ii) ⇔ (iii) ⇔ (iv) for a countable locally finite group and if a countable locally finite group has periodic cohomology after k-steps, then it has periodic cohomology after 1-step.

We also believe that the following may be true :

Conjecture B Let G be a group with periodic cohomology after k-steps.

(i) If G is torsion free, then G has finite cohomological dimension.

(ii) If G is torsion, then G is locally finite.

Conjecture C Let G be a group with periodic cohomology after k-steps, with k minimal. Then for $H^i(G, \text{projective}) = 0$ for $i > k$ and $H^k(G, P_0) \neq 0$ for some $\mathbb{Z}G$-projective module P_0.

References

[B] K.S. Brown, Cohomology of Groups, Graduate Texts in Math. Vol. 87, Springer-Verlag, Berlin-Heidelberg-New York, 1982.

[B, B-C] D.J. Benson and J. Carlson, Products in negative cohomology, J. Pure Appl. Algebra 82 (1992), 107-130.

[C-E] H. Cartan and S. Eilenberg, Homological Algebra, Princeton University Press 1956.

[C-K] J. Cornick and P.H. Kropholler, On complete resolutions, Topology Appl., to appear.

[C-P] F. Connolly and S. Prassidis, Groups which act freely on $\mathbb{R}^n \times S^m$, Topology 28 (1989), 133-148.

[D-D] W. Dicks and M.J. Dunwoody, Groups Acting on Graphs, Cambridge University Press, 1988.

[F] F.T. Farrell, An extension of Tate cohomology to a class of infinite groups, J. Pure Appl. Alg. 10 (1977) 153-161.

[F-S] F.T. Farrell and C.W. Stark, Cocompact spherical-euclidean spaceform groups of infinite vcd, Bull. London Math. Soc. 25 (1993), 189-192.

[G-G] T.V. Gedrich and K.W. Gruenberg ,Complete cohomological functors on groups, Topology Appl. 25 (1987), 203-223.

[G] F. Goichot, Homologie de Tate-Vogel équivariante, J. Pure Appl. Algebra 82 (1992), 39-64.

[G-T] R.M. Guralnick and O. Talelli, A remark on infinite torsion groups with periodic cohomology, Comm. Algebra 20 (1992), 1217-1221 .

[H-R] P.J.Hilton and D. Rees, Natural maps of extension functors and a theorem of R.G. Swan, Proc. Cambridge Philos. Soc. 57 (1961), 489-502.

[H] H. Hopf, Zum Clifford-Kleinschen Raumformen, Math. Ann. 95 (1926), 313-339.

[I] B.M. Ikenaga, Homological dimension and Farrell cohomology, J. Algebra 87 (1984), 422-457.

[J] F.E.A. Johnson, On groups which act freely on $\mathbb{R}^n \times S^m$, J. London Math. Soc. 32 (1985), 370-376.

[Ki] W. Killing, Über die Cliffordl-Kleinschen Raumformen, Math. Ann. 39 (1891), 257-278.

[K] P.H. Kropholler, On groups of type FP_∞, J. Pure Appl. Algebra 90 (1993), 55-67.

[K-T] P.H. Kropholler and O. Talelli, On a properly of fundamental groups of graphs of finite groups, J. Pure Appl. Algebra 74 (1991), 57-59.

[M-T-W] I. Madsen, C.B. Thomas, and C.T.C. Wall, The topological space form problem-II: Existence of free actions, Topology 15 (1976), 375-382.

[Mil] J.W. Milnor, Groups which operate on S^n without fixed points, Amer. J. Math. 79 (1957), 612-623.

[Mis] G. Mislin, Tate cohomology for arbitrary groups via satellites, Topology Appl. 56 (1994), 293-300.

[P] S. Prassidis, Groups with infinite virtual cohomological dimension which act freely on $\mathbb{R}^n \times S^{m-1}$, J. Pure Appl. Algebra 78 (1992), 85-100.

[Se] J.P. Serre, Cohomologie des groupes discrets, Ann. Math. Studies 70 (1971), 77-169.

[Su] M. Suzuki, On finite groups with cyclic Sylow subgroups for odd prime, Am. J. Math. 77 (1955), 657-91.

[Sw$_1$] R.G. Swan, Periodic resolutions for finite groups, Ann. Math. 72 (1960), 267-291.

[Sw$_2$] R.G. Swan, Groups with no Odd dimensional Cohomology, J. Algebra 17 (1971), 401-403.

[T$_1$] O. Talelli, On cohomological periodicity for infinite groups, Comment. Math. Helv. 55 (1980), 178-192.

[T$_2$] O. Talelli, On groups with periodic cohomology after 1-step, J. Pure Appl. Algebra 30 (1983), 85-93.

[T$_3$] O. Talelli, On groups with property $\mathcal{P}_1$, Bull. Soc. Math. Grece (No 5) 29 (1988), 85-90.

[T$_4$] O. Talelli, On groups with $\mathrm{cd}_Q G \leq 1$, J. Pure Appl. Algebra 88 (1993), 245-247.

[T$_5$] O. Talelli, On cohomological periodicity isomorphisms, Bull. London Math. Soc. 28 (1996), 600-602.

[W$_1$] C.T.C. Wall, Periodic projective resolutions, Proc. London Math. Soc. 39 (1979), 509-533.

[W$_2$] C.T.C. Wall, The topological space-form problems, Topology of Manifolds (J.C. Cantrell and C.H. Edwards, Jr. Eds), Proc. Of the Univ. of Georgia Topology of Maniffolds Institute, 1969, Markham, Chicago, 1970, 319-331.

[W$_3$] C.T.C. Wall, Free actions of finite groups on spheres, Proceedings Symposia in Pure Mathematics 32 (Amer. Math. Soc., Providence, R.I, 1978), 145-54.

[J.W] J.A. Wolf, Spaces of constant curvature McGraw-Hill, N.Y. 1967.

[Yo] N. Yoneda, On the homology theory of modules, J. Fac. Sci. Univ. Tokyo Sect. I (1954), 193-227.

ON MODULES OVER GROUP RINGS OF SOLUBLE GROUPS OF FINITE RANK

A. V. TUSHEV

Department of Mathematics, University of Dnepropetrovsk, Prospect Gagarina 72, Dnepropetrovsk 320625, Ukraine

1 Introduction

The study of group rings of soluble groups of finite rank and modules over them was begun with famous papers of P. Hall [13, 14], where he obtained the first results on this subject. Some later these investigations were developed in very important papers of J. Roseblade [22, 23, 24]. These results were followed by many interesting papers on polycyclic group rings and modules over them by other investigators and the list of these papers is too large to be presented here.

Some theorems proved for modules over polycyclic group rings are formulated more naturally for soluble groups of finite rank. In fact, the real cause of some properties of group rings of polycyclic groups is finiteness of rank of these groups. For this and many other reasons, most notably due to the needs of the theory of soluble groups, group rings of soluble groups of finite rank attract attention of many investigators. We must note in particular the following papers of Brookes [2], Brown [7, 8], Musson [19], Wehrfritz [36], Wilson [37], Zaitsev [38]. But we should note that theory of group rings of soluble groups of finite rank and modules over them is not so deep and completed as in the "polycyclic" case. In this survey we consider some techniques, approaches and conjectures which might be useful for further progress in this domain.

In Section 2 we consider control theorems on induced modules. These theorems together with the modernised local approach described in Section 3 give us quite a strong tool for our investigations. In Section 4 we announce results about divisibility in modules over group rings of linear groups which were recently obtained using these techniques, and we also survey some earlier results on this subject. Results on divisibility of modules allow us make our contribution to problems of the Nullstellensatz (Section 5) and the primitivity of group algebras (Section 6) . In Section 7 we describe some progress made in problems of faithful irreducible representations of soluble groups of finite rank over an absolute field. In Section 8 we consider just infinite modules. These modules are connected with some group theory applications, however in our opinion, they are themselves quite interesting for investigations. In the last section we collect some possible group theory applications of our module theoretic results.

In this paper, we try to survey only "post-polycyclic" results, mostly, describing our contribution. Of course, this survey can not be considered as quite complete and I beg pardon if someone's achievements are not reflected here.

2 Control theorems

Let R be a ring, let G be a group and let M be an RG -module. We say that a subgroup H of G is a control subgroup for the module M if there is an RH-submodule U of M such that $M = U \bigotimes_{RH} RG$.

Existence of a control subgroup may be very useful if G has finite torsion-free rank, especially, in the case where

$$r_0(H/C_H(U)) < r_0(G) \qquad (*)$$

because we could use induction on torsion-free rank then. This approach (as many other very fruitful ideas) belongs to Roseblade. Although it is not formulated directly, in fact it was proved in [22] that if A is a finitely generated abelian normal subgroup of a group G, R is a commutative Noetherian domain and M is an RG-module which is not RA-torsion-free then there is an element $0 \neq a \in M$ such that $N_G(Ann_{RA}(a))$ is a control subgroup of aRG.

Let A be an abelian torsion-free group of finite rank acted upon by a group Γ and let I be an ideal of kA, where k is a field. The subgroup $S_\Gamma(I)$ of Γ of elements γ such that $I \cap kB = I^\gamma \cap kB$ for some finitely generated dense subgroup B of A is said to be the standardizer of I (see [2]). The spectrum $Sp(I)$ of I is the set of prime ideals of kA which contain I. The subgroup $Sep_\Gamma(I)$ of $S_\Gamma(I)$ generated by elements γ such that $Sp(I) \cap Sp(I^\gamma) \neq \emptyset$, where $Sp(I)$ is the prime spectrum of I, is said to be the separator of I. In fact, this definition is equivalent to the definition of separator in [33].

In [20] Nabney developed techniques of [22] for the case where A is an abelian torsion-free normal subgroup of finite rank of G replacing $N_G(Ann_{kA}(a))$ by the standardizer $S_G(Ann_{kA}(a))$ of $Ann_{kA}(a)$ in G. But, generally,the relation (*) does not hold if $H = S_G(Ann_{kA}(a))$. To provide (*) we replaced $S_G(Ann_{kA}(a))$ by the separator $Sep_G(Ann_{kA}(a))$ of $Ann_{kA}(a)$ in G (see [33] for details) and this approach was found quite fruitful in the case, where G is a metabelian group of finite rank. However, in the case of soluble groups we have to consider a more general situation where A is not an abelian normal subgroup of G.

Let R be a ring and let M be an R-module. We say that submodules X and Y of M are separated if X and Y have no nonzero isomorphic sections which are isomorphic to a submodule of M.

Let R be a ring and let G be a group with a normal subgroup A. Let M be an RG-module with an RA-submodule W. We say that a subgroup N of G separates W in G if W and Wt are separated RA-modules for any $t \in G$ which is not contained in N. Evidently, a subgroup N separates W if and only if for any $g \in N$ and $t \notin N$ submodules Wg and Wt are separated. It is not difficult to check that the intersection $Sep_{(G,A)}(M, W)$ of all subgroups of G separating W also separates W. The subgroup $Sep_{(G,A)}(M, W)$ will be called the separator of W in G. We should also note that $Sep_{(G,A)}(M, W)$ is the subgroup of G generated by elements $g \in G$ such that W and Wg are not separated. It is also not difficult to show that $Sep_{(G,A)}(akG, akA) \leq Sep_G(Ann_{kA}(a))$ if the subgroup A is torsion-free abelian of finite rank and the element a is chosen as in [33, Theorem 4.2].

Let R be a ring, R-modules X and Y are said to be similar if their injective hulls are isomorphic (see [5]). Let k be a field, let A be a normal subgroup of a group G and let W be a kA-module. By [5, Lemma 3.2], $Stab_G[W] = \{g \in G \mid Wg$ is similar to $W\}$ is a subgroup of G which is called the stabiliser in G of the similarity class $[W]$ of W. We say that a section A/B of W is proper if $A \neq B \neq 0$. It is easy to note that $Stab_G[W] \leq Sep_{(G,A)}(M, W)$ and $Stab_G[W] = Sep_{(G,A)}(M, W)$ if W is uniform and M has no submodule isomorphic to a proper section of W. Control theorems based on stabilisers were considered by Brookes and Brown in [5] and by Brookes in [4]. Some other approaches to control theorems for modules over group rings of infinite groups were developed by Brookes and Brown in [6].

Proposition 1 *Let A be a normal subgroup of a group G. Let R be a ring and let M be an RG-module. Then for any element $a \in M$ $Sep_{(G,A)}(aRG, aRA)$ is a control subgroup of aRG.*

Moreover, the separator is a quite successfully chosen control subgroup because with some additional conditions on A it provides (*).

Theorem A *Let k be a field of characteristic zero and let G be a soluble group of finite torsion-free rank. Let A be a minimax nilpotent torsion-free normal subgroup of G which has no infinite polycyclic G-invariant sections. Let M be a kG -module which is not kA-torsion-free. Then there are a subgroup F of finite index in G such that $A \leq F$, an F-invariant subgroup B of A and an element $a \in M$ such that $akF = akH \bigotimes_{kH} kF$ and $r_0(H/C_H(akH)) < r_0(F)$, where $H = Sep_{(F,B)}(akF, akB)$.*

The proof of this theorem may be found in [35]. In the case where the subgroup A is abelian, our assertion easily follows from [33, Theorem 4.2]. In the not abelian case we developed some ideas introduced by Brookes in [3, Section 5], where the abelian case appears by passing to an appropriate quotient module. In the proof we also use some techniques developed by Brookes and Brown in [5].

3 Local approach

Let RG be the group ring of a locally polycyclic-by-finite (in particular, locally nilpotent) group G over a Noetherian commutative ring R. By local approach we mean techniques developed by Zaitsev [38] and Brown [7] which consist of studying the ring RG and modules over RG by applications of results of Hall to subrings of finitely generated subgroups of G.

In [27] we developed the local approach applying Roseblade's results [22], [24] to modules over group rings of locally polycyclic soluble groups of finite rank.

The most important ingredient for this approach is that the group algebra RG of a locally polycyclic-by-finite group G over a Noetherian ring R is locally Noetherian that is for any finitely generated subgroup H of G the group ring RH is Noetherian. The proof of [38, Theorem 2.3] gives a clear example how it can work.

If G is a finitely generated soluble group and A is the derived subgroup of G then the quotient group G/A is finitely generated abelian. If there exists the partial quotient ring $RG(RA)^{-1}$ with a Noetherian subring $RA(RA)^{-1}$ then arguments of [34, Lemma 11] show that $RG(RA)^{-1}$ is Noetherian. So, there is a hope to modernize the local approach to the general case of soluble groups. The next proposition may be easily proved by the arguments of [34, lemma 11].

Proposition 2 *Let H be a soluble group with a torsion-free nilpotent normal subgroup A and let R be a commutative domain. Let M be an RA-torsion-free RH-module. Then:*

(i) *there exists the partial quotient ring $RH(RA)^{-1}$.*

(ii) *if the quotient group H/A is polycyclic then $RH(RA)^{-1}$ is Noetherian.*

(iii) *there exists an $RH(RA)^{-1}$ -module L such that $M \leq L$.*

This proposition yields us a modernized version of local approach. So, in the case, where M is RA-torsion-free, passing to an $RH(RA)^{-1}$-module we may use the local techniques. If the module M is not RA -torsion-free then we may apply control theorems described above. The embryo of this method may be found in [32] and its offsprings in [34], [35]. It seems to be quite likely that this alternative gives us a very strong tool for future investigations.

4 Divisibility of modules

Let J be a principal ideal domain, let G be a group and let M be a nonzero J-torsion-free JG-module. Following Nabney [20] we say that the module M is sparsely divisible if $M = Mp$ for only finitely many prime elements p of J. In [38], Zaitsev proved that in the case of a locally polycyclic-by-finite group G of finite torsion-free rank any nonzero J-torsion-free finitely generated JG-module M is sparsely divisible.

Conjecture 4.1 Let G be a soluble group of finite torsion-free rank and let J be a principal ideal domain. Let M be a nonzero J-torsion-free JG-module. If M is a nonzero J-torsion-free finitely generated JG-module then M is sparsely divisible.

Moreover, in the case of a nilpotent group G of finite torsion-free rank, if $J = \mathbb{Z}$ or $J = k\langle t\rangle$, where k is a finite field and $\langle t\rangle$ is the infinite cyclic group, Zaitsev obtained some deeper results. In [39, Theorem 3.1] he proved that if M is a finitely generated JG -module of infinite J-rank ($char k \notin SpG$ in the case $J = k\langle t\rangle$) then M/Mp is finite for only finitely many prime elements p of J. If M is a Noetherian JG-module then such a result holds without the restriction on $char k$ [39, Theorem 3.3], see also [17].

In the case of polycyclic groups Zaitsev's theorems follow from much stronger results of Hall [14]. Let π be a set of prime elements of J. We recall that a J-module M belongs to the class $\mathcal{U}(J,\pi)$ if there is a free submodule U of M such that M/U is π-torsion. In [14] Hall proved that if G is a polycyclic group then any finitely generated JG-module belongs to a class $\mathcal{U}(J,\pi)$ for some finite set π. In [28] we proved that if G is an abelian group of finite torsion-free rank then any Noetherian JG-module belongs to a class $\mathcal{U}(J,\pi)$ for a finite set π.

Conjecture 4.2 Let G be a soluble group and let J be $\mathbb{Z}$ or $k\langle t\rangle$, where k is a finite field and $\langle t\rangle$ is the infinite cyclic group. Any Noetherian JG-module belongs to a class $\mathcal{U}(J,\pi)$ for a finite set π if and only if G has finite torsion-free rank.

It easily follows from Zaitsev's result [20, Theorem 2.3] that if G is a locally polycyclic group of finite rank and J is a principal ideal domain with infinite set of prime elements then for any simple JG-module M there is a prime element $p \in J$ such that $Mp = 0$.

Throughout we consider linear groups only over fields of characteristic zero. By Auslander-Swan theorem, any polycyclic-by- finite group is linear. The techniques developed in [35] allow us to prove the following theorem.

Theorem B *Let G be a linear group of finite (Prufer) rank over a field of characteristic zero and let J be a principal ideal domain of characteristic zero with infinite set of prime elements. Then for any simple JG-module M there is a prime element $p \in J$ such that $Mp = 0$.*

Corollary B *Let G be a linear group over a field of characteristic zero and let J be a principal ideal domain of characteristic zero with countable set of prime elements. For any simple JG-module M there is a prime element $p \in J$ such that $Mp = 0$ if and only if G has finite rank.*

In the case $J = k\langle t\rangle$ where k is a field of characteristic zero and $\langle t\rangle$ is an infinite cyclic group the proof of these results may be found in [35].

Conjecture 4.3 Theorem B and Corollary B do not depend on the characteristic of J.

5 The Nullsteilensatz

Let G be a group and let k be a field. We say that the Nullstellensatz holds for kG if $End_{kG}M$ is algebraic over k for any irreducible kG-module M (see [18]).

Various versions of the Nullstellensatz were proved for group rings of polycyclic, locally finite and locally polycyclic-by-finite groups, by Hall [14], Roseblade [22], McConnell [18], Baer [1] and Brown [8].

In [35], due to Theorem B we also made our contribution to this issue.

Theorem C *Let G be a linear group of finite rank and let k be a field of characteristic zero. Then the Nullstellensatz holds for kG.*

So, we may conclude that finiteness of rank is the real reason why the Nullstellensatz holds for polycyclic-by-finite groups in the class of linear groups.

Conjecture 5.1 Theorem C holds for any field k.

We also suspect that in the class of group algebras of soluble groups the Nullstellensatz is also connected with finiteness of torsion-free rank. At least in [34] we proved that the Nullstellensatz holds for group algebras of metabelian groups of finite rank over a field of characteristic zero.

Conjecture 5.2 Let G be a soluble group and let k be a field. If G has finite torsion-free rank then the Nullstellensatz holds for kG. If k is countable then finiteness of torsion-free rank of G is also necessary for the Nullstellensatz.

6 Primitivity of group algebras

In [12] Farkas and Pasman proved that the group algebra kG of a polycyclic-by-finite group G over a field k of characteristic zero is primitive if and only if the FC-centre $\Delta(G)$ of G is trivial and conjectured that triviality of $\Delta(G)$ is also necessary and sufficient condition for primitivity of the group algebra kG of a polycyclic-by-finite group G over an arbitrary non-absolute field k. This assertion was proved by Roseblade in [23]. In [7] Brown conjectured that this result remains true in the case of a soluble group G of finite rank and Brookes [2] proved that if G is a soluble group of finite rank such that $\Delta(G) = 1$ and k is a non-absolute field then kG is primitive. However, it is still unknown whether the triviality of $\Delta(G)$ is also necessary for primitivity of kG.

As an application of Theorem B we proved the next theorem in [35].

Theorem D *Let G be a soluble-by-finite group of finite rank and let k be a field of characteristic zero. The group algebra kG is primitive if and only if $\Delta(G) = 1$.*

In the case of a metabelian group G such a result was proved in [34].

7 Irreducible representations of soluble groups over an absolute field

By the abelian socle $Absoc(G)$ of a group G we mean the characteristic subgroup of G which is generated by the minimal abelian normal subgroups of G. If G has no minimal abelian normal subgroups then $Absoc(G) = 1$. We say that a normal subgroup A of a group G is essential if $A \cap B \neq 1$ for any nontrivial normal subgroup B of G. We let S be the class of groups G for which every minimal abelian normal subgroup is finite and $Absoc(G)$ is essential. The class S is rather large and contains besides finite groups also for instance locally normal groups and torsion soluble groups of finite rank.

By [27, Theorem 3], a locally polycyclic soluble group G of finite rank has a faithful irreducible representation over an absolute field k if and only if G has no torsion-free normal subgroup, the socle $Soc(G)$ of G is a locally cyclic $\mathbb{Z}G$-module, where G acts on $Soc(G)$ by conjugation, and $char k \notin \pi(Soc(G))$. This easily implies that if a soluble locally polycyclic group G of finite rank has a faithful irreducible representation over an absolute field k then $G \in S$.

In [31, Theorem 2] we proved that if $G \in S$ and k is an absolute field then the group G has a faithful irreducible representation over k if and only if $Absoc(G)$ of G is a locally cyclic $\mathbb{Z}G$-module, where G acts on $Absoc(G)$ by conjugation, and $char k \notin \pi(Absoc(G))$. In the case of finite groups this result follows from the Gaschutz criterion [9].

The search of condition for existence of faithful irreducible representations of soluble groups of finite rank over an absolute field has many obstacles. The following conjectures may point us in the right direction on this way.

Conjecture 7.1 Let G be a soluble group of finite rank and let k be an absolute field. If G has an irreducible faithful representation over k then G has no infinite polycyclic normal subgroup and $Soc(G)$ is a locally cyclic $\mathbb{Z}G$ -module such that $char k \notin \pi(Soc(G))$.

Conjecture 7.2 Let G be a soluble group of finite rank and let k be an absolute field. Suppose that G has no abelian normal torsion-free subgroup A such that $char k \in Sp(A)$ and $Soc(G)$ is a locally cyclic $\mathbb{Z}G$-module such that $char k \notin \pi(Soc(G))$. Then the group G has an irreducible faithful representation over k.

8 Just infinite modules

An abelian group M acted upon by a group G is said to be a G-module. A G-module M is said to be faithful if $C_G(M) = 1$.

We recall that a G-module M is said to be just infinite if any proper submodule of M has a finite index in M. Throughout we also assume that just infinite modules are residually finite and faithful. It is easy to note that any just infinite module M is either a torsion-free group or an elementary abelian p-group. We say that M has characteristic 0 in the first case or p in the latter.

At first, Groves [11] considered just infinite modules over finitely generated abelian groups. In [21] Robinson and Wilson described just infinite modules over polycyclic groups. We should note that in the polycyclic case (but only in this case) due to results of Roseblade [24] just infinite modules may be also defined as finitely generated 1-critical modules in sense of Krull dimension.

Just infinite G-modules over nilpotent groups of finite torsion-free rank were considered in [39] by Zaitsev, Kurdachenko and the author. By [39, Theorem 4.1], in this case G is finitely generated and abelian-by-finite.

In [25, 26] we proved that if G is a soluble group of finite torsion-free rank and M is a just infinite G-module of characteristic zero then G is finitely generated and abelian-by-finite. In [16] Karbe and Kurdachenko generalized this result to the class of locally soluble groups.

In [29, Theorem 1] we proved that if G is a soluble group of finite rank and M is a just infinite G-module then G is almost torsion-free. By [29, Theorem 2], in this case G is finitely generated abelian-by-finite if and only if $|\Delta(G)| = \infty$. If the group G is locally polycyclic of finite rank then, by [29, Theorem 3], G is polycyclic.

The case of finitely generated metabelian groups of finite rank was considered in [30]. In this case we obtained a description of just infinite modules which is similar to that obtained in [21].

Theorem E (Cf. [30, Theorem 3]) *Let G be a finitely generated metabelian group of finite rank and let M be a faithful just infinite G-module of characteristic p.*

(i) *If* $|\Delta(G)| = \infty$ *then* G *is a finitely generated abelian-by-finite group.*

(ii) *If* $|\Delta(G)| < \infty$ *then there are a normal subgroup* H *of finite index in* G *and a just infinite* kH*-submodule* $N \leq M$ *such that* $H/C_H(N) = A\lambda T$ *is a semidirect product of a torsion-free abelian group* A *of finite rank and a free abelian group* T *of finite rank such that* A *is a Noetherian* T*-plinth.*

So, we have the following conjecture.

Conjecture 8.1 Theorem E holds for any soluble group G of finite rank.

9 Group theory applications

If A is an abelian normal subgroup of a group G then A may be considered as a $\mathbb{Z}G$-module, where G acts on A by conjugation. So, we can apply Theorem B in the case where $J = \mathbb{Z}$ to group theory.

A group G satisfies the weak minimal condition for normal subgroups if for any descending chain $\{G_n\}$ of normal subgroups of G there is an integer m such that $|G_n/G_{n+1}| < \infty$ for any $n > m$.

A soluble group G is said to be minimax if it has a finite series each of whose factors is either cyclic or quasicyclic. In [15] Zaitsev conjectured that a torsion-free soluble group G satisfies the weak minimal condition for normal subgroup if and only if it is minimax. This conjecture also has affirmative answer due to Theorem B.

Following Groves [10] we say that a group G is just-of-infinite rank if it has infinite rank but any proper quotient group of G has finite rank. Due to theorem B we can prove that any soluble group G just-of-infinite rank is not torsion-free.

Conjecture 9.1 Each soluble group G just-of-infinite rank has an infinite normal elementary abelian subgroup A which is either irreducible or just infinite $\mathbb{Z}G$-module, where G acts on A by conjugation.

Thus, the module theoretic results described above may be very useful in the construction of soluble groups just-of-infinite rank. Some results of this kind were obtained in [30].

Acknowledgements I am deeply grateful to the Dnepropetrovsk branch of International Renaissance Foundation, the organizing committee of "Groups St Andrews 1997 in Bath" and, especially, Geof. Smith for their support of my attending of the conference. I am also deeply grateful to the referee for his helpful notes.

References

[1] R. Baer, "Irreducible groups of automorphisms of abelian groups", *Pacific J. Math.* 14 (1964) 385-406.

[2] C.J.B. Brookes, "Ideals in group rings of soluble groups of finite rank", *Math. Proc. Cambridge. Philos. Soc.* 97 (1985) 27-49.

[3] C.J.B. Brookes, "Modules over polycyclic groups", *Proc. London Math. Soc.* (3) 57 (1988) 88-108.

[4] C.J.B. Brookes, "Stabilisers of injective modules over nilpotent groups", *Proc. of Singapore Group Theory Conference, 1987, de Gruyter* (Berlin: 1989) 275-291.

[5] C.J.C. Brookes, K.A. Brown, "Primitive group rings and Noetherian rings of quotients", *Trans. Amer. Math. Soc.* (2) 288 (1985) 605-623.

[6] C.J.C. Brookes, K.A. Brown, "Injective modules, induction maps and endomorphism rings", *Proc. London Math. Soc.* (3) 67 (1993) 127-158.

[7] K.A. Brown, "Primitive group rings of soluble groups", *Archiv der Math.* 36 (1981) 404-413.

[8] K.A. Brown, "The Nullstellensatz for certain group rings", *J. London Math. Soc.* (2) 26 (1982) 425-434.

[9] W. Gaschutz, "Endliche Gruppen mittreuen absolutirreduziblen Darstellungen", *Math. Nachr.* (3/4) 12 (1954) 253-255.

[10] J.R.J. Groves, "Soluble groups with every proper quotient polycyclic, *Illinois J.Math.* 22 (1978) 90-95.

[11] J.R.J. Groves, "Metabelian groups with finitely generated integral homology", *Quart. J. Math. Oxford* (2) 33 (1982) 405-420.

[12] D.R. Farcas and D.S. Passman, "Primitive Noetherian group rings", *Comm. Algebra* 6 (1982) 301-315.

[13] P. Hall, "Finiteness conditions for soluble groups", *Proc. London Math. Soc.* (3) 4 (1954) 419-436.

[14] P. Hall, "On the finiteness of certain soluble groups", *Proc. London Math. Soc.* (3) 9 (1959) 595-622.

[15] *Kourovka Notebook* (Unsolved problems of group theory) (Novosibirsk, 12th Ed., 1992).

[16] M.I. Karbe, L.A. Kurdachenko, "Just infinite modules over locally soluble groups", *Archiv der Math.* 51 (1991) 401-414.

[17] L.A. Kurdachenko, A.V. Tushev, D.I. Zaitsev, "Noetherian modules over nilpotent groups of finite rank", *Archiv der Math.* 56 (1991) 433-436

[18] J.C. McConnell, "The Nullstellensatz and Jacobson properties for rings of differential operators", *J. London Math. Soc.* 26 (1982) 37-42.

[19] I. Musson, "On the structure of certain injective modules over group algebras of soluble groups of finite rank, *J. Algebra* (1) 85 (1981) 51-75.

[20] I.T. Nabney, *Soluble minimax groups and their representations.* Ph.D. thesis (University of Cambridge, 1989).

[21] D.J.S. Robinson, J.S.Wilson, "Soluble groups with many polycyclic quotients", *Proc. London Math. Soc.* (3) 48 (1984) 193-229.

[22] J.E. Roseblade, "Group rings of polycyclic groups" *J. Pure Appl. Algebra* 3 (1973) 307-328.

[23] J.E. Roseblade, "Prime ideals in group rings of polycyclic groups", *Proc. London Math. Soc.* (3) 36 (1978) 385-447.

[24] J.E. Roseblade, "Applications of the Arthin-Rees lemma to group rings", *Symp. Math.17.* (London: Academic Press, 1976) 471-478.

[25] A.V. Tushev, "Residually finite soluble torsion-free groups with the weak minimal condition for normal subgroups", *18th All-Union Conf. on Alg. Prt.2* (Kishenev, 1985) 175. (in Russian)

[26] A.V. Tushev, "Condition $Min - \infty - N$ and the representations of soluble groups connected with it", *Ukrainian Math. J.* (5) 42 (1990) 599-602.

[27] A.V. Tushev, "Irreducible representations of locally polycyclic groups over an absolute field", *Ukrainian Math. J.* (10) 42 (1990) 1233-1238.

[28] A.V. Tushev, "Noetherian modules over abelian groups of finite torsion-free rank", *Ukrainian Math. J.* (7,8) 43 (1991) 975-981.
[29] A.V. Tushev, "Just infinite modules over locally polycyclic groups of finite rank", *Infinite groups and connected structures.* (Kiev: In-te Math.Acad.Sci.Ukraine, 1993) 312-325. (in Russian)
[30] A.V. Tushev, "On soluble groups all of whose proper quotient groups have finite rank", *Ukrainian Math. J.* (9) 45 (1993) 1430-1437.
[31] A.V. Tushev, "On irreducible representations of locally normal groups", *Ukrainian Math. J.* (12) 45 (1993) 1900-1906.
[32] A.V. Tushev, "On the primitivity of group algebras of certain classes of soluble groups of finite rank", *Sbornik: Mathematics* (3) 186 (1995) 447-463.
[33] A.V. Tushev, "Spectra of conjugated ideals in group algebras of abelian groups of finite rank and control theorems", *Glasgow Math. J.* 38 (1996) 309-320.
[34] A.V. Tushev, "Induced modules over group algebras of metabelian groups of finite rank", *Archiv der Math.*, (to appear).
[35] A.V. Tushev, "The Nullstellensatz for group algebras of linear groups" (to appear).
[36] B.A.F. Wehrfritz, "Invariant maximal ideals in certain group algebras", *J. London Math. Soc.* 46 (1992) 101-110.
[37] J.S. Wilson, "Soluble product of minimax groups, and nearly surjective derivations", *J. Pure Appl. Algebra* 53 (1988) 297-318.
[38] D.I. Zaitsev, "Products of abelian groups", *Algebra i Logica* 19 (1980) 150-172.
[39] D.I. Zaitsev, L.A.Kurdachenko, A.V.Tushev, "Modules over nilpotent groups of finite rank", *Algebra and Logic* 24 (1986) 412-436.

ON SOME SERIES OF NORMAL SUBGROUPS OF THE GUPTA-SIDKI 3-GROUP

ANA CRISTINA VIEIRA

Departamento de Matemática, Universidade Federal de Minas Gerais, Belo Horizonte, M.G., Brazil

1 Introduction

In [3], N. Gupta and S. Sidki constructed one of the simplest examples of recursively defined infinite finitely generated periodic group (*Burnside Groups*) . This is the infinite 3-group of automorphism $\mathcal{G}$ of an infinite one-rooted regular 3-tree $\mathcal{T}$ generated by two automorphisms x and y such that $x^3 = y^3 = 1$ defined in Section 2.

An important property of this group is the fact that all its quotients are finite. This is a motivation to study some series of normal subgroups of $\mathcal{G}$. For other remarkable properties of this group we refer to [4], [5] and [9]. In this paper, we are interested in the behavior of the lower central series $\gamma_k(\mathcal{G})$ and derived series $\mathcal{G}^{(k)}$ of $\mathcal{G}$.

Let $\mathcal{G}_k$, $k \geq 1$, be the subgroup of $\mathcal{G}$ which fixes pointwise the kth level vertices of the tree $\mathcal{T}$. The following theorem of Sidki [10] is the starting point of our work.

Theorem 1 *The subgroup $\mathcal{G}_1$ is the normal closure of $\langle y \rangle$ in $\mathcal{G}$, $\mathcal{G}_2 = \gamma_2(\mathcal{G}_1)\langle y^{[1]} \rangle$ and $\mathcal{G}_{k+1} = \prod_3 \mathcal{G}_k$, $k \geq 2$. Furthermore,*

$$[\mathcal{G} : G_1] = 3,\ [\mathcal{G} : G_k] = 3^{2.3^{k-2}+1},\ k \geq 2.$$

Moreover,

$$\gamma_2(\mathcal{G}_1) = \prod_3 \gamma_2(\mathcal{G}),\ \gamma_2(\mathcal{G}) = \gamma_2(\mathcal{G}_1)\langle [x,y], y^{[1]} \rangle \text{ and } \mathcal{G}/\gamma_2(\mathcal{G}_1) \cong C_3 \mathrm{Wr} C_3,$$

where $y^{[1]} = (y,y,y)$ and for $U \leq \mathcal{G}$, $\prod_3 U$ denotes the product $U \times U \times U = \{(u_1,u_2,u_3) | u_i \in U, 1 \leq i \leq 3\}$.

The main results we shall prove here are relations between the lower central series $\gamma_k(\mathcal{G})$ and the derived series $\mathcal{G}^{(k)}$ of $\mathcal{G}$ involving the chain of the stabilizer subgroups $\mathcal{G}_k$. The order of some important quotients are also determined.

Theorem A (1) *$rank(\gamma_k(\mathcal{G})/\gamma_{k+1}(\mathcal{G})) \leq 2$, for $1 \leq k \leq 9$;*
(2) $[\gamma_2(\mathcal{G}), \gamma_3(\mathcal{G})] = \gamma_5(\mathcal{G})$.

Theorem B (1) $\mathcal{G}^{(2)} = \gamma_5(\mathcal{G})$;
(2) $\mathcal{G}^{(k+1)} \leq \mathcal{G}_{k+1} \leq \mathcal{G}^{(k)} \leq \mathcal{G}_k$, *for all $k \geq 1$;*
(3) $[\mathcal{G} : G'] = 3^2$, $\left[\mathcal{G} : G^{(k)}\right] = 3^{5.3^{k-2}+1}$, *for all $k \geq 2$.*

Theorem C (1) $|\mathcal{G}_k/[\mathcal{G}_k,\mathcal{G}]| = 3^2$, *for all* $k \geq 2$;
(2) $|\mathcal{G}^{(k)}/[\mathcal{G}^{(k)},\mathcal{G}]| = 3^2$, *for all* $k \geq 2$.

In [1], Grigorchuk constructed an infinite 2-group $\mathcal{H}$ which can be interpreted as a group of automorphisms of a binary tree T, generated by τ, which permutes cyclically the vertices of the first level of T, and automorphisms recursively defined by $u = (1,v), v = (\tau,w)$ and $w = (\tau,u)$, satisfying $\tau^2 = u^2 = v^2 = w^2 = 1$ and $uv = w$.

Some interesting properties of this group are given in [2]. Rozhkov calculated in [6] the indices of the stabilizer subgroups and terms of the derived series of $\mathcal{H}$. He proved that:

Theorem 3 *The following equalities hold in* $\mathcal{H}$*:*
(i) $[\mathcal{H}:\mathcal{H}_2] = 2^3, [\mathcal{H}:\mathcal{H}_k] = 2^{2+5.2^{k-3}}, k \geq 3$;
(ii) $[\mathcal{H}:\mathcal{H}'] = 2^3, [\mathcal{H}:\mathcal{H}^{(2)}] = 2^7, [\mathcal{H}:\mathcal{H}^{(k)}] = 2^{2+2^{2k-2}}, k \geq 3$.

Rozhkov published a result concerning the lower central series of $\mathcal{H}$ in [7]:

Theorem 3. *In the lower central series of* $\mathcal{H}$*, we have that*

$$\mathcal{H}/\gamma_2(\mathcal{H}) \cong \mathbb{Z}_2 \oplus \mathbb{Z}_2 \oplus \mathbb{Z}_2,$$
$$\gamma_2(\mathcal{H})/\gamma_3(\mathcal{H}) \cong \mathbb{Z}_2 \oplus \mathbb{Z}_2.$$

Furthermore, if $m = 1, 2, \ldots$

$$\gamma_n(\mathcal{H})/\gamma_{n+1}(\mathcal{H}) \cong \begin{cases} \mathbb{Z}_2 \oplus \mathbb{Z}_2, & if\ 2^m + 1 \leq n < 3.2^{m-1} + 1, \\ \mathbb{Z}_2, & if\ 3.2^{m-1} + 1 \leq n < 2^m + 1. \end{cases}$$

In the case of the group $\mathcal{G}$, by using the group software GAP [8], we can extend our calculations and determine

$$r_k = rank(\gamma_k(\mathcal{G})/\gamma_{k+1}(\mathcal{G})) \leq 4, \text{ for } 1 \leq k \leq 39.$$

In spite of the slow growth of r_k, it seems to us that it is unbounded and there is reason to believe that the average rank of factors $\gamma_k(\mathcal{G})/\gamma_{k+m}(\mathcal{G})$ grows exponentially, for some values of k and m.

2 The construction of $\mathcal{G}$

Let $\mathcal{T}$ be the infinite ternary tree with root $\emptyset$. We label the other vertices of $\mathcal{T}$ by elements of the monoid $\mathcal{M}$ freely generated by $\{0,\ 1,\ 2\}$, on which we define the order relation: $m \leq m' \Leftrightarrow m'$ is as prefix of m. Let $\mathcal{T}_m$ denote the subtree of $\mathcal{T}$ headed by $m \in \mathcal{M}$. If $\mathcal{A} = Aut(\mathcal{T})$, we observe that each automorphism α of $\mathcal{T}$ can be written in the form $\alpha = (\alpha_0, \alpha_1, \alpha_2)\sigma, \sigma \in S_3$, the symmetric group on $\{0,1,2\}$. Thus, $\alpha = (\alpha_0, \alpha_1, \alpha_2)$ is an automorphism fixing the first level vertices and acting as the automorphism α_0 on $\mathcal{T}_0$, α_1 on $\mathcal{T}_1$ and α_2 on $\mathcal{T}_2$.

We denote by $\alpha^{[1]}$ the automorphism (α,α,α) and $\alpha^{[i+1]} = (\alpha^{[i]},\alpha^{[i]},\alpha^{[i]})$, $i \geq 1$. We use $\alpha_{[1]}$ to denote the automorphism $(\alpha,1,1)$ and $\alpha_{[i+1]} = (\alpha_{[i]},1,1)$, $i \geq 1$.

Consider two particular automorphisms of $\mathcal{T}$. The first of them, denoted by x, permutes cyclically the first level vertices of $\mathcal{T}$, $x : iu \to (i+1)u$ (addition is done modulo 3), and the second one, y, is recursively defined by $y = (y, x, x^{-1})$.

The Gupta-Sidki group is the subgroup of $\mathcal{A}$ generated by x and y

$$\mathcal{G} = \langle x, y \rangle .$$

Let $\mathcal{A}_k$, $k \geq 0$, be the subgroup of $\mathcal{A}$ which fixes pointwise the kth level vertices of $\mathcal{T}$. We see that $\mathcal{A}/\mathcal{A}_k$ is the group of automorphisms of the finite tree $\mathcal{T}(k)$ containing all the vertices $u \in M$ of all levels less than or equal to k. So if $\alpha \in \mathcal{A}$, the representative of α in $\mathcal{A}/\mathcal{A}_k$ is determined by its action on the kth level vertices of $\mathcal{T}$.

Define $\mathcal{G}_k = \mathcal{A}_k \cap \mathcal{G}$, $k \geq 0$. We will see that the quotients $\mathcal{G}/\mathcal{G}_k$ play a special role on the study of the lower central series and of the derived series of $\mathcal{G}$. We observe that $\mathcal{G}/\mathcal{G}_k$ is represented on the truncated tree $\mathcal{T}(k)$ and so, $\mathcal{G}/\mathcal{G}_k$ is isomorphically embedded into $\mathrm{Wr}_k C_3$, the iterated wreath product of the cyclic group C_3.

3 Some special products in $\mathcal{G}$

We can note that the set of automorphisms $\mathcal{G} \times \mathcal{G} \times \mathcal{G}$ contains some subsets which are subgroups of $\mathcal{G}$; for instance $\prod_3 \mathcal{G}_k$, $k \geq 2$, as we can see from Theorem 1. Some products like that are very useful in several calculations. We establish now some properties concerning the commutator calculus in $\mathcal{G}$, involving product of subgroups. From now on, we use c to denote the commutator $[x, y] = (xy, y^{-1}x, x)$ and for $U \leq \mathcal{G}$, we denote $[U, n\mathcal{G}] = [[U, \mathcal{G}], (n-1)\mathcal{G}]$, for $n \geq 2$. If $\prod_3 U \leq \mathcal{G}$, we observe the following:

Lemma 1. *If $u \in U$, then*

(i) $[(u, 1, 1), x] = (u^{-1}, u, 1)$;

(ii) $[(u, u^{-1}, 1), x] = u^{-[1]}(1, u^3, 1)$;

(iii) $[(u, 1, 1), y] = ([u, y], 1, 1)$;

(iv) $[(1, u, 1), y] = (1, [u, x], 1)$.

Parts (*i*) and (*ii*) follow from the action of x

$$[(u, 1, 1), x] = (u^{-1}, 1, 1)(1, u, 1), \ [(u, u^{-1}, 1), x] = (u^{-1}, u, 1)(1, u, u^{-1}).$$

Parts (*iii*) and (*iv*) follow from the definition of $y = (y, x, x^{-1})$

$$[(u, 1, 1), y] = ([u, y], 1, 1), \ [(1, u, 1), y] = (1, [u, x], 1). \qquad \square$$

Lemma 2 *If $U^3 \leq [U, \mathcal{G}]$, then $\prod_3 [U, \mathcal{G}] = [\prod_3 U, 3\mathcal{G}]$.*

Proof Since $\mathcal{G} = \mathcal{G}_1\langle x \rangle$ and $\mathcal{G}_1 = \langle y \rangle^{\mathcal{G}}$, the normal closure of $\langle y \rangle$ in $\mathcal{G}$, then:

$$\left[\prod_3 U, \mathcal{G}\right] = \left[\prod_3 U, y\right]\left[\prod_3 U, x\right].$$

By the previous lemma, in $[\prod_3 U, y]$, we obtain the elements $(1, [u, x], 1)$ and $([u, y], 1, 1), \forall u \in U$, producing $\prod_3 [U, \mathcal{G}]$. Moreover, in $[\prod_3 U, x]$, we had $(u, u^{-1}, 1)$ and $u^{-[1]}(1, u^3, 1)$, where $u \in U$. But, since the quotient $U/[U, \mathcal{G}]$ has exponent 3, on defining $U_x = \langle (u, u^{-1}, 1), u^{[1]} \mid u \in U \rangle$, we have:

$$\left[\prod_3 U, \mathcal{G}\right] = (\prod_3 [U, \mathcal{G}])U_x.$$

The commutator of $[\prod_3 U, \mathcal{G}]$ with $\mathcal{G}$ gives us

$$\left[\prod_3 U, 2\mathcal{G}\right] = \left[\prod_3 [U, \mathcal{G}], \mathcal{G}\right] [U_x, \mathcal{G}].$$

But, from the previous calculations, we have

$$\left[\prod_3 [U, \mathcal{G}], \mathcal{G}\right] = (\prod_3 [U, 2\mathcal{G}]) [U, \mathcal{G}]_x,$$

where $[U, \mathcal{G}]_x = \langle (([u, g], [u, g]^{-1}, 1), [u, g]^{[1]} \mid u \in U, g \in \mathcal{G} \rangle$. Furthermore, calculations of $[U_x, x]$ produces $u^{[1]}, \forall u \in U$ and from $[U_x, y]$ we obtain the elements $([u, y], [u^{-1}, x], 1)$, $([u, y], [u, x], [u, x^{-1}])$, where $u \in U$.

Now since

$$[u^{-1}, x] = [u, x]^{-1} \left[[u, x]^{-1}, u^{-1}\right], \quad [u, x^{-1}] = [u, x]^{-1} [u, x^{-1}, x]^{-1},$$

then,

$$([u, y] [u^{-1}, x], 1) \equiv \left([u, y], [u, x]^{-1}, 1\right) \bmod (\prod_3 [U, 2\mathcal{G}]),$$

$$([u, y], [u, x], [u, x^{-1}]) \equiv \left([u, y], [u, x], [u, x]^{-1}\right) \bmod (\prod_3 [U, 2\mathcal{G}]).$$

Thus, since the element $([u, x], [u, x]^{-1}, 1)$ has been generated in $[U, \mathcal{G}]_x$, we have $([u, y], 1, 1)$ also generated and consequently, we are able to produce $([u, x], 1, 1)$. Therefore, we have $\prod_3 [U, \mathcal{G}]$ and then

$$\left[\prod_3 U, 2\mathcal{G}\right] = (\prod_3 [U, \mathcal{G}])U^{[1]},$$

where $U^{[1]} = \langle u^{[1]} \mid u \in U \rangle$. Consequentely,

$$\left[\prod_3 U, 3\mathcal{G}\right] = \left[\prod_3 [U, \mathcal{G}], \mathcal{G}\right] \left[U^{[1]}, \mathcal{G}\right].$$

Using again the previous arguments, we have

$$\left[\prod_3 U, 3\mathcal{G}\right] = (\prod_3 [U, 2\mathcal{G}]) [U, \mathcal{G}]_x \left[U^{[1]}, \mathcal{G}_1\right]$$

and now it is clear that $\prod_3 [U, \mathcal{G}] = [\prod_3 U, 3\mathcal{G}]$. □

We can also note that

Lemma 3 *If U is a subgroup of $\mathcal{G}$ such that $U/[U,\mathcal{G}]$ has exponent 3 and $N_k = \prod_{3^k} U$, $k \geq 1$, is still a subgroup of $\mathcal{G}$ then*

$$\frac{N_k}{[N_k,\mathcal{G}]} \cong \frac{U}{[U,\mathcal{G}]}.$$

Proof If $N_1 = \prod_3 U$, with U in the conditions of the previous lemma, then we observe

$$\frac{N_1}{[N_1,\mathcal{G}]} = \frac{\prod_3 U}{\prod_3 ([U,\mathcal{G}])U_x}.$$

But,

$$\frac{\prod_3 U}{\prod_3 [U,\mathcal{G}]} \cong \prod_3 \frac{U}{[U,\mathcal{G}]}$$

and since that $U_x = \langle (u,u^{-1},1), u^{[1]} | u \in U\rangle$, in the quotient $N_1/[N_1,\mathcal{G}]$ will survive only a copy of $U/[U,\mathcal{G}]$; that is,

$$\frac{N_1}{[N_1,\mathcal{G}]} \cong \frac{U}{[U,\mathcal{G}]}.$$

By using this reasoning in repetition, we have for $N_k = \prod_3 U$

$$\frac{N_k}{[N_k,\mathcal{G}]} \cong \frac{U}{[U,\mathcal{G}]}.$$

□

Lemma 4 *If $i,j \geq 2$, then*

(i) $\prod_3 \gamma_i(\mathcal{G}) \leq \gamma_i(\mathcal{G})$;

(ii) $\prod_3 \gamma_{i+1}(\mathcal{G}) \leq [\prod_3 \gamma_i(\mathcal{G}), 2\mathcal{G}]$;

(iii) *If* $\prod_3 \gamma_i(\mathcal{G}) \leq \gamma_j(\mathcal{G})$ *then* $\{(g,g^{-1},1), g^{[1]} \mid g \in \gamma_i(\mathcal{G})\} \subseteq \gamma_{j+1}(\mathcal{G})$.

Proof Since $\gamma_i(\mathcal{G})/\gamma_{i+1}(\mathcal{G})$ has exponent 3, for all $i \geq 2$, the parts (*i*) and (*ii*) follow from Lemma 2. To prove part (*iii*) it is sufficient to calculate $[(g,1,1),x]$ and $[(g,1,1),2x]$ and use Lemma 1. □

4 The lower central series of $\mathcal{G}$

By Theorem 1, $\gamma_2(\mathcal{G}) = (\prod_3 \gamma_2(\mathcal{G}))\langle c, y^{[1]}\rangle$. We start by giving a similar description for the initial terms of the lower central series of $\mathcal{G}$, which will be useful to establish some identities between the commutator subgroups. It is possible to continue with the other terms of this series using the same reasoning, but the work becomes increasingly cumbersome.

We will determine $\gamma_3(\mathcal{G})$ modulo $(\prod_3 \gamma_3(\mathcal{G}))$. From Lemma 4, we have $\gamma_3(\mathcal{G}) \supseteq \{(c,c^{-1},1), c^{[1]}\}$. But, $[x,y,y] \equiv (c,c,1) \bmod (\prod_3 \gamma_3(\mathcal{G}))$, therefore, $(c,1,1) \in \gamma_3(\mathcal{G})$

and $\prod_3 \gamma_2(\mathcal{G}) \leq \gamma_3(\mathcal{G})$. Since,

$$\begin{aligned} y^{[1]} &\equiv yy^x y^{x^{-1}} \bmod (\prod_3 \gamma_2(\mathcal{G})) \\ &\equiv [x,y][y,x^{-1}] \bmod (\gamma_3(\mathcal{G})) \\ &\equiv 1 \bmod (\gamma_3(\mathcal{G})) \end{aligned}$$

we obtain $y^{[1]} \in \gamma_3(\mathcal{G})$. Thus,

$$\gamma_3(\mathcal{G}) = (\prod_3 \gamma_2(\mathcal{G}))\langle y^{[1]}\rangle.$$

We observe from this that

$$\gamma_2(\mathcal{G}) = \gamma_3(\mathcal{G})\langle c\rangle, \ |\gamma_2(\mathcal{G})/\gamma_3(\mathcal{G})| = 3.$$

Remark 1 From Theorem 1, we note that $\mathcal{G}_2 = \gamma_3(\mathcal{G})$ and thus, $\mathcal{G}_3 = \prod_3 \gamma_3(\mathcal{G})$. Consequently, $\mathcal{G}_k = \prod_{3^{k-2}} \gamma_3(\mathcal{G})$, for $k \geq 3$.

Now, by Lemma 4 we have $\prod_3 \gamma_3(\mathcal{G}) \leq \gamma_4(\mathcal{G})$ and $\{(c, c^{-1}, 1), c^{[1]}\} \subseteq \gamma_4(\mathcal{G})$, then

$$\gamma_4(\mathcal{G}) = (\prod_3 \gamma_3(\mathcal{G}))\left\langle (c, c^{-1}, 1), c^{[1]}\right\rangle.$$

Therefore

$$\gamma_3(\mathcal{G}) = \gamma_4(\mathcal{G})\langle y^{[1]}, c_{[1]}\rangle, \ |\gamma_3(\mathcal{G})/\gamma_4(\mathcal{G})| = 3^2.$$

In order to determine $\gamma_5(\mathcal{G})$ modulo $(\prod_3 \gamma_3(\mathcal{G}))$, we calculate that $[(c, c^{-1}, 1), x] \equiv c^{-[1]} \bmod (\prod_3 \gamma_3(\mathcal{G}))$. Thus,

$$\gamma_5(\mathcal{G}) = (\prod_3 \gamma_3(\mathcal{G}))\left\langle c^{[1]}\right\rangle.$$

And so,

$$\gamma_4(\mathcal{G}) = \gamma_5(\mathcal{G})\left\langle (c, c^{-1}, 1)\right\rangle \text{ and } |\gamma_4(\mathcal{G})/\gamma_5(\mathcal{G})| = 3.$$

The next 5 terms of the lower central series of $\mathcal{G}$ are determined using similar arguments as before. We find

$$\gamma_6(\mathcal{G}) = (\prod_{3^2} \gamma_2(\mathcal{G}))\langle (y^{[1]}, y^{-[1]}, 1), y^{[2]}\rangle, \ |\gamma_5(\mathcal{G})/\gamma_6(\mathcal{G})| = 3^2,$$

$$\gamma_7(\mathcal{G}) = (\prod_3 \gamma_4(\mathcal{G}))\langle (c_{[1]}, c_{[1]}^{-1}, 1), (c_{[1]})^{[1]}, y^{[2]}\rangle, \ |\gamma_6(\mathcal{G})/\gamma_7(\mathcal{G})| = 3^2,$$

$$\gamma_8(\mathcal{G}) = (\prod_3 \gamma_4(\mathcal{G}))\langle (c_{[1]})^{[1]}\rangle, \ |\gamma_7(\mathcal{G})/\gamma_8(\mathcal{G})| = 3^2,$$

$$\gamma_9(\mathcal{G}) = (\prod_3 \gamma_5(\mathcal{G}))\langle ((c, c^{-1}, 1), (c, c^{-1}, 1)^{-1}, 1), (c, c^{-1}, 1)^{[1]}\rangle, \ |\gamma_8(\mathcal{G})/\gamma_9(\mathcal{G})| = 3^2,$$

$$\gamma_{10}(\mathcal{G}) = (\prod_3 \gamma_5(\mathcal{G}))\langle (c, c^{-1}, 1)^{[1]}\rangle, \ |\gamma_9(\mathcal{G})/\gamma_{10}(\mathcal{G})| = 3.$$

From the descriptions we have found, we can draw some conclusions. The first of them corresponds to the next result.

Theorem A(1) $|\gamma_k(\mathcal{G})/\gamma_{k+1}(\mathcal{G})| \leq 3^2$, *for* $1 \leq k \leq 9$.

But the more important conclusion is the following identity

Theorem A(2) $[\gamma_2(\mathcal{G}), \gamma_3(\mathcal{G})] = \gamma_5(\mathcal{G})$.

Proof It follows from the equalities

$$\gamma_2(\mathcal{G}) = (\textstyle\prod_3 \gamma_2(\mathcal{G}))\, \langle c, y^{[1]} \rangle,\ \gamma_3(\mathcal{G}) = (\textstyle\prod_3 \gamma_2(\mathcal{G}))\, \langle y^{[1]} \rangle,$$
$$\gamma_5(\mathcal{G}) = (\textstyle\prod_3 \gamma_3(\mathcal{G}))\, \langle c^{[1]} \rangle$$

that

$$\begin{aligned}[\gamma_2(\mathcal{G}), \gamma_3(\mathcal{G})] &= \textstyle\prod_3 [\gamma_2(\mathcal{G}), \gamma_2(\mathcal{G})] \prod_3 [\gamma_2(\mathcal{G}), y] \cdot \\ &\quad [\textstyle\prod_3 \gamma_2(\mathcal{G}), \langle [x,y], y^{[1]} \rangle] \; [\langle [x,y], y^{[1]} \rangle, \langle y^{[1]} \rangle].\end{aligned}$$

Since $[x,y] = (xy, y^{-1}x, x)$ and $([x,y,y],1,1) \in \prod_3 [\gamma_2(\mathcal{G}), y]$, we have

$$([x,y,x],1,1) \in \left[\prod_3 \gamma_2(\mathcal{G}), \left\langle [x,y], y^{[1]} \right\rangle\right].$$

Then, $\prod_3 \gamma_3(\mathcal{G}) \leq [\gamma_2(\mathcal{G}), \gamma_3(\mathcal{G})]$. Now, $[[x,y], y^{[1]}] = [x, y^{[1]}].([x,y,y],1,1)$ and so $[x, y^{[1]}] \in [\gamma_2(\mathcal{G}), \gamma_3(\mathcal{G})]$. Thus,

$$\gamma_5(\mathcal{G}) \leq [\gamma_2(\mathcal{G}), \gamma_3(\mathcal{G})].$$

Since the other inclusion is always true, the theorem is proved. □

We can observe other identities concerning commutators between the terms of the lower central series of $\mathcal{G}$.

Proposition 1 *The following identities are true in* $\mathcal{G}$:

(i) $[\gamma_3(\mathcal{G}), \gamma_5(\mathcal{G})] = \gamma_8(\mathcal{G})$;

(ii) $[\gamma_2(\mathcal{G}), \gamma_7(\mathcal{G})] = [\gamma_3(\mathcal{G}), \gamma_6(\mathcal{G})] = \gamma_9(\mathcal{G})$;

(iii) $[\gamma_2(\mathcal{G}), \gamma_8(\mathcal{G})] = [\gamma_3(\mathcal{G}), \gamma_7(\mathcal{G})] = \gamma_{10}(\mathcal{G})$.

Proof Since

$$\gamma_3(\mathcal{G}) = \left(\prod_3 \gamma_2(\mathcal{G})\right) \langle y^{[1]} \rangle,\ \gamma_5(\mathcal{G}) = \left(\prod_3 \gamma_3(\mathcal{G})\right) \left\langle c^{[1]} \right\rangle,$$

by the previous theorem, we have:

$$\begin{aligned}[\gamma_3(\mathcal{G}), \gamma_5(\mathcal{G})] &= \left(\prod_3 \gamma_5(\mathcal{G})\right) \left[\prod_3 \gamma_2(\mathcal{G}), \left\langle c^{[1]} \right\rangle\right] \left[\prod_3 \gamma_3(\mathcal{G}), \langle y^{[1]} \rangle\right] \cdot \\ &\quad \left[\left\langle c^{[1]} \right\rangle, \left\langle y^{[1]} \right\rangle\right].\end{aligned}$$

On the other hand, we have

$$\gamma_4(\mathcal{G}) = \gamma_5(\mathcal{G})\left\langle (c, c^{-1}, 1)\right\rangle, \; \gamma_8(\mathcal{G}) = \left(\prod_3 \gamma_4(\mathcal{G})\right) \langle (c_{[1]})^{[1]}\rangle.$$

Since $[y^{[1]}, (y^{[1]}, 1, 1)] \in [\gamma_3(\mathcal{G}), \gamma_5(\mathcal{G})]$ and

$$[y^{[1]}, (y^{[1]}, 1, 1)] = ([y, y^{[1]}], 1, 1) \equiv ((1, c, c^{-1}), 1, 1) \bmod (\prod_3 \gamma_5(\mathcal{G})),$$

we see that $((c, c^{-1}, 1), 1, 1) \in [\gamma_3(\mathcal{G}), \gamma_5(\mathcal{G})]$. Then, $\prod_3 \gamma_4(\mathcal{G}) \leq [\gamma_3(\mathcal{G}), \gamma_5(\mathcal{G})]$.

Now,

$$[c^{[1]}, y^{[1]}] = [c, y]^{[1]} \equiv (c, c, 1)^{[1]} \bmod (\prod_3 \gamma_4(\mathcal{G})),$$

and so, $(c, c, 1)^{[1]} \in [\gamma_3(\mathcal{G}), \gamma_5(\mathcal{G})]$. Since $(c, c^{-1}, 1)^{[1]} \in [\gamma_3(\mathcal{G}), \gamma_5(\mathcal{G})]$, then $(c_{[1]})^{[1]} = (c, 1, 1)^{[1]} \in [\gamma_3(\mathcal{G}), \gamma_5(\mathcal{G})]$. We conclude that $\gamma_8(\mathcal{G}) \leq [\gamma_3(\mathcal{G}), \gamma_5(\mathcal{G})]$, and so (i) is proved.

The proofs of (ii) and (iii) use similar arguments. □

5 The derived series of $\mathcal{G}$

We obtain in this section some results concerning the relationship between the derived series $\mathcal{G}^{(k)}$ and the chain of stabilizers subgroups $\{\mathcal{G}_k \mid k \geq 1\}$ of $\mathcal{G}$. We will also establish the index of $\mathcal{G}^{(k)}$ in $\mathcal{G}$, $k \geq 2$ and determine the terms of the derived series.

It is an elementary fact that for any group G, we have $G^{(i)} \leq \gamma_{2^i}(G)$, $i \geq 0$. For $i = 2$, this result can be improved since that if G is a 2-generated group then $G^{(2)} \leq \gamma_5(G)$. In particular, this is true for our group $\mathcal{G}$.

Now we have

Theorem B(1) $\mathcal{G}^{(2)} = \gamma_5(\mathcal{G})$.

Proof One inclusion is valid from previous remark. By Theorem A(2) we have $[\gamma_2(\mathcal{G}), \gamma_3(\mathcal{G})] = \gamma_5(\mathcal{G})$, and then we have $\gamma_5(\mathcal{G}) \leq [\gamma_2(\mathcal{G}), \gamma_2(\mathcal{G})] = \mathcal{G}^{(2)}$. □

Now we are able to prove:

Proposition 2 $\mathcal{G}^{(k+1)} = \prod_3 \mathcal{G}^{(k)}$, $k \geq 2$.

Proof First we prove the result for $k = 2$. From Theorem B(1), we have $\mathcal{G}^{(2)} = (\prod_3 \gamma_3(\mathcal{G}))\,\langle [x, y]^{[1]}\rangle$. Then,

$$\begin{aligned} \mathcal{G}^{(3)} &= [\mathcal{G}^{(2)}, \mathcal{G}^{(2)}] \\ &= [\textstyle\prod_3 \gamma_3(\mathcal{G}), \prod_3 \gamma_3(\mathcal{G})]\,[\prod_3 \gamma_3(\mathcal{G}), \langle [x, y]^{[1]}\rangle] \\ &= \textstyle\prod_3 [\gamma_3(\mathcal{G}), \gamma_3(\mathcal{G})] \cdot \prod_3 [\gamma_3(\mathcal{G}), [x, y]]\,. \end{aligned}$$

As a consequence, we have

$$\mathcal{G}^{(3)} = \prod_3 \left([\gamma_3(\mathcal{G}), \gamma_3(\mathcal{G})]\,[\gamma_3(\mathcal{G}), [x, y]]\right).$$

Since $\mathcal{G}^{(3)}$ is a normal subgroup of $\mathcal{G}$, we substitute all the conjugates of $[x, y]$ in the above expression by $\gamma_2(\mathcal{G})$. Since $[\gamma_3(\mathcal{G}), \gamma_2(\mathcal{G})] = \gamma_5(\mathcal{G})$, we see that

$$\mathcal{G}^{(3)} = \prod_3 \gamma_5(\mathcal{G}) = \prod_3 \mathcal{G}^{(2)},$$

as desired.

Now, by induction, we have

$$\mathcal{G}^{(k+2)} = \left[\mathcal{G}^{(k+1)}, \mathcal{G}^{(k+1)}\right] = \prod_3 \left[\mathcal{G}^{(k)}, \mathcal{G}^{(k)}\right] = \prod_3 \mathcal{G}^{(k+1)}. \qquad \square$$

Remark 2 From Proposition 2 and Theorem B(1) we can conclude that $\mathcal{G}^{(k)} = \prod_{3^{k-2}} \gamma_5(\mathcal{G})$, for $k \geq 3$.

We note that since the quotients $\mathcal{G}_k/\mathcal{G}_{k+1}$ are abelian, we have $\mathcal{G}_k' \leq \mathcal{G}_{k+1}$, for $k \geq 2$. The next result establishes a relationship between the derived series and the chain of stabilizers in $\mathcal{G}$.

Theorem B(2) (i) *$\mathcal{G}^{(k)} \leq \mathcal{G}_k$, for all $k \geq 1$;*
(ii) *$\mathcal{G}_{k+1} \leq \mathcal{G}^{(k)}$, for all $k \geq 1$.*

Proof (i) By induction on k. For $k = 1$, we know that $\mathcal{G}^{(1)} = \mathcal{G}' \leq \mathcal{G}_1$. Suppose the statement is true for $k > 1$. Since $\mathcal{G}_k' \leq \mathcal{G}_{k+1}$, we have

$$\mathcal{G}^{(k+1)} = \left[\mathcal{G}^{(k)}, \mathcal{G}^{(k)}\right] \leq [\mathcal{G}_k, \mathcal{G}_k] \leq \mathcal{G}_{k+1}.$$

(ii) Again, we use induction on k. For $k = 1$, it is true since $\mathcal{G}_2 \leq \mathcal{G}'$.
Suppose $k > 1$ and use the Theorem 1 and Proposition 2 to conclude:

$$\mathcal{G}_{k+2} = \prod_3 \mathcal{G}_{k+1} \leq \prod_3 \mathcal{G}^{(k)} = \mathcal{G}^{(k+1)}.$$

From Theorem 1, $[\mathcal{G} : \mathcal{G}_1'] = 3^4$ and so $[\mathcal{G} : \mathcal{G}'] = 3^2$. Now, we obtain the index of $\mathcal{G}^{(k)}$ in $\mathcal{G}_k$, $k \geq 2$. $\square$

Theorem B(3) $[\mathcal{G} : \mathcal{G}^{(k)}] = 3^{5.3^{k-2}+1}, k \geq 2.$

Proof It is enough to prove that $\left[\mathcal{G}_k : \mathcal{G}^{(k)}\right] = 3^{3^{k-1}}$, because from Theorem 1, $[\mathcal{G} : \mathcal{G}_k] = 3^{2.3^{k-2}+1}$, for $k \geq 2$. Clearly, this is true for $k = 2$, since $\left[\mathcal{G}_2 : \mathcal{G}^{(2)}\right] = [\gamma_3(\mathcal{G}) : \gamma_5(\mathcal{G})] = 3^3$. Assuming it is true for $k > 2$, we have from Theorem 1 and Proposition 2,

$$\left[\mathcal{G}_{k+1} : \mathcal{G}^{(k+1)}\right] = \left[\prod_3 \mathcal{G}_k : \prod_3 \mathcal{G}^{(k)}\right] = \left[\mathcal{G}_k : \mathcal{G}^{(k)}\right]^3 = 3^{3^k}. \qquad \square$$

We are able to obtain some information about the order of certain quotients $N/[N,\mathcal{G}]$, where N is a normal subgroups of $\mathcal{G}$.

Theorem C *If* $k \geq 2$, *then* $|\mathcal{G}_k/[\mathcal{G}_k,\mathcal{G}]| = |\mathcal{G}^{(k)}/[\mathcal{G}^{(k)},\mathcal{G}]| = 3^2$.

Proof From Lemma 3, and Remarks 1 and 2 (respectively) we have

$$\frac{\mathcal{G}_k}{[\mathcal{G}_k,\mathcal{G}]} \cong \frac{\gamma_3(\mathcal{G})}{\gamma_4(\mathcal{G})}, \text{ and } \frac{\mathcal{G}^{(k)}}{[\mathcal{G}^{(k)},\mathcal{G}]} \cong \frac{\gamma_5(\mathcal{G})}{\gamma_6(\mathcal{G})}.$$

So the theorem is proved since $|\gamma_3(\mathcal{G})/\gamma_4(\mathcal{G})| = |\gamma_5(\mathcal{G})/\gamma_6(\mathcal{G})| = 3^2$. □

As we remarked in the introduction, we cannot say anything in the event that N is a term $\gamma_k(\mathcal{G})$ of the lower central series of $\mathcal{G}$, for $k \geq 40$.

Acknowledgements The author thanks Prof Said Sidki and Prof Norai Rocco for the suggestions and help in the elaboration of this paper.

References

[1] R. I. Grigorchuk, *Burnside's Problem on periodic groups,* Funktsional Anal. i Prilozhen **14** (1980), 53-54; English transl. in Functional Anal.Appl. **14** (1980), 41-43.

[2] R. I. Grigorchuk, *On the Hilbert-Poincaré Series of Graded Algebras Associated with Groups,* Mat. Sb. **180** (1989), 207-225; English transl. in Math. USSR Sbornik **66** (1990), 211-229.

[3] N. Gupta and S. Sidki, *On the Burnside Problem for periodic groups,* Math. Z. **182** (1983), 385-388.

[4] N. Gupta and S. Sidki, *Some infinite p-groups,* Algebra i Logica **22** (1983), 584-589.

[5] N. Gupta and S. Sidki, *Extensions of groups by tree automorphisms,* Contemp. Math. **33** (1984), 232-246.

[6] A. V. Rozhkov, *Centralizers of elements in a group of tree automorphisms,* Izv. Ross. Akad. Nauk. Ser. Mat. **57** (1993), 82-105; English transl. in Russian Acad. Sci. Izv. Math. **43** (1994), 471-492.

[7] A. V. Rozhkov, *Lower Central Series of one Group of Automorphisms of trees,* Mat. Zametki **60**, n$^{\underline{o}}$3 (1996). [In Russian]

[8] M. Shoenert et. al., *GAP "Groups, Algorithms and Programming", User's Manual,* Lehrstuhl D fuer Mathematik, RWTH, Aachen, Germany (1995), 1150pp.

[9] S. Sidki, *On a 2-generated infinite 3-group: the presentation problem,* J. Algebra **110** (1987), 13-23.

[10] S. Sidki, *On a 2-generated infinite 3-group: subgroups and automorphisms,* J. Algebra **110** (1987), 24-55.

Printed in the United States
By Bookmasters